Functional Foods
Nutrients, Health, and Beyond

Editors:
Surajbhan Sevda
Assistant Professor, Department of Biotechnology
National Institute of Technology Warangal
Warangal, India

Meena Krishania
Scientist, Center of Innovative and Applied Bioprocessing
Mohali, India

CRC Press
Taylor & Francis Group
Boca Raton London New York

CRC Press is an imprint of the
Taylor & Francis Group, an **informa** business
A SCIENCE PUBLISHERS BOOK

First edition published 2027
by CRC Press
2385 NW Executive Center Drive, Suite 320, Boca Raton FL 33431

and by CRC Press
4 Park Square, Milton Park, Abingdon, Oxon, OX14 4RN

CRC Press is an imprint of Taylor & Francis Group, LLC

Library of Congress Cataloging-in-Publication Data (applied for)

ISBN: 978-1-041-07407-6 (hbk)
ISBN: 978-1-041-11403-1 (pbk)
ISBN: 978-1-003-65980-8 (ebk)

DOI: 10.1201/9781003659808

Typeset in Times New Roman
by Prime Publishing Services

Preface

Food and nutrition science has undergone a remarkable transformation in recent decades, shifting from a focus on meeting basic dietary requirements to recognizing food as a powerful modulator of health, disease prevention, and overall well-being. This evolution has led to the emergence of *functional foods*—foods that provide health benefits beyond basic nutrition due to the presence of biologically active components. The edited volume *Functional Foods: Nutrients, Health, and Beyond* is conceived to present a comprehensive, multidisciplinary perspective on this rapidly expanding field, integrating advances in nutrition science, chemistry, biotechnology, microbiology, public health, and food technology.

The opening chapters establish a strong scientific foundation. Chapter 1 introduces the concept of functional foods and outlines their nutritional and health benefits, emphasizing the role of bioactive compounds, technological innovations in food processing, and the growing importance of personalized nutrition. Chapter 2 explores the chemical basis of functional foods, linking molecular structure and chemical composition of bioactive constituents with their biological functionality. Chapter 3 focuses on vitamins and minerals in functional foods, discussing their sources, bioavailability, physiological roles, fortification strategies, and their significance in addressing global micronutrient deficiencies. Chapter 4 highlights the role of phytochemicals, detailing their antioxidant, anti-inflammatory, and disease-preventive properties, and their contribution to reducing the risk of chronic lifestyle-related disorders.

Chapters 5 to 7 address functional foods in relation to major physiological systems and disease management. Chapter 5 examines probiotics and prebiotics, emphasizing their role in maintaining gut microbiota balance, modulating the gut–brain and gut–immune axes, and supporting systemic health. Chapter 6 discusses the immunomodulatory effects of functional foods, focusing on nutrients and bioactives that enhance innate and adaptive immune responses. Chapter 7 evaluates the role of functional foods in cancer prevention and management, highlighting bioactive compounds that regulate oxidative stress, inflammation, apoptosis, and immune function, and their potential as complementary approaches in integrative oncology.

The subsequent chapters address technological, regulatory, and societal dimensions. Chapter 8 discusses food fortification and enrichment as public health strategies, outlining methods, benefits, and challenges related to nutrient stability, safety, and consumer acceptance. Chapter 9 focuses on quality control and regulatory considerations, emphasizing analytical validation, labeling, health claims, and the need for harmonized global regulations. Chapter 10 examines consumer awareness and acceptance, analyzing perceptions of health claims, trust in regulatory frameworks, and the balance between naturalness and scientific credibility. Chapter 11 explores marketing strategies for functional foods, addressing opportunities and challenges related to commercialization, branding, digital engagement, and evolving market trends.

The final chapters look toward future innovations. Chapter 12 highlights the pivotal role of biotechnology in the development of functional foods, including genetic engineering, fermentation, microbial technologies, and enzyme optimization, while also addressing ethical and regulatory concerns. Chapter 13 concludes the volume by integrating nutritional insights, emerging bioactive molecules, and future directions such as biofortification, nanoencapsulation, nutrigenomics, personalized nutrition, and sustainable functional food development. Chapter 14 focuses on "Phenolic Bioactives in Functional Foods: Antioxidant and Anti-Inflammatory Health-Promoting Properties", providing a comprehensive overview of phenolic compounds, their natural sources, mechanisms of action, bioavailability, and their significant role in mitigating oxidative stress and inflammation. The chapter further explores technological approaches to enhance stability and efficacy of phenolic bioactives in functional food formulations, emphasizing their therapeutic potential and future applications in health promotion and disease prevention.

This book is intended to serve as a valuable reference for researchers, academicians, postgraduate students, food technologists, nutritionists, healthcare professionals, policymakers, and industry stakeholders. By presenting scientific evidence alongside technological, regulatory, and market perspectives, *Functional Foods: Nutrients, Health, and Beyond* aims to foster interdisciplinary research, support evidence-based innovation, and contribute meaningfully to global efforts in improving nutrition, preventing chronic diseases, and enhancing quality of life.

Acknowledgment: This work was supported by the *Scheme for Promotion of Academic and Research Collaboration* (*SPARC*), *India* (Grant No. SPARC/2024-2025/ENSU/P3267).

Editors
Surajbhan Sevda, Ph.D.
Meena Krishania, Ph.D

Contents

List of Contributors

Akriti Sharma
Indian Council of Agricultural Research-Indian Agricultural Research Institute, New Delhi 110 012.

Alexandros Tsoupras
Hephaestus Laboratory, School of Chemistry, Faculty of Sciences, Democritus University of Thrace, Kavala University Campus, St. Lucas 65404 Kavala, Greece.

Amarjeet Singh
Indian Council of Agricultural Research-Indian Agricultural Research Institute, New Delhi 110 012.

Ankur Ojha
National Institute of Food Technology Entrepreneurship & Management-Kundli, Sonepat, Haryana 131 028.

Ankush Sharma
Department of Physiotherapy. G.D. Goenka University, Gurugram, Haryana-122103, India.

Anurag Singh
Department of Food Technology, Harcourt Butler Technical University, Kanpur-208002.

Ashish Kumar Singh
National Agri-Food and Biomanufacturing Institute (BRIC-NABI), Formerly CIAB, Knowledge City, Mohali 140306, India.

Aysha Sherin K.
Department of Food Engineering and Technology, Institute of Chemical Technology, Mumbai, India-400019.

Balaji, M.
Department of Food Engineering and Technology, Institute of Chemical Technology, Mumbai-400019, India.

Debakanta Tripathy
Department of Chemistry, School of Advanced Sciences, VIT-AP University, Amaravati-522237, Andhra Pradesh, India.

Eleni Koutsia
Hephaestus Laboratory, School of Chemistry, Faculty of Sciences, Democritus University of Thrace, Kavala University Campus, St. Lucas 65404 Kavala, Greece.

Garima Singh
School of Biosciences and Technology. Galgotias University, Greater Noida, Uttar Pradesh-203201, India.

Jannatul Ferdushi Jany
Department of Food Engineering and Technology, Faculty of Agricultural Engineering and Technology, Bangladesh Agricultural University, Mymensingh-2202, Bangladesh.

Khushi Sharma
Department of Food Engineering and Technology, Institute of Chemical Technology, Mumbai, India-400019.

Kritika Bajpai
Department of Food Technology, Harcourt Butler Technical University, Kanpur-208002.

Kurraboyina Sai Venkata
Department of Food Engineering and Technology, Institute of Chemical Technology, Mumbai-400019, India.

Latika Yadav
Vijay Singh Pathik Government PG College, Kairana, Shamli, Uttar Pradesh-247774, India.

Meena Krishania
National Agri-Food and Biomanufacturing Institute (BRIC-NABI), Formerly CIAB, Knowledge City, Mohali 140306, India.

Meena Tippana
Department of Biotechnology, National Institute of Technology Warangal, Warangal, Telangana-506004, India.

Muskaan Duggal
BRIC-National Agri Food and Biomanufacturing Institute (formerly CIAB and NABI), Mohali 140306, India.

Neera Agarwal
BRIC-National Agri Food and Biomanufacturing Institute (formerly CIAB and NABI), Mohali 140306, India.

Nidhi Gossai
Department of Biotechnology, Punjab University, Chandigarh 160014, India.

Nitin Sharma
Indian Council of Agricultural Research-Indian Agricultural Research Institute, New Delhi 110 012.

Pesaru Sai Laya
Department of Biotechnology, National Institute of Technology Warangal, Warangal, Telangana, India-506004.

Pisal Ketali Prakash
Department of Biotechnology, National Institute of Technology Warangal, Warangal, Telangana-506004, India.

Priya Prasad
Department of Biotechnology, National Institute of Technology Warangal, Warangal, Telangana-506004, India.

Priyanka Bairwa
BRIC-National Agri Food and Biomanufacturing Institute (formerly CIAB and NABI), Mohali 140306, India.

Raman M.S.
Indian Council of Agricultural Research-Indian Agricultural Research Institute, New Delhi 110 012.

Roji Waghmare
Department of Food Engineering and Technology, Institute of Chemical Technology, Mumbai-400019, India.

Sadsimrat Kaur
BRIC-National Agri Food and Biomanufacturing Institute (formerly CIAB and NABI), Mohali 140306, India.

Samrudhi Anant Devadiga
Department of Food Engineering and Technology, Institute of Chemical Technology, Mumbai-400019, India.

Satinder Singh
Indian Council of Agricultural Research-Indian Agricultural Research Institute, New Delhi 110 012.

Satya Sravan Kumar
Department of Food Engineering and Technology, Institute of Chemical Technology, Mumbai-400019, India.

Soumya Lipsa Rath
Department of Biotechnology, National Institute of Technology Warangal, Hanamkonda-506004, Telangana, India.

Surabhi Pandey
Department of Food Technology, Harcourt Butler Technical University, Kanpur-208002.

Surajbhan Sevda
Environmental Bioprocess Laboratory, Department of Biotechnology, National Institute of Technology Warangal, Warangal, Telangana 506004, India.

Tania Raheja
BRIC-National Agri Food and Biomanufacturing Institute (formerly CIAB and NABI), Mohali 140306, India.

Tejinder Kaur
BRIC-National Agri Food and Biomanufacturing Institute (formerly CIAB and NABI), Mohali 140306, India.

Theodora Adamantidi
Hephaestus Laboratory, School of Chemistry, Faculty of Sciences, Democritus University of Thrace, Kavala University Campus, St. Lucas 65404 Kavala, Greece.

Yuna Seo
Department of Industrial and Systems Engineering, Tokyo University of Science, 2651, Yamazaki, Noda, Chiba, JPN (278-8510).

Chapter 1

Nutritional and Health Benefits of Functional Foods

Neera Agarwal[1], Muskaan Duggal[2], Tejinder Kaur[3], Priyanka Bairwa[4], Tania Raheja[5], Sadsimrat Kaur[6], Surajbhan Sevda[7*] and Meena Krishania[8*]

1. Introduction

The traditional wisdom of the healing power of food has now been validated by modern nutritional science in our current health-conscious society. Scientific evidence has proved that food is not only required for satisfying hunger but also plays a role in protecting health and preventing diseases (Chaudhary et al., 2025). People are now aware of the link between health and diet and rising healthcare costs, which have driven them to choose personalised health solutions. The easiest way to deal with this is by consuming functional food, which would enhance the well-being and also proactively manage the health of an individual.

Functional foods or medical foods confer additional health benefits to an individual, in addition to their conventional nutritional value. These can be naturally occurring, formulated, or fortified with safe and effective concentrations of certain bioactive ingredients or functional molecules, such as omega-3 fatty acids, dietary fibres, pre/pro biotics, antioxidants, and phytoestrogens. These bioactive compounds, when introduced into the food by technological routes, confer specific health benefits such as shielding against chronic diseases and maintaining healthy vitamin levels in the body (Toy et al., 2025). Moreover, they can help enhance the physical performance, cognitive function, behavioural aspects, and psychological functions of an individual. Functional foods can also be designed or modified to meet specific dietary needs, such as low-sugar, gluten-free, and lactose-free foods (Ahmed et al., 2022). Examples of functional foods include iodised salt, vitamin D-fortified milk, and fortified breads.

The term "functional food" was coined in the 1980s (Ahmed et al., 2022). Japan was also the first country to term these foods legally as Food for Specified Health Use (Yamada et al., 2008). They are also known as medical foods, vita foods, or sometimes nutraceuticals. According to a report published in *Fortune Business Insights*, the market size of functional foods and beverages in 2024 was approximately USD 364.18 billion and is expected to increase up to USD 398.81 billion by 2025. This is because of increasing awareness regarding functional foods and their health benefits among health-conscious consumers.

[1-6,8] BRIC-National Agri Food and Biomanufacturing Institute (formerly CIAB and NABI), Mohali 140306, India.

[7] Environmental Bioprocess Laboratory, Department of Biotechnology, National Institute of Technology Warangal, Warangal, Telangana 506004, India.

* Corresponding author: meena@ciab.res.in

DOI: 10.1201/9781003659808-1

The functional food market was started by Coca-Cola at the beginning of the 20th century, which was created as a non-sweetened, non-carbonated French wine coca tonic that served as a morphine-free pain reliever (Weststrate et al., 2002). This was followed by the development of different functional foods and beverages including hypoallergenic rice, certain fruits and vegetables, seeds, and nuts. Currently, as health and wellness are prioritised across the globe, functional foods are becoming an important part of the daily nutritional needs of an individual's personalised diet. Food scientists are working on producing new functional molecules using the latest processing technologies, which would confer various health benefits to people. This chapter discusses the mechanism of action of vital foods in the body, various categories of functional foods, including plant-based and dairy-based functional foods, and the future direction of food markets. The chapter also discusses emerging innovative food technologies, challenges in the production and marketing of functional foods, and regulation of the functional foods framework, including the standardised labelling and claims. This chapter explores the potential of personalised nutrition, the application of advanced technologies in food processing, and the development of innovative bioactive compounds.

2. Understanding the Science Behind Functional Foods

Functional foods are specially developed products that contain added or naturally occurring bioactive ingredients to promote health or reduce disease risk (Figure 1). These foods surpass the basic nutrition value and deliver targeted physiological benefits.

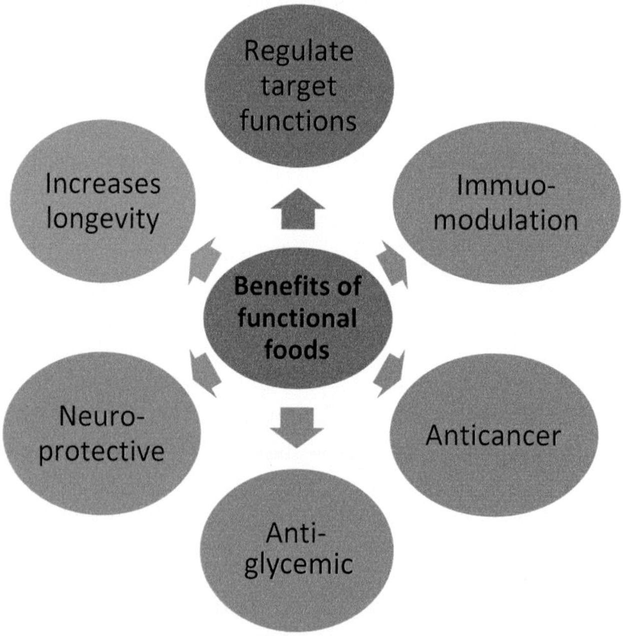

Fig. 1 The benefits of Functional Foods.

2.1 *Bioactive Compounds in Functional Foods*

Bioactive compounds are naturally occurring or synthetically enhanced components of foods that can measurably affect cellular or metabolic processes in humans and animals. They are found in plants and other organisms, and are linked to a decline in the occurrence of chronic diseases, as shown by epidemiological, clinical, and biochemical studies. Flavonoids, vitamins, and carotenoids in fruits and vegetables, which possess anti-inflammatory, antioxidant, anticancer, and antimicrobial properties, are some examples of bioactive molecules. Some of the benefits of the bioactive compounds are listed in Table 1.

Table 1 Biological properties and associated health benefits of key bioactive compounds in functional foods.

Biological Property	*Function/Benefit*	*Examples of Bioactive Compounds*	*References*
Neuroprotective	Reduces neurodegeneration, improves memory	Flavonoids, Phenylphenols, Omega-3	Kerry et al., 2018
Controls Blood Sugar	Enhances insulin sensitivity	Polyphenols, Alkaloids, Terpenoids	Maliha et al., 2025
Anti-inflammatory	Decreases joint pain and swelling	Curcumin, Resveratrol, Quercetin	Pandey et al., 2009
Regulates Endocrine System	Restores Hormonal Balance	Isoflavones, Phytoestrogens	Martirosyan et al., 2022
Decreases Cholesterol	Reduces low-density lipoprotein (LDL), promotes heart health.	Saponins, Plant Sterols, Flavanoids	Martirosyan et al., 2021
Strengthens Immune System	Enhances immunity, fights infections	Beta-glucans, Peptides, Alkaloids	Kerry et al., 2018
Controls High Blood Pressure	Relax blood vessels, improves circulation	Polyphenols, Peptides, Nitric Oxide	Leeuwendaal et al., 2022
Combats Oxidative Stress	Prevents cellular damage, aging	Vitamin C, E, Flavonoids	Zhang et al., 2021
Fights Infection	Antimicrobial, antiviral, antifungal	Tannins, Alkaloids, Essential Oils	Dixit et al., 2021

2.2 *Minerals in Functional Foods*

Minerals are inorganic compounds or metallic elements that act as enzyme cofactors and support biological processes, such as bone development, nerve transmission, and glucose regulation. Commonly fortified minerals include zinc, calcium, iron, magnesium, and copper. Modern techniques like nanoencapsulation include enclosing bioactive compounds in nanostructures (<1000 nm), which can enhance mineral absorption by better dissolution, preservation, and absorption.

Antinutrients (e.g., phytates and tannins) bind to divalent minerals like calcium and iron, hence affecting the bioavailability of minerals. However, the gut microbiota can improve absorption by producing enzymes that release minerals from these complexes. For instance, short-chain fatty acids (SCFAs) from microbial fermentation lower the intestinal pH and increase calcium solubility. Probiotics, such as *Enterococcus faecium* increase phosphorus retention, but excessive iron supplementation can disrupt gut balance and increase infection risk (Dixit et al., 2021). Absorption occurs through three pathways: paracellular transport, where minerals pass through intestinal tight junctions and dependent on mineral solubility. Transcellular transport, where active uptake via specialised proteins (e.g., TRPV5/TRPV6 channels) is regulated by calcitriol (activated vitamin D); and microbial-assisted absorption, where gut bacteria (e.g., *Lactobacillus*) degrade phytic acid to release minerals and SCFAs, boost absorption efficiency. Some food constituents, such as oxalates, tannins, and phytic acid present in large amounts in plant foods bind to iron, calcium, and zinc, thereby decreasing their bioavailability in the body. Soaking, fermentation, and germination of food have been reported to decrease antinutrients and increase mineral availability. For instance, fermentation enhances bioaccessibility of magnesium and calcium in probiotic foods. Functional food minerals can be utilised in several ways. Most of these act as cofactors for enzymes that catalyse crucial biochemical reactions. Manganese, for instance, is a cofactor for manganese superoxide dismutase, an enzyme utilised in the process of protecting cells from oxidative damage (Aguirre and Culotta, 2012). Others, such as phosphorus and calcium, also serve structural functions by being the building components of teeth and bones and providing skeletal rigidity. Minerals, such as magnesium, also control important physiological processes, such as muscle contraction, nerve function, and blood pressure (Gröber et al., 2015). Strategic food combinations can also be used to increase mineral availability. Combining iron-rich plant foods like spinach and beans with vitamin C sources like citrus fruits (oranges, lemons, limes, and grapefruits) peppers significantly improves iron absorption in the body (Moustarah and Daley, 2024). Processing, such as fermentation, also reduces antinutrient content and enhances mineral absorption, as demonstrated by the increased bioavailability of zinc and iron in fermented yogurt.

2.3 Polyols: Properties and Applications

Polyols and sugar alcohols are low-calorie carbohydrates with hydroxyl groups that replace aldehyde/ketone moieties. Naturally occurring examples are erythritol (fruits) and xylitol (birch), whereas synthetic variants such as lactitol are industrially produced. Their sweetness ranges from 30% to 100% sucrose and they are popular sugar substitutes in chocolates and pharmaceuticals. Polyols improve product shelf-life and texture by changing the osmolality, viscosity, and freezing points. For instance, erythritol provides a cooling sensation in breath mints, whereas xylitol smooths out ice cream (Poka et al., 2025). Polyols are partially absorbed in the small intestine and the unabsorbed portion is fermented by colonic bacteria. This fermentation produces SCFAs (e.g., acetate and propionate) that feed on beneficial microbes, such as *Bifidobacteria*, and some polyols are prebiotics (Ríos-Covián et al., 2016). However, excessive intake of polyols can cause bloating or diarrhoea due to gas production. In irritable bowel syndrome (IBS) patients, low-FODMAP diets (excluding fermentable polyols) are recommended to alleviate symptoms related to gut microbiota shifts. Polyols occur naturally in fruits and vegetables and are produced by microorganisms, such as lactic acid bacteria (LAB). Recent biotechnological developments have investigated the bioproduction of polyols using LAB as a substitute for conventional industrial processes. Certain LAB can naturally generate mannitol and erythritol when grown in specific ways, but sorbitol and xylitol production require genetic modification. Polyol addition to functional foods not only improves their organoleptic appeal, but also contributes to functionalities of health. For example, xylitol and glycerol have been found to induce LAB, i.e., *Lactobacillus plantarum*, to produce antifungal compounds, including cyclic dipeptides such as cyclo (Leu-Leu), cyclo (Pro-Gly), and cyclo (Val-Phe) (StröM et al., 2002). These compounds are found to inhibit fungi such as *Aspergillus carbonarius*, which is a frequent contaminant of foods like nuts, coffee, cereals, and grapes, thus diminishing the production of toxic mycotoxins like ochratoxin A (Banahene et al., 2024). Such antifungal activity is suggestive of the prospective application of polyols to enhance food safety and improve shelf-life. Moreover, polyol supplementation through diet as functional foods can lead to alchemically good control and oral health as the polyols are metabolized distinctively as opposed to customary sugar and are not supportive of dental caries.

2.4 Omega-3 Fatty Acids: Roles and Clinical Insights

Omega-3 fatty acids, including alpha-linolenic acid (ALA), eicosatetraenoic acid (EPA), and docosahexaenoic acid (DHA), contribute significantly to cell membrane integrity, inflammation regulation, and heart health. ALA found in flaxseeds and walnuts is converted into EPA and DHA *in vivo*, but not efficiently. Omega-3 present in flaxseeds helps in lipid metabolism, which was proven by a trial in which 30 hemodialysis patients showed reduced CRP, LDL-C, total cholesterol, and triglycerides when consuming 40 g/day of flaxseed for 8 weeks (Nattagh-Eshtivani et al., 2025). Another is glucose regulation, which was proven by lowered fasting blood glucose in obese, insulin-resistant adults when they were supplemented with flaxseed for 12 weeks (Kavyani et al., 2023). Similarly, chia seeds also act as an anti-inflammatory agent.

Omega-3 fatty acids act as metabolic "managers", helping the body to burn fat efficiently while reducing fat storage. They suppress the activity of genes involved in fat production (lipogenesis) and increase the enzymes that break down fats (β-oxidation). Omega-6 fatty acids like arachidonic acid act as building blocks for inflammatory molecules. Consuming EPA and DHA help reduce inflammation by displacing omega-6s in cell membranes, thus increasing the production of anti-inflammatory compounds. This tug-of-war helps regulate immune responses and reduces chronic inflammation linked to heart disease and arthritis. Emerging research also suggests that omega-3 may play a role in cancer prevention. They activate cellular repair pathways (AMPK/SIRT) and slow down tumor growth in certain cancers (Kim and He, 2013). In the brain, DHA integrates into nerve cell membranes, protects against Alzheimer's disease by blocking toxic protein buildup (amyloid-beta plaques) (Xiao et al., 2022). These fats also support eye health by maintaining blood vessel integrity in the retina, potentially delaying age-related vision loss. Omega-3 fatty acids are essential polyunsaturated fatty acids that are a component of human well-being. Although they are better known in connection with fatty fish, they are also present in functional foods such as fruits, vegetables, nuts, seeds, yogurt, dairy products, and fortified food. These necessary fatty acids, primarily ALA in plant sources and EPA and DHA in animal sources and food products fortified with these substances, contribute to health in numerous ways via multiple mechanisms.

Plant-based foods that contain omega-3s like walnuts, chia seeds, flaxseeds, and hemp seeds contain ALA, which the body can transform, to some extent, into EPA and DHA. Spinach, Brussels sprouts, and seaweed also have trace amounts of omega-3s. Nuts and seeds not only provide these fats but also contain fibre, antioxidants, and polyphenols that improve cardiovascular well-being through controlling inflammation and lipid metabolism. Yogurt and fortified dairy foods are excellent sources of omega-3s because fortification with EPA and DHA increases their nutritional value, improving brain function and heart health (McCowen et al., 2010).

The action involves their effect on cell membrane fluidity, anti-inflammatory pathways, and hormonal control. The interference of omega-3 fatty acids with omega-6 fatty acids in the enzymatic pathways leads to the production of eicosanoids, which produce anti-inflammatory molecules. It is of most utility in preventing chronic disease risk, including cardiovascular disease, metabolic syndrome, and neurodegenerative disease. Additionally, dairy omega-3s and yogurt contribute to maintaining gut health by interacting with probiotics, which can affect gut microbiota balance and inhibit inflammation. The synergy between omega-3s and other bioactive nutrients interacts with cognition and supports metabolism in seeds and nuts. The ingestion of omega-3-rich nuts and seeds, leafy greens, yogurt, and dairy products provide a means by which individuals can sustain optimal brain function, heart health, and overall well-being. Their inclusion in the diet enhances anti-inflammatory processes, hormonal balance, and cellular integrity and makes them integral parts of an optimal and health-enhancing nutritional regimen.

2.5 Probiotics: The Gut's Guardians

Probiotics, according to the WHO (World Health Organization), are live microorganisms that when ingested in specific amounts, can improve health. Common strains include *Lactobacillus* (in yogurt) and *Bifidobacterium* (in fermented foods like kimchi), which aid digestion, strengthen immunity, and may even influence mood through the gut-brain axis (Kerry et al., 2017). Probiotics fortifies the gut barrier by reinforcing the intestinal lining, acting like a "security system" to block harmful bacteria and toxins from leaking into the bloodstream. It crowds out bad bacteria by probiotics competing for space and nutrients, making it harder for pathogens like *E. coli* to grow (Figure 2).

In immune system coaching, immune cells are trained to respond appropriately by reducing excessive inflammation (associated with allergies) while enhancing defence against infections. Gut-brain communication, in which neurotransmitters like serotonin and gamma-aminobutyric acid, probiotics may influence mood, stress responses, and even neurological health (Ashique et al., 2024). For example, after a stroke, gut imbalances (dysbiosis) and a "leaky" intestinal lining can worsen the recovery. Probiotics help restore microbial balance, which may improve outcomes by reducing inflammation and supporting nerve function (Das et al., 2024). Scientists are developing advanced probiotic strains with targeted health benefits, such as *Christensenella minuta*, associated with improved blood sugar regulation, and *Faecalibacterium prausnitzii*, which exhibits anti-inflammatory properties and potential protective effects against colon cancer.

Combining probiotics with prebiotics (like inulin) enhances their effects by lowering cholesterol levels and improving blood vessel function. Specific strains, such as *Lactobacillus rhamnosus*, reduce gut inflammation and metabolic disorders by blocking harmful bacterial toxins (Leeuwendaal et al., 2022).

The gut microbiome, a complex community of microorganisms residing in the gastrointestinal tract, plays a vital role in human health; its imbalance has been linked to conditions such as atherosclerosis and diabetes. For instance, patients with heart disease often have too much *Collinsella* (a microbe that promotes inflammation) and disrupted ratios of key bacterial groups (Nesci et al., 2023). Probiotics counter this by increasing mucus-producing cells (goblet cells) to protect the gut lining, boosting proteins (mucins) that seal intestinal gaps, and promoting anti-inflammatory microbes like *Prevotella*, which produce compounds that calm overactive immune responses. Probiotics may help stroke patients with digestive problems by influencing the communication pathway between the gut and the brain. This pathway, known as the gut-brain axis, allows gut bacteria to affect nerve signals and overall inflammation in the body (Agirman et al., 2021). Probiotics in functional foods are crucial in sustaining gut health, immunity, and overall well-being. Vegetables like fermented food (kimchi, sauerkraut, and pickles) also contain naturally occurring probiotics that facilitate smooth digestion and also contribute to enhanced nutrient supply. Raw fruits like bananas and apples even serve as a prebiotic as they encourage the growth of good gut microbes with fibre supplementation (Aziz et al., 2024). Fermented legumes like tempeh and miso add probiotic strains that benefit gut and immune health. Seeds and nuts like flaxseeds, almonds, and chia seeds are carriers of prebiotics and probiotics that enhance microbial balance in the gut and lipid metabolism (Sugizaki and Naves, 2018). Whole grains like oats, barley, and quinoa provide beta-glucans that assist in the survival of probiotics through digestion, resulting in enhanced gut colonisation and metabolic activity (Arena et al., 2014). Fermented foods and dairy products are also good sources of probiotics, where yogurt and fortified dairy products contain strains such as *Lactobacillus acidophilus* and *Bifidobacterium bifidum* that improve lactose digestion, support the gut barrier, and strengthen the immune system. Milk fermentation of kefir provides a probiotic drink in high demand because of its ease of digestion, anti-inflammatory properties, and cardiovascular protection (Hertzler and Clancy, 2003). Probiotic-enriched cheeses and fortified milk have been created to enhance gastrointestinal function, reduce cholesterol, and increase calcium absorption for healthier bones. Functional foods such as tea and coffee have also appeared in the form of probiotic-rich beverages, with kombucha, a fermented tea that contains probiotics and organic acids to promote health in the gut, liver and detoxification (De Oliveira et al., 2023). Probiotic teas, including green tea and herbal tea, combine beneficial microorganisms with antioxidants to neutralise inflammation and promote metabolic well-being. Probiotic coffee items containing *Lactobacillus* strains have also been proposed as an alternative for traditional dairy-based probiotic foods that support digestion and equilibrium in gut microbiota (Shah et al., 2024). The mechanisms through which probiotics act are manifold, beginning with gut microbiota modulation, whereby they preserve microbial homeostasis, hence suppressing disorders like IBS and inflammatory bowel disease. Immune function of the body is also enhanced using probiotics, which stimulates the immune cells, hence promoting the production of immunoglobulins and cytokines, which in turn protect against infections and reduce inflammation. Probiotics facilitate the breakdown of complex carbohydrates, proteins, and fats, thereby improving the bioavailability of essential nutrients such as calcium, magnesium, and vitamin B12. They also

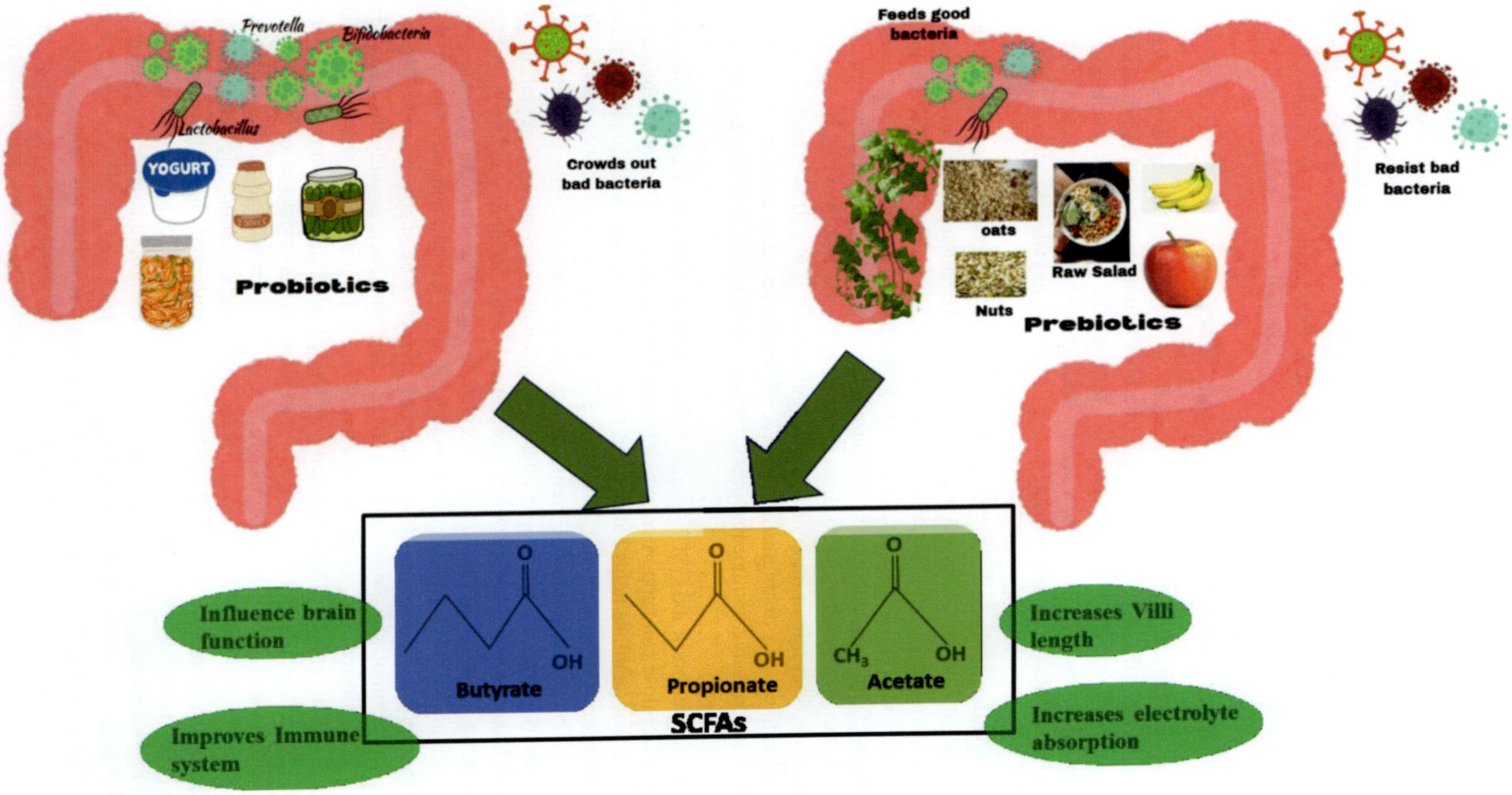

Fig. 2 Synergistic effect of probiotics and prebiotics in maintaining gut microbiome balance and improving host health via short-chain fatty acid metabolism.

produce SCFAs—including butyrate, acetate, and propionate—that nourish gut cells, modulate inflammation, and support metabolic health. Probiotics also generate antimicrobial substances like bacteriocins, hydrogen peroxide, and organic acids that suppress the growth of pathogenic bacteria like *Salmonella* and *Escherichia coli.* In addition to gut health, probiotics also help in stabilising mood and lowering stress of an individual by helping in synthesis of neurotransmitters like serotonin and gamma-aminobutyric acid. Overall, the incorporation of probiotics into functional foods containing fruits, vegetables, legumes, nuts, seeds, yogurt, whole grains, fortified milk, tea, and coffee gives consumers ample health advantages that enhance their function in preventive nutrition and health.

2.6 Prebiotics: The Gut's Natural Fertilizers

Prebiotics are specialised "food" for good bacteria in the gut. Officially defined as selectively fermented ingredients that boost beneficial gut microbes to improve health, they work by surviving digestion by resisting stomach acid and enzymes to reach the colon intact. Feeding good bacteria, fermented by microbes such as *Bifidobacterium* and *Lactobacillus,* helps them thrive and blocks bad bacteria by crowding out harmful bacteria (e.g., *Clostridium*) while producing health-promoting compounds.

The fermentation of prebiotics by gut bacteria produces SCFAs, bioactive metabolites that play a pivotal role in gut health and overall well-being (Figure 2). Among them, butyrate is particularly important, as it reinforces the intestinal barrier and protects against increased permeability. Additionally, SCFAs lower colonic pH, thereby creating an unfavorable environment for pathogenic microorganisms. They also function as "postbiotics", actively regulating metabolism and reducing inflammation, further contributing to digestive and systemic health. In addition to gut health, butyrate plays a pivotal role in immune system regulation (Ji et al., 2023). It aids dendritic cells—key immune sentinels—in distinguishing between harmful invaders and benign substances. This function prevents excessive immune responses, reducing the likelihood of allergies and autoimmune disorders.

Through the gut-brain axis, SCFAs also influence brain function. Research has shown that prebiotics like lactulose can support cognitive health. For instance, studies on Alzheimer's disease have suggested that lactulose helps balance gut microbiota associated with improved memory and neurological function (Lei et al., 2025). Moreover, SCFAs contribute to metabolic health, particularly in conditions like obesity and diabetes. Prebiotics enhance insulin sensitivity, thereby preventing the harmful effects of prolonged high sugar intake. Additionally, they assist in repairing gut barrier damage, often linked to metabolic disorders. Probiotics are important for gut health, immunity, and overall well-being. Plant-based foods such as fermented vegetables—including pickles, sauerkraut, and kimchi—naturally provide probiotics that support digestion and enhance nutrient absorption. Raw fruits like apples and bananas also function as prebiotics by supplying fibre that feeds good gut bacteria and promotes probiotic growth. Seeds and nuts such as almonds, chia seeds, and flaxseeds serve as substrates for the delivery of probiotics and prebiotics, promoting a balance of the gut microbes with enhanced lipid metabolism. Oats, barley, and quinoa are whole grains that contain beta-glucans, which can resist digestion to allow the probiotics to colonise (Maqsood et al., 2025). Fermented dairy foods and milk are also good sources of probiotics. Milk fermented into kefir produces a probiotic beverage popular for supporting digestion, alleviating inflammation, and maintaining cardiovascular health. Probiotic-fortified milk and cheese have been created to enhance the function of the gastrointestinal system, reduce cholesterol levels, and improve calcium assimilation for bone health. Tea and coffee are also becoming probiotic-enriched functional foods, including fermented tea or kombucha containing organic acids and probiotics, which benefit the digestive system and liver. Prebiotic tea, such as green tea and herbal tea, brings together healthy microbes and antioxidants to counteract inflammation and bring about metabolic balance. In the same way, probiotic coffee beverages containing *Lactobacillus* strains have been presented as a substitute for conventional dairy-based probiotic foods for better digestion and gut flora balance. The methods by which probiotics work vary, beginning with gut microbiota modulation, through which they influence microbial equilibrium by suppressing harmful bacteria and favouring the multiplication of good bacteria, thereby treating such conditions as IBS and inflammatory bowel disease (Latif et al., 2023). Probiotics also stimulate the production of immunoglobulins and cytokines that combat infection and inflammation. Probiotics aids in digestion and absorption of nutrients and stabilizes mental health.

2.7 Carotenoids: Nature's Colourful Shields

Carotenoids are naturally occurring plant pigments with strong antioxidant properties, functioning as protective agents against cellular damage. Since humans cannot synthesise these compounds, they must be obtained through a diet rich in colourful fruits and vegetables. There are many plant sources containing these protective substances, red and orange foods such as tomatoes, producing lycopene, while carrots and sweet potatoes produce beta-carotene, a vitamin A precursor. Yellow foods like corn are lutein-rich, and saffron is a source of characteristic carotenoids that give it its rich gold color. Leafy vegetables like spinach and kale are rich in lutein and zeaxanthin which are specifically important for eye well-being. Purple foods like grapes and blueberries even have carotenoid chemicals that promote the overall cellular function (Abdel-Aal et al., 2013).

Carotenoids are important for health and disease prevention. They are components of the body's antioxidant defence system that counteract free radicals by reacting with and "disarming" toxic oxygen species. Carotenoid pigments also boost the functions of antioxidants like vitamins C and E. For example, the carotenoid capsanthin in chili peppers is synergistic with vitamin C to enhance its antioxidant effect. Increased carotenoid intake through diet lowers the incidence of chronic diseases, including heart disease, particular cancers, and age-related macular degeneration. To enhance the health benefits of carotenoids, it is necessary to consume them in adequate amounts through a well-nourished diet (Villa-Rivera and Ochoa-Alejo, 2020). A good amount of beta-carotene, approximately four milligrams per day, can be obtained by consuming a large carrot. A minimum of 3.3 milligrams of lutein can also be obtained from two cups of spinach. The intake of half a sweet potato also gives over 0.6 milligrams of alpha-carotene. These dietary choices help to ensure adequate carotenoid levels and promote overall well-being (Ranard et al., 2017).

Besides carotenoids, prebiotics maintain gut health by serving as "microbiome fertiliser", feeding healthy bacteria within the digestive tract. Avoiding supplements in favour of whole foods offers a superior and more natural method of supporting digestive function. Carotenoids also promote an "eat the rainbow" philosophy, with their natural colour serving as a marker of their individualistic protective benefits. The interaction of these compounds is responsible for maximising health as they are optimally utilised together. Combining prebiotic vegetables with foods rich in carotenoids maximises their combined advantages. Carotenoids are present in various functional foods including vegetables, fruits, nuts, yogurt, seeds, whole-grain cereals, tea, and coffee. These compounds are best known for their remarkable antioxidant activity, wherein they are the primary components that scavenge free radicals and offer protection against cell damage that occurs due to oxidative reactions. Beta-carotene, lutein, and zeaxanthin are the major carotenoids present in plant-based foods. Leafy greens such as kale and spinach are dense lutein and zeaxanthin; carotenoids associated with eye health and protection against age-related macular degeneration.

Although nuts and seeds contain lower levels of carotenoids compared to fruits and vegetables, they are rich in compounds that enhance the absorption and utilisation of these pigments. Their healthy fats improve carotenoid bioavailability for uptake in body tissues. Additionally, carotenoids can be incorporated into yogurt as an added source of antioxidants. However, whole grains, tea, and coffee contain a low concentration of carotenoids. The ways in which carotenoids have positive impacts include neutralisation of free radicals, influencing gene expression, including antioxidant enzymes, and influencing cell signalling pathways involved in inflammation regulation and cell survival.

2.8 Flavonoids: Nature's Multitasking Protectors

Flavonoids are colourful, plant-based molecules found in high quantities in fruits, vegetables, and herbs, which give them bright colours and impressive health-promoting properties. With more than 6,000 types known, these natural compounds help in combating oxidative stress, lowering inflammation, and even offering antiviral protection. Their strong effects are a result of their distinctive molecular structure, which has three strategically located hydroxyl groups that work like "defence stations", enabling flavonoids to counteract dangerous threats (Jomova et al., 2025).

Tea, citrus fruits, and berries contain flavones and catechins, which protects the cells from damage by acting as antioxidants. Isoflavones present in soybeans and lentils lowers the risk of breast cancer. Anthocyanins, which give blueberries and red cabbage their rich colours, play a role in heart health. Flavanols, found in onions and kale, maintain immune function and fight chronic inflammation, on the other hand.

Even with all the benefits that flavonoids have to offer, they have one major drawback: poor body absorption. The daily flavonoid consumption is estimated to range between 50–800 mg, of which quercetin accounts for nearly 75% (Zahra et al., 2024). To address this limitation, researchers have developed nanoencapsulation techniques that enclose flavonoids within microscopic carriers, such as liposomes or chitosan, thereby enhancing their stability and bioavailability. Studies have indicated that non-encapsulated polyphenols, for instance, are more successful in controlling diabetes than their naked counterparts (Aryaeian et al., 2017). Flavonoids play a crucial role as antioxidants and act as cellular bodyguards to protect against oxidative stress. Their hydroxyl groups play a number of different roles: they neutralise free radicals by giving up electrons and halting chain reactions that can harm DNA and proteins, they sequester toxic metals such as iron and copper to inhibit oxidative stress, and they augment the body's internal defences by activating antioxidant enzymes like glutathione while at the same time inhibiting pro-oxidant enzymes. In certain instances, flavonoids have been proven to be even more effective than vitamins C and E (Ullah et al., 2020). However, their efficacy varies with structure; smaller, less polar molecules such as aglycones are better absorbed from the gut, whereas larger, sugar-conjugated forms known as glycosides need to be cleaved by gut microbes first.

In addition to their antioxidant properties, flavonoids regulate the immune system and function as central modulators of immune responses. They trigger "alarm systems" by binding to immune receptors (pattern recognition receptors, or PRRs) which recognise pathogens and initiate protective actions (Gugliesi et al., 2021). They also mobilize immune cells by potentiating natural killer (NK) cells that attack infected cells or cancerous cells and attract dendritic cells to assist in training T-cells while maximising macrophages to clear injurious debris (Yang et al., 2023). Flavonoids also modulate overactive immune systems and prevent autoimmune disease. Some flavonoids have also shown potential to hinder the COVID-19 virus's spike proteins, hence preventing COVID-19 infection.

Increased flavonoid consumption has been associated with important health benefits. For example, flavonoid-rich Asian diets have shown to reduce the risk of breast cancer to 60% (Ahn-Jarvis et al., 2019). To obtain the maximum number of flavonoids, it is crucial to "eat the rainbow" and a variety of brightly coloured plant foods. Red coloured foods like tomatoes and strawberries contain lycopene, whereas purple and blue foods like eggplant and blackberries contain anthocyanins. Green foods, such as spinach and matcha tea, provide a range of flavonoids, while white foods like garlic and onions have flavanols that boost immune function. Flavonoids have antioxidant, anti-inflammatory, and cardioprotective properties. Fruits like blueberries, strawberries, and blackberries are full of anthocyanins, flavonoids that give these fruits their deep colour and high antioxidant activity. Citrus fruits like grapefruits, oranges, and lemons contain flavanones, which are associated with cardiovascular health. Leafy greens and onions are great sources of flavanols, which are anti-inflammatory. Seeds and nuts, including walnuts and flaxseeds, contain flavonoids that maintain metabolic well-being and limit oxidative stress. Whole grains, including quinoa and oats, also contribute to the dietary contribution of flavonoids, which lower the risk of chronic disease. Teas, particularly green tea and black tea, contain high levels of catechins and theaflavins, flavonoids that improve vascular function and possess neuroprotective properties. Coffee is also a significant source of several flavonoids contributing to the antioxidant capacity of coffee.

3. Key Functional Food Categories and Their Benefits

Functional foods encompass both naturally occurring and fortified or modified products that provide health benefits beyond basic nutrition. Broadly, they can be categorised into plant-based, dairy-based, meat-based, and other specialised foods. Plant-based functional foods include fruits, vegetables, nuts, seeds, legumes, and whole grains, offering vitamins, minerals, fibre, healthy fats, and bioactive compounds that support metabolic and digestive health. Dairy-based products, such as fortified milk and probiotic yogurt, supply calcium, vitamin D, and beneficial microbes that enhance bone and gut health. Meat-based and other functional foods provide proteins, peptides, and bioactive molecules, collectively contributing to overall wellness and disease prevention. Each category is discussed in detail.

3.1 Plant-based Functional Foods

The plant-derived foods that provide health benefits along with basic nutrition come under the category of plant-based functional foods. The bioactive compounds present in plant-based diets include micronutrients, carbohydrates, healthy fats, carotenoids, polyphenols, vitamins, minerals, dietary fibres, biogenic amines, and peptides (Samtiya et al., 2021). Naturally occurring, plant-derived functional foods can be classified into four main categories: fruits and vegetables, whole grains, legumes, and nuts and seeds. These foods are rich in bioactive compounds that play a crucial role in promoting overall health and preventing disease (Table 2).

3.1.1 Fruits and Vegetables

Fruits like apples, kiwis, cherries, plums, and blueberries are rich sources of hydroxycinnamic acids (HCA) with varying concentrations of approximately 0.5–2 g/ kg of fresh weight. Caffeic acid is the most prevalent form of HCA, accounting for approximately 75% of the total HCA content of fruits (Samtiya et al., 2021). Anthocyanins, responsible for the vibrant colors of many fruits and vegetables, possess antioxidant properties that promote several health benefits, including the regulation of neurological disorders, cardiovascular diseases, diabetes, and cancer. Other important bioactive compounds in plant-based foods include epicatechin, catechin, gallocatechin, tannins, and related polyphenols, which similarly support overall health and disease prevention. Bananas are rich in biogenic amines like serotonin, dopamine, norepinephrine and phenolic compounds like gallic acid, vanillic acid, hydrobenzoic acid, salicylic acid, vitamin E, and vitamin C. These antioxidants regulate the immune system, reduce DNA damage, and prevent lipid peroxidation (Baroni et al., 2024). Red pepper is rich in luteolin and apigenin. Berries, brinjal, and plums are rich in cyanidin, delphinidin, and pelargonidin (Samtiya et al., 2021). Apples, onions, and tea are rich in a flavonoid, quercetin. Grapes and citrus fruits contain naringenin, which regulates cognitive functions. Allium (garlic, onion) is rich in isoalliin (a thiosulphinate) and S-methylcysteine sulfoxide, which confer anti-cancer and anti-diabetic properties (Samtiya et al., 2021). Onions contain organosulphur compounds that contribute to the prevention of diabetes, obesity, and other metabolic disorders. Additionally, vitamin K, found in green leafy vegetables, serves as a cofactor for enzymes that facilitate calcium deposition in bones, supporting skeletal health.

Dietary fibres, found in vegetables, fruits, wheat bran, oats, and other plant-based foods, are non-starchy carbohydrates that are poorly digestible yet offer numerous health benefits. Based on their water solubility, water-holding capacity, and viscosity, dietary fibres can be classified into soluble and insoluble types (Al Ali et al., 2021). Fibres like beta-glucans, fructans, and hemicellulose are soluble viscous fibres and reduce the speed of gastric emptying. Insoluble fibres are non-viscous, water-insoluble fibres which accelerate gastric emptying time and provide relief from constipation. A high-fibre diet helps in treating inflammatory bowel diseases like Crohn's Disease and Ulcerative colitis (Al Ali et al., 2021). Bioactive compounds present in these foods possess anti-inflammatory, anti-oxidant, and anti-cancer effects, hence regulating immune function, improving cardiovascular health, and impacting both physical and mental health (Samtiya et al., 2021).

Table 2 Bioactive components and associated health benefits of different plant-based functional foods.

Food Category	*Examples*	*Key Bioactive Compounds*	*Health Benefits and Function*	*References*
Fruits and Vegetables	Apples, Blueberries, Bananas, Red Pepper, Onions, Garlic, Green Leafy Veg.	Hydroxycinnamic Acids (HCAs), Anthocyanins, Quercetin, Naringenin, Biogenic Amines (Serotonin, Dopamine), Organosulfur Compounds (Isoalliin), Vitamin K, Dietary Fibres (Beta-glucans, Fructans)	Regulate neural disorders, support cardiovascular health, anti-cancer, anti-diabetic, boost immunity, prevent lipid peroxidation, reduce DNA damage, provide relief from constipation (insoluble fibre), slow gastric emptying (soluble fibre).	Samtiya et al., 2021; Baroni et al., 2024; AlAli et al., 2021
Whole Grains	Wheat, Oats, Rye, Barley, Rice, Maize	Arabinoxylans (prebiotic), Ferulic Acid (antioxidant), Methyl Donors, Beta-glucans, Zeaxanthin, Lutein, Dietary Fibre, essential Minerals (Mn, Fe, Mg, Se)	Prebiotic effect, antioxidant property, neurotransmitter synthesis, lower blood cholesterol, improve glycaemic index (Oats, Barley, Rye), promote satiety and weight management (Rye), reduce risk of Type 2 Diabetes, support eye health (Maize), enhance insulin sensitivity, decrease inflammation.	Samtiya et al., 2021
Nuts and Seeds	Almonds, Walnuts, Pistachios, Cashews, Hazelnuts, Flaxseeds, Chia Seeds	Monounsaturated and Polyunsaturated Fats, Omega-3 Fatty Acids, Polyphenols, Vitamin E, Magnesium, Arginine, Phytosterols, Chlorophyll	Reduce LDL and increase HDL cholesterol, regulate blood sugar, produce short-chain fatty acids (gut health), prevent plaque formation in arteries, lower blood pressure, improve cognitive functions, support endothelial function (vasodilation), anti-inflammatory effects (reduce TNF-alpha, CRP).	Li et al., 2009; Singar et al., 2024
Legumes	Kidney Beans, Chickpeas, Lentils, Black Beans, Soybeans	Proteins (Soy protein is complete), Soluble Fibres, Antioxidants (Polyphenols, Flavonoids, Anthocyanins, Saponins), Resistance Starch, Folate, Magnesium, Potassium, Isoflavones	Reduce LDL cholesterol and promote cardiovascular health, weight management (satiety), lower oxidative stress, improve insulin sensitivity, regulate blood sugar (low GI), reduce hypertension, support bone health (Soybeans), reduce menopausal symptoms.	Capurso et al., 2021; Al Ali et al., 2021

3.1.2 *Whole Grains*

Whole grains consist of the entire grain kernel, excluding inedible parts such as hulls and husks. Their main components include the starchy endosperm, germ, and bran. Wheat, for example, contains arabinoxylans that exhibit prebiotic effects, ferulic acid with antioxidant properties, and methyl donors such as betaine and glycine, which are important for neurotransmitter synthesis and overall metabolic health (Neyrinck et al., 2012). Wheat is rich in fibre and helps to regulate digestion and provides several essential B vitamins. Spelt (or dinkel wheat) contains higher levels of soluble polymeric proteins and a greater proportion of gliadins compared to common wheat. Oats are unique among cereals in containing a legume-like protein, avenalin. They are also rich in beta-glucans, which contribute to lowering blood cholesterol, improving glycemic control, and supporting gut health. Rye has a lower gluten content than wheat. It is beneficial for controlling the glycaemic index. It contains a high amount of non-cellulosic polysaccharides, which help in promoting satiety and weight management. Rye also lowers the risk of Type 2 diabetes by improving glucose metabolism in the body (Iversen et al., 2022). Barley contains the highest number of β-glucans among the grains and helps in supporting cardiovascular health and stabilising blood sugar. Rice is low in fat, but some varieties contain high levels of resistant starch, which is beneficial for gut health. Maize contains zeaxanthin, lutein, and high lipid content, which supports eye health (Abdel-Aal et al., 2013b). Whole grains supply essential minerals such as manganese, iron, magnesium, and selenium, which are vital for metabolic and immune functions. Regular consumption of whole grains has been associated with reduced obesity prevalence, improved insulin sensitivity, and decreased inflammation.

3.1.3 *Nuts and Seeds*

Nuts are dry fruits with edible seeds and hard shells. These are categorised into four types: classic nuts, drupes, gymnosperm seeds, and angiosperm seeds (Li et al., 2009). Globally, almonds and hazelnuts are the most consumed nuts. Peanuts are technically legumes but due to their similar nutritional profile and use, they are categorised in the category of nuts. Nuts are different from seeds as they are larger and have a harder shell. Moreover, nuts have a single seed or kernel, whereas seeds are present in large quantities within a fruit. Nuts have higher fat and calorie content than seeds, and seeds have higher protein and fibre content than nuts. Examples of nuts include walnuts, almonds, and hazelnuts, while seeds comprise flaxseeds, chia seeds, and pumpkin seeds (Li et al., 2009). Almonds are rich in monounsaturated and polyunsaturated fats, protein, and dietary fibre, and provide significant amounts of magnesium, potassium, vitamin E, and polyphenols (Singar et al., 2024). Almonds reduce LDL cholesterol and increase high-density lipoprotein levels, thereby improving the overall lipid profile of an individual. It regulates blood sugar by reducing the insulin resistance in patients with Type 2 diabetes. It also produces the short-chain fatty acids, which enhance the gut microflora (Singar et al., 2024). Walnuts are rich in healthy fats, protein, dietary fibre, magnesium, selenium, and bioactive compounds such as polyphenols and ellagitannins. Regular consumption helps prevent arterial plaque formation, reducing the risk of atherosclerosis, heart attacks, and strokes. Walnuts also support cardiovascular health by lowering blood pressure and improving cholesterol profiles. The presence of omega-3-fatty acids further improves the cognitive functions and reduces the inflammatory cytokines (Li et al., 2009). Pistachios are rich in potassium, phytosterols, and L-arginine which reduce the LDL cholesterol and improve epithelial function. They also help to regulate blood sugar and body weight. Pistachios also contain chlorophyll and L-arginine. Chlorophyll is conferred for its green colour and anti-oxidant properties whereas L-arginine supports nitric oxide production, which serves as a vasodilator (Singar et al., 2024). Another nut, cashews, is a rich source of magnesium and copper and has a lower glycaemic index (Gonçalves et al., 2023). Hazelnuts are excellent sources of tocopherols (vitamin E), magnesium, and manganese. They promote skin, cardiovascular, and brain health by reducing oxidative stress. Magnesium contributes to the prevention of metabolic syndrome and Type 2 diabetes, while plant sterols in nuts help lower intestinal cholesterol absorption.

Nuts also exhibit anti-inflammatory effects. They reduce the inflammatory markers like C-reactive protein, ICAM-1, TNF-αwith the help of polyphenols, vitamin E, and magnesium (Li et al., 2009). Anti-inflammatory properties help in preventing vascular damage, arterial stiffness and insulin resistance. Nuts like walnuts and almonds have arginine which acts as a precursor of nitric oxide and enhances the blood flow. Production of nitric oxide increases in our body by consuming these nuts, and this improves the endothelial function of blood vessels, promoting vasodilation. This prevents hypertension, stroke risk, and improves the blood circulation (Li et al., 2009).

3.1.4 Legumes

Legumes are nutrient-rich foods containing essential vitamins, minerals, fibres, proteins, and carbohydrates. They are naturally low in fat and are cholesterol-free. They are excellent sources of proteins and therefore help in muscle growth and weight management. Soy protein reduces LDL cholesterol and promotes cardiovascular health (Capurso et al., 2021). Soluble fibres in legumes bind to bile salts and lower the blood cholesterol. Intake of high fibres contributes to satiety, reduction of overall calorie intake, and helps in weight management (Al Ali et al., 2021). Antioxidants such as polyphenols, flavonoids, and anthocyanins in legumes contribute to cardiovascular health by reducing oxidative stress. Kidney beans are rich in magnesium, iron, zinc, phosphorus, and resistant starch, which promotes gut microbiota and improves insulin sensitivity. Regular consumption of kidney beans helps lower LDL cholesterol and triglycerides, and aids in blood sugar regulation due to their low glycaemic index, typically ranging from 10 to 40 (Capurso et al., 2021). Chickpeas are good sources of folate, iron, and magnesium. They contain saponins, having anti-oxidant and anti-inflammatory properties. They help in weight management and also improve glycaemic control in diabetes. Lentils are rich in Vitamins (B), iron, potassium, and magnesium. Due to high potassium and magnesium content, they reduce hypertension. By lowering cholesterol, they prevent cardiovascular diseases. These do not require soaking before cooking. Black beans are rich in anthocyanins and flavonoids. These are one of the highest antioxidant-containing legumes. They scavenge the free radicals and prevent lipid peroxidation to prevent various chronic diseases. They reduce the oxidative stress and inflammation. They also help in weight loss. Since soybeans contain all the essential amino acids, they are considered as a complete protein (Hernández-Ruiz et al., 2025). They are also high in isoflavones. They have a good amount of magnesium and calcium which support bone health. They also help in reducing the menopausal symptoms and lowering the risk of cardiovascular diseases (Capurso et al., 2021).

To maximise their health benefits, legumes can be consumed in diverse forms such as sprouts, cooked preparations, fermented products, roasted snacks, or blended into functional food formulation (Table 3). Whole cooked legumes are preferred for retaining fibre content for enhanced digestion and heart health, preserving proteins for muscle maintenance and antioxidants to reduce oxidative stress. Sprouted forms enhance the nutrient absorption and digestibility. These enhance the bioavailability of vitamins, reduce antinutrients like lectins and phytic acids, which interfere in mineral absorption (Ikram et al., 2021). Fermented legumes improve gut health and reduce bloating. Tempeh, miso soup, and fermented black bean paste are a few examples of fermented legumes. Roasted or dry-roasted forms have enhanced crunch and flavour. Legumes are rich in resistant starch, contributing to improved gut health and metabolic regulation. Commercially, roasted forms such as chickpeas, peanuts, and soybeans are widely available. Legume-based flours serve as gluten-free, high-protein alternatives to wheat flour, while blended legume dips and spreads provide healthier substitutes for butter and mayonnaise. Hummus is prepared by blending chickpeas, tahini, lemon juice, and olive oil. Black bean dip is prepared with garlic and lime juice. Legumes are also consumed in liquid forms, such as smoothies and shakes, for example soy milk, or lentil and peanut butter shakes. These beverages serve as high-protein, high-fibre dietary options, supporting satiety and overall nutritional balance (Capurso et al., 2021).

Table 3 Different legume preparation methods and their key health benefits.

Consumption Form	*Preparation Method*	*Primary Benefits*
Whole Cooked	Cooked without modification	Retains maximum fibre, preserves protein, maintains antioxidants, enhanced digestion and heart health.
Sprouted	Legumes are germinated	Enhanced nutrient absorption and digestibility, increased bioavailability of vitamins, reduced anti-nutrients (lectins, phytic acids).
Fermented	Processed through fermentation (e.g., Tempeh, Miso, Black Bean Paste)	Improved gut health, reduced bloating.
Roasted/Dry-Roasted	Dry-roasted (e.g., Chickpeas, Peanuts, Soybeans)	Enhanced crunch and flavour, high resistant starch.
Flour	Ground into flour (e.g., chickpea flour)	Gluten-free and high-protein alternative to wheat flour.
Blended Dips/ Spreads	Blended with other ingredients (e.g., Hummus, Black Bean Dip)	Healthy alternatives for high-fat spreads, high in fibre.
Liquid Forms	Made into beverages (e.g., Soy milk, shakes)	High-protein and high-fibre liquid diets.

3.2 Dairy-based Functional Foods

Dairy-derived foods constitute a major source of biologically active compounds, including antioxidants, organic acids, proteins, and other bioactive constituents that support a wide range of physiological and metabolic processes. Their established nutritional value, combined with extensive global consumption, makes dairy products an integral component of a balanced human diet. In recent years, growing attention has been directed toward functional foods, as these products offer benefits beyond basic nutrition by supporting health maintenance, enhancing physiological functions, and reducing disease risk. Given the widespread consumption of dairy foods, they have emerged as one of the most promising platforms for the development of functional food products. Consequently, the dairy industry has increasingly focused on value-added formulations that incorporate functional ingredients such as bioactive peptides, probiotics, and prebiotics, thereby supporting product diversification while simultaneously contributing to the reduction of food-chain waste (Lamsar and Abhilasha, 2023). Several conventional dairy products are now recognized for their functional attributes, including milk, cream, condensed and powdered milk, yogurt, butter, kefir, ricotta, casein, cheese, fermented dairy beverages, infant formula, and ice cream. Beyond their nutritional contributions, these products contain specific chemical constituents that positively influence human health and overall well-being. Dairy foods provide a wide spectrum of essential nutrients, notably proteins, vitamins, and minerals; however, certain formulations may exhibit limitations such as reduced bioavailability of active compounds or nutritional imbalances. A representative example is milk fortified with calcium and vitamin D, which addresses deficiencies commonly observed in unfortified products. Notably, dairy foods account for approximately 60–70% of total calcium intake in the typical Western diet (Burckhardt, 2015).

Fermented dairy products exert health benefits through two principal mechanisms. The first is a probiotic effect, arising from the consumption of viable beneficial microorganisms, while the second is a biogenic effect linked to microbial metabolites synthesised during fermentation, including vitamins, peptides, oligosaccharides, proteins, and organic acids. These products provide an effective vehicle for delivering probiotic bacteria to the gastrointestinal tract. Due to their substantial content of organic acids—such as lactic, acetic, butyric, and citric acids—fermented products including yogurt, kefir, koumiss, cheese, acidophilus–bifidus milk, cream, and other cultured beverages create favourable conditions for probiotic survival and colonisation (Widyastuti et al., 2021). The health-

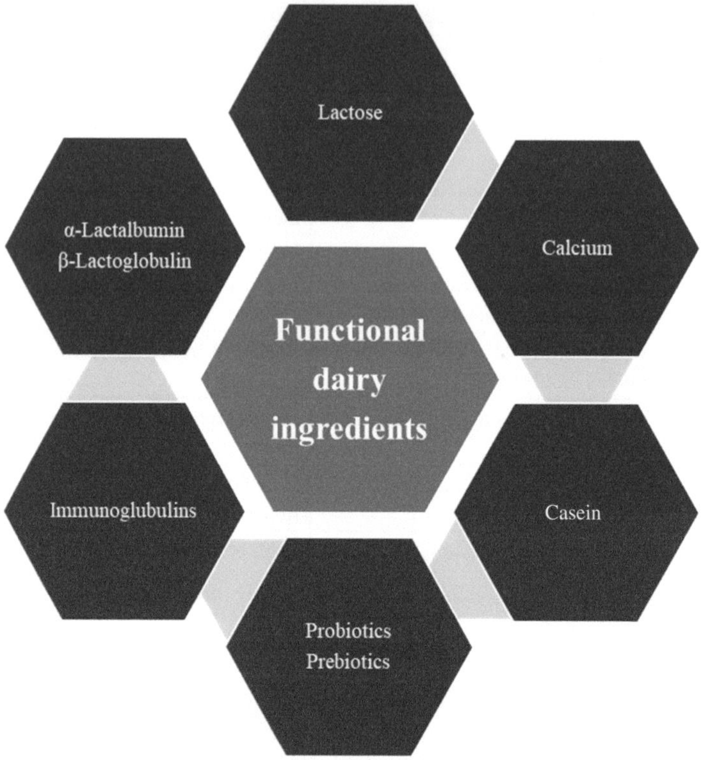

Fig. 3 Functional bioactive compounds used for fortification of dairy-based food products.

promoting action of probiotics is largely attributed to their capacity to modulate immune responses by stimulating innate immunity and influencing the production of cytokines and antimicrobial peptides (Biancone et al., 2002). In addition, whey proteins present in dairy products exhibit a range of biological activities, including immunomodulatory, antibacterial, anticarcinogenic, and general health-enhancing effects. Beyond supplying essential nutrients and energy, dairy-based functional foods also contribute to the maintenance of gut epithelial integrity (Cena and Calder, 2020). Collectively, the functional components present in dairy products have been shown to assist in lowering blood pressure, reducing cholesterol absorption, regulating appetite and satiety, improving metabolic disturbances linked to obesity, and potentially exerting antimicrobial effects (Figure 3) (Anastasova et al., 2018).

3.2.1 Probiotic Yogurt

Yogurt is a fermented dairy product produced through the action of two thermophilic, homofermentative lactic acid bacteria, namely *Lactobacillus delbrueckii* subsp. *bulgaricus*, and *Streptococcus thermophilus* (Ozcan-Yilsay et al., 2007). It is widely recognised as a functional dairy food due to its ability to supply essential nutrients along with biologically active components that contribute to human health and longevity. In recent years, yogurt consumption has increased markedly, largely driven by growing awareness of its health-related benefits. Owing to its balanced composition of fats, proteins, and key micronutrients—including calcium, magnesium, potassium, zinc, iodine, and several vitamins—yogurt is often described as an almost complete food (Zhu et al., 2021). The fermentation process enhances the nutritional quality of yogurt by improving the bioavailability of minerals and vitamins such as calcium, magnesium, riboflavin (vitamin B2), and cobalamin (vitamin B12), while also generating bioactive peptides. Furthermore, yogurt proteins exhibit high digestibility, facilitating efficient nutrient absorption. Several studies have demonstrated that regular yogurt intake supports lactose digestion, strengthens immune responses, and reduces the risk of gastrointestinal disorders (Anastasova et al., 2018). Yogurt also represents an effective medium for probiotic delivery, as it promotes the survival and activity of beneficial microorganisms within the gut. The incorporation of probiotic strains into yogurt offers a practical, cost-effective, and consumer-acceptable approach to producing functional fermented dairy products. The combined presence of probiotics and prebiotic components in yogurt has been associated with improved cardiovascular health outcomes, including reduced blood pressure, enhanced weight management, and lower serum cholesterol levels (Khalaf et al., 2021).

Based on microbial composition, yogurt can be broadly categorised into conventional culture yogurt and probiotic yogurt. Traditional yogurt relies solely on *S. thermophilus* and *L. bulgaricus*, whereas probiotic yogurt includes additional beneficial microorganisms, most commonly strains of *Bifidobacterium* and *Lactobacillus acidophilus* (Hadjimbei et al., 2022). Probiotics may be introduced either independently or alongside starter cultures; when combined, starter cultures enhance probiotic activity, resulting in synergistic health benefits. A wide variety of probiotic-enriched yogurts containing diverse microbial strains have been developed globally, and extensive research has focused on maintaining probiotic viability during fermentation and refrigerated storage (Mani-López et al., 2014).

Yogurt is commercially available throughout the year and is offered in multiple forms, including set, stirred, and drinkable varieties, with variations in fat content, texture, and flavour. It may be consumed as a standalone snack or incorporated into meals in sweet or savoury preparations, with formulations that include fruit, cereals, or natural flavours. This versatility, together with its reputation as a nutritious and wholesome food, has contributed to its widespread acceptance across all age groups. Notably, yogurt combining prebiotics (dietary fibres) and probiotics such as *Lactobacillus rhamnosus* GG has been shown to help maintain healthy fecal microbiota in individuals suffering from inflammatory bowel syndrome. Beyond probiotic delivery, yogurt serves as an ideal base for fortification with a broad range of bioactive compounds. Its nutritional value can be further enhanced through supplementation with cereal grains, nuts, and fresh or dried fruits (Anastasova et al., 2018). For example, Khalil et al. (2022) developed a functional probiotic yogurt fortified with white sapote fruit pulp at concentrations of 5%, 10%, and 15% in combination with *Bifidobacterium longum*. During refrigerated storage at 5°C for 21 days, the fortified yogurt samples exhibited improved physicochemical, microbiological, and sensory characteristics. Compared to control samples, fruit-enriched yogurts showed higher contents of protein, carbohydrates, total solids, and ash. The incorporation of white sapote pulp also influenced syneresis and water-holding capacity, while significantly increasing levels of ascorbic acid, vitamins, minerals, total phenolics, and antioxidant activity (Khalil and Elkot, 2020). For individuals with lactose intolerance, yogurt offers a distinct advantage over other dairy products. The metabolic activity of yogurt cultures facilitates lactose breakdown, thereby improving digestion and tolerance in individuals with reduced lactase activity. Additionally, the yogurt matrix protects bacterial lactase from gastric acidity, enhancing its functional efficacy. Yogurt has also demonstrated considerable potential as a nutritional intervention in HIV-infected populations (Hemsworth et al., 2010). In resource-limited settings, yogurt consumption can substantially improve dietary protein intake and overall nutritional adequacy, which are critical for immune system function. Moreover, yogurt serves as a delivery vehicle for probiotics, prebiotics, and micronutrient fortification, all of which have been shown to mitigate microbial translocation and chronic inflammation associated with HIV infection (Anastasova et al., 2018).

Clinical studies have reported that probiotic yogurt fortified with *Lactobacillus rhamnosus* GR-1 enhances immune parameters, including CD4 cell counts, improves tolerance to antiretroviral therapy, and supports overall nutritional status and productivity in HIV-positive individuals. Additional investigations indicate that yogurt enriched with *L. rhamnosus* GR-1 and selected micronutrients may slow HIV disease progression, although its impact on intestinal microbiota composition appears to be limited (Sarkar, 2019).

3.2.2 Fortified Milk

Milk has long occupied a central position in human nutrition and is often described as one of nature's most complete foods. It serves as the primary source of nourishment for newborns, supplying essential macronutrients such as proteins, lipids, and carbohydrates, along with vital minerals required for early growth and development. Beyond infancy, milk continues to play an important role in the diets of adults and elderly populations by providing a readily available and balanced supply of nutrients. Its classification as a physiologically functional food arises from its rich content of essential minerals, including calcium, phosphorus, and magnesium, as well as vitamins such as riboflavin and β-carotene, which are present in highly bioavailable forms that are efficiently absorbed by the human body.

During early life, milk supports physiological adaptation by facilitating digestion, promoting growth, and delivering bioactive components necessary for healthy development. It contributes to the maturation of lymphoid tissues and encourages the establishment of beneficial intestinal microbiota. Increasing evidence suggests that the diversity, concentration, and biological relevance of milk-derived bioactive compounds are greater than previously assumed. Because dairy products are widely consumed, affordable, and nutritionally dense, they represent an effective platform for mineral fortification. Milk, in particular, allows the incorporation of additional nutrients with minimal impact on sensory characteristics while preserving nutrient stability and bioavailability (Ocak and Rajendram, 2013).

Owing to these attributes, milk has been extensively investigated as a base material for the development of functional foods enriched with probiotics and other bioactive ingredients. Among the most extensively studied approaches is the fortification of milk with probiotics and prebiotics (Sazawal et al., 2010). During fermentation, probiotic microorganisms metabolize milk components to generate unique bioactive substances known as probioactives, which originate from casein hydrolysis and are exclusive to fermented foods. This characteristic makes dairy matrices more effective probiotic carriers than dietary supplements (Champagne et al., 2018).

Vitamins are organic micronutrients required in small quantities for normal physiological functioning and must be obtained through diet due to the body's inability to synthesise them. Vitamin D plays a critical role in calcium absorption and bone mineralisation; its deficiency leads to rickets in children and osteomalacia in adults (Sahay and Sahay, 2012). Vitamin A is essential for vision, immune function, growth, reproduction, and epithelial tissue maintenance, with deficiency potentially causing night blindness, xerophthalmia, epithelial keratinization, and, in severe cases, mortality. To mitigate these deficiencies, milk is frequently fortified with vitamins A and D, as dairy products provide an ideal medium for vitamin delivery. In several countries, such fortification is mandated by regulatory authorities. Additionally, β-carotene is added to certain dairy products, such as butter, to enhance visual appeal.

Calcium in milk exists primarily as a calcium caseinate–phosphate complex, which is readily released during digestion. Given the importance of calcium intake for skeletal health, regular consumption of calcium-fortified milk, combined with an active lifestyle, may help reduce the risk of osteoporosis and bone fractures. Powdered milk is commonly enriched with vitamins A and D, as well as minerals such as calcium and iron. The choice of mineral fortificant depends on several factors, including sensory impact, bioavailability, safety, and cost. Iron fortification poses particular challenges, as certain iron compounds can induce undesirable colour changes and off-flavours in dairy products. Soluble iron salts may promote lipid oxidation, resulting in metallic tastes, while compounds such as ferrous sulfate can impart greenish or grey discolouration to infant foods (Kaur et al., 2022). These issues can be minimised through encapsulation techniques, such as coating iron salts with ethyl cellulose or hydrogenated oils (Jackson and Lee, 1991).

Iron-fortified milk has been developed as a nutritional strategy to address iron-deficiency anaemia, with multiple studies demonstrating its effectiveness in improving haemoglobin levels (Rajasekaran and Kalaivani, 2013). A notable advantage of milk-based iron fortification is the relatively high bioavailability of iron delivered through this matrix. In contrast, milk contains only trace amounts of selenium, prompting interest in selenium fortification strategies. One promising approach involves fermenting milk supplemented with selenium, which may enhance dietary intake of organic selenium compounds. Consumption of selenium-enriched milk has been shown to increase systemic selenium levels and may contribute to a reduced risk of certain cancers (Anastasova et al., 2018).

Zinc is another essential micronutrient naturally present in milk, although research on zinc fortification remains limited. Available studies indicate that adolescent females consuming zinc-fortified milk at recommended intake levels exhibit improved zinc absorption and nutritional status. Sustained consumption for periods exceeding 27 days has been suggested to achieve physiological requirements and maintain adequate zinc levels (Méndez et al., 2012). The success of milk fortification strategies depends on factors such as the nutritional needs of the target population, the sensory stability of the final product, and nutrient retention during processing and storage. One of the major challenges in fortification is the risk of sensory deterioration and nutrient loss—particularly vitamin C degradation—when iron salts are incorporated.

Recent advances in micro- and nanotechnology provide innovative solutions to these challenges by enabling controlled nutrient delivery and improved bioavailability. Techniques such as microencapsulation have gained attention as effective methods for incorporating sensitive bioactive compounds into milk while preserving product quality and functionality (Kwak et al., 2014).

3.3 Functional Beverages

Drinks containing bioactive molecules like vitamins, minerals, probiotics, herbs, and antioxidants that confer various health benefits like digestion, immunity, energy levels, mental clarity and stress relief, come under the category of functional beverages. Recently these functional beverages are increasing in demand for health-conscious customers looking for products that are not only palatable but also help in attaining a healthy lifestyle. Functional drinks constitute probiotic beverages, energy drinks, herbal teas, sports, and electrolyte drinks and various detox beverages (Gupta et al., 2023). Probiotic beverages like kombucha and yogurt drinks consist of live microorganisms that help improve digestion and gut health. Energy drinks provide a boost to the energy level of an individual by incorporating caffeine, sugar, and vitamins but must be consumed in moderation to avoid potential side effects. Herbal teas like chamomile tea, green tea, peppermint tea contain herbs with medicinal properties known for its benefits such as relaxation, stress reduction, and improved digestion (Sugajski et al., 2023).

3.3.1 Tea as a Functional Beverage

Tea is the most popular and widely consumed largely all across the world due to its known stimulating properties. Tea can be categorised into different types on the basis of its source of origin and fermentation. Green tea and black tea are produced from *Camellia sinensis* plant, I known as non-herbal teas, whereas tea produced from other medicinal plant species are known as herbal teas (Ruchika and Sehgal, 2021). To make herbal tea, fresh or dried roots, leaves, flowers, fruits, seeds, stems, barks, or whole plants are brewed or boiled. Although, some teas like green tea can also be prepared by soaking the plant part directly in cold water for a definite time. Herbal teas, like oolong tea and black tea, are prepared by semi-fermentation and complete fermentation processes, respectively (Sugajski et al., 2023).

Yerba Mate tea is a functional tea that is prepared by infusing the leaves of *Ilex paraguariensis*. This tea is consumed widely as in South America a non-alcoholic beverage due to its active ingredient in dietary supplements. This tea has been used for centuries by indigenous people as a medicinal beverage (Heck and De Mejia, 2007). Yerba tea is hypocholesterolemic, hepatoprotective, a nervous system stimulant, diuretic, thus conferring various health benefits. It is also beneficial for the cardiovascular system and obesity management. Studies suggested that this tea prevents oxidative DNA damage and *in vitro* lipoperoxidation of low-density lipoprotein (Xu et al., 2025). Yerba tea has been reported to prevent different types of cancers.

Liang cha, also known as cooling tea, is produced from water infusion of a variety of plant species from tea plant *Camellia sinensis*; Theaceae (Liu et al., 2013). This tea is produced from plant materials like fresh or dried leaves, pollens, nuts, flowers, fruits, seeds, barks from single as well as multiple species. The Lingnan region of China experiences a health condition known as internal heat due to extreme conditions of damp humidity and high heat levels of heat in this region (Tang et al., 2025). Ling cha provides a cooling effect, reduces internal heat, and treats associated health disorders (Liu et al., 2013). Green tea is another example of herbal tea most commonly used as a functional food beverage. The extracts of green tea contain a unique set of catechins that is rich in antioxidants and has biological activity in angiotensinogenesis with antiproliferative assays that has significant potential in the treatment of different types of cancer (Cheng et al., 2020). *Castanopsis lamontii* also known as "Shaji" is a functional herbal tea popular in Southwest China produced from buds of plants of the Fagaceae family (Gao et al., 2019). The members of this family contain functional components like polyphenols, triterpenoids, and flavones and extracts of these plants also show anti-inflammatory and antioxidant properties. This tea is commonly used in Southwest China to prevent oral inflammation and refresh breath (Gao et al., 2020). *Jasonia glutinosa* or rock tea is a popular herbal tea in the Iberian Peninsula used as a traditional medicine due to its tremendous digestive properties. This tea contains essential oil-rich compounds like camphor and borneol. The plant and components of this tea show significant antioxidant, antiprotozoal and anti-inflammatory activity (Valero et al., 2013). Chamomile tea is another popular single-ingredient herbal tea produced from dried flower heads and is widely used for medicinal purposes. The medicinal properties of this tea are attributed due to its constituents of the flowers that contain various phenolic compounds like flavonoids apigenin, patuletin, quercetin, luteolin and their glucosides (McKay and Blumberg, 2006a). Chamomile shows antioxidant, antimicrobial, and significant antiplatelet activity. Peppermint tea is rich in rosmarinic acid, eriocitrin, hesperidin, and luteolin and shows significant antioxidant, antiviral, and antimicrobial activities. It offers a calming effect on the stomach and intestines, while also producing an anesthetic effect on the body's nervous system (McKay and Blumberg, 2006b).

3.3.2 Coffee as a Functional Beverage

Coffee is one of the most widely consumed beverages globally and is produced extensively in approximately 60 tropical and subtropical countries. The primary sources of coffee are the beans of *Coffea arabica* (Arabica) and *Coffea canephora* (Robusta) (Kemsley, 1995). Traditionally, the beneficial effects were attributed solely due to the presence of caffeine, but recent studies have shown presence of other potential compounds contributing to its functional properties. The composition of coffee beans and roasting process determine the functional properties of coffee. Coffee is rich in bioactive compounds such as caffeine, chlorogenic acids, diterpenes like kahweol, cafestol, and polyphenols. Coffee acts as a stimulant and improves alertness due to high caffeine content and also shows significant antioxidant and anti-inflammatory properties. Recent studies have shown that coffee helps in weight management as it enhances metabolism and contains fat oxidation properties. Coffee consumption is also helpful in the maintenance of gut microbiota, thus improving digestion and overall gastrointestinal health (Jaquet et al., 2009)

Coffee can be classified into several types based on the source of the beans and the brewing process, as described below:

3.3.3 Green Coffee

Green coffee is produced from unroasted coffee beans from Coffea species grown in subtropical and tropical countries like Brazil, Ethiopia, Colombia, and Vietnam (Da Cruz Correia et al., 2024). The harvesting process includes handpicking up of coffee cherries on ripening followed by extraction of beans from the cherries by the dry (natural) method or wet (washed) method. These beans are further dried to remove the moisture content and milled to obtain green coffee. The main component of green coffee is chlorogenic acids, which aids in weight loss by boosting metabolism that helps to reduce the body fat and also helps in regulating blood sugar levels (Kanchanasurakit et al., 2023). It also includes other health benefits since it is rich in antioxidants, it helps to reduce blood pressure, lower LDL (bad cholesterol), and improve alertness by reducing inflammation in the brain.

3.3.4 Decaffeinated Coffee

Decaffeinated coffee is primarily produced from the Arabica (Coffea arabica) coffee beans due to its lower caffeine content (1.2% content) (Silvarolla et al., 2000). As its name suggests, this coffee is low in caffeine, maintaining its flavour and beneficial properties. The coffee beans are decaffeinated by several methods like ethyl acetate process, Swiss water process, and the carbon dioxide process. The ethyl acetate process uses ethyl acetate to extract the caffeine from the coffee beans while preserving the coffee's original flavour. The Swiss water process is a chemical-free method based on the osmosis process in which beans are soaked in hot water to dissolve the caffeine followed by filtration to remove the dissolved caffeine from the water further reintroduced to the beans which enables it to reabsorb the remaining flavour compounds. In the carbon dioxide process, a supercritical extraction method is performed in which carbon dioxide is used as a solvent to remove the caffeine (Hiraga et al., 2025). Decaffeinated coffee is rich in antioxidants and other bioactive compounds like chlorogenic acids, thus conferring health benefits like reducing risk of Type 2 diabetes and colorectal, improving liver health, head, and neck cancer, and also helps in improving mental health.

3.3.5 Monsooned Coffee

Monsooned coffee is produced from Arabica and Robusta coffee beans, mostly in the Malabar coast of India, especially in Karnataka, Kerala, and Goa (Menon et al., 2022). The coffee is produced by harvesting ripe coffee cherries followed by drying in the sun to remove the moisture content. The dried beans are further stored in warehouses until the monsoon season; afterwards in the months of June to September, these are exposed to the moist monsoon winds (Murthy and Manonmani, 2009). On exposure to moist monsoon winds, the beans absorb moisture, swell, and turn into a pale golden hue colour along with intense, mellow, and musty flavour. This coffee is rich in antioxidative, anti-inflammatory properties with enhanced cognitive functions. Research studies suggest that this coffee is the well-known and distinctive coffee exported from India to Europe, Asia, Africa, and North America (Jagadeesh et al., 2024). This coffee is mostly used in blended form with other more acidic coffees to provide mellow flavour and impart richness to their texture (Farah, 2012).

3.3.6 Instant Coffee

Instant coffee is one the most convenient beverages produced from ground and roasted coffee beans by treating them with hot water, followed by high-pressure water extraction to obtain the soluble material. The soluble material is further heated, concentrated, and spray dried or lyophilized to obtain the soluble powder form. This process helps to preserve the flavour and aroma of coffee, resulting in a product which can be easily dissolved in hot water (Ishwarya and Anandharamakrishnan, 2014). Instant coffee is rich in bioactive compounds like caffeine and antioxidants. It also contains diterpenes compounds such as cafestol and kahweol, which confer anti-inflammatory and anticancer effects on the body (Zanetti et al., 2021).

3.4 Other Beverages

Apart from tea and coffee, popular functional beverages include probiotic- and prebiotic-fortified yogurt drinks, omega-3-enriched juices, functional milk, and vitamin- and mineral-fortified drinks.

These beverages offer a range of health benefits, such as reducing the risk of breast cancer, enhancing immune function, and improving mental and physical well-being. They may also help alleviate stress, slow the aging process, and provide anti-inflammatory and antioxidant effects (Tolun and Altintas, 2019). The different types of fortified functional beverages are described in detail below.

3.4.1 Phyto-active Fortified Drinks

These drinks are fortified with various macro and micronutrients and bioactive compounds like phytosterols, isoflavones, arabinose, tocopherol, β-glucan, β-sitosterol, and many other vitamins (Munekata et al., 2020).

3.4.2 Prebiotic and Probiotic Drinks

These beverages dominate the functional beverage consumer market. These drinks are fortified with neutral milk oligosaccharides, which help in maintaining the beneficial gut microflora. Moreover, these are fortified with fructo-oligosaccharides and other prebiotics like maltodextrin, pectin, and mannitol that help in the development of probiotics (Rivas and Cabral, 2021).

3.4.3 Beauty Drinks

Beauty drinks, primarily developed by the cosmetic industry, typically contain a combination of hydrolysed collagen powder and natural fruit juices. These beverages, often marketed as beauty elixirs, are intended to support skin health and overall aesthetic wellness. These drinks help in improving skin hydration, retaining moisture, decreasing dermal collagen fragmentation and preventing wrinkles (Polonini et al., 2021).

3.4.4 Sports Drinks

Sports drinks, also referred to as carbohydrate–electrolyte beverages, often contain bioactive compounds such as the alkaloid caffeine and the flavonoid catechin as key ingredients. Other functional ingredients include ginseng, guarana, acai, vitamin B, and ginkgo. Sports drinks help in hydrating the body of athletes (Berry et al., 2022).

3.4.5 Cognitive and Immune System Enhancers

These beverages, often referred to as psychobiotic drinks, are used in the management of psychiatric disorders. In addition to supporting mental health, they also enhance immune function and promote alertness. These drinks, on ingestion in adequate amounts of neuroactive compounds like gamma-aminobutyric acid, dopamine, norepinephrine, and serotonin, act as regulators of the central nervous system (de Araújo et al., 2020).

3.5 Functional Meat Products

Meat products contain many bioactive compounds like vitamins, minerals, proteins, and fatty acids. The advancements in the food sector and growing awareness about the health benefits of functional foods has led to the development of healthy meat products. Recent studies reported that meat products are supplemented with functional constituents like fatty acids, vitamins, minerals, probiotics, bioactive peptides, probiotics, and dietary fibres that impart various health benefits to the consumers (Pogorzelska-Nowicka et al., 2018). Meat products are fortified with omega-3 fatty acids like EPA and DHA which act as anti-inflammatory agents and help in restoring heart health. The most common method used for meat fortification is microencapsulation, which additionally helps in enhancing the shelf- life of the products by preventing oxidation of fatty acids (Feizollahi et al., 2018).

3.5.1 Omega-3-fatty Acids Fortified Meat Products

Omega-3 fatty acids are essential fatty acids that cannot be synthesised by the human body and must be obtained through diet. Flaxseed, chia seeds, and fish oil are rich in these fatty acids. Fatty acids can be incorporated into meat products by feeding the livestock with omega-3-fatty acid rich foods. The studies showed that feeding chickens with flaxseeds enhanced the omega-3 levels in their meat and eggs (Whittle et al., 2024). Omega-3 oils can be directly incorporated into meat products such as sausages and burgers during processing. Fortification with omega-3 fatty acids provides multiple health benefits, including lowering triglyceride levels, reducing the risk of cardiac arrhythmias, and promoting overall cardiovascular health. DHA, a type of omega-3 fatty acids, also helps in the reduction of risks of neurodegenerative disorders (Feizollahi et al., 2018).

3.5.2 Fortification of Meat Products with other Nutrients

Meat products can also be fortified with essential nutrients such as iron, selenium, calcium, magnesium, and iodine. Detailed information about the roles and benefits of these nutrients is provided below:

3.5.2.1 Iron

Iron is a divalent metal and a critical component of human nutrition. It supports several physiological functions, including oxygen transport, enzyme synthesis, energy production, and immune system regulation. Meat can be fortified with iron but fortification of iron in meat leads to rapid oxidation of lipids and deterioration of color (Radlowski and Johnson, 2013). Encapsulation of iron in liposomes before addition to meat can overcome the problem of lipid oxidation. The addition of iron to meat patties may improve the nutritional profile of meat and meat products.

3.5.2.2 Selenium

Selenium is an essential component of selenoproteins, playing a key role in regulating redox reactions and neutralizing free radicals. It also plays a role in regulating the immune system, metabolism of thyroid hormone, and the prevention of diseases like diabetes, cardiovascular diseases and certain cancers. Supplementation of animal feed with selenium can be used for fortification of meat of animals with selenium (Hadjimbei et al., 2015). Selenium-rich animal feed increases the stability of lipid oxidation by decreasing microbial growth. It also makes the meat firmer and chewier.

3.5.2.3 Calcium and magnesium

Calcium is one of the most important cations in the body and plays an important role in muscle functions, intracellular transmission, nerve transmission, vascular contraction, and dilation. Magnesium is essential for regulating protein and nucleotide synthesis and serves as a cofactor for numerous enzymatic reactions. Since meat products are typically low in magnesium, fortification can help address this nutritional gap. The concentration of calcium and magnesium can be increased by supplementation of animal feed with these minerals in the form of salts or brine to meats. The study showed that higher concentration of calcium and magnesium was found in meat obtained from animals fed with calcium and magnesium-enriched diets (Dikeman et al., 2003).

3.5.2.4 Iodine

Iodine is an essential element for thyroid hormone synthesis. The diet deficient in iodine leads to health problems like deformations in skeletal muscles, retarded physical and mental growth, goitre, and decreased fertility. The symptoms are collectively referred to as iodine deficiency disorder (Kapil, 2007). The iodine content can be increased in meat by increasing the iodine content in animal feed.

3.6 Functional Food Regulations: Standardized Labelling, Verified Claims, and Consumer Safety

The regulatory landscape of functional foods encompasses several aspects such as consumer protection, health claims, and labelling requirements. A robust regulatory framework is required to avoid inconsistent labels and claims which might cause confusion and misinformation to consumers. Creating rules for food labelling in the US, including functional foods, was an essential step by the Nutrition Labelling and Education Act (NLEA) of 1990 (Ross, 2000). In order to empower customers by making educated diet decisions, the legislation sought to incentivise producers to give accurate and clear nutritional information. The prevalence of nutrition labelling rose from 65.9% to 75.5% after the NLEA was passed, yet despite these efforts, there is still much improvement required in terms of consumer awareness and compliance (Wang et al., 2011). The absence of precise definitions and guidelines for what qualifies as a functional food makes the regulatory landscape more difficult to navigate. Labelling the functional foods with health claims impacts consumer behaviour and serves as a primary reason for purchasing functional foods. Kirwan et al. (2016) suggested tailoring the labelling format to accommodate consumer preference and comprehension where nutrient values are stated per portion size instead of nutrient values per 100 g (Kirwan et al., 2016). Poor dietary choices can be made if consumers lack understanding of food labels and nutritional information, which further necessitates standardised labelling. The Food Allergen Labelling and Consumer Protection Act (FALCPA) in the U.S. mandates that packaged foods clearly indicate the presence of major allergens, yet compliance remains inconsistent (Gendel and Zhu, 2013). A lack of clarity in product information can undermine consumer trust and impede the growth of the functional food market. Allergen labelling presents an additional regulatory challenge, as inadequate labelling poses serious health risks for individuals with food allergies. Inconsistent or inaccurate labelling can create dangerous situations for consumers who depend on precise information to avoid allergens. Regulatory frameworks for nutraceuticals vary significantly across regions, impacting market entry and product safety. Regulatory frameworks for nutraceuticals and dietary supplements vary significantly across jurisdictions, impacting product safety, market access, and global trade. The FDA has an important role in guaranteeing that food label health claims are supported by sound scientific data, generally in the form of robust clinical trials and epidemiologic studies (Méndez and Stratton, 2023). Food regulation varies dramatically around the world. For instance, Australia applies the WHO's Nutrient and Promotion Profile Model for assessing infant and toddler foods, which emphasises the necessity of well-defined policies to protect public health from commercial forces (Chung et al., 2024). This highlights the importance of developing regulations that not only address labelling, but also consider the broader implications of food marketing and consumer health. In Japan, the "Foods for Specified Health Uses" system exemplifies a structured approach to functional food regulation, allowing products that meet specific health criteria to be marketed with health claims (Tousen et al., 2020). Regulatory approaches for food products vary considerably across regions: the United States emphasises post-market surveillance, India operates a relatively simpler registration system, and Europe mandates pre-market approval and verification of health claims. Such disparities pose challenges for businesses seeking to operate internationally. Greater alignment of global food regulations is needed to enhance safety, facilitate trade, and strengthen consumer confidence (Bose and Sharma, 2025). EU health claim regulations require robust scientific backing (Jost et al., 2025). While effective, challenges persist in defining healthy populations, conducting food trials, establishing biomarkers, and ensuring product consistency. Consumer comprehension and potential misinterpretations are also concerns, prompting calls for better communication on commercial claim authorisation (Ashwell et al., 2023). Furthermore, the introduction of front-of-package labelling systems, such as traffic light labels, has been shown to influence consumer choices positively by making nutritional information more accessible (Zhang et al., 2020). By addressing the ambiguous definition of "functional food", a standardised definition can be proposed and guidelines for efficacy testing, which are crucial for navigating diverse international regulatory landscapes and ensuring truthful product labelling.

Providing consumers with clear and accurate information and labelling can help in reducing the rising rates of obesity and related health issues. The implementation of physical activity calorie equivalent labels encouraged healthier food choices by contextualising calorie content in terms of physical activity (Daley et al., 2020). The need for a comprehensive regulatory framework is further emphasised by the increasing prevalence of non-communicable diseases globally. Proper food labelling is a crucial strategy in addressing public health concerns related to diet and nutrition. In the complexities of novel food regulation, the necessity of balance between oversight and innovation is essential. Despite the potential of nutraceutical markets, insufficient information and substantiation of health claims necessitate collaborative efforts to develop regulations ensuring safety, efficacy, and innovation across diverse regulatory landscapes. Probiotic regulation requires global harmonisation to address inconsistencies impacting consumers, professionals and industry. Accurate labelling and robust safety assessments, especially for vulnerable groups, are crucial. According to Gerke (2025), proper regulatory frameworks are required for labelling amidst rapid growth of artificial intelligence/machine learning (AI/ML)-based medical devices, as the consumer needs transparency and essential information access (Gerke 2025). "AI Facts labels" and "front-of-package AI labelling systems", which are comparable to food labels, can improve user comprehension and help make well-informed dietary choices (Bramstedt, 2025). This includes a suggested app and labelling specifications for material created by artificial intelligence. Luthra et al. (2025) investigated barriers to green personal care product purchase, using innovation resistance theory, and revealed limited availability, inadequate labelling, certification, poor product performance, and lack of government regulation as key impediments (Luthra et al., 2025). Zandonadi et al. (2025) reviewed cell-based meat production, focusing on hazards, regulations, and critical control points (Zandonadi et al., 2025). While production protocols are established, regulatory frameworks are still developing. Potential hazards in functional foods include cell line instability, microbiological and chemical contamination, and nutritional risks. To mitigate these risks, rigorous safety assessments such as "Hazard Analysis and Critical Control Points" are essential, ensuring systematic monitoring and control throughout production and processing.

3.7 Challenges and Opportunities: Functional Foods

As the functional food market continues to expand, regulatory frameworks also evolve to address challenges associated with these products. Government and regulatory bodies focus on establishing guidelines for health claims, labelling, and safety assessments to protect consumers and ensure the integrity of the functional food market. Navigating through the regulatory landscape is a major challenge in the functional food industry. Consumer preferences are inclining more towards healthier and sustainable options, compelling the food industry to adapt and innovate. However, the absence of proper regulations and for health claims associated with functional food has complicated the approval process for these products, and with the lengthy approval process, innovative products get significantly delayed to market. Due to this uncertainty, without proper guidelines for functional foods and their health claims, investments in new product development waver, becoming a hindrance to innovation. The regulatory landscape is crucial in fostering consumer trust and encouraging innovation within the industry. Although functional foods and nutraceuticals are gaining popularity, the lack of standardised global definitions and regulations continues to pose a challenge. Concurrently, consumer acceptance remains a critical barrier to the widespread adoption of functional foods amidst increasing awareness due to skepticism about the efficacy of health claims made by manufacturers. Additionally, the increased cost of functional foods as compared to the conventional options deters price-sensitive consumers, further complicating market penetration strategies. This skepticism is often rooted in a lack of understanding of the scientific evidence supporting these claims, which can lead to hesitance in purchasing decisions.

The nutraceutical market is expanding globally, driven by population-based trends, while challenges remain in product adaptation to regional preferences and regulations. The Indian nutraceutical market presents distinct challenges for new entrants. Market complexities arise from the diverse product categories and ingredient options, necessitating specialised expertise. Consumer preferences mandate product adaptation to local tastes. Significant hurdles include establishing foundational infrastructure, maximising market penetration, competition from established domestic entities, and inconsistent raw material sourcing. Pharmaceutical sector dominance within the nutraceutical market is a prevailing trend, likely to continue due to information asymmetry for potential investors.

Technological integration in the functional food sector is challenged by the need to develop and implement advanced processing techniques capable of incorporating bioactive compounds into food products while preserving functionality and maintaining product safety and quality. Knowing how the microbiome influences health, in addition to the fact that functional foods affect individuals differently, makes this field particularly significant. While micro- and nanoencapsulation techniques show promise for enhancing probiotic stability and bioavailability, their high cost can limit smaller manufacturers' competitiveness against larger companies (Gutiérrez-Álzate et al., 2024). The development of functional beverages shows the ongoing challenge of balancing consumer acceptance with scientific formulation. Replacing ingredients such as sucrose with natural sweeteners while maintaining sensory appeal remains a significant challenge. The broader functional food sector faces a complex convergence of scientific, regulatory, and commercial issues, including establishing robust scientific consensus, navigating intricate regulatory frameworks, addressing consumer skepticism, and managing rising competition. These trends underscore the critical need for ongoing research, innovation, and collaboration to overcome these multifaceted challenges and fully harness the potential of functional foods to improve public health outcomes. Despite these challenges, the functional food industry is ripe with opportunities for growth and innovation. So, there is an expanding market for functional foods that cater to specific health needs, such as heart health, digestive health, and weight management. The food industry is focusing more and more on functional foods in its new product launches, fuelled by the growing global health and wellness market. Functional food product development presents complexities distinct from traditional new product development, requiring more integrative and innovative approaches. Rising prevalence of lifestyle-related diseases, such as obesity and diabetes, has further fuelled consumer interest in health-promoting foods This is particularly pronounced among aging populations, who seek out products that can support their health and well-being. Social media platforms provide an effective channel for engaging with consumers and building brand loyalty where digital marketing strategies can help overcome some of the barriers to consumer acceptance by providing accessible information and fostering community around functional food products. Co-branding initiatives combine the strengths of different stakeholders and enhance product visibility and consumer trust, ultimately leading to increased market share. The integration of sustainability into product development can enhance brand reputation and appeal to health-conscious consumers who prioritise ethical sourcing and production practices. Collaborative partnerships between ingredient suppliers, manufacturers, and research institutions are emerging as a key strategy for driving innovation in the functional food sector. It can facilitate knowledge sharing and resource pooling, enabling companies to leverage cutting-edge research and technology to develop new products. The increasing focus on nutrient-dense, health-promoting foods, including nutraceuticals, spurred by growing health consciousness and recent global health events, presents significant market opportunities for nutraceuticals, particularly in regions like India, but also poses challenges related to formulation, regulation, and quality control. Natural biopreservatives, such as bacteriocins, offer promising alternatives for enhancing food safety, though further research is needed to address potential safety concerns and optimise their production and application. Microbial biotransformation is another technique where microorganisms can modify food components, enhancing their functional properties and nutritional content, and presents another avenue for improving food quality and nutritional value, but requires careful management of safety and scalability. Controlled environment agriculture like vertical farming, hydroponics, greenhouse systems, and automation, holds potential for sustainable food production, but its widespread adoption is hampered by high initial costs, gaps in technical knowledge, resource utilisation, scalability, crop profitability, and unstable energy infrastructure (Nwanojuo et al., 2025). Innovation in traditional fermented foods, like kefir, gained attention due to its distinctive sensory attributes, low lactose content, and more health benefits. The development of novel beverages from pseudocereals such as quinoa, buckwheat, and amaranth offer opportunities for creating functional and nutritious products, exhibiting promising health benefits, including antioxidant, hypoglycaemic, antidiabetic, and antihypertensive effects. Moreover, advancements in food science and technology also pave the way for the development of innovative and novel functional food products such as plant-based ingredients sourced from microalgae and functional proteins derived from food by products. Personalised nutrition (PN), with the aim of tailored dietary advice driven by AI, ML and personal data, is the objective, but safety, data protection, and regulatory challenges need to be overcome. Adapting existing regulatory frameworks to accommodate PN's evolving technologies, including new biomarkers and AI/ML integration, is crucial for ensuring user safety, transparency, and trust (Donovan et al., 2025). Synthetic biology offers powerful tools for producing food components and additives, but its application in the food industry requires careful assessment of potential risks and benefits.

3.8 The Future of Functional Foods

The landscape of functional foods, products, and technologies is rapidly evolving, driven by a confluence of consumer demand for health-enhancing foods, advancements in food science, and innovative processing technologies (Figure 4). This shift in consumer perception led to a budding market that emphasised focus on functional food products that incorporate bioactive compounds, probiotics and other health-promoting ingredients. Overtime shift in focus from food's fundamental role in energy provision and structural development to the nuanced impact of bioactive food components on human health was observed. The functional food concept was formalised with active contribution of consumer interest in food's contribution to well-being, longevity, and chronic disease prevention. Significant market growth observed in probiotics, known for their potential to positively modulate gut microbiota and overall health, particularly non-dairy probiotic options driven by dietary preferences, health concerns (milk cholesterol) and lactose intolerance. Therefore, non-dairy probiotic product development is the key area of focus and a complex task for researchers in both commercial and academic settings. Analysis of technological trends in functional food development revealed a reliance on traditional methods, especially in the period from 2000 to 2010, but also a growing interest in PN approaches thereafter (Betoret et al., 2011). As a vital economic sector, the food industry recognises innovation, particularly in functional foods as crucial for competitive advantage and meeting evolving consumer demands, driven by factors such as rising healthcare costs, increasing life expectancy, and a desire for improved quality of life. Beverages emerged as the most dynamic category within functional foods due to their convenience and adaptability to consumer preferences regarding packaging and storage while considering safety concerns. Furthermore, they served as an exceptional mode for delivering nutrients and bioactive compounds like plant extracts, fibre, vitamins, omega-3 fatty acids, minerals, antioxidants, probiotics, and prebiotics. Nevertheless, safety concerns remained a key area of focus for functional beverages development. Bioactive compounds, derived from diverse organisms such as bacteria, microalgae, and fungi, along with their metabolic products, hold potential as natural alternatives to pharmaceuticals for mitigating adverse drug effects. The epidemiological and clinical evidence supported the positive impact of consuming specific foods on health, wellness, and disease risk reduction, driving anticipated future product innovation with robust health claims. The functional foods market, especially the products targeting health and mental well-being, spurred increased research and development within the food

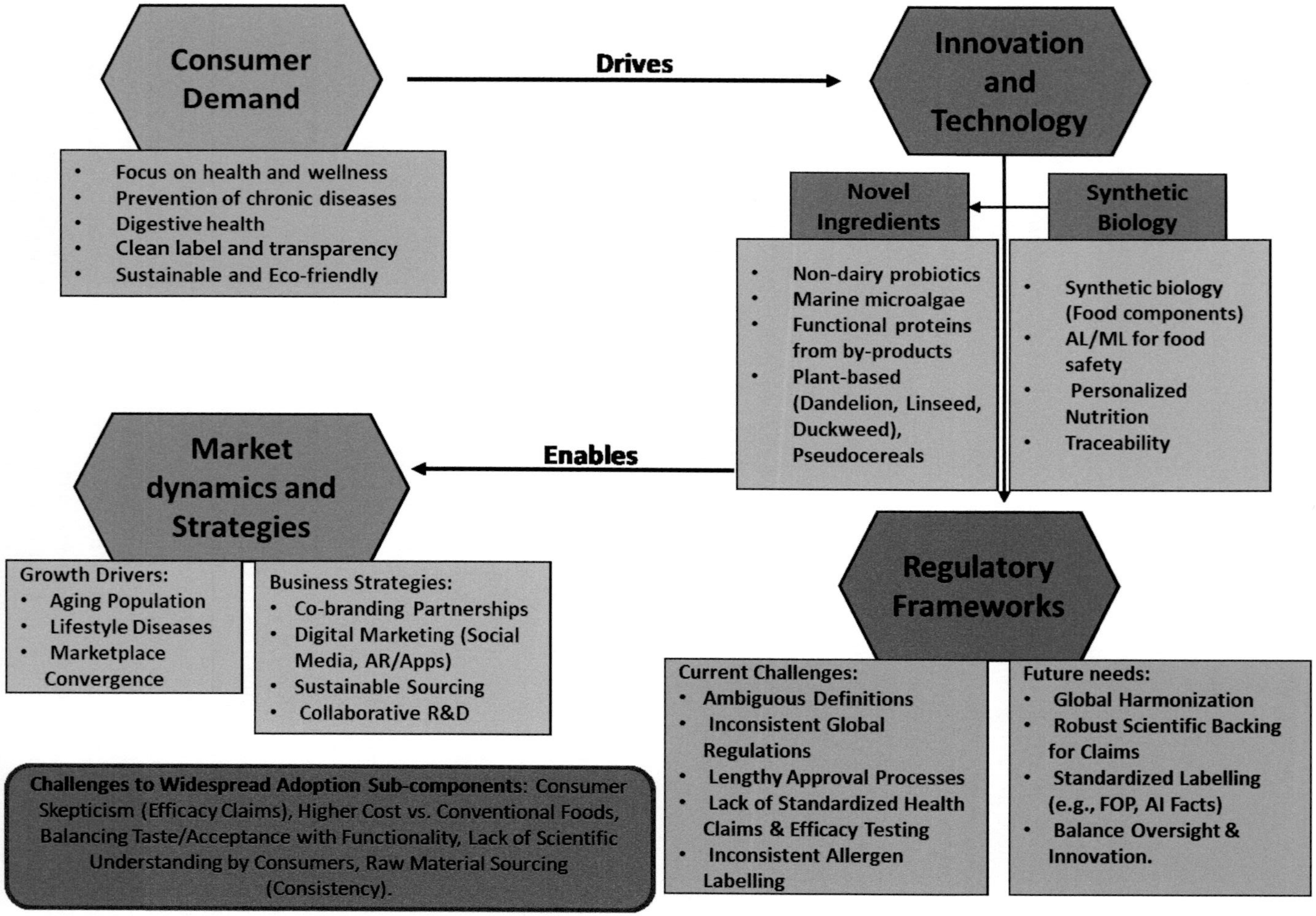

Fig. 4 Factors defining the future of functional foods.

industry, navigating the complexities of evolving regulations in a rapidly expanding global market. This growth is evident in the functional food and beverage market, which has experienced an annual growth rate exceeding 10%, driven by key trends such as marketplace convergence. This accelerated the growth of functional foods and beverages, co-branding partnerships, increased focus on scientific validation, demand for sustainable and eco-friendly products, the rise of active nutrition, innovative delivery forms, and the influence of millennial consumers on health and wellness paradigms (Hilton, 2017). Marine macroalgae emerged as a promising source of functional food ingredients, owing to their unique bioactive compounds, such as phlorotannin, which demonstrated potential protective effects against hyperglycaemia, hyperlipidaemia, inflammation, and oxidative stress. Using food by products as functional food ingredients to promote sustainability and reduce waste in the food industry.

Apart from product development, research studies focus on safety and efficacy of functional ingredients while actively exploring health benefits of bioactive compounds. This ensures to support the informed choices of consumers about their dietary options and health claims related to functional foods. Plant-based products particularly vegetables gained importance due to their low calorific value high phytochemical and fibre content, and also as a source of carotenoids and flavonoids which contribute to disease prevention through multiple and often synergistic mechanisms like antioxidant activity and cell signalling modulation. Incorporation of new vegetables, like dandelions which possess unique and valuable nutraceutical profile, expands the dietary options. Linseed crop is recognised as a superfood due to high omega-3 fatty acid, dietary fibres, proteins, and other bioactive compounds, which spurred an increase in research focusing on cultivation, processing, and diverse applications. Due to this, linseed crops have the potential to manage mindfulness, blood sugar levels, bone health, skin disorders, gastrointestinal issues, and heart health (Langyan et al., 2025). Duckweed is used as a sustainable resource for producing high-value bio-products, with a major focus on optimizing cultivation, harvesting, and biorefining processes. Synthetic biology explored for producing important food components, functional additives, and nutritional chemicals that offer potential solutions to challenges in the food industry.

Nowadays, consumer awareness regarding the functional food market is increasing which is helping in the acceptance of these products (Figure 4). With the rise in health-conscious consumers, the demand for functional foods that address health concerns like chronic diseases and digestive health increased. It was further supported by the rise of clean-label products which emphasized transparency and absence of artificial ingredients and aligned with consumer preferences for natural and wholesome foods. The rise of digital technology and data analytics helps in marketing and distribution of products. Blockchain technology has allowed the consumers to verify quality and source of functional foods purchased, hence ensuring traceability and transparency in the food supply chain (Sundarakani and Ghouse, 2024). Furthermore, integration of augmented reality and mobile applications in food marketing enhances the consumer engagement and educates about the functional food benefits (Abdul Rani and Ramlie, 2023). AI is also found to have various applications across food safety, quality, and security, as well as integrating with advanced food processing techniques.

Innovative processing techniques, such as microencapsulation, 3D food printing, and non-thermal methods, preserve the nutritional and functional properties of foods without compromising quality. These approaches also enhance the stability, bioavailability, and sensory attributes of functional food ingredients. The incorporation of functional components—such as plant-based proteins, probiotics, and bioactive compounds from fruits, vegetables, and dairy—into conventional foods has become a notable trend, improving nutritional profiles and meeting the rising consumer demand for plant-based and functional food options.

4. Conclusion

Functional foods have taken a forefront of nutritional research and consumer interest, as the world is more focussed on health management, overall well-being, and disease prevention. The exploration of functional foods has revealed different types of functional foods, including naturally occurring foods like legumes, dairy-based food, and different kinds of tea. These foods impart health benefits beyond the basic nutritional requirements of an individual. Functional foods can also be designed by incorporating specific bioactive compounds into traditional foods, allowing them to interact with the body at the molecular level and provide targeted health benefits. Bioactive compounds encompass phytochemicals, antioxidants, probiotics, prebiotics, and other functional ingredients. Advancements in understanding their mechanisms of action, alongside insights from metabolomics and nutrigenomics, are facilitating the design of novel foods aimed at enhancing overall human health and vitality. Functional foods include both plant- and dairy-based sources that provide essential nutrients and bioactive compounds. Plant-based options—such as fruits, vegetables, nuts, seeds, legumes, and whole grains—offer vitamins, minerals, healthy fats, dietary fibre, and bioactive molecules that support metabolic health, regulate blood sugar, and promote satiety. Dairy-based functional foods, including fortified milk and probiotic yogurt, supply calcium and can be enriched with vitamin D to support bone health, while probiotics in yogurt enhance gut microbiota and overall digestive function. Moreover, bioactive peptides, which are produced through the fermentation of probiotic yogurt, are helpful to gut microflora and immune system function.

Food scientists are now looking to expand the era of functional foods by innovating and identifying novel bioactive functional molecules, developing advanced processing technologies and creating personalised nutrition solutions. Attempts are being made to enhance the bioavailability and stability of functional biomolecules by regulating encapsulation, fermentation, and using nanotechnology approaches. Despite their potential, the functional food industry faces ongoing challenges in ensuring product safety, efficacy, and regulatory compliance. Robust guidelines are needed to govern product development, manufacturing, and marketing. Establishing consumer trust requires standardised labelling, transparent health claims, and scientific validation. Equally important is raising consumer awareness to promote informed choices and dispel misconceptions about functional foods. The growth of the functional food market is reinforced by increasing consumer awareness of their benefits. Incorporating vital foods into the diet can help individuals manage their health and well-being. With advancing scientific understanding of their mechanisms, functional foods are poised to play a pivotal role in promoting health, preventing disease, and enhancing overall quality of life.

References

Abdel-Aal, E., Akhtar, H., Zaheer, K. and Ali, R. 2013. Dietary sources of lutein and zeaxanthin carotenoids and their role in eye health. *Nutrients*, *5*(4), 1169–1185.

Abdel-Aal, E., Akhtar, H., Zaheer, K. and Ali, R. 2013b. Dietary sources of lutein and zeaxanthin carotenoids and their role in eye health. *Nutrients*, *5*(4), 1169–1185.

Abdul Rani, M.N.H.A.R. and Ramlie, M.K. 2023. The Utilization of Augmented Reality (AR) Applications as Packaging Design Enhancement. *International Journal of Academic Research in Business and Social Sciences*, *13*(7), 251–274.

Agirman, G., Yu, K.B., and Hsiao, E.Y. 2021. Signaling inflammation across the gut-brain axis. *Science*, *374*(6571), 1087–1092.

Ahmed, M.H., Vasas, D., Hassan, A. and Molnár, J. 2022. The impact of functional food in prevention of malnutrition. *PharmaNutrition*, *19*, 100288.

Ahn-Jarvis, J., Parihar, A. and Doseff, A. 2019. Dietary flavonoids for immunoregulation and Cancer: Food Design for Targeting Disease. *Antioxidants*, *8*(7), 202.

AlAli, M., Alqubaisy, M., Aljaafari, M.N., AlAli, A.O., Baqais, L., Molouki, A., Abushelaibi, A., Lai, K.-S., Lim, S.-H.E. 2021. Nutraceuticals: Transformation of Conventional Foods into Health Promoters/Disease Preventers and Safety Considerations. *Molecules*, *26*, 2540.

Anastasova, L., Petreska Ivanovska, T., Petkovska, R. and Petrusevska-Tozi, L. 2018. Concepts, benefits, and perspectives of functional dairy food products. *Macedonian Pharmaceutical Bulletin*, *64*(2), 73–83.

Aguirre, J.D. and Culotta, V.C. (2012). Battles with Iron: Manganese in Oxidative Stress Protection. *Journal of Biological Chemistry*, *287*(17), 13541–13548.

Arena, M., Caggianiello, G., Fiocco, D., Russo, P., Torelli, M., Spano, G. and Capozzi, V. 2014. Barley B-Glucans-Containing food enhances probiotic performances of beneficial bacteria. *International Journal of Molecular Sciences*, *15*(2), 3025–3039.

Aryaeian, N., Sedehi, S. K. and Arablou, T. 2017. Polyphenols and their effects on diabetes management: A review. *Medical Journal of the Islamic Republic of Iran*, *31*(1), 886–892.

Ashique, S., Mohanto, S., Ahmed, M. G., Mishra, N., Garg, A., Chellappan, D. K., Omara, T., Iqbal, S. and Kahwa, I. 2024. Gut-brain axis: A cutting-edge approach to target neurological disorders and potential synbiotic application. *Heliyon*, *10*(13), e34092.

Ashwell, M., Hickson, M., Stanner, S., Prentice, A. and Williams, C. M. (2023). Nature of the evidence base and strengths, challenges, and recommendations in the area of nutrition and health claims: A position paper from the Academy of Nutrition Sciences. *British Journal of Nutrition*, *130*(2), 221–238.

Aziz, T., Hussain, N., Hameed, Z. and Lin, L. 2024. Elucidating the role of diet in maintaining gut health to reduce the risk of obesity, cardiovascular, and other age-related inflammatory diseases: Recent challenges and future recommendations. *Gut Microbes*, *16*(1), 2297864.

Banahene, J. C. M., Ofosu, I. W., Odai, B. T., Lutterodt, H. E., Agyemang, P. A. and Ellis, W. O. 2024. Ochratoxin A in Food Commodities: A Review of Occurrence, Toxicity, and Management Strategies. *Heliyon*, *10*(20), e39313.

Baroni, L., Rizzo, G., Galchenko, A.V., Zavoli, M., Serventi, L. and Battino, M. 2024. Health Benefits of Vegetarian Diets: An Insight into the Main Topics. *Foods*, *13*, 2398.

Berry, C.W., Murray, B. and Kenney, W.L. 2022. Scientific basis for a milk permeate-based sports drink: A critical review. *International Dairy Journal*, *127*, 105296.

Betoret, E., Betoret, N., Vidal, D. and Fito, P. 2011. Functional foods development: Trends and technologies. *Trends in Food Science and Technology*, *22*(9), 498–508.

Biancone, L., Monteleone, I., Del Vecchio Blanco, G., Vavassori, P. and Pallone, F. 2002. Resident bacterial flora and immune system. *Digestive and Liver Disease*, *34*, S37–S43.

Bose, A. and Sharma, S. 2025. Global regulatory trends and comparative insights: Nutraceuticals in the USA, India, and Europe. *PharmaNutrition*, *31*, 100430.

Bramstedt, K.A. 2025. Artificial Intelligence (AI) Facts Labels: An Innovative Disclosure Tool Promoting Patient-Centric Transparency in Healthcare AI Systems. *J. Med. Syst.*, *49*, 78.

Burckhardt, P. 2015. Calcium revisited, part III: Effect of dietary calcium on BMD and fracture risk. *BoneKEy Reports*, *4*, 708.

Capurso, C. 2021. Whole-Grain Intake in the Mediterranean Diet and a Low Protein to Carbohydrates Ratio Can Help to Reduce Mortality from Cardiovascular Disease, Slow Down the Progression of Aging, and to Improve Lifespan: A Review. *Nutrients*, *13*, 2540.

Cena, H. and Calder, P. C. 2020. Defining a Healthy Diet: Evidence for The Role of Contemporary Dietary Patterns in Health and Disease. *Nutrients*, *12*(2), 334.

Champagne, C.P., Gomes da Cruz, A. and Daga, M. 2018. Strategies to improve the functionality of probiotics in supplements and foods. *Current Opinion in Food Science*, *22*, 160–166.

Chaudhary, D., Guleria, D., Aggarwal, H., Mishra, V., Chauhan, A., Dufossé, L. and Joshi, N.C. 2025. Nutrigenomics and Personalized Diets: Tailoring nutrition for optimal health. *Applied Food Research*, *5*(1), 100980. https://doi.org/10.1016/j.afres.2025.100980.

Cheng, Z., Zhang, Z., Han, Y., Wang, J., Wang, Y., Chen, X., Shao, Y., Cheng, Y., Zhou, W., Lu, X. and Wu, Z. 2020. A review on anti-cancer effect of green tea catechins. *Journal of Functional Foods*, *74*, 104172.

Chung, A., Torkel, S., Myers, J. and Skouteris, H. 2024. Assessment of foods for infants and toddlers in Australia against the World Health Organization's Nutrient and Promotion Profile Model for food products for infants and young children. *Public Health Nutrition*, *27*(1), e201.

Da Cruz Correia, P.F., Reis, J.G.M.D., Amorim, P.S., Da Costa, J.S. and Da Silva, M.T. 2024. Impacts of Brazilian green coffee production and its logistical corridors on the international coffee market. *Logistics*, *8*(2), 39.

Daley, A.J., McGee, E., Bayliss, S., Coombe, A. and Parretti, H.M. 2020. Effects of physical activity calorie equivalent food labelling to reduce food selection and consumption: Systematic review and meta-analysis of randomised controlled studies. *J Epidemiol Community Health*, *74*(3), 269–275.

Das, T.K., Pradhan, S., Mal, S. and Ghosh, K. 2024. Probiotics as a Therapeutic Approach for Non-infectious Gastric Ulcer Management: A Comprehensive Review. *Probiotics and Antimicrobial Proteins*, *16*(1), 1–15.

de Araújo, F.F. and de Paulo Farias, D. 2020. Psychobiotics: An emerging alternative to ensure mental health amid the COVID-19 outbreak? *Trends in Food Science and Technology*, *103*, 386.

De Oliveira, P.V., Da Silva Júnior, A.H., De Oliveira, C.R.S., Assumpção, C.F. and Ogeda, C.H. 2023. Kombucha benefits, risks, and regulatory frameworks: A review. *Food Chemistry Advances*, *2*, 100288.

Dikeman, M.E., Hunt, M.C., Addis, P.B., Schoenbeck, H.J., Pullen, M., Katsanidis, E. and Yancey, E.J. 2003. Effects of postexsanguination vascular infusion of cattle with a solution of saccharides, sodium chloride, and phosphates or with calcium chloride on quality and sensory traits of steaks and ground beef. *Journal of Animal Science*, *81*(1), pp.156–166.

Dixit, V., Joseph, S.W., et al., 2021. The influence of gut microbiota on the bioavailability of oral drugs. *British Journal of Pharmacology*, *178*(20), 4129–4146.
Donovan, S.M., Abrahams, M., Anthony, J.C., Bergia, R., Blander, G., Brisbois, T.D., Keck, A.-S., Moore, E.G., Morck, T.A., Nieman, K.M., Ordovas, J.M., Steiber, A., Winters, B.L. and Wu, T. 2025. Perspective: Challenges for Personalized Nutrition in the Current U.S. Regulatory Framework and Future Opportunities. *Advances in Nutrition*, 100382.
Farah, A. 2012. Coffee constituents. In: *Coffee: Emerging Health Effects and Disease Prevention*, (Vol. 1, pp. 22–58). ResearchGate.
Feizollahi, E., Hadian, Z. and Honarvar, Z. 2018. Food fortification with omega-3 fatty acids: Microencapsulation as an addition method. *Current Nutrition and Food Science*, *14*(2), 90–103.
Gao, Y., Wang, J.Q., Fu, Y.Q., Yin, J.F., Shi, J. and Xu, Y.Q. 2020. Chemical composition, sensory properties, and bioactivities of Castanopsis lamontii buds and mature leaves. *Food Chemistry*, *316*, 126370.
Gao, Y., Zhang, X., Yin, J., Du, Q., Tu, Y., Shi, J. and Xu, Y. 2019. Castanopsis lamontii Water Extract Shows Potential in Suppressing Pathogens, Lipopolysaccharide-Induced Inflammation, and Oxidative Stress-Induced Cell Injury. *Molecules*, *24*(2), 273.
Gendel, S.M. and Zhu, J. 2013. Analysis of U.S. Food and Drug Administration Food Allergen Recalls after Implementation of the Food Allergen Labeling and Consumer Protection Act. *Journal of Food Protection*, *76*(11), 1933–1938.
Gerke, S. 2025. The Need for "Nutrition Facts Labels" and "Front-Of-Package Nutrition Labeling" For Artificial Intelligence/Machine Learning-Based Medical Devices: Lessons Learned From Food Labeling. *SSRN Electronic Journal, 74*(5), 1297.
Gjerlaug-Enger, E., Haug, A., Gaarder, M., Ljøkjel, K., Stenseth, R.S., Sigfridson, K., Egelandsdal, B., Saarem, K. and Berg, P. 2015. Pig feeds rich in rapeseed products and organic selenium increased omega-3 fatty acids and selenium in pork meat and backfat. *Food Science and Nutrition*, *3*(2), 120–128.
Gonçalves, B., Pinto, T., Aires, A., Morais, M.C., Bacelar, E., Anjos, R., Ferreira-Cardoso, J., Oliveira, I., Vilela, A. and Cosme, F. 2023. Composition of Nuts and Their Potential Health Benefits: An Overview. *Foods*, *12*(5), 942.
Gröber, U., Schmidt, J. and Kisters, K. 2015. Magnesium in prevention and therapy. *Nutrients*, *7*(9), 8199–8226.
Gugliesi, F., Pasquero, S., Griffante, G., Scutera, S., Albano, C., Pacheco, S.F.C., Riva, G., Dell'Oste, V. and Biolatti, M. 2021. Human cytomegalovirus and autoimmune diseases: Where are we? *Viruses*, *13*(2), 260.
Gupta, A., Sanwal, N., Bareen, M.A., Barua, S., Sharma, N., Olatunji, O.J., Nirmal, N.P. and Sahu, J.K. 2023. Trends in functional beverages: Functional ingredients, processing technologies, stability, health benefits, and consumer perspective. *Food Research International*, *170*, 113046.
Gutiérrez-Álzate, K., Beltrán-Cotta, L.A., dos Santos Rekowsky, B.S., Cavalheiro, C.P. and Pereira da Costa, M. 2024. Micro- and Nanoencapsulation of Probiotics: Exploring Their Impact on Animal-Origin Foods. *ACS Food Science and Technology*, *4*(12), 2799–2812.
Hadjimbei, E., Botsaris, G. and Chrysostomou, S. 2022. Beneficial Effects of Yoghurts and Probiotic Fermented Milks and Their Functional Food Potential. *Foods*, *11*(17), 2691.
Heck, C. and De Mejia, E. 2007. Yerba Mate Tea (Ilex paraguariensis): A Comprehensive Review on Chemistry, Health Implications, and Technological Considerations. *Journal of Food Science*, *72*(9),138–151.
Hemsworth, J., Hekmat, S. and Reid, G. 2010. The development of micronutrient supplemented probiotic yogurt for people living with HIV: Laboratory testing and sensory evaluation. *Innovative Food Science and Emerging Technologies*, *12*(1), 79–84.
Hernández-Ruiz, R.G., Olivares-Ochoa, X.C., Salinas-Varela, Y., Guajardo-Espinoza, D., Roldán-Flores, L.G., Rivera-Leon, E.A. and López-Quintero, A. 2025. Phenolic compounds and anthocyanins in legumes and their impact on inflammation, oxidative stress, and metabolism: Comprehensive review. *Molecules*, *30*(1), 174.
Hertzler, S.R. and Clancy, S.M. 2003. Kefir improves lactose digestion and tolerance in adults with lactose maldigestion. *Journal of the American Dietetic Association*, *103*(5), 582–587.
Hilton, J. 2017. Growth patterns and emerging opportunities in nutraceutical and functional food categories: Market overview. In: D. Bagchi & S. Nair (Eds.), *Developing New Functional Food and Nutraceutical Products* (pp. 1–28). Academic Press.
Hiraga, Y., Yoshida, C., Aoki, K. and Watanabe, M. 2025. Efficient decaffeination of green coffee beans using pressure swing supercritical CO_2 extraction. *Innovative Food Science and Emerging Technologies*, 104154.
Ikram, A., Saeed, F., Afzaal, M., Imran, A., Niaz, B., Tufail, T., Hussain, M. and Anjum, F.M. 2021. Nutritional and end-use perspectives of sprouted grains: A comprehensive review. *Food Science and Nutrition*, *9*(8), 4617–4628.
Ishwarya, S.P. and Anandharamakrishnan, C. 2014. Spray-Freeze-Drying approach for soluble coffee processing and its effect on quality characteristics. *Journal of Food Engineering*, *149*, 171–180.
Iversen, K.N., Jonsson, K. and Landberg, R. 2022. The effect of Rye-Based foods on postprandial plasma insulin concentration: The Rye factor. *Frontiers in Nutrition*, *9*, 868938.
Jackson, L.S. and Lee, K. 1991. Microencapsulated Iron for Food Fortification. *Journal of Food Science*, *56*(4), 1047–1050.
Jagadeesh, M.S., Jahnavi. A.P., Vinay, H.T, Abhishek, G.J., Abhishek, S.S. and Chikkalaki, A.S. (2024.). Export Competitiveness of Indian Coffee: Analysing trade potential in the global market. *Journal of Scientific Research and Reports*, *30*(9), 51–61.

Jaquet, M., Rochat, I., Moulin, J., Cavin, C. and Bibiloni, R. 2009. Impact of coffee consumption on the gut microbiota: A human volunteer study. *International Journal of Food Microbiology*, *130*(2), 117–121.

Ji, J., Jin, W., Liu, S., Jiao, Z. and Li, X. 2023. Probiotics, prebiotics, and postbiotics in health and disease. *MedComm*, *4*(6), e420.

Jomova, K., Alomar, S.Y., Valko, R., Liska, J., Nepovimova, E., Kuca, K. and Valko, M. 2025. Flavonoids and their role in oxidative stress, inflammation, and human diseases. *Chemico-Biological Interactions*, 111489.

Jost, S., Herzig, C. and Birringer, M. 2025. A Balancing Act—20 years of nutrition and health claims regulation in Europe: A Historical perspective and reflection. *Foods*, *14*(9), 1651.

Kanchanasurakit, S., Saokaew, S., Phisalprapa, P. and Duangjai, A. 2023. Chlorogenic acid in green bean coffee on body weight: A systematic review and meta-analysis of randomized controlled trials. *Systematic Reviews*, *12*(1),163.

Kapil, U. 2007. Health consequences of iodine deficiency. *Sultan Qaboos University Medical Journal*, *7*(3), 267–272.

Kavyani, Z., Pourfarziani, P., Kakhki, A.M.J., Ahrabi, S.S., Moridpour, A.H., Mollaghasemi, N., Musazadeh, V. and Faghfouri, A. H. 2023. The effect of flaxseed supplementation on glycemic control in adults: An updated systematic review and meta-analysis of randomized controlled trials. *Journal of Functional Foods*, *110*, 105816.

Kaur, N., Agarwal, A. and Sabharwal, M. 2022. Food fortification strategies to deliver nutrients for the management of iron deficiency anaemia. *Current Research in Food Science*, *5*, 2094–2107.

Kemsley, E. 1995. Discrimination between Coffea arabica and Coffea canephora variant robusta beans using infrared spectroscopy. *Food Chemistry*, *54*(3), 321–326.

Kerry, R.G., Patra, J.K., Gouda, S., Park, Y., Shin, H.S. and Das, G. 2018. Benefaction of probiotics for human health: A review. *Journal of Food and Drug Analysis*, *26*(3), 927–939.

Khalaf, A.T., Wei, Y., Alneamah, S., Al-Shawi, S.G., Kadir, S.Y.A., Zainol, J. and Liu, X. 2021. What Is New in the Preventive and Therapeutic Role of Dairy Products as Nutraceuticals and Functional Foods? *BioMed Research International*, 8823222.

Khalil, O.S.F., Ismail, H.A. and Elkot, W.F. 2022. Physicochemical, functional and sensory properties of probiotic yoghurt flavored with white sapote fruit (Casimiroa edulis). *Journal of Food Science and Technology*, *59*(9), 3700–3710.

Khalil, R. A. and Elkot, W.F. 2020. Functional Properties and Nutritional Quality of Processed Cheese Spreads Enriched with Black Rice Powder. *Egyptian Journal of Food Science*, *48*(2), 281–289.

Kim, I. and He, Y. (2013). Targeting the AMP-Activated Protein kinase for cancer prevention and therapy. *Frontiers in Oncology*, *3*,175.

Kirwan, L., Walsh, M.C., Brennan, L., Gibney, E.R., Drevon, C.A., Daniel, H., Lovegrove, J.A., Manios, Y., Martínez, J.A., Saris, W.H.M., Traczyk, I., Mathers, J.C. and Gibney, M. 2016. Comparison of the portion size and frequency of consumption of 156 foods across seven European countries: Insights from the Food4ME study. *European Journal of Clinical Nutrition*, *70*(5), 642–644.

Kwak, H.S., Mijan, M. Al and Ganesan, P. 2014. Application of Nanomaterials, Nano- and Microencapsulation to Milk and Dairy Products. In: H. S. Kwak (Ed.), *Nano- and Microencapsulation for Foods* (1st Edn., pp. 273–300). John Wiley & Sons, Ltd.

Lamsar, H. and Abhilasha, A. 2023. Dairy-based functional food products. In: *Industrial Application of Functional Foods, Ingredients, and Nutraceuticals: Extraction, Processing, and Formulation of Bioactive Compounds* (pp. 127–170). Research Gate.

Langyan, S., Yadava, P., Khan, F.N., Sharma, S., Singh, R., Bana, R.S., Singh, N., Kaur, V., Kalia, S. and Kumar, A. 2025. Trends and advances in pre- and post-harvest processing of linseed oil for quality food and health products. *Critical Reviews in Food Science and Nutrition*, *65*(4), 746–769.

Latif, A., Shehzad, A., Niazi, S., Zahid, A., Ashraf, W., Iqbal, M.W., Rehman, A., Riaz, T., Aadil, R.M., Khan, I.M., Özogul, F., Rocha, J.M., Esatbeyoglu, T. and Korma, S.A. 2023. Probiotics: Mechanism of action, health benefits, and their application in food industries. *Frontiers in Microbiology, 14*.

Leeuwendaal, N.K., Stanton, C., O'Toole, P.W. and Beresford, T.P. 2022. The Most Promising Next-Generation Probiotic Candidates: Impact on Human Health. *Fermentation*, *8*(9), 444.

Lei, W., Cheng, Y., Liu, X., Gao, J., Zhu, Z., Ding, W., Xu, X., Li, Y., Ling, Z., Jiang, R. and Chen, X. 2025. Gut microbiota-driven neuroinflammation in Alzheimer's disease: From mechanisms to therapeutic opportunities. *Frontiers in Immunology*, *16*, 1161211.

Li, T. Y., Brennan, A.M., Wedick, N.M., Mantzoros, C., Rifai, N. and Hu, F.B. 2009. Regular Consumption of Nuts is Associated with a Lower Risk of Cardiovascular Disease in Women with Type 2 Diabetes. *The Journal of Nutrition*, *139*(7), 1333–1338.

Liu, Y., Ahmed, S. and Long, C. 2013. Ethnobotanical survey of cooling herbal drinks from southern China. *Journal of Ethnobiology and Ethnomedicine*, *9*(1), 1530012.

Liu, Yujing, Selena Ahmed and Chunlin Long. 2013. Ethnobotanical survey of cooling herbal drinks from southern China. *Journal of Ethnobiology and Ethnomedicine*, *9*, 1–8.

Luthra, C., Deshwal, P., Kushwah, S. and Gokarn, S. 2025. Exploring barriers towards green personal care products purchase: An integrated ISM-MICMAC approach. *International Journal of Pharmaceutical and Healthcare Marketing*, *19*(3), 618–642.

Maliha, S. and Hayat, U. 2025. Preventive and Therapeutic Role of Bioactive Food Components and Functional Foods in Diabetes. *Anatomy Physiol. Biochem. Int. J.*, *7*(5), 555722.

Mani-López, E., Palou, E. and López-Malo, A. 2014. Probiotic viability and storage stability of yogurts and fermented milks prepared with several mixtures of lactic acid bacteria. *Journal of Dairy Science*, *97*(5), 2578–2590.

Maqsood, S., Bordiga, M. and Xu, B. 2025. The role of fiber in gut health and chronic diseases. *Quality Assurance and Safety of Crops and Foods*, *17*(2), 165–199.

Martirosyan, D. and Stratton, S. 2023. Comparing the FDA's process for health claim submission and the FFC's guidelines for regulating functional foods. *Functional Foods in Health and Disease*, *13*(9), 424–436.

McCowen, K.C., Ling, P.R., Decker, E., Djordjevic, D., Roberts, R., Coupland, J. and Bistrian, B.R. 2010. A simple method of supplementation of omega-3 polyunsaturated fatty acids. *Nutrition in Clinical Practice*, *25*(6), 641–645.

McKay, D.L. and Blumberg, J.B. 2006a. A Review of the bioactivity and potential health benefits of chamomile tea (*Matricaria recutita* L.). *Phytotherapy Research*, *20*(7), 519–530.

McKay, D.L. and Blumberg, J.B. 2006b. A review of the bioactivity and potential health benefits of peppermint tea (*Mentha piperita* L.). *Phytotherapy Research: An International Journal Devoted to Pharmacological and Toxicological Evaluation of Natural Product Derivatives*, *20*(8), 619–633.

Méndez, R.O., Galdámez, K., Grijalva, M.I., Quihui, L., García, H.S. and de la Barca, A.M.C. 2012. Effect of micronutrient-fortified milk on zinc intake and plasma concentration in adolescent girls. *Journal of the American College of Nutrition*, *31*(6), 408–414.

Menon, Rekha and Vasanthakumari, P. 2022. Implications of GI on Indian Specialty Coffee: A Case Study on 'Monsooned Malabar Arabica Coffee'. *Journal of Intellectual Property Rights*, *27*, 333–338.

Moustarah, F. and Daley, S.F. 2024. Dietary iron. In: *StatPearls, Treasure Island* (*FL*). StatPearls Publishing, PMID: 31082013.

Munekata, P.E., Domínguez, R., Budaraju, S., Roselló-Soto, E., Barba, F.J., Mallikarjunan, K., Roohinejad, S. and Lorenzo, J.M. 2020. Effect of innovative food processing technologies on the physicochemical and nutritional properties and quality of non-dairy plant-based beverages. *Foods*, *9*(3), 288.

Murthy, P.S. and Manonmani, H.K. 2009. Physico-chemical, antioxidant, and antimicrobial properties of Indian monsooned coffee. *European Food Research and Technology*, *229*(4), 645–650.

Nattagh-Eshtivani, E., Barghchi, H., Hatami, A., Barati, M., Heydari, A., Moghaddam, Z.A., Fard, M.V., Mohammadhasani, K., Miri, H., Talebi, S. and Pahlavani, N. 2025. The effects of flaxseed supplementation on metabolic syndrome risk factors: A Comprehensive review. *Clinical Nutrition Open Science*, *60*, 262–285.

Nesci, A., Carnuccio, C., Ruggieri, V., D'Alessandro, A., Di Giorgio, A., Santoro, L., Gasbarrini, A., Santoliquido, A. and Ponziani, F.R. 2023. Gut microbiota and cardiovascular disease: Evidence on the metabolic and inflammatory background of a complex relationship. *International Journal of Molecular Sciences*, *24*(10), 9087.

Neyrinck, A.M., Van Hée, V.F., Piront, N., De Backer, F., Toussaint, O., Cani, P.D. and Delzenne, N.M. 2012. Wheat-derived arabinoxylan oligosaccharides with prebiotic effect increase satietogenic gut peptides and reduce metabolic endotoxemia in diet-induced obese mice. *Nutrition and Diabetes*, *2*(1), e28.

Nwanojuo, M.A., Anumudu, C.K. and Onyeaka, H. 2025. Impact of Controlled Environment Agriculture (CEA) in Nigeria: A Review of the Future of Farming in Africa. *Agriculture*, *15*(2), 117.

Ocak, E. and Rajendram, R. 2013. Fortification of milk with mineral elements. In: *Handbook of Food Fortification and Health: From Concepts to Public Health Applications* (Vol. 1, pp. 213–224). Humana Press.

Ozcan-Yilsay, T., Lee, W.J., Horne, D. and Lucey, J.A. (2007). Effect of trisodium citrate on rheological and physical properties and microstructure of yogurt. *Journal of Dairy Science*, *90*(4), 1644–1652.

Pandey, K.B. and Rizvi, S.I. 2009. Plant polyphenols as dietary antioxidants in human health and disease. *Oxidative Medicine and Cellular Longevity*, *2*, 270–278.

Pogorzelska-Nowicka, E., Atanasov, A.G., Horbańczuk, J. and Wierzbicka, A. 2018. Bioactive compounds in functional meat products. *Molecules*, *23*(2), 307.

Poka, M.S., Milne, M., Wessels, A. and Aucamp, M. 2025. An Investigation into the Effect of Maltitol, Sorbitol and Xylitol on the Formation of Carbamazepine Solid Dispersions through Thermal Processing. *Pharmaceutics*, *17*(3), 321.

Polonini, H., Dijkers, E. and Ferreira, A.O. 2021. Beauty from within: A review of the science behind YulivTM collagen drink: An anti-aging nutraceutical. *Journal of Cosmetics, Dermatological Sciences, and Applications*, *11*(3), 263–278.

Radlowski, E.C. and Johnson, R.W. 2013. Perinatal iron deficiency and neurocognitive development. *Frontiers in Human Neuroscience*, *7*, 585.

Rajasekaran, A. and Kalaivani, M. 2013. Designer foods and their benefits: A review. *Journal of Food Science and Technology*, *50*(1), 1–16.

Ranard, K.M., Jeon, S., Mohn, E.S., Griffiths, J.C., Johnson, E.J. and Erdman, J.W. 2017. Dietary guidance for lutein: Consideration for intake recommendations is scientifically supported. *European Journal of Nutrition*, *56*(S3), 37–42.

Ríos-Covián, D., Ruas-Madiedo, P., Margolles, A., Gueimonde, M., De Los Reyes-Gavilán, C.G. and Salazar, N. 2016. Intestinal Short Chain Fatty Acids and their Link with Diet and Human Health. *Frontiers in Microbiology*, *7*, 185.

Rivas, J.C., Cabral, L.M.C. and Rocha-Leão, M.H.M.D., 2021. Microencapsulation of guava pulp using prebiotic wall material. *Brazilian Journal of Food Technology*, *24*, e2020213.

Ross, S. 2000. Functional foods: The Food and Drug Administration perspective. *American Journal of Clinical Nutrition*, *71*(6), 1735S–1738S.

Ruchika, N. and Sehgal, A. 2021. An update on disease preventing potential of green tea in comparison with some tisanes. *South African Journal of Botany*, *144*, 92–96.

Sahay, M. and Sahay, R. 2012. Rickets-vitamin D deficiency and dependency. *Indian Journal of Endocrinology and Metabolism*, *16*(2), 164.

Samtiya, M., Aluko, R.E., Dhewa, T. and Moreno-Rojas, J.M. 2021. Potential Health benefits of Plant Food-Derived Bioactive Components: An Overview. *Foods*, *10*(4), 839.

Sarkar, S. 2019. Potentiality of probiotic yoghurt as a functional food: A review. *Nutrition and Food Science*, *49*(2), 182–202.

Sazawal, S., Dhingra, U., Hiremath, G., Sarkar, A., Dhingra, P., Dutta, A., Verma, P., Menon, V.P. and Black, R.E. 2010. Prebiotic and Probiotic Fortified Milk in Prevention of Morbidities among Children: Community-Based, Randomized, Double-Blind, Controlled Trial. *PLoS ONE*, 5(8), e12164.

Shah, A.B., Baiseitova, A., Zahoor, M., Ahmad, I., Ikram, M., Bakhsh, A., Shah, M.A., Ali, I., Idress, M., Ullah, R., Nasr, F.A. and Al-Zharani, M. 2024. Probiotic significance of Lactobacillus strains: A comprehensive review on health impacts, research gaps, and future prospects. *Gut Microbes*, *16*(1), 2431643.

Silvarolla, M.B., Mazzafera, P. and De Lima, M.M.A. 2000. Caffeine content of Ethiopian Coffea arabica beans. *Genetics and Molecular Biology*, *23*(1), 213–215.

Singar, S., Kadyan, S., Patoine, C., Park, G., Arjmandi, B., Nagpal, R. 2024. The Effects of Almond Consumption on Cardiovascular Health and Gut Microbiome: A Comprehensive Review. *Nutrients*, , *16*, 1964.

StröM, K., SjöGren, J., Broberg, A. and SchnüRer, J. 2002. Lactobacillus plantarum MiLAB 393 Produces the Antifungal Cyclic Dipeptides Cyclo(l-Phe-l-Pro) and Cyclo(l-Phe-trans-4-OH-l-Pro) and 3-Phenyllactic Acid. *Applied and Environmental Microbiology*, *68*(9), 4322–4327.

Sugajski, M., Buszewska-Forajta, M. and Buszewski, B., 2023. Functional beverages in the 21st century. *Beverages*, *9*(1), 27.

Sugizaki, C.S.A. and Naves, M.M.V. 2018. Potential prebiotic properties of nuts and edible seeds and their relationship to obesity. *Nutrients*, *10*(11), 1645.

Sundarakani, B. and Ghouse, A. 2024. A Systematic Literature Review and Bibliometric Analysis of Blockchain Technology for Food Security. *Foods*, *13*(22), 3607.

Tang, H., Huang, G., Hu, K., Wang, J., Huang, C. and Yang, X. 2025. Compound extreme events and health risks in China: A review. *Atmospheric and Oceanic Science Letters*, *18*(5), 100647.

Tolun, A. and Altintas, Z., 2019. Medicinal properties and functional components of beverages. In: A. M. Grumezescu & A. M. Holban (Eds.), *Functional and Medicinal Beverages* (Vol. 11, pp. 235–284). Academic Press.

Tousen, Y., Kondo, T., Chiba, T. and Ishimi, Y. 2020. Regulation of the Food Labelling Systems for Health and Nutrition in Japan and Associated Role of the National Institute of Health and Nutrition. *The Japanese Journal of Nutrition and Dietetics*, *78*(Supplement), S80–S90.

Toy, J.Y.H., Yu, L.L. and Huang, D. 2025. Functional Foods for Health: Evidence and Insights. In: *Royal Society of Chemistry eBooks* (pp. 1–15).

Ullah, A., Munir, S., Badshah, S.L., Khan, N., Ghani, L., Poulson, B.G., Emwas, A. and Jaremko, M. 2020. Important flavonoids and their role as therapeutic agent. *Molecules*, *25*(22), 5243.

Valero, M.S., Berzosa, C., Langa, E., Gomez-Rincon, C. and Lopez, V., 2013. Jasonia glutinosa DC (rock tea): Botanical, phytochemical, and pharmacological aspects. *Bol Latinoam Caribe Plant Med. Aromat*, *12*(Supl 6), 543–557.

Villa-Rivera, M.G. and Ochoa-Alejo, N. 2020. Chili Pepper carotenoids: Nutraceutical properties and mechanisms of action. *Molecules*, *25*(23), 5573.

Wang, S., Chen, Y., Liu, M., Hong, Z., Sun, D., Du, Y., Su, M., Yu, C., Liu, Q., Ren, Y., Lv, J. and Li, L. 2011. The Changes of Nutrition Labeling of Packaged Food in Hangzhou in China during 2008-2010. *PLoS ONE*, *6*(12), e28443.

Weststrate, J.A., Van Poppel, G. and Verschuren, P.M. 2002. Functional foods, trends, and future. *British Journal of Nutrition*, *88*(S2), S233–S235.

Whittle, R.H., Kiarie, E.G., L, D.W., MA. and Widowski, T.M. 2024. Feeding flaxseed to chicken hens changes the size and fatty acid composition of their chicks' brains. *Frontiers in Physiology*, *15*, 1400611.

Widyastuti, Y., Febrisiantosa, A. and Tidona, F. 2021. Health-Promoting Properties of Lactobacilli in Fermented Dairy Products. *Frontiers in Microbiology*, *12*, 673890.

Xiao, M., Xiang, W., Chen, Y., Peng, N., Du, X., Lu, S., Zuo, Y., Li, B., Hu, Y. and Li, X. 2022. DHA ameliorates cognitive ability, reduces amyloid deposition, and nerve fiber production in Alzheimer's disease. *Frontiers in Nutrition*, *9*, 852433.

Xu, Z., Tao, Z. and Guo, Y. 2025. The role of tea in managing cardiovascular risk factors: Potential benefits, mechanisms, and interventional strategies. *Frontiers in Nutrition*, *12*.

Yamada, K., Sato-Mito, N., Nagata, J. and Umegaki, K. 2008. Health claim evidence requirements in Japan. *Journal of Nutrition*, *138*(6), 1192S–1198S.

Yang, Y., Yang, F., Huang, Z., Li, Y., Shi, H., Sun, Q., Ma, Y., Wang, Y., Zhang, Y., Yang, S., Zhao, G. and Xu, F. 2023. T cells, NK cells, and tumor-associated macrophages in cancer immunotherapy and the current state of the art of drug delivery systems. *Frontiers in Immunology*, *14*, 1125205.

Zahra, M., Abrahamse, H. and George, B.P. 2024. Flavonoids: Antioxidant powerhouses and their role in nanomedicine. *Antioxidants*, *13*(8), 922.

Zandonadi, R.P., Ramos, M. C., Elias, F.T.S. and Guimarães, N.S. 2025. Global Insights into Cultured Meat: Uncovering Production Processes, Potential Hazards, Regulatory Frameworks, and Key Challenges: A Scoping Review. *Foods*, *14*(1), 129.

Zanetti, W.A.L., Marques, M.D.S., do Amaral, A.M.S., da Silva, A.B., Barcelos, J., Goés, B.C. and Putti, F.F., 2021. Analysis of the Technological Evolution of Coffee Production in Brazil. *J. Agric. Stud.*, *9*, 352–362.

Zhang, X., Liu, Y., Gu, Y., Wang, S. and Chen, H. (2020). Red for "Stop": "Traffic-Light" Nutrition Labels Decrease Unhealthy Food Choices by Increasing Activity and Connectivity in the Frontal Lobe. Nutrients, *12*(1), 128.

Zhu, Y., Jain, N., Holschuh, N. and Smith, J. (2021). Associations between frequency of yogurt consumption and nutrient intake and diet quality in the United Kingdom. *Journal of Nutritional Science*, *10*, e85.

CHAPTER 2

Chemistry of Functional Foods

Debakanta Tripathy[1*] and Soumya Lipsa Rath[2]

1. Introduction

The concept of "functional food" was introduced through large-scale research conducted in Japan under the sponsorship of the Ministry of Education, Science, and Culture during 1984–1991 (Arai 1996). The primary objective of developing functional food was to reduce the health care cost of the ageing population in the country. The term *functional food* was first used by Swinbanks and O'Brien in the year 1993 (Swinbanks and O'Brien, 1993). As such, there is no universally accepted definition of functional food available to date. However, different government regulatory bodies of different countries and different research organisations have proposed various definitions for functional food. The most recent definition for functional food was proposed by N. J. Temple:

> *"Functional foods are novel foods that have been formulated so that they contain substances or live microorganisms that have a possible health-enhancing or disease-preventing value, and at a concentration that is both safe and sufficiently high to achieve the intended benefit. The added ingredients may include nutrients, dietary fiber, phytochemicals, other substances, or probiotics."* (Temple, 2022)

Japan is the first and only country that recognises functional food under the Foods for Specified Health Uses (FOSHU) category (Intrasook et al., 2024). The term *functional food* is very often confused with *nutraceuticals*, which is derived from two different terms such as nutrients and pharmaceuticals (Mukherjee et al., 2025; Puri et al., 2022). The fundamental difference between functional food and nutraceuticals lies in the way they are ingested. Functional food is taken as a whole as food, whereas nutraceuticals are food extracts that are taken in dosages (Shahidi, 2009). However, the physiological effect of both is due to the bioactive compounds present in them. The bioactive compounds present in a particular functional food depend on its source. The sources of functional foods are broadly classified into two types: (i) plant-based functional foods, and (ii) animal-based functional foods (Negi et al., 2025; Frumuzachi et al., 2025). This article discusses the chemical (structural and functional) aspects of various types of bioactive components present in functional foods derived from different sources. Based on the chemistry, the bioactive components present in functional foods can be classified into bioactive compounds based on phytochemicals, bioactive compounds based on carbohydrates, lipid-based bioactive compounds, bioactive peptides, vitamins, and minerals.

[1] Department of Chemistry, School of Advanced Sciences, VIT-AP University, Amaravati-522237, Andhra Pradesh, India.
[2] Department of Biotechnology, National Institute of Technology Warangal, Hanamkonda-506004, Telangana, India.
* Corresponding author: debakanta.tripathy@vitap.ac.in

DOI: 10.1201/9781003659808-2

2. The Bioactive Compounds Based on Phytochemicals

Phytochemicals are non-nutrient, plant-derived bioactive compounds that provide health benefits beyond basic nutrition (Hossain, 2025). They play important roles in functional foods by exhibiting antioxidant, anti-inflammatory, antimicrobial and disease-preventive properties (Shahidi and Ambigaipalan 2015). Common examples include polyphenols, terpenoids, glucosinolates, and alkaloids.

2.1 Polyphenols

Polyphenols are a diverse group of plant-based antioxidants whose molecular weight ranges from 500 to 4000 Da (Bolat et al., 2024). Based on their core chemical structure, they can be sub-classified into phenolic acids, stilbenoids, flavonoids, lignans, tannins, coumarins and phenolic amides (Dirimanov and Högger, 2019) (Figure 1). These compounds show antioxidant, anti-inflammatory, cardioprotective, anticancer, and neuroprotective properties (Rana, 2022).

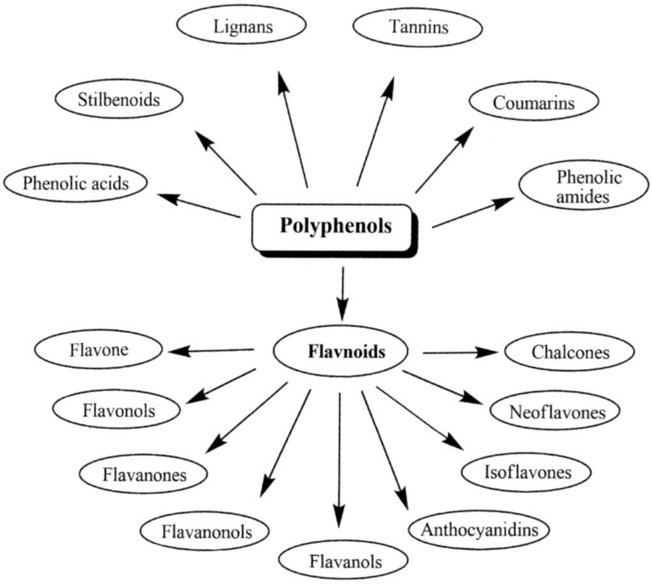

Fig. 1 Classification of plant-based polyphenolic compounds
(This figure is adopted from reference Dirimanov and Högger, 2019).

2.1.1 Phenolic Acids

Phenolic acids are aromatic phenolic compounds with an acid group directly or indirectly attached to the aromatic ring. They are classified into benzoic acid derivatives and cinnamic acid derivatives (Pereira et al., 2009). Their antioxidant and anti-inflammatory activities are attributed to the hydroxy group(s) attached to the aromatic ring, which help reduce oxidative stress and cardiovascular risk (Kumar and Goel, 2019). Phenolic acid compounds that are commonly found in plant-based foods are given in Figure 2.

2.1.2 Stilbenoids

Stilbenoids are a broad class of stilbene derivatives with two benzene rings linked by an ethylene group as core structure (Oh et al., 2021). Based on the type of substitution, stilbenoids can be classified into simple stilbenes, prenylated, geranylated stilbenes, 2-phenyl-benzofuran derivatives

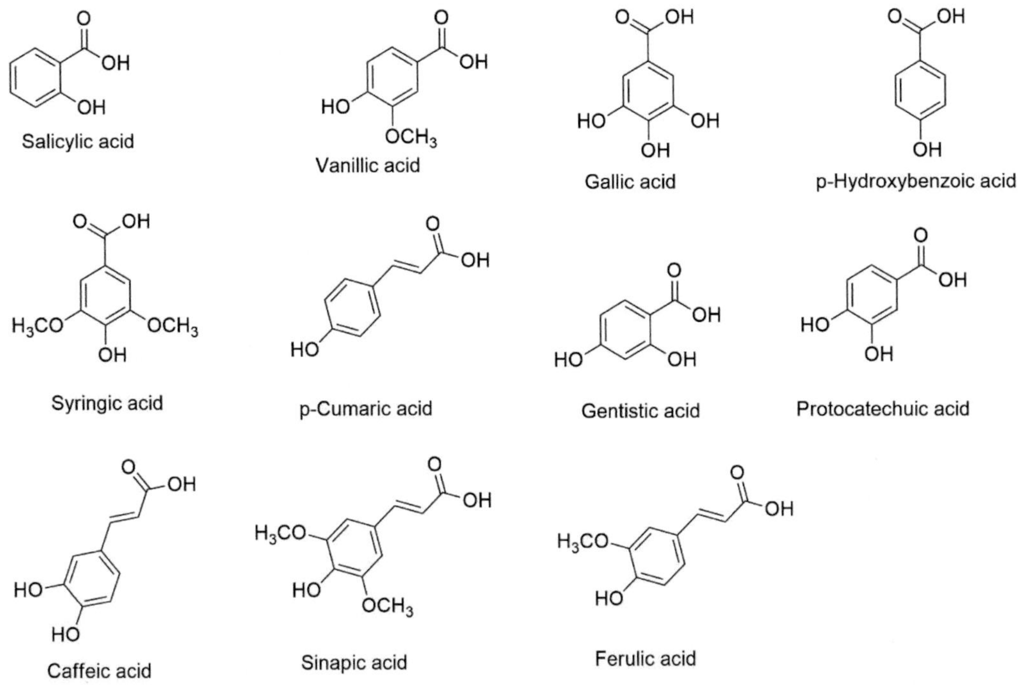

Fig. 2 Chemical structures of commonly found phenolic acids.

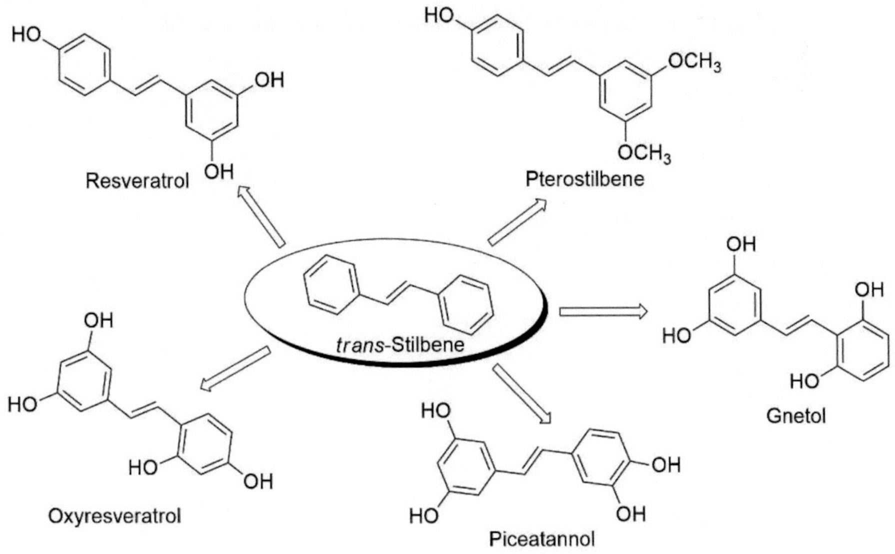

Fig. 3 Chemical structure of some simple stilbene derivatives.

and carbon substituted stilbenes. Most commonly available stilbenoids in plant-based foods are resveratrol, pterostilbene, gnetol, piceatannol, and oxyresveratrol. Among all these the biological activities of resveratrol is well documented in the literature. Resveratrol is especially known for vascular and metabolic benefits (Springer and Mocom, 2019). Chemical structure of some simple stilbenoids, prenylated and geranylated stilbenes are given in Figures 3, 4, and 5, respectively. Chemical structure of dimeric, trimeric, tetrameric stilbenoids, and glycosidic derivatives are given in Figure 5. Like other polyphenolic compounds, stilbenoids show anti-oxidant, anti-inflammatory

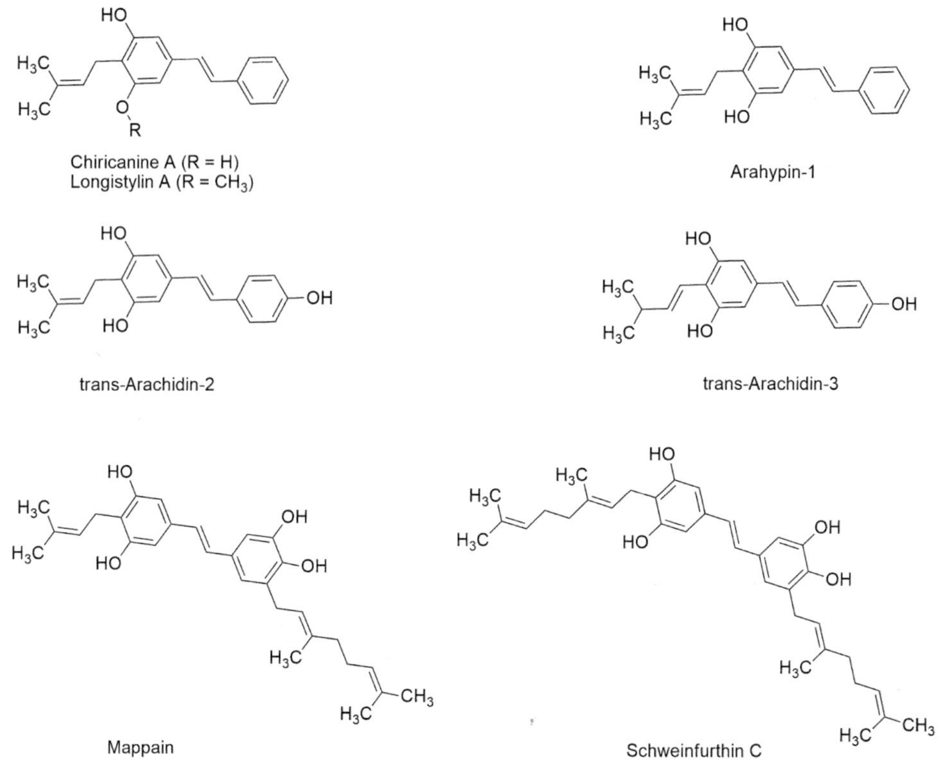

Fig. 4 Chemical structure of some prenylated and geranylated stilbenes.

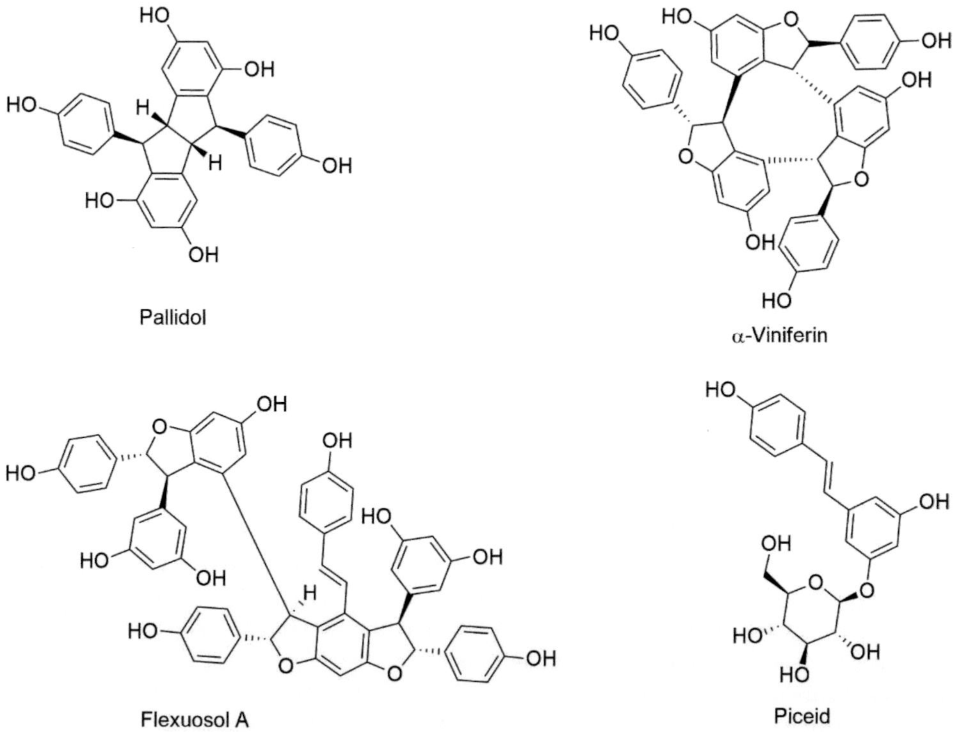

Fig. 5 Chemical structure of dimeric, trimeric, tetrameric stilbenoids, and glycosidic derivatives.

properties. Apart from that, stilbenoids also have anti-obesity and anti-diabetic effect (Dvorakova and Landa, 2017).

2.1.3 *Flavonoids*

Flavonoids are a class of polyphenolic compounds known for their various health benefits such as antioxidant, anti-inflammatory and anti-cancer activities (Duarte et al., 2025). For example, quercetin and kaempferol are linked to reduced risk of coronary heart disease, while anthocyanidins support eye health (Dabeek and Marra, 2019). The general chemical structure of flavonoids comprises two benzene rings linked by a three-carbon chain that forms a pyran ring (Safe et al., 2021) (Figure 6). Therefore, the core structure of flavonoids are very often referred to as C6-C3-C6. The ring B can bind to any of the available positions of ring C. Depending upon which position of ring C is attached to ring B along with the functionalization of all three (A, B, and C) and degree of unsaturation of ring C, flavonoids can be subclassified into flavones, flavonols, flavanononones, flavanonols, flavanols, anthocyanidins, isoflavones, neoflavonoids, chalcones (Figure 6).

2.1.3.1 Flavones

Flavones are an important subclass of flavonoids and are present in leaves, flowers, and fruits of many plants (Hostetler et al., 2017). Some examples include apigenin (parsley, chamomile, celery, onions, oranges, apple, grape), tangeretin (mainly found in peels of tangerine and oranges), baicalein (primarily from the roots of Scutellaria baicalensis), and chrysin (found in blue passionflower, honey, and propolis) (Figure 7). Apigenin has documented antioxidant, anti-inflammatory, anticancer, neuroprotective, anxiolytic, and blood-pressure-modulating effects (Salehi et al., 2019). Tangeretin exhibits anticancer, antioxidant, anti-inflammatory, and metabolic modulation including lowering LDL, improving diabetic parameters (Ashrafizadeh et al. 2020). Baicalein has been shown to have antioxidant, anti-inflammatory, anticancer, neuroprotective effects, and even antiviral activity (Hu et al., 2022). Chrysin demonstrates anticancer, anti-inflammatory, antioxidant, neuroprotective, and anti-allergy effects (Mani and Natesan, 2018).

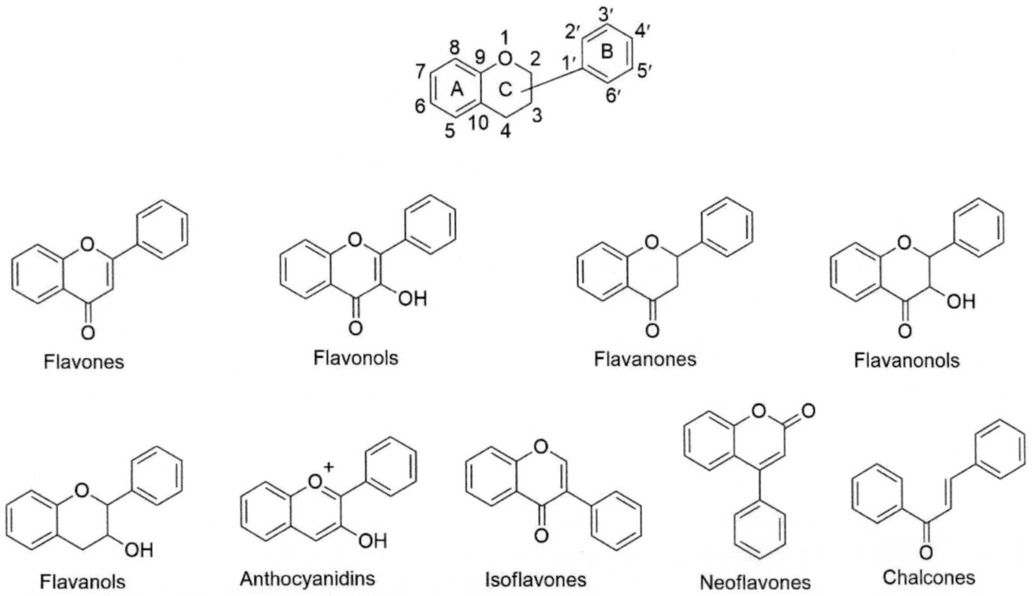

Fig. 6 Core structure and subclasses of flavonoids.

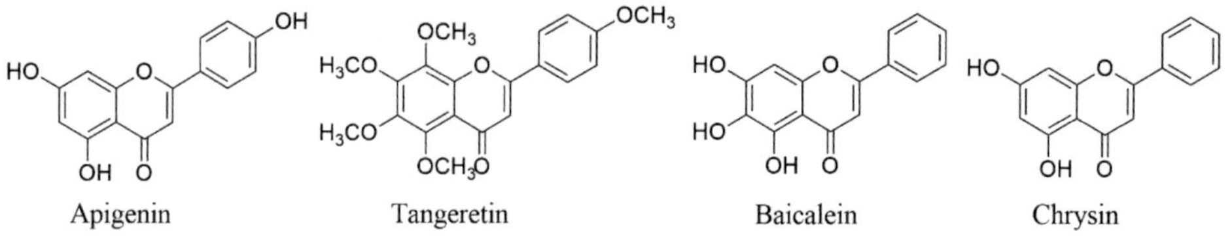

Fig. 7 Chemical structures of representative flavones illustrating their structural diversity.

2.1.3.2 Flavonols

These are another important class of flavonoids (Przybylski et al., 2025). They are most commonly found in fruits and vegetables. One of the well-known examples of flavonols is quercetin (Figure 8). It is found in apples, onions, berries, and citrus fruits. Quercetin has a wide variety of health benefits including reduction in inflammation, blood pressure, and blood sugar level. It may also have anticancer, anti-allergy, and brain-protecting effects (Yang et al., 2020). Another example of flavonol is morin. It is found in various plant products such as almonds, figs, Indian guava, tea, coffee, cereals, fruits, and red wine. Like any other members of the flavonoid family, it also shows anticancer, antioxidant, anti-inflammatory, and neuroprotective effects. Apart from this, morin also shows the ability to protect the heart by reducing oxidative stress (Jain et al., 2025). Kaempferol is also another example of flavonols. These are found in grapes, kale, broccoli, and tea. Kaempferol shows anticancer, anti-inflammatory effects along with antidiabetic and hepatoprotective pharmacological effects (Alrumaihi et al., 2024). Common sources of myricetin are blueberries, oranges, spinach, cauliflower, tea, and red wine. Apart from anticancer, anti-inflammatory, and neuroprotective effects, myricetin also protects cardiovascular health (Nadalin et al., 2023).

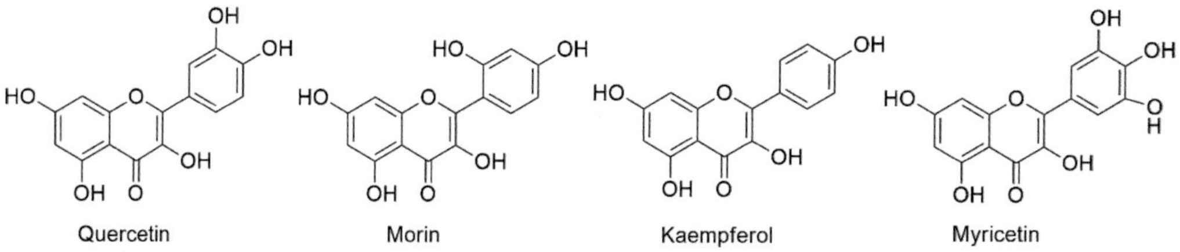

Fig. 8 Representative chemical structures of flavonols.

2.1.3.3 Flavanones

These are another important class of flavonoids mainly found in citrus fruits (Khan et al., 2014). Some examples of flavanones are eriodictyol, hesperetin, hesperidin, and naringenin (Figure 9). The core chemical structure of flavanones is very similar to that of flavones except that there is no double bond between the C2 and C3 of ring B, and the sp^3-hybridisation of C2 of ring B leads to various stereoisomers. Naringenin is primarily found in citrus fruits such as grapefruits. It has anticancer, anti-inflammatory, and antioxidant properties. Also, various studies suggest that it might be beneficial in controlling lipid levels and blood pressure (Shilpa et al., 2023). Eriodictyol has a very similar chemical structure to naringenin with an additional hydroxy group at the 3-position of the B ring. This compound is found in food sources such as lemons, oranges, grapes and tomatoes. It has antioxidant, anti-inflammatory, anticancer properties. It may also have neuroprotective effects, particularly beneficial in Alzheimer's disease. Hesperetin is a methylated version of eriodictyol (the hydroxy group at the 4-position of ring B is methylated). Hesperidin is the glycated form of hesperetin (the hydroxy group at the 7-position of ring A is glycated) and shares many of its benefits as a flavanone. Hesperetin shows neuroprotective effects in Alzheimer's disease models (Song et al., 2024).

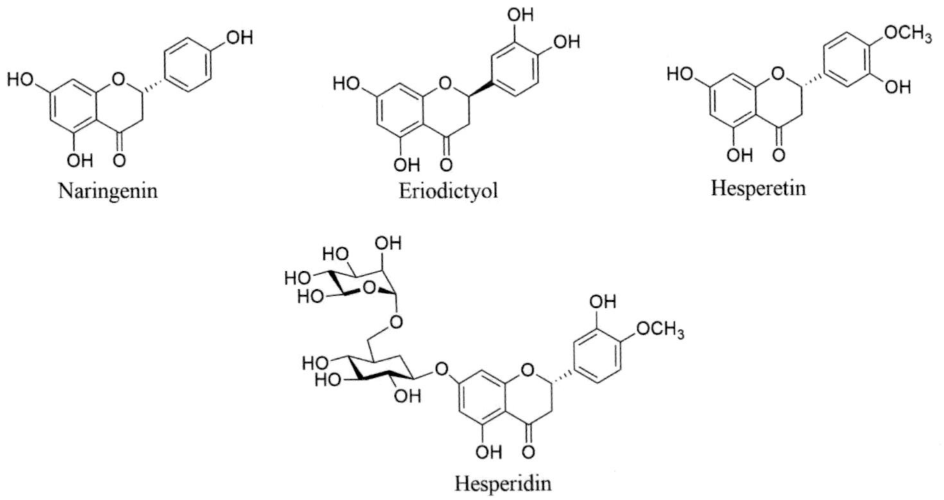

Fig. 9 Chemical structure of selected flavanones.

2.1.3.4 Flavanonols

These are flavonoids with a structure like that of flavanones. The only difference is the additional hydroxyl group present at the 3-position of ring C. Some examples of flavanonols are aromadendrin, taxifolin and engeletin (Figure 10). Aromadendrin has shown antioxidant, anti-inflammatory, anticancer, cardioprotective effects (especially against cardiac hypertrophy), glucose-uptake enhancing activity, and liver-protective effects (Patel and Patel, 2024; El-Shiekh et al., 2025). Taxifolin exhibits anticancer activity, strong antioxidant and anti-inflammatory properties, neuroprotective effects, and has been studied for its potential to protect against metabolic disorders (Sunil and Xu, 2019).

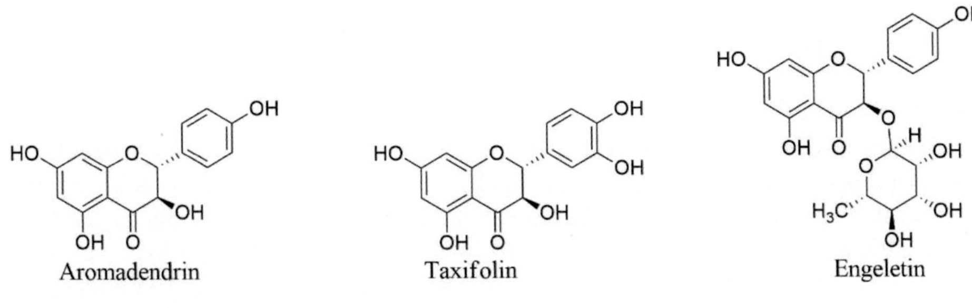

Fig. 10 Representative chemical structure of some selected flavanonols.

2.1.3.5 Flavanols

Flavanols, sometimes referred to as flavan-3-ols, are flavan derivatives that possess a 2-phenyl-3,4-dihydro-2H-chromen-3-ol skeleton (Figure 11). The core structure of flavanols is close to that of flavanonols. The difference is that the flavanol structure lacks the unsaturation at the 2-position and the carbonyl group at the 4-position of the pyran ring (ring C). The most popular example of a flavanol is catechin, which can exist in both stereoisomeric forms (R-catechin and S-epicatechin). Other examples are catechin derivatives (gallocatechin, catechin gallate) and oligomeric or polymeric forms of catechin (procyanidins) (Figure 11). These compounds have been shown to exhibit strong antioxidant, anti-inflammatory, cardioprotective, antidiabetic, anti-atherosclerotic, and neuroprotective effects (Bae et al., 2020; Baranwal et al., 2022; Sheng et al., 2023).

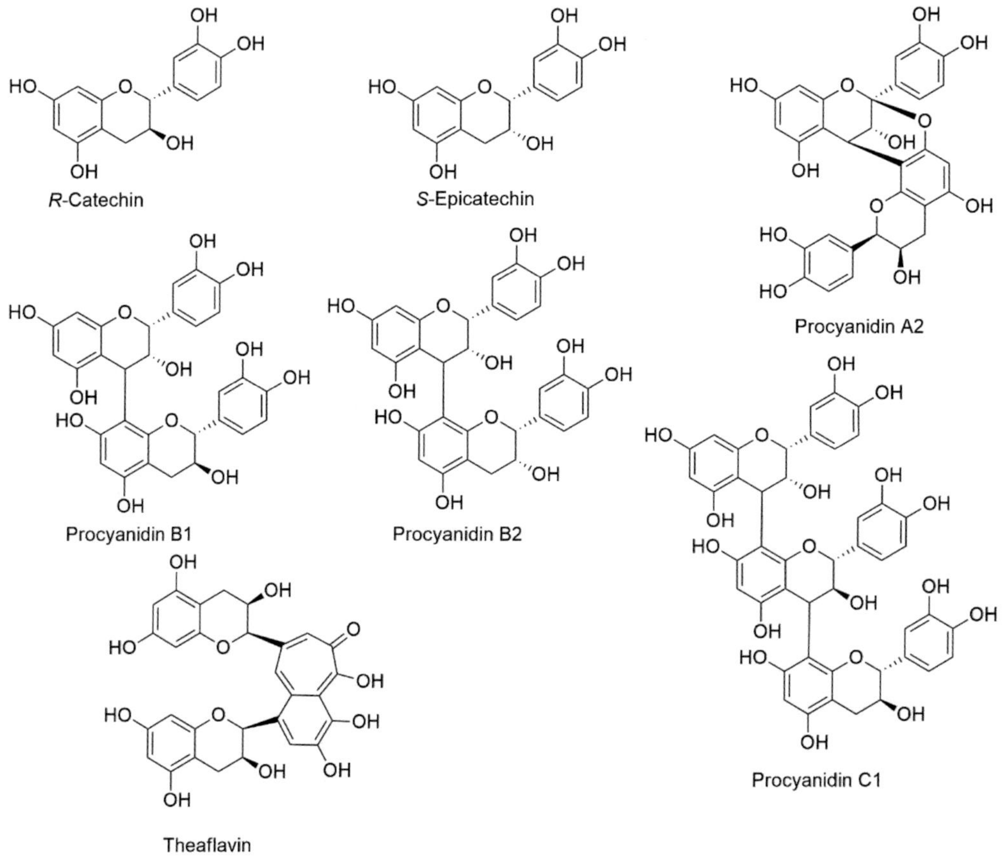

Fig. 11 Chemical structure of common flavanols in functional foods.

2.1.3.6 Anthocyanidins

Anthocyanidins are one of the key components responsible for the blue, purple, pink and red coloration of different parts of plants (Khoo et al., 2017). The core structure of anthocyanidins resembles flavanols but differs in the degree of unsaturation of the pyran ring (ring C). In anthocyanidins, there exist double bonds at the 1- and 3-positions of the pyran ring, giving rise to a hypervalent oxygen atom in the core structure. Chemical structures of some predominantly found anthocyanidins are given in Figure 12. These compounds have demonstrated antioxidant, anti-inflammatory, anti-obesity, cardioprotective, antidiabetic, anticancer, and vision-protective effects (Khoo et al., 2017; Safdar et al., 2022; Kooshki et al., 2025).

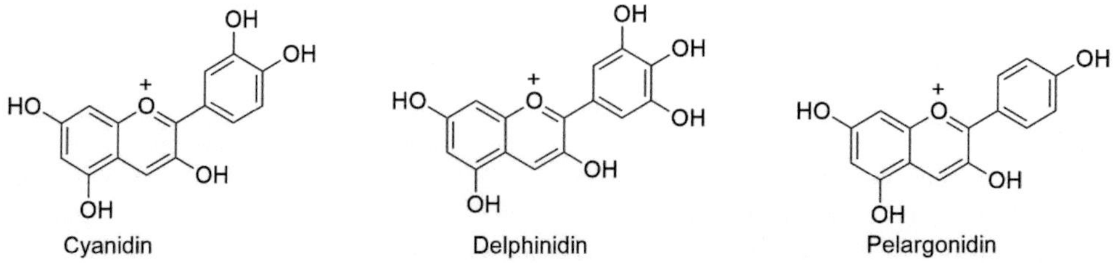

Fig. 12 Chemical structure of some anthocyanidins.

2.1.3.7 Isoflavones

Isoflavones differ from flavones in their core chemical structure with respect to the attachment of ring B: in isoflavones the ring B is attached to the 3-position of ring C. Examples include genistein, daidzein and glycitein (Křížová et al., 2019) (Figure 13). Isoflavones have shown benefits in reducing risk of hormone-related cancers, improving bone health, cardiovascular protection, antioxidant and anti-inflammatory effects, and amelioration of metabolic syndrome (Miadoková, 2009; Wang et al., 2025).

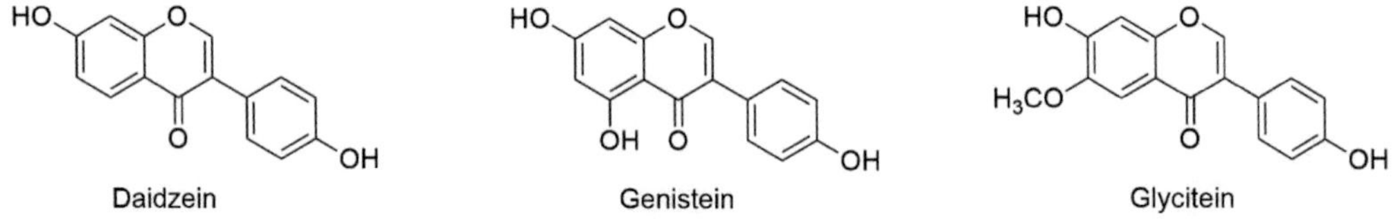

Fig. 13 Chemical structure of some isoflavones.

2.1.3.8 Neoflavones

Neoflavones are flavonoids in which the B-ring is attached to ring C at the 4-position, and the carbonyl group is at the 2-position—a pattern opposite to flavones. Some important examples of neoflavones are shown in Figure 14. While less studied, neoflavones appear to share many of the benefits of other flavonoids including antioxidant, anti-inflammatory, and potential anticancer effects (Miadoková, 2009, Kumar et al., 2020).

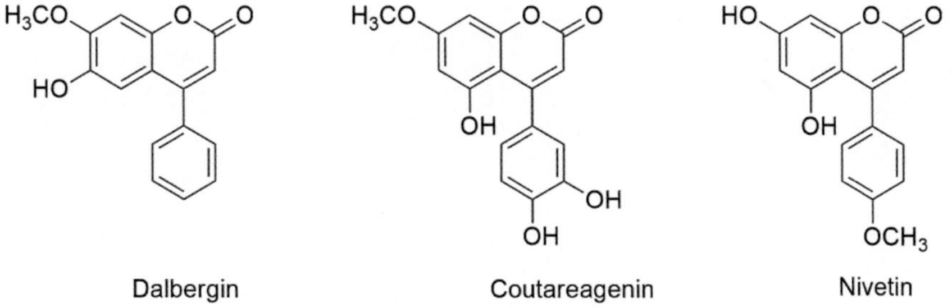

Fig. 14 Chemical structure of some neoflavones.

2.1.3.9 Chalcones

Chalcones are basically open-chain flavonoids and often called 1,3-diaryl-2-propen-1-ones (Zhuang et al., 2017). They possess two aromatic rings but unlike flavones, isoflavones, and neoflavones, chalcones do not have the closed heterocyclic pyran ring. They are considered important intermediates for flavonoid biosynthesis (Figure 15). Chalcones have been found to exhibit anti-inflammatory, antioxidant, antimicrobial, and anticancer activities (Adhikari et al., 2025).

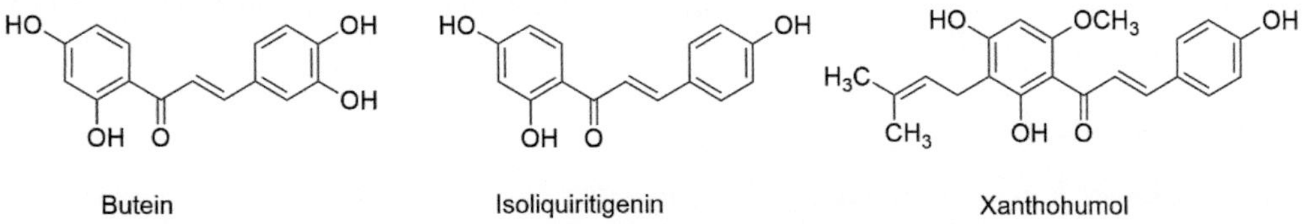

Fig. 15 Chemical structure of chalcones.

2.1.4 Lignans

Lignans are a group of naturally occurring polyphenolic compounds found mostly in plants—particularly in seeds, whole grains and vegetables (Berenshtein et al., 2024). Structurally distinct from flavonoids and chalcones, lignans typically occur as dimers, though some are trimers or tetramers. If the monomers are linked at the 8-8′ (β-β′) position the lignans are termed classical lignans; otherwise, they are neolignans (Li et al., 2018) (Figure 16). Lignans have been associated with hormone metabolism modulation, antioxidant and anti-cancer effects, cardiovascular health benefits, and may help in reducing risk of hormone-dependent cancers (Mukhija et al., 2022; Zhou et al., 2021).

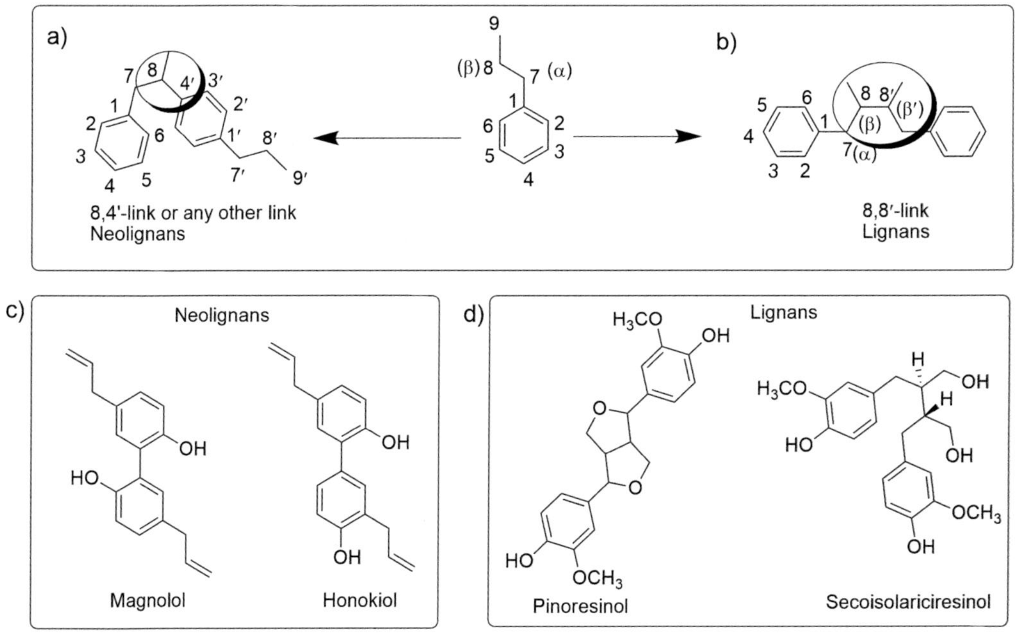

Fig. 16 Linking of phenylpropane units in (a) neolignans, (b) lignans, and examples of (c) neolignans and (d) lignans.

2.1.5 Tannins

Tannins are the largest and most complex group of polyphenols (Chang et al., 2019). They are generally classified into hydrolysable and condensed tannins. Gallotannins are hydrolysable tannins made of gallic acid units esterified to sugars; ellagitannins are another group. Proanthocyanidins are condensed tannins, formed by polymerization of flavanol units. Chemical structures of some condensed and hydrolysable tannins are shown in Figures 11 and 17. Tannins display strong antioxidant, anti-inflammatory, anti-microbial, cardioprotective, antidiabetic, and gut health promoting properties (Pizzi, 2021).

2.1.6 Curcuminoids

Curcuminoids are a group of polyphenolic compounds belonging to the diarylheptanoid subclass of non-flavonoid (Priyadarsini, 2014). The most abundant example is curcumin, which constitutes almost 70–80% of total curcuminoids. Other less common examples are demethoxycurcumin and bisdemethoxycurcumin (Figure 18). Curcuminoids are mostly obtained from turmeric and are well known for their antioxidant activity, anti-inflammatory action, anticancer potential, neuroprotective effects, and metabolic regulation (Hewlings and Kalman, 2017).

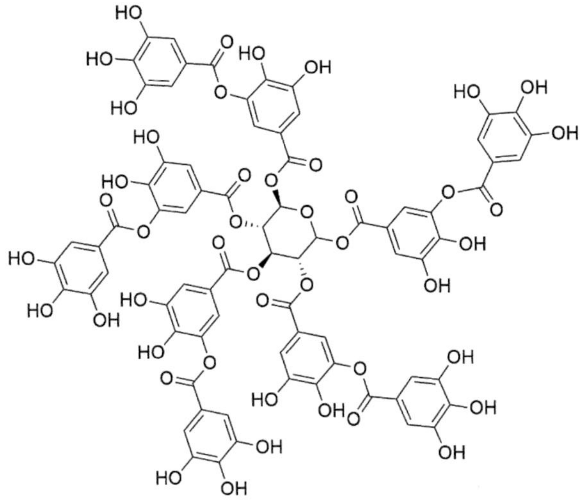

Tannic acid

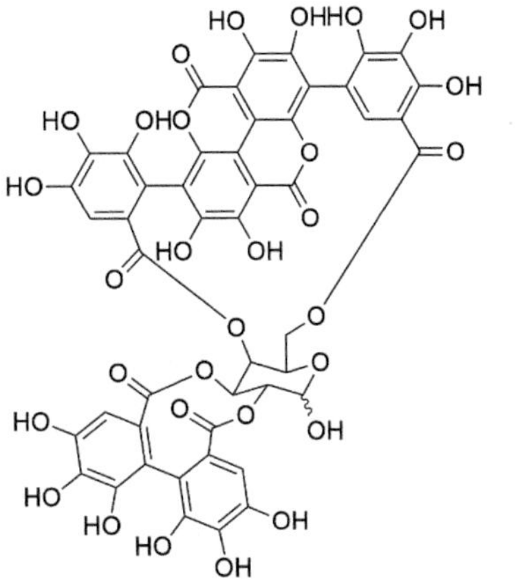

Punicalagin

Fig. 17 Chemical structure of hydrolysable tannins.

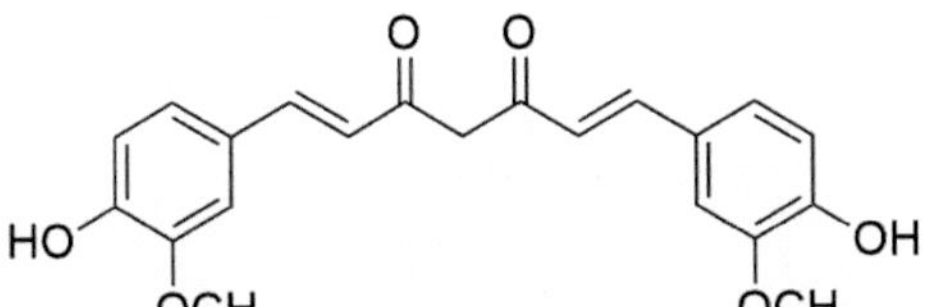

Curcumin

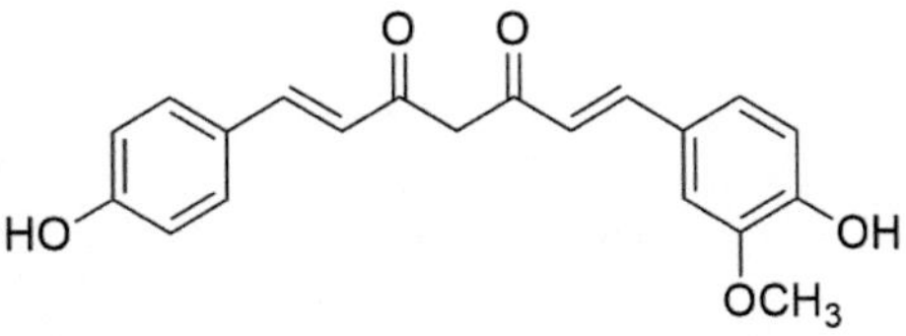

Demethoxycurcumin

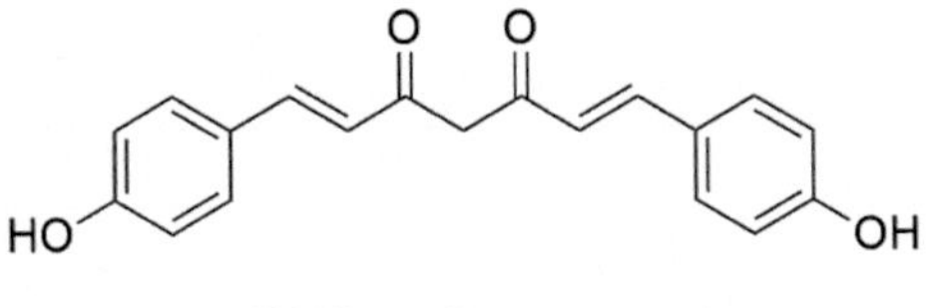

Bisdemethoxycurcumin

Fig. 18 Chemical structure of curcuminoids.

Apart from the above discussed classes of polyphenols, there are a few minor classes of polyphenols such as phenolic amides and coumarin. The majority of phenolic amides and coumarin contain a single hydroxy group but when these contain more than one hydroxy groups, they can be included under the polyphenol category (Shahidi and Ambigaipalan, 2015).

2.2 *Terpenoids*

Terpenoids are oxygenated derivatives of terpenes which are basically made up of two or more isoprene units. Generally, isoprene (2-methyl-1,3-butadiene) unit is a linear conjugated diene with molecular formula C_5H_8 having two double bonds at 1,3-positions and a methyl group at 2-position (Figure 19). Terpenoids can be classified into mono, di, tri, etc., depending upon the number of isoprene units present; the name can be accordingly (Yadav et al., 2014). The chemical structure of some representative terpenoids is given in Figure 19. Table 1 gives the classification details of terpenoids based on the number of isoprene units and examples from each class. Terpenoids are known to possess antioxidant, anti-inflammatory, and antimicrobial properties (Jahangeer et al., 2021).

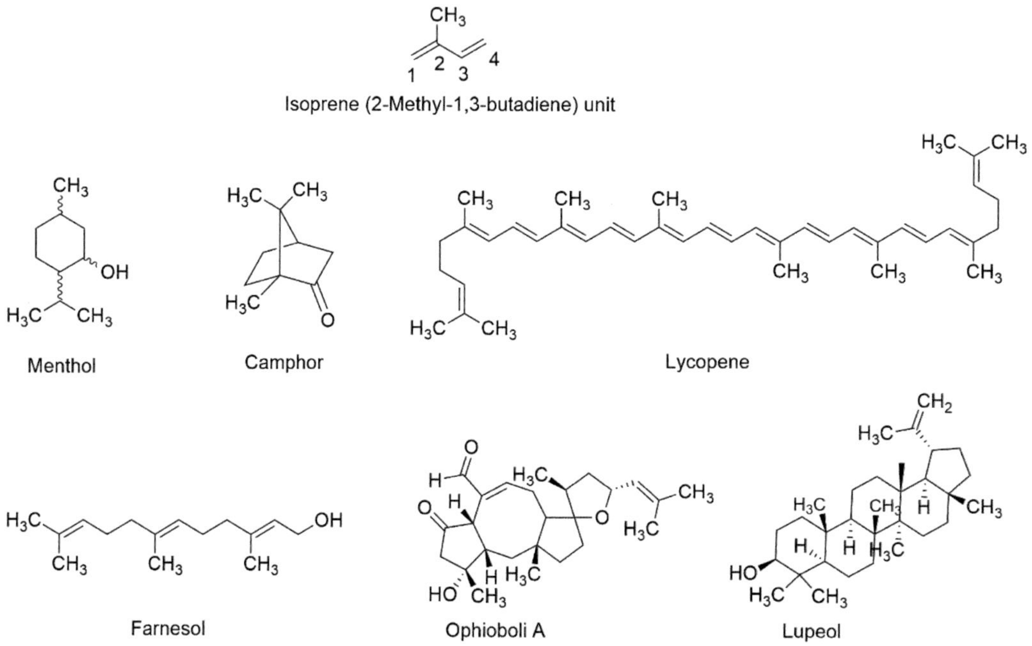

Fig. 19 Structure of isoprene unit and some well-known terpenoids.

Table 1 List of terpenoids with examples.

Class	*Number of Isoprene Units*	*Number of Carbon Atoms*	*Examples*
Monoterpenoids	2	10	Menthol, Limonene, Camphor
Sesquiterpenoids	3	15	Artemisinin, Farnesol
Diterpenoids	4	20	Taxol (Paclitaxel), Retinol
Sesterterpenoids	5	25	Ophiobolins
Triterpenoids	6	30	Squalene, Lupeol, Beta-amyrin
Tetraterpenoids	8	35	β-Carotene, Lycopene, Canthaxanthin, Capsanthin

2.3 Glucosinolates

Glucosinolates are a class of plant secondary metabolites containing both sulphur and nitrogen (Abdel-Massih et al., 2023). The core chemical structure of glucosinolates consists of a β-D-glucopyranose ring connected to a thiohydroximate group with a variable side chain derived from amino acids. Well-known examples include glucoraphanin, sinigrin, and glucobrassicin, which are mainly found in cruciferous vegetables (Barba et al., 2016). The chemical structures of some glucosinolates are shown in Figure 20. These compounds are recognized for their antioxidant, anti-inflammatory, anticancer, and detoxification-promoting properties (Choi et al., 2023; Baldelli et al., 2025). Table 2 lists some examples of glucosinolates along with their dietary sources and health benefits.

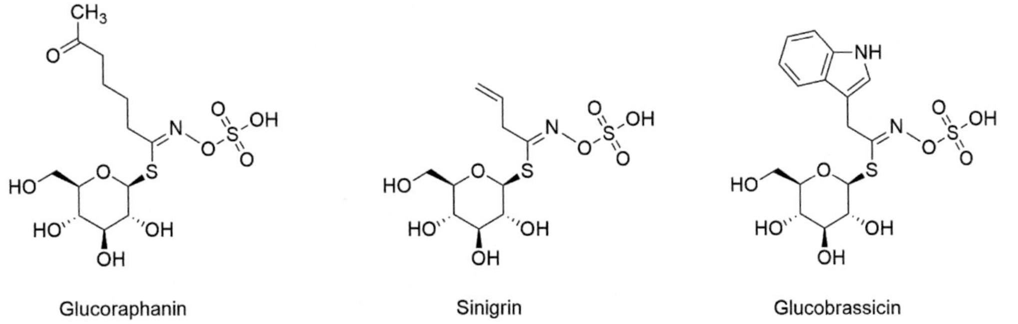

Fig. 20 Chemical structure of glucosinolates.

Table 2 Glucosinolates: Sources and health benefits.

Glucosinolates	*Food Source*	*Health Benefits*
Glucoraphanin	broccoli sprouts, kale, cabbage, and cauliflower	Antioxidants, anti-cancer, anti-inflammatory, and detoxification effects
Sinigrin	mustard seeds, horseradish, wasabi, cabbage, broccoli, and Brussels sprouts	Antimicrobial, antioxidant, anti-inflammatory, and anticancer properties, and also plays a role in promoting detoxification enzymes
Glucobrassicin	broccoli, cabbage, kale, and Brussels sprouts	Anti-inflammatory, antioxidant, and anticancer properties, and protection against cardiometabolic diseases and neurodegenerative disorders

2.4 Alkaloids

Alkaloids are primarily secondary plant metabolites. Their core structure consists of a nitrogen-containing heterocyclic ring system, often derived from amino acids, with diverse substituents that contribute to their wide structural variety and biological activity (Bhambhani et al., 2021). Common examples include nicotine, caffeine, ephedrine, and morphine. Alkaloids exhibit multiple health benefits such as neurostimulatory, anticancer, antimicrobial, anti-inflammatory, and analgesic effects (Debnath et al., 2018). The chemical structures of these alkaloids are shown in Figure 21, and Table 3 lists important alkaloids along with their food sources and health benefits.

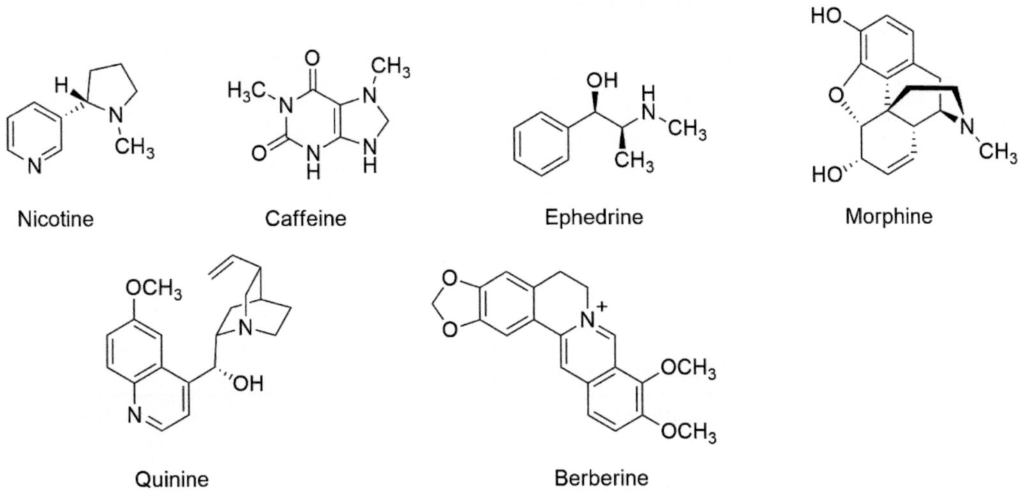

Fig. 21 Chemical structure of some alkaloids.

Table 3 Food source of important alkaloids along with their health benefits.

Compound	*Food Source*	*Health Benefits*
Nicotine	Tobacco plant; trace amounts in eggplant, tomato, potato, green pepper	Acts as a stimulant (enhances alertness, focus), appetite suppression, may have neuroprotective effects in Parkinson's disease
Caffeine	Coffee beans, tea leaves, cacao (chocolate), kola nut, guarana, yerba maté	Central nervous system stimulant, improves alertness, mood, and cognitive performance; enhances physical endurance, may reduce risk of Parkinson's and Alzheimer's in moderate intake
Ephedrine	Ephedra sinica (traditional Chinese medicinal plant)	Bronchodilator (used in asthma, bronchitis), decongestant, stimulates metabolism and thermogenesis
Morphine	Opium poppy	Strong analgesic (pain relief), used in surgery, trauma, cancer pain, induces sedation and euphoria
Quinine	Bark of Cinchona tree	Antimalarial agent, reduces fever, muscle relaxant for leg cramps, bitter compound used in tonic water
Berberine	Found in Berberis species (barberry, tree turmeric), goldenseal, Oregon grape	Antimicrobial (effective against bacteria, fungi, parasites), supports blood sugar regulation, improves cholesterol levels, anti-inflammatory and antioxidant effects

3. Carbohydrate-based Bioactive Compounds.

Dietary fibers, including resistant starch, prebiotics, and β-glucans, are the major carbohydrate-based bioactive compounds with a wide range of health benefits. These indigestible carbohydrates resist digestion in the small intestine and are partially or fully fermented in the colon by gut microbiota (Slavin, 2013; Roberfroid, 2007). Resistant starch, composed of glucose polymers amylose and amylopectin, is fermented in the colon to produce short-chain fatty acids (SCFAs) such as acetate, propionate, and butyrate, which support gut health, modulate glucose metabolism, and reduce inflammation (Birt et al., 2013). Prebiotics are non-digestible food ingredients, typically oligosaccharides (fructo- and galacto-oligosaccharides) or some polysaccharides, that selectively stimulate the growth of beneficial gut bacteria and promote a healthy microbiome (Gibson et al., 2017). β-Glucans are soluble dietary fibers composed of glucose polymers with β-glycosidic linkages. They form viscous solutions in the gut, reduce cholesterol absorption, slow glucose uptake, and improve cardiovascular and metabolic health (Volman et al., 2008). Table 4 lists food sources of different dietary fibers along with their health benefits.

Table 4 Dietary fibers: Sources and health benefits.

Dietary fibers	*Food Source*	*Health Benefits*
Resistant starch	Raw potato, green banana, legumes, cooked and cooled rice, pasta, potato maize, high-amylose corn	Improves colonic health via SCFA production, enhances satiety and weight control, lowers postprandial blood glucose and insulin response, supports bowel regularity
Prebiotics	Chicory root, onions, garlic, leek, legumes, soybeans	Selective stimulation of Bifidobacteria and Lactobacilli, strengthens gut barrier and immune response, enhances mineral absorption (Ca, Mg), improves digestion and reduces risk of colon disorders
β-glucans	Oats, barley, mushrooms, yeast cell wall	Cholesterol-lowering, blood glucose regulation, modulation of gut microbiota, immune system activation

4. Lipid-based Bioactive Compounds

Lipid-derived molecules obtained from various food sources have significant health benefits beyond basic nutrition (Peña-Vázquez et al., 2024). Important examples include omega-3 fatty acids, short-chain fatty acids "SCFAs", sterols, and stanols.

4.1 Omega-3 Fatty Acids

Omega-3 fatty acids are polyunsaturated fatty acids (PUFAs) essential for human health. The three main types are α-linolenic acid (ALA), eicosapentaenoic acid (EPA), and docosahexaenoic acid (DHA) (Figure 22). ALA is primarily found in plant-based sources such as flaxseeds, chia seeds, walnuts, hazelnuts, spinach, kale, and soybeans. EPA, commonly obtained from fatty fish such as salmon, mackerel, sardines, herring, trout, and fish oil supplements. DHA, abundant in the same marine sources, is crucial for eye and retinal health, supports brain development and function, lowers triglyceride levels and blood pressure, and contributes to mental health and infant neural development. Thus, most omega-3 PUFAs must be obtained from dietary sources, as humans cannot efficiently synthesize them (Saini and Keum, 2018). These compounds support cardiovascular health, anti-inflammatory responses, neuroprotection, and metabolic regulation (Patted et al., 2024). Table 5 lists its dietary sources and associated health benefits.

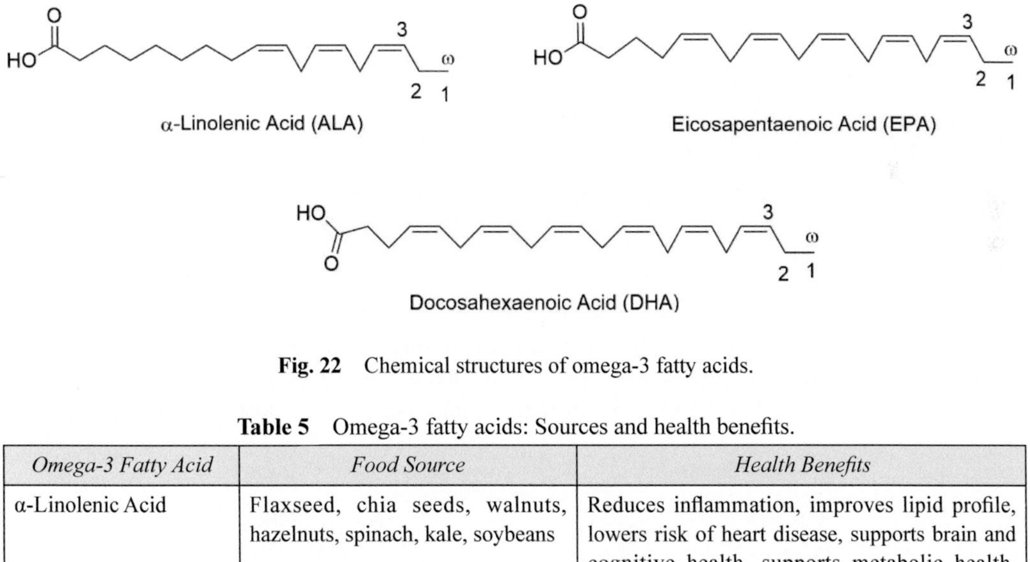

Fig. 22 Chemical structures of omega-3 fatty acids.

Table 5 Omega-3 fatty acids: Sources and health benefits.

Omega-3 Fatty Acid	*Food Source*	*Health Benefits*
α-Linolenic Acid	Flaxseed, chia seeds, walnuts, hazelnuts, spinach, kale, soybeans	Reduces inflammation, improves lipid profile, lowers risk of heart disease, supports brain and cognitive health, supports metabolic health, improve skin and hair health
Eicosapentaenoic Acid (EPA)	Salmon, mackerel, sardines, herring, trout, fish oil supplements	Anti-inflammatory, improves mental health and cognitive function, improves metabolic health
Docosahexaenoic Acid (DHA)	Salmon, mackerel, sardines, herring, trout	Beneficial for eye and retinal health, anti-inflammatory, improves heart health by reducing triglycerides and blood pressure, improves mental health, helps in infant development

4.2 Short Chain Fatty Acids

SCFAs are organic acids produced mainly by gut microbiota (e.g., *Lactobacillus and Bifidobacterium spp.*) during the fermentation of dietary fibers and resistant starches in the colon (Silva et al., 2020). Acetate, propionate, and butyrate are the major SCFAs, while formate, valerate, and caproate are minor components (Figure 23). Other organic acids such as lactate, succinate, pyruvate, citrate, and malate are also produced in small amounts. SCFAs contribute to gut health, energy metabolism, anti-inflammatory effects, and immune modulation (Sankarganesh et al., 2025). Table 6 lists major short chain fatty acids, their sources, and health benefits.

Table 6 SCFAs: Sources and health benefits.

SCFA	*Probiotics*	*Health Benefits*
Formate	Bifidobacterium breve, B. infantis	Early-life gut health
Acetate	Bifidobacterium longum, B. breve, L. acidophilus	Pathogen inhibition, barrier strengthening
Propionate	Propionibacterium freudenreichii, L. plantarum	Cholesterol reduction, satiety, glucose regulation
Butyrate	Clostridium butyricum, Bifidobacterium adolescentis	Colonocyte fuel, anti-inflammatory
Valerate/ Caproate	Clostridium spp.	Amino acid fermentation by-products

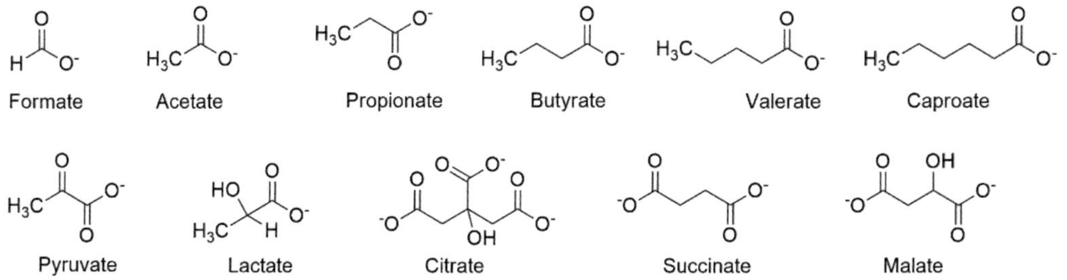

Fig. 23 Chemical structure of SCFAs.

4.3 Sterols and Stanols

Sterols are a subgroup of steroids found in plants and animals, characterized by a four-fused-ring structure with a hydroxyl group. Plant-derived sterols, or phytosterols, include β-sitosterol, campesterol, and stigmasterol, while cholesterol is the major animal sterol (Plat et al., 2019). Stanols are the hydrogenated form of sterols, e.g., sitostanol, campestanol, and ergostanol (Figure 24). These

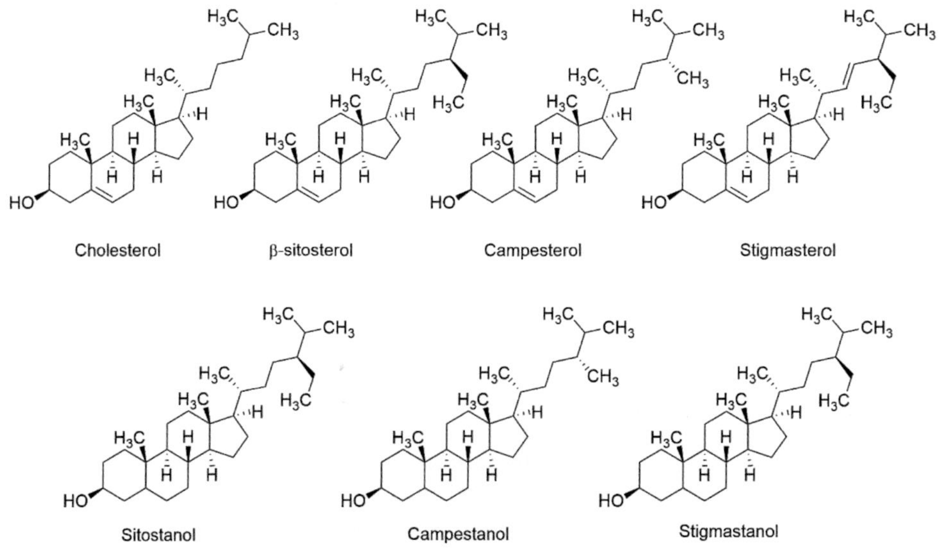

Fig. 24 Chemical structures of sterols and stanols.

compounds lower LDL cholesterol, support cardiovascular health, and modulate inflammatory responses (Barkas et al., 2023). Table 7 provides dietary sources of some sterols and stanols along with their health benefits.

Table 7 Sterols and stanols: sources and health benefits.

Sterols and Stanols	*Food Source*	*Health Benefits*
Cholesterol	Egg yolk, shrimp, crab, beef and chicken liver, cheese, butter, milk	Precursor for steroid hormones, helps in digestion via bile acid production, acts a precursor for vitamin D, essential for brain and nervous system
β-sitosterol	Corn, soybean and canola oil, almonds, peanuts, pistachios, pumpkin, sesame and sunflower seeds, legumes, whole grains	Supports heart health by lowering LDL, beneficial for prostate health, anti-inflammatory and anticancer effect, also have antioxidant potential
Campesterol	Vegetable oils, almonds, peanuts, cashews; sesame, pumpkin, sunflower seeds; Legumes; whole grains; broccoli, Brussels sprouts	Supports cardiovascular health, anti-inflammatory and anticancer potential
Stigmasterol	Vegetable oils, almonds, peanuts, cashews; sesame, pumpkin, sunflower seeds; Legumes; whole grains; spinach, cauliflower, soy sprouts	Supports cardiovascular health, anti-inflammatory and anticancer potential, beneficial for bone health
Sitostanol	Yogurts, milk, margarine; corn, rapeseed and soybean oil; almonds, sunflower seeds, Legumes and cereals	Reduces the intestinal absorption of dietary cholesterol, improves cardiovascular health
Campestanol	Margarine, spreads, yogurts, milk, vegetable oils, cereals and whole grains, Nuts and seeds, legumes	Helps in cholesterol management, improves cardiovascular health.
Ergostanol	Mushrooms, yeast, fermented foods	Cholesterol reducing effect, helps in improving cardiovascular health

5. Bioactive Peptides

Bioactive peptides are oligomeric amino acid chains produced during enzymatic hydrolysis of proteins (Korhonen and Pihlanto, 2006). They are classified into two main types based on origin: endogenous and exogenous (Akbarian et al., 2022) (Figure 25). Endogenous bioactive peptides are generated within cells by enzymatic cleavage of body proteins or via direct biosynthesis, whereas exogenous peptides are obtained from food proteins or supplements through enzymatic hydrolysis or microbial fermentation. Some exogenous peptides are directly consumed as functional foods or nutraceuticals. Another variety of bioactive peptides produced by bacteria through ribosomal synthesis are termed bacteriocins (Akbarian et al., 2022). Table 8 presents categories of bioactive peptides with representative examples and associated health benefits.

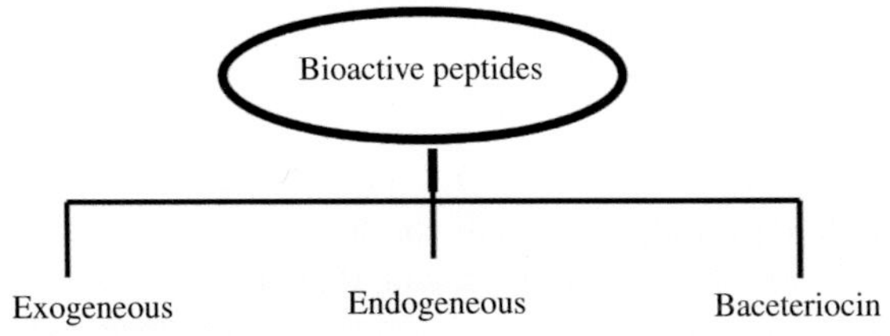

Fig. 25 Classification of bioactive peptides.

Table 8 Bioactive peptides: Categories and health benefits.

Types of Bioactive Peptides	*Examples*	*Health Benefits*
Endogenous	Angiotensin II	Blood pressure regulation, fluid and electrolyte balance, support of renal function
	Endorphins	Pain Relief, stress reduction, mood enhancement, immune system support
	Thymosins	Antiviral and antimicrobial effects, anti-cancer supportive role, wound healing and tissue repair
Exogenous	Lactoferricin	Antiviral and antimicrobial effects, anti-inflammatory properties, gut health benefits
	Hemorphins	Pain regulation and analgesic effect, stress and mood regulation, cognitive benefits
	Casomorphins	Pain relief and relaxation, mood regulation, gastrointestinal health
Bacteriocin	Nisin	Antimicrobial, anti-inflammatory, and anticancer properties
	Enterocin A and B	Antimicrobial property
	Helveticin J	Antimicrobial and probiotic

6. Vitamins

Vitamins are organic micronutrients which are essential for growth and development of living organisms and are considered as bioactive compounds in many functional foods (Barker, 2023). Vitamins A, C, and E have antioxidant properties however vitamin B acts as coenzyme. These micronutrients are majorly classified based on their water solubility as well as their origin. Vitamins A, D, E, and K are fat soluble vitamins (Youness et al., 2022) whereas B and C are water soluble vitamins (Figure 26).

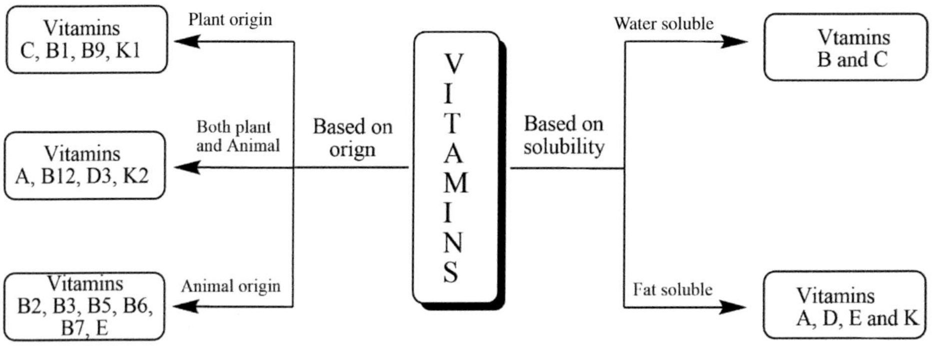

Fig. 26 Classification of vitamins based on solubility and origin.

6.1 *Vitamin A*

Vitamin A includes fat-soluble retinoids (retinol, retinal, retinoic acid) derived from plant and animal sources. β-carotene, a tetraterpenoid, is a metabolic precursor of vitamin A. Vitamin A acts as an antioxidant and supports vision, growth, development, immune function, and reproduction (Youness et al., 2022). Chemical structure of β-carotene and other active forms of vitamin A are given in Figure 27. Major dietary sources and health benefits of vitamin A are given in Table 9.

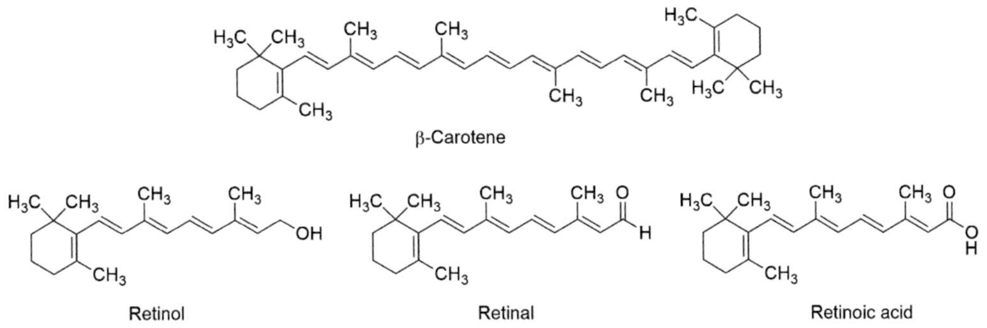

Fig. 27 Structure of β-carotene and active forms of vitamin A.

6.2 Vitamin B

A group of eight water-soluble vitamins essential for metabolism. Most of those act as coenzymes or coenzyme precursors in enzymatic reactions. Depending upon their chronology of discovery from vitamin B complex they are named as B1 (thiamine), B2 (riboflavin), B3 (niacin), B5 (pantothenic acid), B6 (pyridoxine), B7 (biotin), B9 (folate) and B12 (cobalamin) (Kennedy, 2016). All these are chemically distinct compounds and there is no structural similarity among them. B1 and B9 are obtained from plant sources and B2, B3, B5, B6, B7 are obtained from animal sources however B12 can be obtained from both plant and animal sources. Vitamin B essentially keeps the nervous system healthy, helps release energy from food and makes healthy blood cells (Hanna et al., 2022). Chemical structure of all forms of vitamin B are given in Figure 28 and major dietary sources and health benefits of vitamin B are given in Table 9.

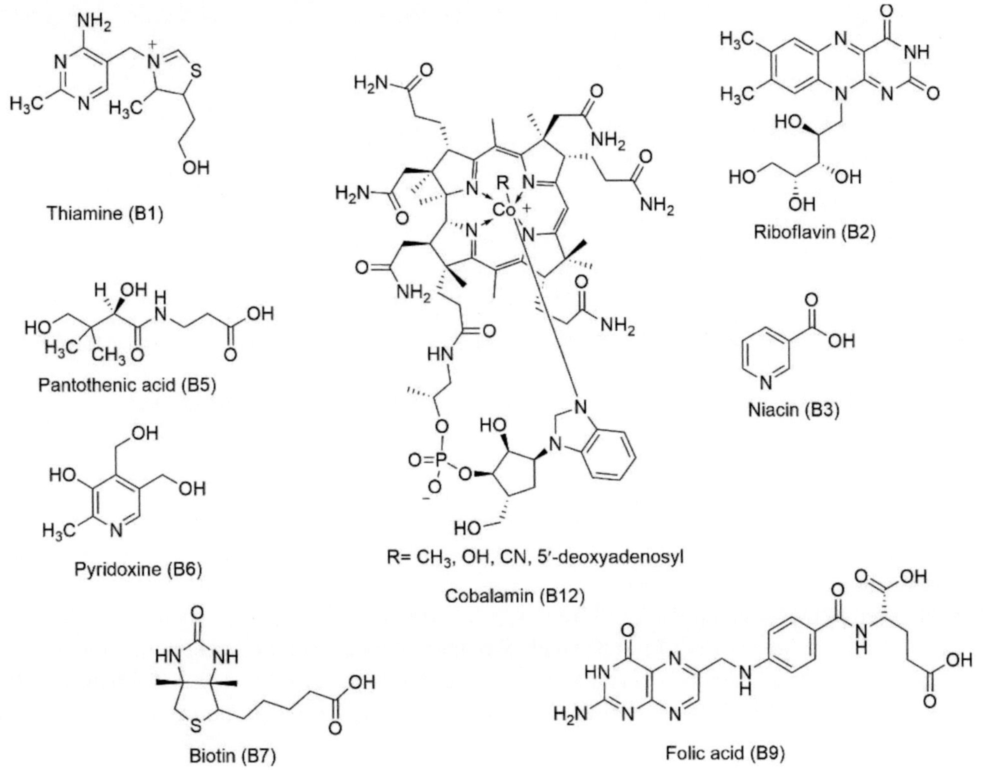

Fig. 28 Chemical structure of vitamin B complex.

6.3 Vitamin C

Vitamin C which is chemically known as L-ascorbic acid (Figure 29) is another water-soluble vitamin obtained only from plant sources. The other isomer, i.e., D-ascorbic acid is biologically inactive and is not considered as a vitamin. Vitamin C acts as an antioxidant, helps in collagen synthesis, wound healing and supports immune function (Carr and Maggini, 2017). Major dietary sources and health benefits of vitamin C are given in Table 9.

6.4 Vitamin D

Vitamin D (also referred as calciferol), like vitamin A is a fat-soluble vitamin mostly obtained from animal sources. There are three different forms of vitamin D, i.e., D1 (no longer used in nutrition science), D2 (Ergocalciferol) and D3 (Cholecalciferol). Vitamin D helps in calcium and phosphate absorption, bone mineralization, and is essential for cardiovascular and brain health (Holick, 2007). Figure 29 presents the structure of various forms of vitamin D and its important metabolites such as calcidiols and calcitriols which are the metabolites of vitamin D (Bikle, 2014). Information about the dietary sources and health benefits are given in Table 9.

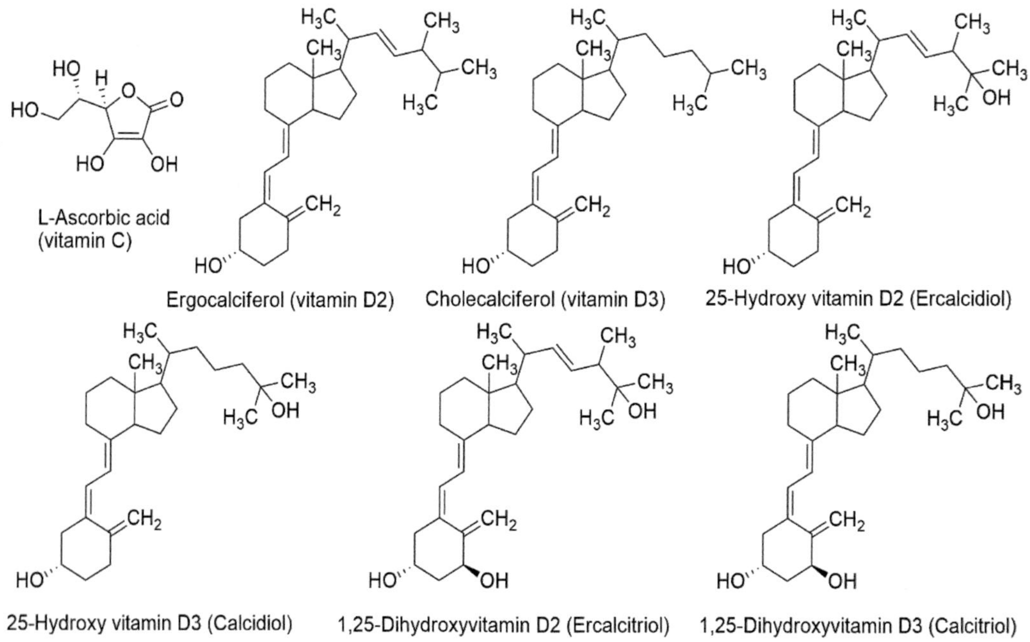

Fig. 29 Chemical structure of vitamin C and vitamin D (forms and metabolites).

6.5 Vitamin E

Vitamin E is a fat-soluble vitamin obtained primarily from plant and some animal sources. It exists in eight natural forms, divided into two classes: tocopherols (α, β, γ, δ) and tocotrienols (α, β, γ, δ) (Galli et al., 2022; Figure 30). Among these, α-tocopherol is the most biologically active and is primarily used to meet human vitamin E requirements. Vitamin E functions as a potent antioxidant, protecting cell membranes from lipid peroxidation, maintaining red blood cell membrane integrity, preventing oxidation of vitamin A, and supporting immune function (Traber and Atkinson, 2007, Rizvi et al., 2014). More details about the health benefits and dietary sources of vitamin E are summarized in Table 9.

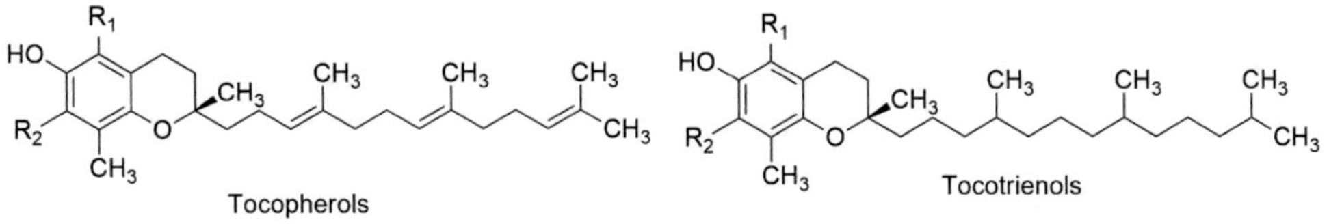

Isomer	R1	R2
α	CH_3	CH_3
β	CH_3	H
γ	H	CH_3
δ	H	H

Fig. 30 Chemical structure of vitamin E classes.

6.6 Vitamin K

Vitamin K, like vitamins A, D, and E, is a fat-soluble vitamin obtained mainly from dietary sources (Booth, 2012). It exists in three biologically active forms: K1 (Phylloquinone), K2 (Menaquinone), and K3 (Menadione) (Figure 31), of which K1 and K2 are naturally occurring and biologically useful for humans. Vitamin K plays a crucial role in regulating blood clotting, maintaining bone health, and supporting vascular function (Simes et al., 2020). More information about the health benefits of vitamin K and dietary sources can be found in Table 9.

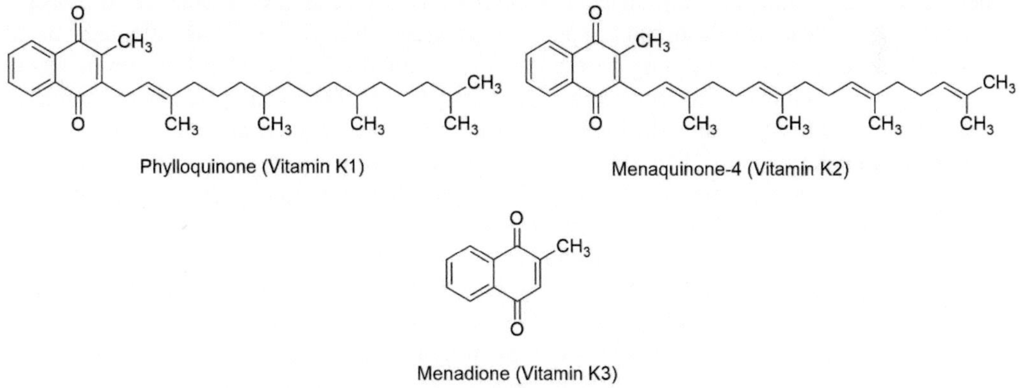

Fig. 31 Chemical structure of different forms of vitamin K.

Table 9 Vitamins: Sources and health benefits.

Vitamin	*Food Sources*	*Key Health Benefits*
A (Retinol, Beta-carotene)	Liver, fish oils, dairy, eggs, carrots, sweet potatoes, spinach, mango	Vision (especially night vision), immune function, skin and cell growth
B1 (Thiamine)	Whole grains, legumes, nuts, pork, sunflower seeds	Carbohydrate metabolism, nerve and muscle function
B2 (Riboflavin)	Milk, eggs, lean meat, leafy greens, almonds	Energy production, antioxidant role, skin and eye health
B3 (Niacin)	Poultry, fish, peanuts, mushrooms, fortified cereals	Energy metabolism, DNA repair, skin and nervous system health

Contd.

Table 9 *Contd.*

Vitamin	*Food Sources*	*Key Health Benefits*
B5 (Pantothenic acid)	Eggs, avocados, broccoli, whole grains, chicken, beef	Coenzyme A synthesis, energy release, hormone production
B6 (Pyridoxine)	Poultry, fish, bananas, potatoes, chickpeas, nuts	Amino acid metabolism, hemoglobin synthesis, neurotransmitter production, immune support
B7 (Biotin)	Egg yolk, nuts, seeds, legumes, whole grains, sweet potatoes	Fatty acid & glucose metabolism, healthy hair, skin & nails
B9 (Folate / Folic acid)	Leafy greens, legumes, citrus fruits, beets, fortified grains	DNA & RNA synthesis, red blood cell formation, fetal neural tube development
B12 (Cobalamin)	Meat, fish, eggs, dairy, fortified cereals	Red blood cell formation, nerve function, DNA synthesis, prevents anemia
C (Ascorbic acid)	Citrus fruits, strawberries, kiwi, bell peppers, broccoli, tomatoes	Antioxidant, collagen synthesis, wound healing, iron absorption, immune defense
D (Calciferol)	Fatty fish, egg yolk, fortified dairy, sunlight exposure	Calcium absorption, bone and teeth health, immune regulation
E (Tocopherol)	Nuts, seeds, vegetable oils, spinach, broccoli	Antioxidant, protects cell membranes, supports skin and heart health
K (Phylloquinone, Menaquinone)	Leafy greens (kale, spinach), broccoli, Brussels sprouts, fermented foods	Blood clotting, bone metabolism, cardiovascular protection

7. Minerals

Minerals, though traditionally recognized as essential nutrients, can also be considered bioactive compounds due to their significant roles in physiological regulation and promotion of human health (Mazumder et al., 2025). Depending on the quantity required by the human body, minerals are classified into macrominerals and microminerals (Farag et al., 2023). Examples of macrominerals include calcium, chloride, magnesium, phosphorus, potassium, sodium, and sulfur, with a daily recommended intake of more than 100 mg. Microminerals (trace elements) include chromium, copper, fluoride, iodine, iron, manganese, molybdenum, selenium, and zinc, with daily recommended intake of less than 100 mg (Farag et al., 2023). Table 10 lists macrominerals and microminerals along with their corresponding roles in human physiological functions.

Calcium (Ca), obtained from milk, cheese, yogurt, leafy greens, and fortified foods, is vital for bone and teeth health, muscle contraction, and blood clotting. Chloride (Cl^-), found in table salt, seaweed, rye, celery, and olives, helps maintain electrolyte balance and supports the production of gastric acid (HCl). Magnesium (Mg), present in nuts, seeds, whole grains, legumes, and spinach, functions as an enzyme cofactor and aids in muscle and nerve function as well as energy metabolism. Phosphorus (P), sourced from meat, poultry, fish, dairy, legumes, and nuts, contributes to bone and teeth structure and is essential for ATP and nucleic acid synthesis. Potassium (K), abundant in bananas, potatoes, oranges, leafy greens, and beans, regulates fluid balance, nerve signaling, and muscle contraction. Sodium (Na), commonly consumed through table salt, processed foods, cheese, and bread, maintains fluid balance, regulates blood pressure, and enables nerve impulse transmission. Sulphur (S), derived from protein foods like meat, fish, eggs, legumes, garlic, and onion, is a component of amino acids and vitamins and aids detoxification processes (Farag et al., 2023, Ye et al., 2018, Ali, 2023).

Among the microminerals, Chromium (Cr), present in whole grains, broccoli, grapes, meat, and brewer's yeast, enhances insulin action and supports glucose metabolism. Copper (Cu), found in shellfish, liver, nuts, seeds, and whole grains, plays a role in iron metabolism, antioxidant defense, and connective tissue formation. Fluoride (F^-), available in fluoridated water, tea, and seafood, strengthens bones and teeth. Iodine (I^-), obtained from iodized salt, seaweed, fish, dairy, and eggs, is necessary for thyroid hormone synthesis. Iron (Fe), present in red meat, poultry, fish, legumes, spinach, and fortified cereals, is essential for oxygen transport and electron transfer in cells. Manganese (Mn), sourced from whole grains, nuts, leafy vegetables, and tea, functions as an enzyme cofactor supporting bone health and antioxidant activity. Molybdenum (Mo), found in legumes, whole grains, nuts, and milk, acts as a cofactor in various metabolic pathways. Selenium (Se), obtained from Brazil nuts, seafood, meat, eggs, and cereals, serves as a potent antioxidant and supports thyroid function. Zinc (Zn), found in meat, shellfish, legumes, nuts, seeds, and dairy, is vital for cell growth, immune response, and wound healing. Finally, Cobalt (Co), present in meat, fish, dairy, and fortified cereals, forms a component of vitamin B_{12} (cobalamin), which is essential for red blood cell formation and neurological function (Rahman et al., 2025, Farag et al., 2021).

Table 10 Macrominerals and microminerals: Roles and recommended intake.

Minerals	*Food Source*	*Role in Basic Function of Human Body*
Calcium (Ca)	Milk, cheese, yogurt, leafy greens, fortified foods	Bone and teeth health, muscle contraction, blood clotting
Chloride (Cl^-)	Table salt (NaCl), seaweed, rye, celery, olives	Electrolyte balance, gastric acid (HCl) production
Magnesium (Mg)	Nuts, seeds, whole grains, legumes, spinach	Enzyme cofactor, muscle and nerve function, energy metabolism
Phosphorus (P)	Meat, poultry, fish, dairy, legumes, nuts	Bone and teeth structure, ATP, DNA/RNA
Potassium (K)	Bananas, potatoes, oranges, leafy greens, beans	Fluid balance, nerve signalling, muscle contraction
Sodium (Na)	Table salt, processed foods, cheese, bread	Fluid balance, blood pressure regulation, nerve impulses
Sulphur (S)	Protein foods (meat, fish, eggs, legumes), garlic, onion	Component of amino acids and vitamins, detoxification
Chromium (Cr)	Whole grains, broccoli, grapes, meat, brewer's yeast	Enhances insulin action, glucose metabolism
Copper (Cu)	Shellfish, liver, nuts, seeds, whole grains	Iron metabolism, antioxidant enzymes, connective tissue
Fluoride (F^-)	Fluoridated water, tea, seafood	Strengthens teeth and bones
Iodine (I^-)	Iodized salt, seaweed, fish, dairy, eggs	Necessary for thyroid hormone production
Iron (Fe)	Red meat, poultry, fish, legumes, spinach, fortified cereals	Essential for oxygen transport in the blood, electron transport
Manganese (Mn)	Whole grains, nuts, leafy vegetables, tea	Enzyme cofactor, bone health, antioxidant defence
Molybdenum (Mo)	Legumes, whole grains, nuts, milk	A cofactor in various metabolic processes
Selenium (Se)	Brazil nuts, seafood, meat, eggs, cereals	An antioxidant that protects cells, thyroid function
Zinc (Zn)	Meat, shellfish, legumes, nuts, seeds, dairy	Important for cell growth and immune function
Cobalt (Co)	Meat, fish, dairy, fortified cereals	Component of vitamin B12 (cobalamin)

8. Conclusions

Functional foods, originally conceptualized to support healthy aging, have emerged as a vital dietary intervention with the potential to prevent disease and enhance overall health. They are consumed in natural or minimally processed forms and provide beneficial effects through diverse bioactive compounds. These compounds namely, polyphenols, terpenoids, glucosinolates, alkaloids, bioactive peptides, lipids, fibers, vitamins, and minerals, offer antioxidant, anti-inflammatory, cardioprotective, neuroprotective, and metabolic regulatory functions. Their chemical diversity shows the complex physiological effects, emphasizing the importance of understanding their structures, sources, and mechanisms of action. Integrating functional foods into diets represents a scientifically grounded strategy to promote wellness, prevent chronic diseases, and improve the overall quality of life.

References

Abdel-Massih, R.M, E. Debs, L. Othman, J. Attieh, and F.M Cabrerizo. 2023. Glucosinolates, a natural chemical arsenal: More to tell than the myrosinase story. *Front. Microbiol.*, *14*, 1130208.

Adhikari, S., Nath, P., Deb, V.K., Das, N, Banerjee, A, Pathak, S., et al., 2025. Pharmacological potential of natural chalcones: A recent studies and future perspective. *Front. Pharmacol.*, *16*, 1570385.

Ali, A.H.A. 2023. Overview of the vital roles of macro minerals in the human body. *J. Trace Elem. Miner.*, *4*, 100076.

Akbarian, M., A. Khani, S. Eghbalpour, and V.N. Uversky. 2022. Bioactive Peptides: Synthesis, Sources, Applications, and Proposed Mechanisms of Action. *Int. J. Mol. Sci.*, *23*(3), 1445.

Alrumaihi, F., S.A. Almatroodi, H.O.A. Alharbi, W.M. Alwanian, F.A. Alharbi, A. A. lmatroudi, et al., 2024. Pharmacological Potential of Kaempferol, a Flavonoid in the Management of Pathogenesis via Modulation of Inflammation and Other Biological Activities. *Molecules*, *29*(9), 2007.

Arai, S. 1996. Studies on functional foods in Japan. *Biosci. Biotechnol. Biochem.*, *60*(1), 9–15.

Ashrafizadeh M., Z. Ahmadi, R. Mohammadinejad, and E. Ghasemipour Afshar. 2020. Tangeretin: A mechanistic review of its pharmacological and therapeutic effects. *J. Basic Clin. Physiol. Pharmacol.*, *31*(4). /j/jbcpp.2020.31.issue-4/jbcpp-2019-0191/jbcpp-2019-0191.xml.

Bae, J., N. Kim, Y. Shin, S.-Y. Kim, and Y.-J. Kim. 2020. Activity of catechins and their applications. *Biomed. Dermatol.*, *4*, 8.

Baldelli, S., M. Lombardo, A. D'Amato, S. Karav, G. Tripodi, and G. Aiello. 2025. Glucosinolates in Human Health: Metabolic Pathways, Bioavailability, and Potential in Chronic Disease Prevention. *Foods*, *14*(6), 912.

Baranwal, A., P. Aggarwal, A. Rai, and N. Kumar. 2022. Pharmacological Actions and Underlying Mechanisms of Catechin: A Review. *Mini Rev. Med. Chem.*, *22*(5), 821–833.

Barba, F.J., N. Nikmaram, S. Roohinejad, A. Khelfa, Z. Zhu, and M. Koubaa. 2016. Bioavailability of Glucosinolates and Their Breakdown Products: Impact of Processing. *Front. Nutr.*, *3*, 24.

Barkas, F., E. Bathrellou, T. Nomikos, D. Panagiotakos, E. Liberopoulos, M.D. Kontogianni. 2023. Plant Sterols and Plant Stanols in Cholesterol Management and Cardiovascular Prevention. *Nutrients*. *15*(13), 2845.

Barker, T. 2023. Vitamins and Human Health: Systematic Reviews and Original Research. *Nutrients*, *15*(13), 2888.

Berenshtein, L., Z. Okun, and A. Shpigelman. 2024. Stability and Bioaccessibility of Lignans in Food Products. *ACS Omega*, 9(2), 2022–2031.

Bhambhani, S., K.R. Kondhare, and A.P. Giri. 2021. Diversity in Chemical Structures and Biological Properties of Plant Alkaloids. *Molecules*, *26*(11), 3374.

Bikle, D.D. 2014. Vitamin D Metabolism, Mechanism of Action, and Clinical Applications. *Chemistry & Biology*, *21*(3), 319–329.

Birt, D.F., T. Boylston, S. Hendrich, J.-L. Jane, J. Hollis, L. Li, et al., 2013. Resistant starch: Promise for improving human health. *Adv. Nutr.*, *4*(6): 587–601.

Bolat, E., S. Sarıtaş, H. Duman, F. Eker, E. Akdaşçi and S. Karav, 2024. Polyphenols: Secondary Metabolites with a Biological Impression. *Nutrients*, *16(*15), 2550.

Booth, S.L. 2012. Vitamin K: Food composition and dietary intakes. *Food Nutr. Res.*, *2012*, 56.

Carr, A.C., and S. Maggini. 2017. Vitamin C and immune function. *Nutrients*, *9*(11), 1211.

Chang, Z., Q. Zhang, W. Liang, K. Zhou, P. Jian, G. She, et al., 2019. A Comprehensive Review of the Structure Elucidation of Tannins from *Terminalia* Linn. *Evid. Based Complement. Alternat. Med.*, *2019*, 8623909.

Choi, H., H. Kim, S. Han, H.W. Park, I.J. Ha, and J.S. Kim et al., 2023. Antioxidant and Anti-Inflammatory Activities of High-Glucosinolate-Synthesis Lines of *Brassica rapa*. *Antioxidants* (*Basel*), *12*(9), 1693.

Dabeek, W.M., and M.V. Marra. 2019. Dietary Quercetin and Kaempferol: Bioavailability and Potential Cardiovascular-Related Bioactivity in Humans. *Nutrients*, *11*(10), 2288.

Debnath, B., W.S. Singh, M. Das, S. Goswami, M.K. Singh, D. Maiti, et al., 2018. Role of plant alkaloids on human health: A review of biological activities. *Mater. Today Chem.*, 9, 56–72.

Dirimanov, S., and P. Högger. 2019. Screening of Inhibitory Effects of Polyphenols on Akt-Phosphorylation in Endothelial Cells and Determination of Structure-Activity Features. *Biomolecules*, *9*(6), 219.

Duarte, M., S.S. Pedrosa, P.R. Khusial, and A.R. Madureira. 2025. The biological potential and health-benefits of flavonoids: A review and development opportunities. *Chem. Biol. Interact.*, *421*, 111755.

Dvorakova, M., and P. Landa. 2017. Anti-inflammatory activity of natural stilbenoids: A review. *Pharmacol. Res.*, *124*, 126–145.

El-Shiekh, R.A., M.H. Radi, and E. Abdel-Sattar. 2025. Unveiling the therapeutic potential of aromadendrin (AMD): A promising anti-inflammatory agent in the prevention of chronic diseases. *Inflammopharmacol.*, *33*, 1209–1220.

Farag, M.A., S. Hamouda, S. Gomaa, A.A. Agboluaje, M.L.M. Hariri, and S.M. Yousof. 2021. Dietary Micronutrients from Zygote to Senility: Updated Review of Minerals' Role and Orchestration in Human Nutrition throughout Life Cycle with Sex Differences. *Nutrients*, *13*(11), 3740.

Farag, M.A., B. Abib, Z. Qin, X. Ze, and S.E. All. 2023. Dietary macrominerals: Updated review of their role and orchestration in human nutrition throughout the life cycle with sex differences. *Curr. Res. Food Sci.*, *6*, 100450.

Frumuzachi, O., A. Flanagan, S. Rohn, and A. Mocan. 2025. The dichotomy between functional and functionalized foods: A critical characterization of concepts. *Food Res. Int.*, *208*, 116173.

Galli, F., M. Bonomini, D. Bartolini, L. Zatini, G. Reboldi, G. Marcantonini, et al., 2022. Vitamin E (Alpha-Tocopherol) Metabolism and Nutrition in Chronic Kidney Disease. *Antioxidants*, *11*(5), 989.

Gibson, G.R., R. Hutkins, M.E. Sanders, S.L. Prescott, R.A. Reimer, S.J. Salminen, et al., 2017. Expert consensus document: The International Scientific Association for Probiotics and Prebiotics (ISAPP) consensus statement on the definition and scope of prebiotics. *Nat. Rev. Gastroenterol. Hepatol.*, *14*(8), 491–502.

Hanna, M., E. Jaqua, V. Nguyen, and J. Clay. 2022. B Vitamins: Functions and Uses in Medicine. *Perm J.*, *26*(2), 89–97.

Hewlings, S.J., and D.S. Kalman. 2017. Curcumin: A Review of Its Effects on Human Health. *Foods*, *6*(10), 92.

Holick, M.F. 2007. Vitamin D deficiency. *N. Engl. J. Med.*, *357*(3), 266–281.

Hossain, M.S., M.A. Wazed, S. Asha, M. R. Amin, and I. M. Shimul. 2025. Dietary Phytochemicals in Health and Disease: Mechanisms, Clinical Evidence, and Applications: A Comprehensive Review. *Food Sci Nutr.*, *13*(3), e70101.

Hostetler G.L., R.A. Ralston, and S.J. Schwartz. 2017. Flavones: Food Sources, Bioavailability, Metabolism, and Bioactivity. *Adv. Nutr.* *8*(3), 423–435.

Hu, Z., Y. Guan, W. Hu, Z. Xu, and M. Ishfaq. 2022. An overview of pharmacological activities of baicalin and its aglycone baicalein: New insights into molecular mechanisms and signaling pathways. *Iran J. Basic Med. Sci.*, *25*, 14–26.

Intrasook, J., T.W. Tsusaka, and A.K. Anal. 2024. Trends and current food safety regulations and policies for functional foods and beverages containing botanicals. *J. Food Drug Anal.*, *32*(2), 112–139.

Jahangeer, M., R. Fatima, M. Ashiq, A. Basharat, S. A. Qamar, and M. Bilal. 2021. Therapeutic and Biomedical Potentialities of Terpenoids: A Review. *J. Pure Appl. Microbiol.*, *15*(2), 471–483.

Jain, S., Y. Yadav, and M. Sharma. 2025. Morin: A Powerful Potential Flavonoid for Human. *J. Drug Delivery Ther.*, *15*(8), 237–241.

Kennedy, D. O. 2016. B vitamins and the brain: Mechanisms, dose, and efficacy: A review. *Nutrients*, *8*(2), 68.

Khan, M.K., Z.-E. Huma, and O. Dangles. 2014. A comprehensive review on flavanones, the major citrus polyphenols. *J. Food Compos. Anal.*, *33*(1), 85–104.

Khoo, H.E, A. Azlan, S.T. Tang, and S.M. Lim. 2017. Anthocyanidins and anthocyanins: colored pigments as food, pharmaceutical ingredients, and the potential health benefits. *Food Nutr. Res.*, *61*(1), 1361779.

Kooshki, L., S. Fakhri, F. Abbaszadeh, A. Kiani, M.H. Farzaei, E. Mohammadi-Noori, et al., 2025. Pelargonidin improves functional recovery and attenuates neuropathic pain following spinal cord injury in rats: Relevance to its neuroprotective, antioxidant, and anti-inflammatory effects. *Front. Pharmacol.*, *16*, 1547187.

Korhonen, H., and A. Pihlanto. 2006. Bioactive peptides: Production and functionality. *Int. Dairy J.*, *16*(9), 945–960.

Křížová, L., K. Dadáková, J. Kašparovská, T. Kašparovský. 2019. Isoflavones. *Molecules*, *24*(6), 1076.

Kumar, Naresh, and N. Goel. 2019. Phenolic acids: Natural versatile molecules with promising therapeutic applications. *Biotechnol. Rep* (*Amst*)., *24*, e00370.

Kumar, P., T. Ahamad, D.P. Mishra, and M.F. Khan. 2020. Plant Neoflavonoids: Chemical Structures and Biological Functions. *In*: Swamy, M. (Eds.), *Plant-derived Bioactives* (pp. 35–57). Springer, Singapore.

Li, Y., S. Xie, J. Ying, W. Wei, and K. Gao, 2018. Chemical Structures of Lignans and Neolignans Isolated from Lauraceae. *Molecules*, *23*(12), 3164.

Mani R. and V. Natesan. 2018. Chrysin: Sources, beneficial pharmacological activities, and molecular mechanism of action. *Phytochemistry*, *145*, 187–196.

Mazumder, K., A. Aktar, B. Biswas, and M.E. Hossain, 2025. Proteins, Vitamins, Minerals, Cereal, Vegetables, and Beverages as Functional Foods. In: B. Mukherjee (Ed.), *Dietary Supplements and Nutraceuticals* (pp. 89–128). Springer, Singapore.

Miadoková E. 2009. Isoflavonoids – an overview of their biological activities and potential health benefits. *Interdiscip. Toxicol.*, *2*(4), 211–218.

Mukherjee B., M. Ray, A.A. Hoque, S. Mallick, A. Mukherjee, M. Barman, et al., 2025. General Concept on Nutraceuticals, Functional Foods, and Dietary Supplements. In: Mukherjee, B. (Eds.), *Dietary Supplements and Nutraceuticals* (pp. 1–37). Springer, Singapore.

Mukhija, M., B.C. Joshi, P.S. Bairy, A. Bhargava, and A N. Sah. 2022. Lignans: A versatile source of anticancer drugs. *Beni-Suef Univ. J. Basic Appl. Sci.*, *11*, 76.

Nadalin, P., J.K. Kim, and S.U. Park. 2023. Recent studies on myricetin and its biological and pharmacological activities. *EXCLI J.*, *22*, 1223–1231.

Negi R., B. Sharma, T. Jan, T. Kaur, S.S. Khan, N. Yadav, et al., 2025. Bioactive compounds as plant-based functional foods for human health: Current scenario and future challenges. *J. App. Biol. Biotech.*, *13*(1), 1–23.

Oh, W.Y., Y. Gao and F. Shahidi. 2021. Stilbenoids: Chemistry, occurrence, bioavailability, and health effects: A review. *J. Food Bioact.*, *13*, 20–31.

Patel, K. and D.K. Patel. 2024. Biological potential of aromadendrin against human disorders: Recent development in pharmacological activities and analytical aspects. *Pharmacol. Res. Mod. Chin. Med.*, *11*, 100424.

Patted, P.G., R.S. Masareddy, A.S., R.R. Patil, Kanabargi and C.T. Bhat. 2024. Omega-3 fatty acids: A comprehensive scientific review of their sources, functions, and health benefits. *Futur. J. Pharm. Sci*, *10*, 94.

Peña-Vázquez, G.I., A. Arredondo-Arenillas, S.N. Serrano-Sandoval and M. Antunes-Ricardo. 2024. Functional foods lipids: Unraveling their role in the immune response in obesity. *Crit. Rev. Food Sci. Nutr.*, *65*(21), 4043–4064.

Pereira, D.M., P. Valentão, J.A. Pereira and P.B. Andrade. 2009. Phenolics: From Chemistry to Biology. *Molecules*, *14*, 2202–2211.

Pizzi, A. 2021. Tannins medical/pharmacological and related applications: A critical review. *Sustainable Chem. Pharm.*, *22*, 100481.

Priyadarsini, K.I. 2014. The chemistry of curcumin: from extraction to therapeutic agent. *Molecules*, *19*(12), 20091–20112.

Plat, J., S. Baumgartner, T. Vanmierlo, D. Lütjohann, K.L. Calkins, D.G. Burrin, et al., 2019. Plant-based sterols and stanols in health & disease: "Consequences of human development in a plant-based environment?" *Prog. Lipid Res.*, *74*, 87–102.

Przybylski, T., J. Czerniel, J. Dobrosielski and M. Stawny, 2025. Flavonol Technology: From the Compounds' Chemistry to Clinical Research. *Molecules*, *30*(15), 3113.

Puri, V., M. Nagpal, I. Singh, M. Singh, G. A. Dhingra, K. Huanbutta, et al., 2022. A Comprehensive Review on Nutraceuticals: Therapy Support and Formulation Challenges. *Nutrients*, *14*(21), 4637.

Rahman, A., M. Rao, A. Aldughpassi, R. Jallad and L. Shaban. 2025. Blood levels of copper, manganese, selenium, and zinc are positively associated with cognitive function and academic performance in adolescents. *Front. Nutr.*, *12*, 1638283.

Rana, A., M. Samtiya, T. Dhewa, V. Mishra, and R.E. Aluko. 2022. Health benefits of polyphenols: A concise review. *J. Food Biochem.*, *46*(10), e14264.

Rizvi, S., S.T. Raza, F. Ahmed, A. Ahmad, S. Abbas and F. Mahdi. 2014. The role of vitamin E in human health and some diseases. *Sultan Qaboos Univ. Med. J.*, *14*(2), e157–e165.

Roberfroid, M. 2007. Prebiotics: The concept revisited. *J. Nutr.*, *137*(3 Suppl 2), 830S–837S.

Safdar, M.A., R.M.N. Aslam, A. Shakeel, Shiza, M. Waqar, A. Jmail, et al., 2022. Cyanidin as potential anticancer agent targeting various proliferative pathways. *Chem. Biol. Drug Des.*, *101*(2), 438–452.

Safe, S., A. Jayaraman, R.S. Chapkin, M. Howard, K. Mohankumar and R. Shrestha. 2021. Flavonoids: Structure-function and mechanisms of action and opportunities for drug development. *Toxicol. Res.*, *37*(2), 147–162.

Saini, R.K. and Y.-S. Keum. 2018. Omega-3 and omega-6 polyunsaturated fatty acids: Dietary sources, metabolism, and significance: A review. *Life Sci.*, *203*, 255–267.

Salehi, B., A. Venditti, M. Sharifi-Rad, D. Kregiel, J. Sharifi-Rad, A. Durazzo, et al., 2019. The Therapeutic Potential of Apigenin. *Int. J. Mol. Sci.*, *20*(6), 1305.

Sankarganesh, P., A. Bhunia, A. Ganesh Kumar, A. Surendra Babu, S.T. Gopukumar and E. Lokesh. 2025. Short-chain fatty acids (SCFAs) in gut health: Implications for drug metabolism and therapeutics. *Med. Microecol.*, *25*, 100139.

Shahidi, F. 2009. Nutraceuticals and functional foods: Whole versus processed foods. *Trends Food Sci. Technol.*, *20*(9), 376–387.

Shahidi, F. and P. Ambigaipalan. 2015. Phenolics and polyphenolics in foods, Beverages and Spices: Antioxidant Activity and Health Effects: A Review. *J. Funct. Foods*, *18*, 820–897.

Sheng, Y., Y. Sun, Y. Tang, Y. Yu, J. Wang, F. Zheng, et al., 2023. Catechins: Protective mechanism of antioxidant stress in atherosclerosis. *Front. Pharmacol.*, *14*, 1144878.

Shilpa V.S., R. Shams, K.K. Dash, V.K. Pandey, A.H. Dar, S. Ayaz Mukarram, et al., 2023. Phytochemical Properties, Extraction, and Pharmacological Benefits of Naringin: A Review. *Molecules*, *28*(15), 5623.

Silva, Y.P., A. Bernardi and R.L. Frozza. 2020. The Role of Short-Chain Fatty Acids From Gut Microbiota in Gut-Brain Communication. *Front. Endocrinol.*, *11*, 25.

Simes, D.C., C.S.B. Viegas, N. Araújo and C. Marreiros. 2020. Vitamin K as a Diet Supplement with Impact in Human Health: Current Evidence in Age-Related Diseases. *Nutrients*, *12*(1), 138.

Slavin, J. 2013. Fiber and prebiotics: Mechanisms and health benefits. *Nutrients*, *5*(4), 1417–1435.

Song B., M. Hao, S. Zhang, W. Niu, Y. Li, Q. Chen, et al., 2024. Comprehensive review of Hesperetin: Advancements in pharmacokinetics, pharmacological effects, and novel formulations. *Fitoterapia*, *179*, 106206.

Springer, M. and S. Moco. 2019. Resveratrol and Its Human Metabolites: Effects on Metabolic Health and Obesity. *Nutrients*, 11(1):143.

Sunil, C. and B. Xu. 2019. An insight into the health-promoting effects of taxifolin (dihydroquercetin). *Phytochemistry*, *166*, 112066.

Swinbanks, D. and J. O'Brien. 1993. Japan explores the boundary between food and medicine. *Nature*, *364*, 180.

Temple, N.J. 2022. A rational definition for functional foods: A perspective. *Front. Nutr.*, *9*, 957516.

Traber, M.G. and J. Atkinson, 2007. Vitamin E, antioxidant, and nothing more. *Free Radic. Biol. Med.*, *43*(1), 4–15.

Volman, J.J., J.D. Ramakers and J. Plat. 2008. Dietary modulation of immune function by beta-glucans. *Physiol. Behav.*, *94*(2), 276–284.

Wang, M., T. Liu, J. Gan, C. Xu, Y. Zhong, Y. Cao, et al., 2025. Glycitein: A comprehensive review of its bioactivities, molecular mechanisms, and therapeutic potential. *Phytomedicine*, *148*, 157430.

Yadav. N., R. Yadava and A. Goyal. 2014. Chemistry of terpenoids. *Int. J. Pharm. Sci. Rev. Res.*, *27*(2), 272–278.

Yang, D., T. Wang, M. Long and P. Li. 2020. Quercetin: Its Main Pharmacological Activity and Potential Application in Clinical Medicine. *Oxid. Med. Cell Longev.*, *2020*, 8825387.

Ye, Q., N. Georges and C. Selomulya. 2018. Microencapsulation of active ingredients in functional foods: From research stage to commercial food products. *Trends Food Sci. Technol.*, *78* 167–179.

Youness, R.A., A. Dawoud, O. ElTahtawy and M.A. Farag. 2022. Fat-soluble vitamins: Updated review of their role and orchestration in human nutrition throughout life cycle with sex differences. *Nutr. Metab.*, (*Lond.*)., *19*, 60.

Zhou, Y., L. Men, Y. Sun, M. Wei and X. Fan. 2021. Pharmacodynamic effects and molecular mechanisms of lignans from Schisandra chinensis Turcz. (Baill.): A current review. *Eur. J. Pharmacol.*, *892*, 173796.

Zhuang, C., W. Zhang, C. Sheng, W. Zhang, C. Xing and Z. Miao. 2017. Chalcone: A Privileged Structure in Medicinal Chemistry. *Chem. Rev.*, *117*(12), 7762–7810.

CHAPTER 3

Vitamins and Minerals in Functional Foods

Sources and Health Benefits

Samrudhi Anant Devadiga[1], Balaji, M.[2], Venkata, Kumar Kurraboyina Sai Satya Sravan[3] and Roji Waghmare[4*]

1. Introduction

The idea of incorporating healthy food additives began in Japan during the 1970s, and the phrase "functional foods" was first used in 1984 (Hamed et al., 2015). According to the Food and Nutrition Board of the National Academy of Sciences, a functional food is defined as "any modified food or food ingredient that may provide health advantages beyond the standard nutrients it contains" (Committee on Opportunities in the Nutrition and Food Sciences, 1994) Thomas and Earl, (1994). These are becoming an increasingly significant segment of the global food industry. Vitamins are essential, naturally occurring nutrients required in small amounts for different biological functions, like wound healing, growth and development, bone and tissue maintenance, and immune system function (Alsaqqa et al., 2011). These organic molecules play diverse roles in maintaining overall health. Since humans cannot produce vitamins, they are obtained through diet (Ottaway, 1993). Adequate micronutrient intake, including both vitamins and minerals, is crucial for overall well-being (Lanham-New et al., 2019) Certain vitamins also function as nutraceuticals and manage many chronic conditions like cancer, macular degeneration, diabetes, heart disease, muscular degeneration and aging-related disorders (Mamede et al., 2011; Thomas, 2006). Vitamin deficiencies remain a global health concern, having an adverse effect on both industrialized and developing nations. This concern is due to the diet falling short, absorption being poor, or the body's needs being higher than normal. To combat deficiencies, many strategies have been introduced, including vitamin and mineral fortification of foods and the usage of dietary supplements. For optimal health benefits, foods should not only contain a high quantity of vitamins but also ensure bioavailability, meaning the body can efficiently absorb and utilize them after consumption (Tan and McClements, 2021).

Minerals are chemical compounds that cannot be generated within the body but can be received through diet. They are naturally found in soil and water. Some are vital to living creatures, whereas others are highly harmful. Plants absorb a large number of minerals from their surroundings and typically pass them down the food chain to animals. A lack of such vitally necessary minerals frequently proves fatal. They are necessary in the formation and activity of key biomolecules in the

[1-5] Department of Food Engineering and Technology, Institute of Chemical Technology, Mumbai-400019, India.
* Corresponding author: rosewaghmare@gmail.com

DOI: 10.1201/9781003659808-3

body. Although minerals are not an energy source in the body, they are required for the maintenance of normal metabolic processes (Zhao et al., 2016).

Minerals are generally classified into two groups, macrominerals and trace minerals. The primary minerals include phosphorus (P), sodium (Na), calcium (Ca), chloride (Cl), magnesium (Mg), potassium (K), and sulfur (S). In contrast, trace minerals consist of copper (Cu), selenium (Se), iodine (I), molybdenum (Mo), iron (Fe), chromium (Cr), zinc (Zn), manganese (Mn), fluoride (F), cobalt (Co), and boron (B). A nutritious diet should incorporate a mix of plant-based foods, such as vegetables and fruits, along with animal-based sources to supply essential nutrients. Although the body's requirement for trace minerals is less than that for major minerals, this does not diminish their importance. Consequently, a balanced diet can fulfill the body's mineral needs (Lukaski, 2004).

Vitamins and minerals play crucial roles in regulating metabolic and cellular activities, promoting overall health, reproduction, and development, while also preventing various illnesses (Lešková et al., 2006).Vitamins are important for various biological processes, like energy metabolism, though they do not directly provide energy. Instead, they support metabolic functions that become more demanding during physical exertion (Lukaski, 2004). Vitamin deficiencies result in serious diseases like scurvy, night blindness, and beriberi. To secure adequate vitamin intake on a global scale, researchers have developed fortification strategies to enhance basic foods with essential vitamins.

Minerals are fundamental to different physiological functions, including bone formation and nerve impulse transmission, contributing to overall health and longevity (Gharibzahedi & Jafari, 2017). They also regulate hormone production and heart rate, playing a significant role in maintaining homeostasis. Different minerals have specific structural and functional roles in the body. For instance, bone and teeth health relies on calcium (Ca), phosphorus (P), and fluoride (F) for strong teeth, while bone structure benefits from calcium (Ca), magnesium (Mg), manganese (Mn), phosphorus (P), boron (B), and fluoride (F). Magnesium (Mg), copper (Cu), iron (Fe), manganese (Mn), zinc (Zn), and selenium (Se) are essential structural elements in various enzymes (Gharibzahedi and Jafari, 2017). They support enzymatic functions, enhance physiological well-being, promote overall health, and improve metabolic efficiency, particularly when intake of these minerals and vitamins is ensured through diet or fortification.

2. Market Trends and Consumer Demand for Functional Foods

A multitude of research has identified factors that shape consumer acceptance of functional foods (Sgroi et al., 2022). Nevertheless, the functional food industry encounters difficulties in formulating effective communication strategies and promotional content to enhance consumer acceptance of their products (Baker et al., 2022). Academic literature classifies consumer preferences for functional foods categorized into five categories: Behavioral attributes, Product characteristics, Socio-demographic factors, Physical aspects and psychological traits (Küçükgöz and Trząskowska, 2022). This study sought to examine the impact of socio-demographic factors on the intake of functional foods. These factors have an important function in the adoption of functional foods by consumers (Hadjimbei et al., 2022; Salvatore et al., 2022).

Empirical studies have explored various characteristics, including marital status, age, nationality, gender, education level, geographic location, and household attributes (Precup et al., 2022; Rai et al., 2023). Studies suggest that older adults are more inclined to consume functional foods (Femi-Oladunni et al., 2023). Specifically, individuals over 65 tend to prefer functional foods more, likely due to a greater focus on health compared to younger generations (Kombanda et al., 2022; Ozen et al., 2013). Conversely, research also indicates that those youths, particularly those under 25, strongly prefer functional foods since they are open and willing to experiment with new food products. The outcomes highlight the fact that young adults are a main group turning to functional foods. Verbeke (2006) said that young people between their late teens to early 30s are a key group regarding functional food. Urala (2003) also suggests that those in the 19–30 age range are the ones who will likely drive consumer trends. Beyond demographics, consumer perceptions of the qualities of functional foods are essential.

The size of the global market of functional food was estimated at USD 337.85 billion in 2024 and is expected to expand at a compound annual growth rate of 6.5% from 2025 to 2033 (Figure 1), from USD 359.81 billion in 2025 to USD 595.49 billion. (https://www.imarcgroup.com/function).

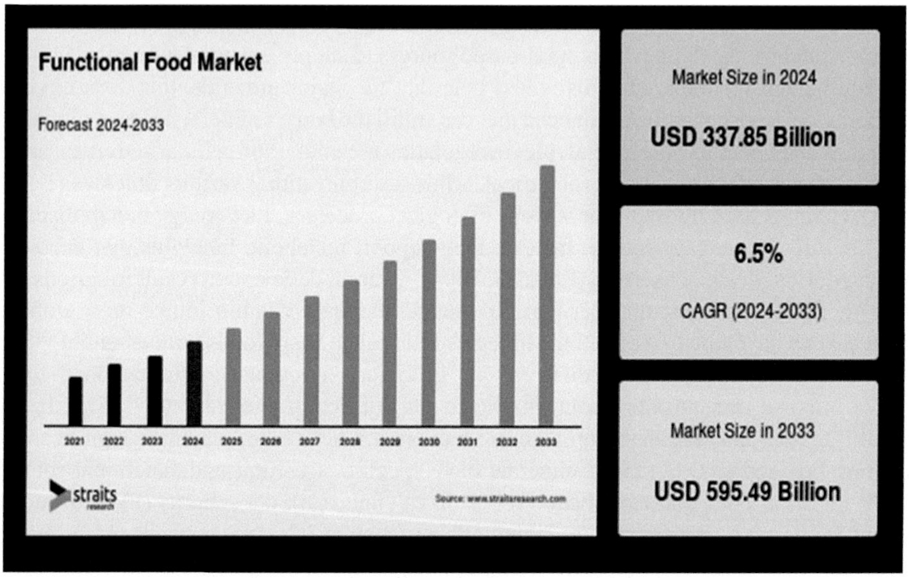

Fig. 1 Functional food market forecast (2024–2033). Adapted from IMARC Group. (2024). Functional Food Market Report. Retrieved from https://www.imarcgroup.com/function.

3. RDA for Various Vitamins and Minerals

The recommended dietary allowances (RDA) specify the levels of essential nutrients required to meet the nutritional needs of healthy individuals on a daily basis. Table 1 shows the RDA values for adult men according to the ICMR-NIN 2020 guidelines.

Action : The word is revised in line no. 130.

Table 1 RDA values for adult men (ICMR-NIN).

Nutrients	*ICMR-NIN*
	RDA 2020
Calcium (mg)	1000
Magnesium (mg)	440
Iron (mg)	19
Zinc (mg)	17
Iodine (μg)	150
Thiamine (mg)	1.8
Riboflavin (mg)	2.5
Niacin (mg)	18
Vitamin B6 (mg)	2.4

Contd.

Table 1 *Contd.*

Nutrients	*ICMR-NIN*
	RDA 2020
Folate (µg) DFE	300
Vitamin B12 (µg)	2.2
Vitamin C (mg)	80
Vitamin A (µg)	1000
Vitamin D (IU)	600

As a result of physiological and metabolic nutrients, females have different nutritional requirements. Table 2 shows the RDA for adult women as per the ICMR-NIN 2020 recommendations.

Table 2 RDA values for adult Females (ICMR-NIN).

Nutrients	*ICMR-NIN (RDA 2020)*
Calcium (mg)	1000
Magnesium (mg)	370
Iron (mg)	29
Zinc (mg)	13.0
Iodine (µg)	150
Thiamine (mg)	1.7
Riboflavin (mg)	2.4
Niacin (mg)	14
Vitamin B6 (mg)	1.9
Folate (µg) DFE	220
Vitamin B12 (µg)	2.2
Vitamin C (mg)	65
Vitamin A (µg)	840
Vitamin D (IU)	600

Numerous plant-derived functional foods include bioactive compounds with potential roles in health promotion and disease prevention are referred as functional foods. Table 3 summarizes these sources, their beneficial constituents, and corresponding health effects.

Table 3 Functional foods of plant origin, key sources, bioactive components, and health benefits.

Functional Food	*Bioactive Compound(s)*	*Health Benefit*	*Reference*
Garlic	Organosulfur compounds	Reduces both total and LDL cholesterol	(Hasler, 2002)
Jerusalem artichoke, garlic, asparagus, onion, chicory, sugar beet	Fructooligosaccharides	Prebiotics	(Sangeetha et al., 2005)
Cruciferous vegetables	Glucosinolates, indoles	Reduces the risk of specific types of cancer	(Hasler, 2002)

Contd.

Table 3 *Contd.*

Functional Food	*Bioactive Compound(s)*	*Health Benefit*	*Reference*
Dark chocolate	Flavonoids	Anti-inflammatory and antioxidant qualities reduce the risk of cardiovascular disease	(Shi et al., 2016)
Tomatoes and processed tomato products	Lycopene	Reduces the risk of prostate cancer	(Hasler, 2002)
Collard greens, kale, spinach	Lutein/zeaxanthin	Reduces the risk of age- related macular degeneration.	(Hasler, 2002)
Saw palmetto	Fatty acids and phytosterols	Improves sexual vigor, reduces prostatic enlargement and boosts sperm production.	(Halster, 2003)
Tea leaves	Caffeic acid, theophylline, p-coumaric acid, catechins, theobromine, caffeine, chlorogenic acid	Neuroprotective, antidiabetic, antioxidative, hepatoprotective, anti-inflammatory, cardioprotective, antimutagenic, central nervous system stimulant, anti-obesity, immunomodulatory	(Bag et al., 2022; Engelhardt, 2020; Samanta, 2022)
Walnuts	Omega-3 fatty acids	Prevents cardiovascular diseases	(Jiménez-Colmenero et al., 2010)
Red beetroot	Betalains	Antioxidant and anticancer	(Gengatharan et al., 2015)
Lemongrass	Luteolin, myrcene, citral, apigenin glycosides, geraniol, citronellol, derivatives of caffeic, chlorogenic, and p-coumaric acids	Anxiolytic, antihyperlipidemic, antidiabetic, antibacterial, anti-inflammatory, antihelmintic, antioxidant, antiproliferative.	(Majewska et al,. 2019; Mukarram et al., 2021; Sousa et al., 2021)

In addition to plant sources, various animal-based foods serve as functional foods due to the presence of health-promoting bioactive substances. Table 4 presents selected animal sources, their active components, and documented health benefits.

Table 4 Functional foods of animal origin, key sources, bioactive components, and health benefits.

Functional Food	*Bioactive Component(s)*	*Health Benefit*	*Reference*
Beef, lamb, turkey, dairy products	Conjugated linoleic acid	Prevents breast cancer	(Hasler, 2002)
Whey protein	Methionine and cysteine	Prevents colon cancer	(Hasler, 2002)
Meats and fish are examples of low-fat foods included in a low-fat diet	Low in total fat or saturated fat	Reduce the risk of cancer and heart disease	(Bang and Dyerberg, 1972)
Fatty fish	(n-3) Fatty acids	Prevent fatal/non-fatal myocardial infarction, reduce triglycerides, and lower the risk of cardiac mortality and heart disease	(Hasler, 2002)

Contd.

Table 4 *Contd.*

Omega-3 enriched eggs	Omega-3 fatty acids	Reduced diastolic and systolic blood pressures, platelet aggregation, and plasma triglyceride levels	(Farrell, 1998)
Fermented dairy products	Probiotics	Support gastrointestinal health and increase immunity	(Hasler, 2002)

4. Sources of Vitamins in Functional Foods

The traditional classification of vitamins in human nutrition is based on their solubility, categorizing them as either fat-soluble or water-soluble. Figure 2 presents the graphical representation of vitamin classification. This classification is determined by a vitamin's ability to be absorbed in either fat or water (Bhagavan and Ha, 2015), also its specific role within the body (Delaney Kerri, 2017). Its solubility depends on both internal and external variables. Vitamins, such as vitamins A, D, E, and K, are lipophilic molecules. These vitamins are typically retained in the liver instead of being eliminated from the body. As opposed to being kept in the body, water-soluble vitamins are lipophobic in nature, which means they dissolve in water and are easily eliminated through urine. This category comprises the B-complex vitamins, which include Biotin, Thiamine, Cobalamin, Riboflavin, Pantothenic acid, Folate, Niacin, Pyridoxine; Ascorbic acid (vitamin C). Small amounts of these nutrients are required each day. Vitamins such as vitamin D, biotin (B7), and vitamin K are not considered technically "essential" for proper physiological activities because they can be produced inside the body. Each of these vitamins has a unique chemical composition. They are shown in Figure 2.

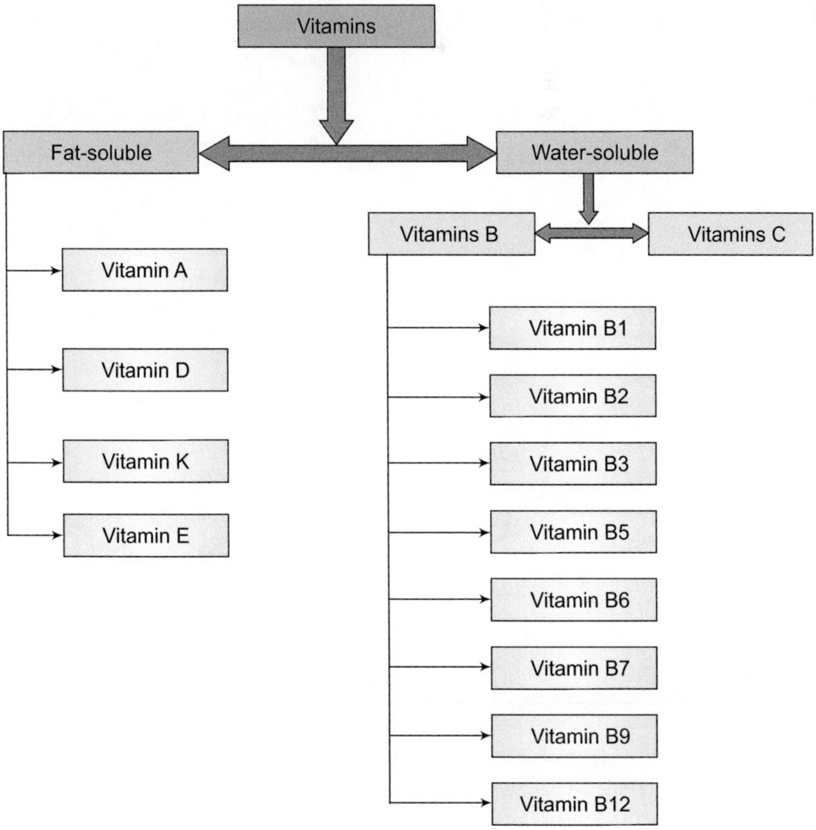

Fig. 2 Classification of vitamins (Ofoedu et al., 2021).

The chemical structure of the vitamins is given in Figure 3.

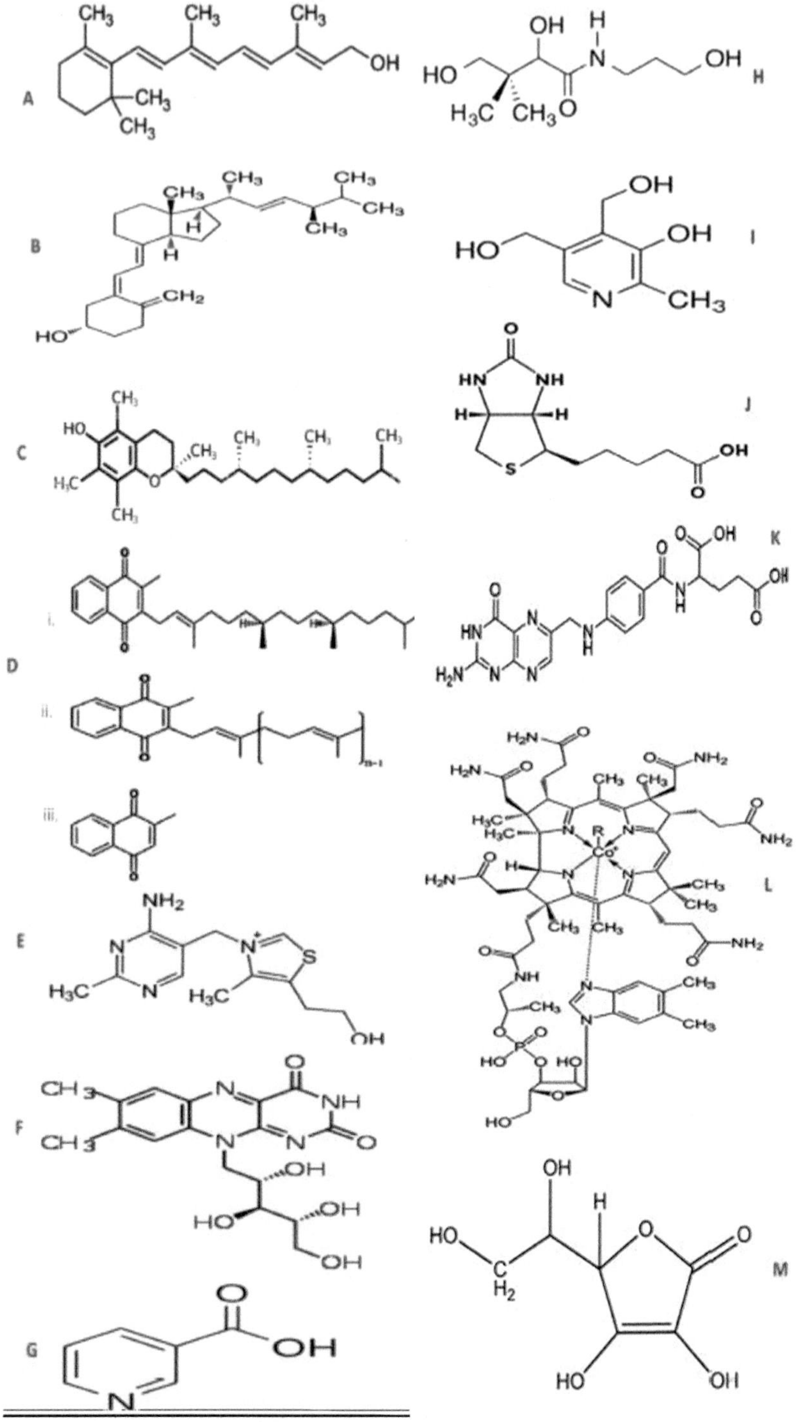

Fig. 3 *Classification and Functions of Vitamins.* Includes Vitamin A (Retinol), Vitamin D (Ergocalciferol), Vitamin E (α-Tocopherol), Vitamin K (Phylloquinone, Menaquinone, Menadione), Vitamin B1 (Thiamine), Vitamin B2 (Riboflavin), Vitamin B3 (Niacin), Vitamin B5 (Pantothenic acid), Vitamin B6 (Pyridoxine), Vitamin B7 (Biotin), Vitamin B9 (Folate), Vitamin B12 (Cobalamin), and Vitamin C (Ascorbic acid).
(*Source:* Ofoedu et al., 2021)

4.1 Fat Soluble Vitamins

The term Vitamin A refers to a group of essential fat-soluble substances known as retinoids. It is categorized within the diet as either provitamin A or pre-vitamin A. Provitamin A is synthesized by plants and found in plant foods only. Animals form pre-vitamin A, usually referred to as vitamin A. Found in Organ meat, spinach, sweet potatoes, milk, fruits, liver, eggs, and carrots. (Ross, 2010). Made up of fat-soluble substances derived from cholesterol, vitamin D is a non-essential one, as ergocalciferol and cholecalciferol, its active forms, are produced in the body (Teegarden, 2014). Ultraviolet B radiation from the sun stimulates vitamin D in the skin. It is a non-essential micronutrient since it can be produced by the body under adequate exposure to sunlight (Carr and Maggini, 2017) However, vitamin D2 is obtainable from plant foods, specifically ergosterol. Vitamins D2 and D3 need to undergo metabolic activation by the body in order to be biologically active. Received through Milk, egg yolks, sunlight, and fatty fish (Carr and Maggin, 2017). (α) tocopherol is the most efficient type of vitamin that suits the needs of the body, as noted by (Delaney Kerri, 2017). Unlike the other vitamins with specific functions such as cofactors, hormones or cellular metabolism, Vitamin E is a lipid-soluble vitamin that possesses strong antioxidant activity that helps defend cells from injury. It acts as a watchdog, finding and eliminating harmful free radicals to stop fats from damaging our cells. Major food sources of Vitamin E are sunflower seeds, spinach, hazelnuts, dandelion greens, avocado, and almonds (Lips, 2006). As a fat-soluble vitamin and member of the isoprenoid quinone group (Shearer & Newman, 2014), vitamin K acts as a cofactor of the enzyme γ-glutamyl carboxylase (Ball, 2004). It can be synthesized both by plants and microbes and exists in two main forms. The most important dietary form is vitamin K1 (phylloquinone), derived from plant matter and possessing a phytyl side chain. On the other hand, bacteria synthesize vitamin K2 as menaquinones, characterized by side chains composed of repeating unsaturated 5-carbon prenyl units. Found in broccoli, cabbage, ground beef, and spinach (Niki and Traber, 2012).

4.2 Water-Soluble Vitamins

Vitamin B1 falls under B-complex vitamins called thiamine, a vitamin containing pyrimidine and thiazole rings connected by the assistance of methylene bridge (Bender D. A, 2003). Found in all foods, especially in yeast, pork, fish, and whole grains; Fortified in some foods like cereals, bread, milk, and infant formulas, as some thiamine is destroyed during processing (Frank, 2015; Gasperi et al., 2019). Also found in legumes (Fattal-Valevski, 2011). Vitamin B2, or riboflavin, which is common in foods, is only associated with proteins as flavin mononucleotide and flavin adenine dinucleotide (Bates et al., 1997; Combs and McClung, 2008; Decker, 1993). Apart from being easily destroyed by ultraviolet light, vitamin B2 plays a significant role in the body as a coenzyme that supports the normal functioning of proteins. It is essential for the formation of flavin mononucleotide and flavin adenine dinucleotide, which participate in energy metabolism, amino acid metabolism, and the protection of proteins against oxidative stress. Vitamin B2 is naturally present in mushrooms, meat, green vegetables, enriched rice, eggs, almonds, dairy products, fish, and legumes, and it is also available as a nutritional supplement (Plantone et al., 2021; Ashoori and Saedisomeolia, 2014). Vitamin B3(niacin): It is of two types: nicotinic acid and nicotinamide. In the body, it gets converted into its active forms, namely nicotinamide adenine dinucleotide and nicotinamide adenine dinucleotide phosphate (Combs & McClung, 2008). Niacin is present in various plant and animal foods, like seeds, legumes, grains, soy, and nuts. Moreover, some cereal foods such as cereals, infant foods, and bread are fortified with niacin to make them more nutritional (*Dietary Reference Intakes for Thiamin, Riboflavin, Niacin, Vitamin B6, Folate, Vitamin B12, Pantothenic Acid, Biotin and Choline 2000*); Other foods that contain the nutrient are fish, eggs, meat, peaches, and dates (Denu, 2005). A water-soluble vitamin required for energy metabolism is pantothenic acid. Coenzyme A, acts as the principal carrier of carbon molecules in cells. Found in nearly all foods, but with greater concentrations in fortified cereals, sweet potatoes, chicken, lentils, mushrooms, sunflower seeds, beef, eggs, dried foods, avocados, infant formulas, fish, and pork (*Dietary Reference Intakes for Thiamin, Riboflavin, Niacin, Vitamin B6, Folate, Vitamin B12, Pantothenic Acid, Biotin, and Choline,* 2000), also, in peanuts and liver (Ismail et al., 2020). Vitamin B6 is part of the family of water-soluble vitamins and occurs naturally in three forms: pyridoxine, pyridoxamine, and pyridoxal, varying according to their chemical alcohol, amine, and aldehyde structures. Found in fortified cereals, starchy vegetables, beef, non-citrus fruits, and poultry (Wendt et al., 2021),Bananas, fish, nuts, meat, and chickpeas are some other sources of vitamin B6 (da Silva and Gregory, 2020). Vitamin B7 also referred to as biotin or vitamin H. It occurs naturally in foods such as seeds, organ meats, soybeans, eggs, nuts and fish. The nutrient exists in dietary as well as supplement form (Hanna et al., 2022); other sources include oysters, almonds, dairy, salmon, legumes, and nuts (Zempleni et al., 2009). Vitamin B9 exists in two forms: folate and folic acid. Vitamin B9's natural form is folate, having the base molecule 5,6,7,8-tetrahydropteroyl glutamate or pteroylglutamic acid and tetrahydrofolate. Found in Brussels sprouts, poultry, grains, milk products, beans, meat, and nuts (*Dietary Reference Intakes for Thiamin, Riboflavin, Niacin, Vitamin B6, Folate, Vitamin B12, Pantothenic Acid, Biotin, and Choline* 2000) (Wendt et al. 2021). Other sources include citrus fruits, carrots, green leafy vegetables, okra, beans, asparagus, celery and broccoli (Guilland and Aimone-Gastin, 2013). Vitamin B12 (cobalamin) is the sole water-soluble vitamin containing cobalt (Delaney Kerri, 2017). Cobalamins are a class of corrinoid with a dimethyl benzimidazole nucleotide attached to the D ring, according to which they coordinate the central cobalt ion. Natural vitamin B12 occurs in two forms hydroxocobalamin and aqua cobalamin that are differentiated by the hydroxyl or water group attached to the cobalt center. Available in fortified foods and in animal foods (Wang et al., 2018); some other sources include milk, poultry, eggs, meat and shellfish (Oh and Brown, 2003).There are two forms of vitamin C: ascorbic acid (reduced) and dehydro-L-ascorbic acid (oxidized) (Said, 2011). Most animals are capable of producing a six-carbon lactone from glucose but not human beings because they lack one enzyme (*FAO/WHO Framework for the Provision of Scientific Advice on Food Safety and Nutrition*, 2007) (Combs & McClung, 2008). Found naturally in papaya, parsley, Brussels sprouts, pineapple, broccoli, guava, citrus fruits, kiwi, and other vegetables and fruits.

5. Sources of Minerals in Functional Foods

Minerals are classified into macro, micro, and ultra-trace minerals (Figure 4).

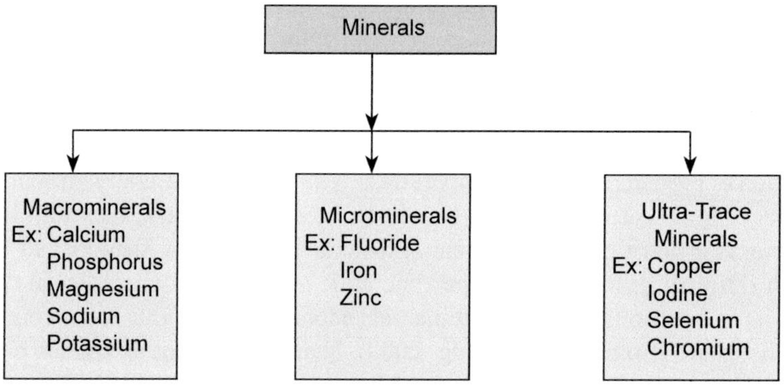

Fig. 4 Classification of minerals.

5.1 Macro Minerals

The macro minerals are needed in comparatively large daily quantities, from a few tenths of a gram to over a gram, and consist of calcium, phosphorus, magnesium, sodium, potassium, and chloride, all of which play crucial roles in a wide array of physiological functions. Calcium is deposited mainly in bones and teeth and has approximately 1% circulating in the blood and found in all tissue cells, where it controls metabolic processes. Besides ensuring skeletal structure, calcium facilitates blood clotting, nerve conduction, intracellular signaling and contraction of smooth muscles, with significant sources being legumes, mustard greens, broccoli, canned fish with bones like sardines and salmon, cheese, milk, soy milk fortified with nutrients, and tofu (Gharibzahedi and Jafari, 2017). Phosphorus, stored in bones and teeth as calcium phosphate crystals and found in a metabolically available pool within cells and extracellular fluid, is part of phospholipids in cell membranes. Its deficiency is rare owing to its richness in Brazil nuts, lentils, beans, pumpkin and squash seeds, soy foods and processed foods such as soft drinks, and animal sources including chicken, non-fat yogurt, lean beef, milk, scallops, salmon, eggs, and Romano cheese (Gharibzahedi and Jafari, 2017). Magnesium, which is predominantly stored in bone and secondarily in muscle, body fluids, and soft tissues, serves as a cofactor for more than 300 enzymatic reactions required for glucose phosphorylation and for the synthesis of fatty acids and proteins. Low dietary intake of magnesium has been associated with heart failure, metabolic syndrome, insulin resistance, and hypertension (Jee et al., 2002; Moore-Schiltz et al., 2015), and major sources are artichokes, beans, leafy green vegetables, nuts, seeds, chocolate, "hard" drinking water, and milk, cheese, and shellfish. Sodium, predominantly derived from sodium chloride or table salt, is needed for extracellular fluid balance, normal nerve conduction, muscle contraction, and acid–base balance. While the daily need is barely 200 mg, intakes on a daily basis are typically well above this level, and this has led to hypertension and raised cardiovascular risk (He and MacGregor, 2002, 2009). Chloride, generally ingested with sodium, occurs in tomatoes, seaweed, olives, lettuce, celery, and rye with small residues in meat and milk. Potassium, the principal intracellular cation also found in trace amounts extracellularly, acts synergistically with sodium to regulate acid–base balance, osmotic balance and fluid balance in addition to serving a principal part in neuromuscular function together with calcium. Raising potassium consumption by consuming more fruits, vegetables, whole grains, legumes, dried fruits, acorn squash, baked potatoes, poultry, and meat may further increase sodium expulsion and decrease blood pressure.

5.2 Micro Minerals

Microminerals are nutrients needed by the body in infinitesimally small quantities ranging from micrograms to milligrams per day to continue essential physiological functions and overall well-being. Fluoride, iron, and zinc are some of these essential microminerals. Fluoride, which occurs naturally in drinking water, soil and foods, is involved in the skeletal health of bones and teeth, although it is not yet solidly entrenched as part of human metabolic needs. Its absorption is significantly based on exposure to the environment and habits, e.g., ingestion of fluoridated water or accidental use of fluoridated dentifrices. Its sources are tea, processed cereal products, fluoridated or naturally fluoridated beverages, juice, wine, beer, mechanically deboned fish, canned shellfish, fish and other seafood (Gharibzahedi and Jafari, 2017). Iron, deposited primarily as ferritin, hemosiderin, and transferrin, is stringently conserved by the organism, which recycles approximately 90% of its iron daily and loses only trace amounts, mostly via bile. Since free iron has the potential to react with oxygen and create compounds that are toxic to DNA (Deoxyribonucleic acid) and cell membranes, it is bound to proteins for protection (Beard, 2001). Iron is critical in oxygen transport via hemoglobin, muscle function through myoglobin, brain function, immune function, and energy production by cytochrome enzymes. Sources of its foods are dried fruits, lentils, white beans, nuts (almonds, hazelnuts, cashews, peanuts, pine nuts), pumpkin and squash seeds, whole grains, pulses, dark leafy greens (Swiss chard, spinach), bran, cocoa powder, tofu, iron-fortified breads and cereals, dark chocolate, and animal foods like lean beef, lamb roast, mussels, poultry, chicken liver, clams, oysters, and eggs. Zinc, a cofactor for more than 300 enzymes that are involved in carbohydrate, fat, nucleic acid, and protein metabolism, also plays structural roles in proteins and is necessary for gene expression, insulin action, immune protection, and cell transport. Although zinc supplements such as lozenges are frequently marketed to prevent and treat the common cold, studies indicate that they have little to no effect. Food sources of zinc are cashews, cooked chickpeas, spinach, pumpkin and squash seeds, toasted whole grains, white mushrooms, wheat germ, cocoa powder, chocolate, and foods from animal sources like lean beef short ribs, lamb, seafood (particularly oysters), poultry, hog shoulder, and fish.

5.3 Ultra-Trace

Copper, iodine, selenium and chromium are essential microminerals that participate in a wide range of physiological processes, with characteristic dietary sources and health outcomes. Copper maintains crucial enzymatic activities in energy metabolism and antioxidant protection, with deficiencies uncommon in adults because of hepatic storage, although reduced intake can compromise immune function. Its rich food sources are prunes, whole grains, avocado, cashew nuts, cooked shiitake mushrooms, raw kale, sesame seeds, beans, chickpeas, fermented soy foods like tempeh, drinking water, and meats. Examples of such meats. are soft goat cheese, oysters, and organ meats (Gharibzahedi and Jafari, 2017). Iodine, deposited mostly in the thyroid gland (20–30 mg), is needed for thyroid hormone synthesis governing body temperature, metabolism, growth, and reproduction; its deficiency leads to goiter and, in infancy, cretinism, whereas excessive intake can induce thyroid dysfunctions in susceptible individuals (Di Matola et al., 2014). Dietary iodine is determined by soil composition, with higher amounts found in coastal areas, the primary sources being iodized salt, canned tuna, baked potatoes, navy beans, cranberries, prunes, milk products like yogurt and Cheddar cheese, and seafood like lobster, cod, and shrimp (Gharibzahedi and Jafari, 2017). Selenium, found largely in the forms of selenomethionine and selenocysteine, has synergistic antioxidant activity with vitamin E, while insufficient intake can increase cancer risk and excess intake may result in toxicity, as indicated in China and Australia (Brozmanová et al., 2010). Some of the most important selenium-dense foods are Brazil nuts, lima and pinto beans, flaxseed, broccoli, chia, cabbage, sunflower, spinach, sesame, whole grains, mushrooms, whole-wheat bread, and animal sources such as tuna, oysters, chicken, beef, lamb, turkey, and selenium-enriched yeast. Chromium is involved in carbohydrate, lipid, and protein metabolism by facilitating insulin action, although its function in glucose tolerance factor is controversial; deficiency can lead to lipid abnormalities and insulin resistance, while high-dose chromium picolinate supplementation has caused toxicity in athletes. Chromium is supplied by bananas, green peppers, wheat germ, molasses, spinach, black pepper, apples, nuts, whole grains, brewer's yeast, oysters, beef, liver, cheese, chicken, eggs, and butter (Gharibzahedi and Jafari, 2017). Collectively, these microminerals play central roles in metabolic regulation, endocrine function, and antioxidant defense, such that balanced consumption is essential to avoid deficiency-related disorders and toxicity.

6. Fortification of Foods with Vitamins and Minerals

Consumers today prefer delicious and convenient foods that offer additional benefits like improved health, vigor, and longevity. While minerals tend to withstand industrial processing better than vitamins, they can still be affected by exposure to moisture, light, heat, or oxygen during food processing and storage. Therefore, proper handling and processing strategies are critical to preserving the nutrients of food products (Gharibzahedi and Jafari, 2017).

6.1 Impact of Food Processing on Mineral Bioavailability

Getting enough micronutrients, both vitamins and minerals, is vital for our overall health and well-being (Lanham-New et al., 2019). Vitamins are organic compounds that support many important biochemical processes essential for our bodies to function and perform. Inadequate intake of certain vitamins can lead to serious health problems, including pellagra, scurvy, osteomalacia, beriberi, night blindness, and rickets (Lanham-New et al., 2019). Beyond preventing deficiencies, some vitamins also act as nutraceuticals, helping to protect against chronic illnesses such as diabetes, heart disease, age-related eye problems, cancer, and the effects of aging. Despite knowing how important vitamins are, many people around the world, even in developed countries, still suffer from vitamin shortages. These shortages can happen for several reasons, including not eating enough nutritious foods, poor absorption of nutrients, or increased needs due to health or lifestyle factors.

6.2 Strategies to Address Vitamin and Mineral Deficiencies

Different techniques have been developed to fight against vitamin and mineral deficiencies. One of the most effective approaches is food fortification, which ensures that essential minerals and vitamins typically not produced in sufficient amounts by the body are added to commonly consumed foods. Vitamin supplementation has long been used as a preventive measure against deficiencies (Zhou et al., 2015). However, not only is it necessary for foods to include sufficient vitamins, but they must also be bioavailable.

6.3 Fortification of Foods with Vitamins

Food fortification with essential vitamins is a widely adopted strategy to combat micronutrient deficiencies at the population level. Table 5 highlights common food items fortified with specific vitamins along with their supporting references.

Table 5 Fortification of foods with vitamins.

Sl. No	*Food Type*	*Fortified With*	*Reference*
1	Food grains	Folic acid	(Choi et al., 2014)
2	Milk	Vitamin A and D	(Osendarp et al., 2018)
3	Sugar	Vitamin A	(Mejia et al., 2019)

6.4 Fortification of Foods with Minerals

Mineral fortification is a crucial strategy in combating global micronutrient deficiencies. Table 6 provides an overview of food types fortified with various essential minerals, including iron, zinc, and selenium, along with relevant studies.

Table 6 Fortification of foods with minerals.

Sl.No	*Food Type*	*Fortified With*	*Reference*
1	Ready-to-eat milk-based	Se: selenate/selenite	(Van Dael et al., 2001)
2	Yogurt	Fe: $FeCl_3$	(Hekmat and McMahon, 1997)
3	Cheddar cheese	Zn: $ZnSO_4$	(Kahraman and Ustunol, 2012)
	Finger millet flour	Fe:$C_4H_2FeO_4$; $Fe_4O_21P_6$	(Tripathi and Platel, 2010 2011)
4	Whole wheat	Zn/Fe: ZnO, Elemental Fe, $ZnSO_4$, NaFeEDT	(Akhtar et al., 2008)
5	Edam cheese	Zn: $Zn(2CCH_3)_2 (H_2O)$ $ZnCl_2$, $ZnSO_4$	(Sami Abd-Rabou et al., 2010)

7. Health benefits of Vitamins in Functional foods

7.1 Vision Health and Skin Care (Vitamin A)

Vitamin A is a group of unsaturated organic compounds, with retinol being the most notable member. This vitamin appears in various forms, which include natural derivatives like 9-cis-retinoic acid (alitretinoin), Retinal, 13-cis-retinoic acid (isotretinoin) and all-trans retinoic acid (tretinoin). Also, this category includes synthetic variants like etretinate, trifarotene, acitretin, adapalene, tazarotene, bexarotene (Carazo et al., 2021). Vitamin A is able to convert between differing forms through specific enzymatic actions that underscore its adaptability, along with its importance as a nutrient. Of vitamin A, there are two primary forms that are biologically active, like 11-cis-retinal and all-trans-retinoic acid, It is the most typical of retinol. In the human body, the principal active form is in all of its trans-retinoic acid form. All trans-retinoic acid regulates the transcription of genes, serving a vital role in cell functions. These cell activities include apoptosis, along with proliferation as well as differentiation. This compound is vital since it supports many bodily functions like reproduction, growth, and embryonic development. It is important because it keeps brain function optimal, makes the immune system more efficient, strengthens bone growth, and helps hair and nails grow. It also aids in the synthesis of RNA (Ribonucleic acid) within epithelial cells, which include the ones in both the conjunctiva and also the cornea, and this contributes to eye health to a large degree (Blaner et al., 2016; Carazo et al., 2021).

Visual perception largely relies upon 11-cis-retinal, a slightly oxidized variant of vitamin A. This compound fundamentally allows the retina to function when processing visual information and relaying it to the brain. Rods as well as cones are vision within the retina as structures that are critical. Cones detect intense light, and rods excel under low-light conditions. They contribute in the direction of night vision in addition to color perception. Retinal's special ionone ring-based structure allows switches between cis-trans isomeric configurations, a key process for rhodopsin formation, plus the visual pigment key for dim-light vision. Insufficient retinol levels impede rhodopsin synthesis, resulting in night blindness (nyctalopia), a condition reversible upon normalization of plasma retinol concentrations (Maiani et al., 2009). Beyond their visual function, retinoids demonstrate significant benefits in dermatological applications, particularly in preventing various skin disorders. The cosmeceutical industry extensively employs vitamin A, its derivatives, and retinoids for their efficacy in treating photo-damaged skin. Studies have shown that topical application of tretinoin can increase the appearance of sun-damaged skin by promoting collagen synthesis and accelerating cellular turnover. Vit A, its derivatives, and retinoids for their efficacy in treating photo damaged skin. Studies show that topical application of tretinoin can increase the appearance of sun-damaged skin by promoting collagen synthesis and accelerating cellular turnover (Kligman et al., 1986).

The liver plays an important function in processing orally consumed vitamin A by transforming retinyl esters and beta-carotene into retinol. This retinol is subsequently either stored as esters or released into the bloodstream. It binds to the retinol-binding protein and enters the dermis through capillaries, where it is dispersed to the skin. Keratinocytes, which are the main cell type in human skin, have the ability to convert retinol into retinoic acid. These transformations are vital because retinoic acid is the form of vitamin A responsible for many of its beneficial effects on the skin. Notably, skin cells can also metabolize beta-carotene, a vitamin A precursor. Studies involving human keratinocyte cultures exposed to radiolabeled beta-carotene demonstrated increased retinol levels, highlighting its significance for skin health. Although the skin's absorption of retinoids is limited, consistent application of tretinoin in various formulations has not resulted in notable changes in plasma tretinoin levels or notable excretion through urine and feces. This conveys that little amount of the active ingredient enters the systemic circulation (Lehman et al., 1988; Schaefer, 1993). Despite these low levels, retinoids continue to have a substantial impact on skin health.

7.2 Role in Immunity (Vitamin C, D)

7.2.1 Vitamin C

Ascorbic acid, commonly called vitamin C, is an essential water-soluble nutrient that humans must obtain from their diet, as the body cannot synthesize it. This vitamin is highly regarded for its antioxidant properties, which are crucial in neutralizing reactive oxygen species (ROS) (Lee and Burckart, 1998). This process is key to shielding biomolecules such as proteins, lipids, and nucleotides from oxidative stress and potential dysfunction. By mitigating oxidative stress, vitamin C acts as a cornerstone in maintaining a pivotal role in maintaining cellular health, particularly during periods of stress or infection. A notable feature of vitamin C is its high concentration in white blood cells (leukocytes), where it is present at levels 50–100 times higher than in plasma. During infections, leukocytes rapidly deplete their vitamin C stores, underscoring its importance in the immune response by directly supporting immune cells during inflammation and infection. When the balance between antioxidant defenses and oxidant production is disrupted, various signaling pathways can be affected (Lee and Burckart, 1998). In particular, elevated levels of oxidants can stimulate nuclear factor κB (NF-κB), an inflammatory transcription factor. Activation of this factor initiates a cascade of signaling that results in enhanced production of oxidative species and inflammatory mediators. NF-κB is involved in regulating inflammatory responses and is associated with the development of several diseases, including viral infections. Consequently, inhibiting NF-κB has been suggested as a potential therapeutic strategy against viral infections (Lee and Burckart, 1998). Vitamin C is widely recognized for its protective effects against infectious diseases. It fortifies respiratory defense mechanisms, contributing to a reduced susceptibility to viral infections and mitigating both their duration and severity. Additionally, its antihistaminic properties may help alleviate associated flu-like symptoms by dampening the histamine response. A significant decline in plasma vitamin C levels is commonly observed in patients suffering from acute respiratory infections, including pneumonia and tuberculosis. Studies reveal that Vitamin C supplements have been shown to aid in the recovery from pneumonia, especially in older adults. Hunt et al. (1994) reported that providing elderly patients with acute respiratory infections with vitamin C substantially improved their well-being and contributed to reducing the duration and severity of the condition (Hunt et al., 1994).

7.2.2 Vitamin D

The process of Vitamin D, a fat-soluble steroid hormone precursor, begins in the skin. Ultraviolet-B radiation causes the conversion of 7-dehydrocholesterol in the epidermis into cholecalciferol that is distributed into the bloodstream. Cholecalciferol in the liver is converted by a process of hydroxylation to become 25-hydroxyvitamin D. This intermediate is then sent to the kidneys, which convert it into the biologically active 1, 25-hydroxyvitamin D (1,25(OH)2D), a hormone with myriad physiological consequences. The importance of vitamin D extends beyond bone health because it is essential for immune system function, strengthens the body's defenses against infection, and helps to fortify the general immune system. Vitamin D promotes cellular defense in innate immunity by triggering the production of antimicrobial peptides like defensins and cathelicidin. These peptides play a role in maintaining the integrity of tight and gap junctions between cells, binding to cellular barriers, and activating antioxidant gene expression. This action is especially essential in fighting viral infections like influenza that can disrupt epithelial tight junctions, making one more susceptible to pulmonary edema. Vitamin D helps to support these junctions, protecting epithelial surfaces from viral attack, as well as pulmonary health (Figure 5).

Furthermore, cellular equilibrium gets the contributions from Vitamin D. Overproduction of claudin-2, a protein that is associated with inflammation, has been bound to the reduced expression of these vitamin receptors. People who are deficient should supplement with Vitamin D to alleviate this inflammation. Supplementing also restores balance in regard to the immune system. Experiments with mice initially lacking Vitamin D have shown that supplementation can diminish lung inflammation as pulmonary epithelium integrity is minimally affected, highlighting Vitamin D's role in managing inflammatory responses and protecting tissues (Gorman et al., 2017).

The role that vitamin D plays in immune function is complex. The ability of vitamin D to facilitate monocyte differentiation into macrophages raises superoxide production, phagocytic activity, and bacteriocidal potential. As a result, the innate immune response to infection is strengthened. Additionally, vitamin D suppresses the activity of Thelper type-1 cells, which are implicated in the induction of inflammation. Vitamin D reduces the production of pro-inflammatory cytokines, including IL-2 and interferon-gamma, thus reducing inflammation. In parallel, Vitamin D acts to induce Th2 cells to secrete anti-inflammatory cytokines, leading to a balanced and protective immune system and not one that is overly inflammatory. Vitamin D also plays a critical role in controlling the regulation of developing suppressive regulatory T cells (Tregs), which enable the immune system to recognize the body's own cells and avoid destroying them by regulating overall immune system function Vitamin D aids in transitioning immune cells from a pro-inflammatory to an anti-inflammatory state, which is necessary for long-term immune regulation (Jeffery et al., 2009).

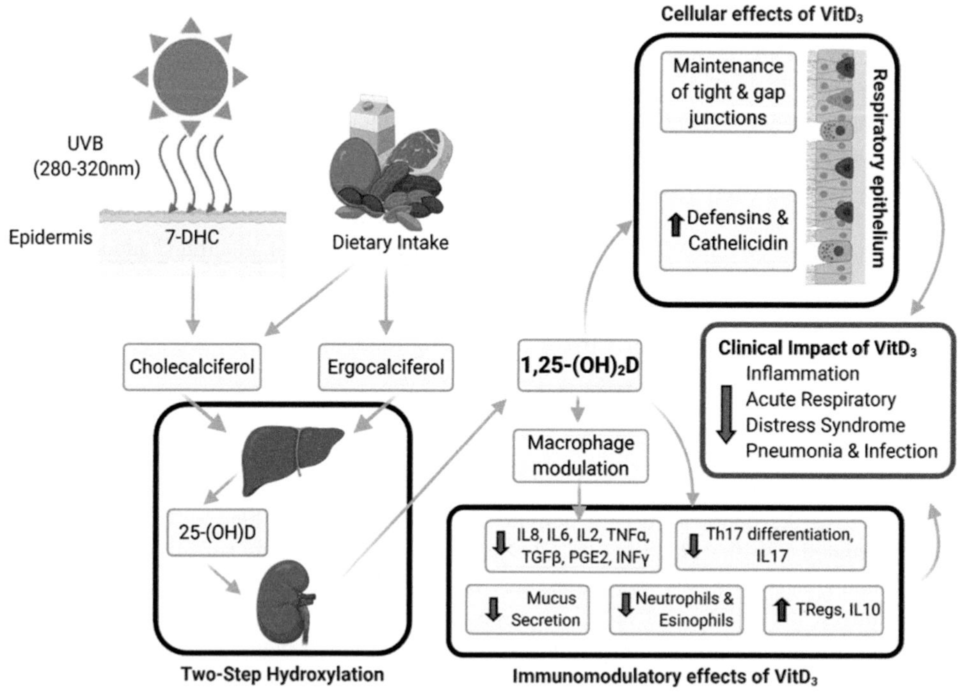

Fig. 5 Immunomodulatory actions of vitamin D. IL: interleukin; TNF: Tumor necrosis factor; IFN: Interferon; Th: T-Helper; 7-DHC: 7-Dehydrocholesterol; PGE2: Prostaglandin E2.
(*Source:* Reproduced with the permission from Shakoor et al., 2021. Copyright © 2021,Elsevier)
Source: Shakoor et al., 2021

7.3 Antioxidant Activity (Vitamin E)

Vitamin E shows various biological activities, this group includes tocopherols and tocotrienols, categorized by antioxidant properties into several isoforms. Eight vitamin E isoforms, each potentially functioning distinctly inside the body, were discovered by scientists. Alpha tocopherol is one of these compounds. It has the capacity for antioxidant potential. Unique abilities are shown by other forms, such as gamma-tocopherol and they neutralize reactive nitrogen species. These molecules do contribute to nitrosative stress, which is a condition similar to oxidative stress. Cellular death and damage can also stem from this condition (Szymańska et al., 2020). For reactive nitrogen species, which amass within the body to a detrimental effect, there is an outstanding effective counteraction from gamma-tocopherol. This particular form of vitamin E, since it works in tandem alongside alpha-tocopherol, creates a thorough defense system that protects against oxidative plus nitrosative stress. These properties are greatly highlighted in contemporary research as means for reducing cellular damage, especially in malignancies and in chronic inflammatory conditions.

The function of vitamin E can go well past only its antioxidant abilities. It may be within some bigger set of tocochromanols. Tocochromanols also include both tocopherols and tocopherolic acids, plus tocopherol phosphate, as well as the tocopherol esters. These substances share a common thread because each fulfills distinct biological roles, combating oxidative damage. Alpha-tocopherol has powerful antioxidant properties, so it is an important element for safeguarding cells against harm (Szymanska et al., 2020).

7.4 Cognitive Function and Mental Health (B Complex Vitamins)

B-complex vitamins are an essential nutrient with roles in several biological processes within the human organism. Such essential vitamins, which need to be obtained through diet from either plant or animal material, include riboflavin (B2), biotin (B7), cobalamin (B12), thiamine (B1), niacin (B3), folate (B9), pantothenic acid (B5), pyridoxine (B6) (Kennedy, 2016). In addition to their life-supporting functions, these vitamins are part of basic physiological processes, including embryogenesis, cellular reproduction, synthesis of DNA/RNA, and processes of cellular repair (Friso et al., 2020). To function well, the central nervous system requires B vitamins. For instance, thiamine (B1) is reported to function as a neuromodulator, i.e., in the synaptic transmission of acetylcholine, a neurotransmitter involved in memory and cognition that is mediated by calcium (Hirsch and Parrott, 2012). Moreover, B5 (pantothenic acid) and B7 (biotin), which get absorbed via the blood-brain barrier, are crucial for the function of the brain (Friso et al., 2020). B12 is a vitamin that is highly crucial for brain function. This vitamin aids in the construction of myelin, a sheath of protection surrounding nerves, enabling them to conduct signals correctly. This is necessary to maintain the brain and nervous system functioning effectively and aid in clear thinking and coordination (Rizzo and Laganà, 2020). But the main work of certain B vitamins, particularly B6, B9, and B12, are as cofactors in methionine and homocysteine metabolism, the essential amino acids that play a role in maintaining a precarious biochemical balance. Homocysteine, a metabolic intermediate synthesized from methionine, needs to be regulated to ensure both physical and mental health. One of the main mechanisms of controlling Hcy levels is remethylation, which is contingent on their methylation, which is in turn dependent on the bioavailability of folate (B9) and B12, which is one the primary processes that regulate Hcy levels. The enzyme 5-methyltetrahydrofolate reductase transfers a methyl group to Hcy during this process, changing it back into methionine. Bioavailability of folate (B9) and B12. In this reaction, a methyl group is transferred to Hcy by the enzyme 5-methyltetrahydrofolate reductase,which restores methionine (Selhub, 1999). This delicate balance helps maintain normal levels of homocysteine and prevents the harmful buildup of this compound in the blood.

8. Health Benefits of Minerals in Functional Foods

8.1 Calcium

Calcium is an indispensable nutrient that significantly influences a wide range of extracellular and intracellular functions, making it vital for the human body's optimal performance. It has a main function in maintaining cellular stability, regulating enzymes, and facilitating neural conduction through ion channels. Furthermore, calcium is fundamental for the development, growth, and maintenance of teeth and bone. In an average adult, approximately 1200 grams of calcium are present, constituting about 2% of body weight, with 99% of this calcium stored in bones and teeth as hydroxyapatite, crucial for tissue mineralization (Beto, 2015).

The demand for calcium fluctuates throughout life, especially during growth periods such as childhood, adolescence, pregnancy, and lactation (Fischer et al., 2018). Adequate calcium intake is crucial for the skeletal system's ability to retain calcium during bone formation, influencing the attainment of peak bone mass in early adulthood. Sufficient calcium consumption during critical growth phases is necessary for proper bone development; inadequate intake during these times can elevate the risk of osteoporotic fractures later in life (Zhu et al., 2008). Additionally, calcium contributes to enhancing bone mineral density ; research indicates that increased calcium intake during growth can enhance bone mineral accretion by 1–5%, depending on the measurement site (Cashman and Flynn, 1999). The body maintains calcium homeostasis through complex regulatory mechanisms involving hormones such as 1,25-dihydroxyvitamin D_3 (1,25$(OH)_2D_3$), parathyroid hormone, fibroblast growth factor 23 (FGF23). These hormones work together to keep blood calcium levels in a very narrow range. Parathyroid hormone promotes calcium reabsorption in the kidneys and osteoclastic bone resorption to bring calcium levels back up when they get too low (Weber et al., 2003; Yamazaki et al., 2002). Negative feedback mechanisms like calcitonin also help regulate calcium levels and maintain homeostasis by inhibiting bone resorption and reducing calcium absorption (Shimada et al., 2001). And the importance of calcium intake throughout life is evident because this whole system of calcium regulation is important for many biological processes, bone health, and mineral balance.

8.2 Magnesium

Magnesium (Mg) is a necessary component for the body so it can operate healthily and is then involved in many cellular metabolic processes (Maguire and Cowan, 2002). The importance of magnesium extends further than just normal cellular activity. Its presence is vital to bone formation and mineralization. Magnesium directly affects osteoblasts' activity because it increases bone production enzymes (Leidi et al., 2011). For preserving bone strength and density, magnesium is needed by these osteoblasts to perform their critical tasks. Magnesium, therefore, also helps strong healthy bone structure development.

Magnesium deficiency directly affects bone density. Several studies have shown that it worsens osteoporosis, a condition that weakens bones and increases fracture risk. Magnesium deficiency in animals results in fragile, easily fractured ones full of trabecular microcracks. The significant impairment of these bones' mechanical properties shows how important magnesium is for bone integrity. Magnesium deficiency can also affect the cortical bone around implants and result in a significant decrease in tibial cortical thickness (Belluci et al., 2011). Magnesium benefits for bone health go beyond bone density and structural integrity. It is essential for mineral and bone homeostasis because it controls bone cells' function and promotes hydroxyapatite crystal development, the main mineral component of bone. In bones, magnesium is stored in both a fixed and dynamic pool, accounting for about 60% of the body's total magnesium content. If there is a deficiency, serum magnesium can be replenished quickly from this magnesium pool (Havaldar et al., 2013).

The function of this dynamic magnesium pool varies throughout life, but it is essential for preserving bone health. About half of the magnesium in the body is deposited in the bones during the early stages of adolescence. But as people become older, this percentage falls, falling to 33% for adults and then to roughly 10% for the elderly (Havaldar et al., 2013). This reduction in magnesium storage emphasizes how crucial it is to consume enough magnesium throughout life, particularly as people age and their natural magnesium stores start to deplete. Magnesium is mainly stored in dense bone, where it sits either on the surface of hydroxyapatite crystals or within the water layer surrounding them. In this position, it helps strengthen the bone's structure from the inside out, making bones more resistant to fractures and improving overall bone health.

8.3 *Phosphorus*

Phosphorus, the second most important component of bone tissue after calcium, plays a critical role. It is the second most abundant mineral in bone after calcium and plays a key role in keeping the skeleton strong and healthy. In the human body, it ranges between 550 and 770 grams, with around 85% stored in the bones and teeth as part of hydroxyapatite crystals and phosphoproteins (Butusov and Jernelöv, 2013). The balance of phosphorus in the body is mainly managed by three hormones: fibroblast growth factor 23, parathyroid hormone, and active Vitamin D [1,25$(OH)_2D_3$], which is released by bone cells called osteocytes (Bergwitz and Jüppner, 2010). Keeping phosphorus levels within a healthy range is essential because it supports the function of osteoblasts and osteocytes, the cells that build and maintain bone, especially during the mineralization process that gives bones their hardness and structure (Magne et al., 2003). The balance between calcium and phosphorus in the diet is equally vital for proper bone development, with current recommendations suggesting a Ca:P molar ratio of 1:1 or 1.3:1 in weight units (Pastore et al., 2012). In physiologically normal adults, phosphorus deficiency results in the release of calcium from the skeleton and causes hypercalciuria, although hypercalcemia does not typically occur. However, in cases of genetic hypophosphatemia and long-term phosphorus deficiency, conditions such as rickets or osteomalacia often develop. When diets are high in phosphorus but low in calcium, complexes form that reduce serum calcium levels, prompting the secretion of PTH. This hormone then triggers bone resorption to restore serum calcium to normal levels. Elevated phosphorus consumption has been shown to contribute to bone loss in animal models, and a 100 mg increase in intake raises fracture risk by 9%. Over the years, researchers have explored the adverse effects of excessive phosphorus on bone metabolism, revealing that high-phosphorus diets can induce hormonal changes like mild hyperparathyroidism. This, in turn, lowers 1,25$(OH)_2$ D levels, disturbing calcium homeostasis. These findings highlight the prevalence of nutrient deficiencies related to bone health in the population and underscore the correlation between increased phosphorus intake and a higher incidence of bone fractures.on bone metabolism, revealing that high-phosphorus diets can induce hormonal changes like mild hyperparathyroidism. This, in turn, lowers 1,25$(OH)_2$ D levels, disturbing calcium homeostasis. These findings highlight the prevalence of nutrient deficiencies related to bone health in the population and underscore the correlation between increased phosphorus intake and a higher incidence of bone fractures.

8.4 *Hemoglobin Production and Oxygen Transport (Iron)*

An essential component of the body, iron plays important roles in DNA synthesis, cell energy production, and oxygen transport in the bloodstream. There are two main ways that iron enters the bloodstream: through the gastrointestinal tract during post-meal digestion, and through liver cells and macrophages that recycle old red blood cells. Iron in the bloodstream is bound to a protein known as transferrin, which facilitates its distribution throughout the body.

The hormone hepcidin is crucial in iron regulation through control of absorption and release. At high levels of hepcidin, iron is sequestered in the liver and macrophages, keeping too much iron out of the bloodstream. On the other hand, more iron enters the bloodstream and is bound to transferrin when hepcidin levels are low (Afsar et al., 2021).

After entering the bloodstream, iron must reach the cells that require it for processes like Adenosine triphosphate production in heart cells and hemoglobin production in red blood cells. Transferrin enters these cells by binding to the TfR1(Transferrin receptor protein 1) receptor and being internalized through endocytosis. Certain enzymes within the cell convert iron from its ferric form (Fe^{3+}) to a more reactive ferrous form (Fe^{2+}). The mitochondria use this Fe2+ to create heme and iron-sulfur clusters for energy production after it has been transported there (Richardson et al.,

2010). Fe^{2+} is highly reactive and can damage cells if not controlled. To prevent this, cells regulate the amount of Fe^{2+} in the cytosol (the cell's fluid). When the level of Fe^{2+} drops, the cell senses this change through proteins called iron regulatory proteins. These proteins increase the expression of TfR1, encouraging more iron to be taken up by the cell. Additionally, in heart cells, when iron stores are low, hepcidin is upregulated to prevent the export of iron from the cells, helping to conserve iron even when levels are low in the rest of the body (Kozłowska et al., 2022). The heart is highly sensitive to disruptions in iron homeostasis. If proteins like TfR1, iron regulatory proteins, or hepcidin are deleted or malfunction in heart cells, it leads to iron deficiency, which impairs the production of heme and Adenosine triphosphate production. This can cause heart muscle dysfunction, resulting in cardiomyopathy, a disease of the heart muscle (Lakhal-Littleton et al., 2016; Xu et al., 2015).

8.5 Thyroid Health (Iodine, Selenium)

8.5.1 Iodine

Iodine, a trace element found naturally in the environment, is primarily absorbed by mammals in the form of iodide (I^-) (Zbigniew, 2017). It is essential for the production of thyroid hormones, which play an important function in regulating metabolism, supporting cell differentiation, and ensuring proper fetal development (Mullur et al., 2014). Dietary sources of iodine are mainly in the form of iodide or iodate (IO_3^-), often combined with sodium or potassium in iodized salt (Patrick, 2008). The process of iodine absorption begins in the intestine, where it is transported across epithelial cells by the sodium/iodide symporter (NIS), a membrane protein that leverages the sodium gradient to actively transport iodide into the body (Nicola et al., 2009). Once in the bloodstream, NIS also facilitates the uptake of iodide into critical tissues such as the thyroid and lactating mammary glands (Portulano et al., 2014).

The principal physiological role of iodine is in thyroid hormone synthesis. This begins with the active uptake of iodide into thyroid cells, a rate-limiting process carried out by NIS (Ravera et al., 2017) Iodide from within the cell is transferred to the apical membrane and then into the colloid space with Pen and other specific transporters (Bizhanova and Kopp, 2009; Zbigniew, 2017) Iodide is also bound to the thyroid gland's major protein framework in the colloid, thyroglobulin. Thyroid peroxidase and dual oxidase 2 are enzymes that catalyze iodination (Di Jeso and Arvan, 2016). Iodinated Tg After iodination, iodinated Tg is recycled into thyroid cells through endocytosis. It fuses with lysosomes, in which Tg is hydrolyzed by enzymes, and active thyroid hormones T3, and T4 are released. These are then secreted into the blood by exocytosis, which acts on many essential functions in the body (Ahad and Ganie, 2010; Becker et al., 2001).

8.5.2 Selenium

Selenium (Se) is essential for thyroid function, especially for the production and metabolism of thyroid hormones. Thyroid tissue contains the highest quantities of it, highlighting its importance for thyroid function (Liu et al., 2013). Through a broad set of proteins called selenoproteins, which are essential to many bodily functions, selenium carries out its biological activities. Thioredoxin reductase, selenoprotein K, glutathione peroxidase, iodothyronine deiodinase, and selenoprotein P are a few of them. Antioxidant defense, immunological responses, thyroid hormone metabolism, DNA synthesis, and other processes all depend on these selenoproteins (Duntas, 2018; S. Gupta et al., 2013).

Although iodine is well-known to have a significant role in thyroid illness and to affect thyroid size, there is mounting evidence that selenium also affects thyroid size (Köhrle, 2013; Liu et al., 2013). For thyroid cells and vascular follicular units, which are involved in the manufacture and storage of thyroid hormones, to operate normally, an adequate selenium intake is necessary. Through the inhibition of Th1-dependent immune responses, it can lessen inflammation and thyroid gland damage. In patients with autoimmune thyroiditis on levothyroxine, selenium supplementation has been seen to decrease thyroid peroxidase antibody levels after three, six, and twelve months of treatment. Supplementing with selenium may lower the malignancies, including those of the breast, pancreas, liver, colon, and esophagus according to research on the subject (Davis et al., 2012; Renko et al., 2009). More precisely, there is evidence linking low selenium levels to a risk of thyroid cancer, mainly papillary thyroid carcinoma (Combs et al., 2005; Glattre et al., 2012). Iodothyronine deiodinase is one such protein that catalyzes the conversion of T4 into the more active T3, helping to maintain stable T3 levels Iodothyronine deiodinase, along with glutathione peroxidases, thioredoxin reductases, provides antioxidant protection for the thyroid gland by scavenging ROS. The thyroid stores significant amounts of selenium, which is vital for generating critical selenoproteins (de Oliveira Maia et al., 2020).

8.6 Cardiovascular Health (Potassium, Magnesium)

8.6.1 Potassium

Potassium has several ways that it helps control blood pressure and support cardiovascular health. Its main function is to decrease intravascular volume, in part by lowering sodium reabsorption, which raises sodium excretion in the urine. The long-term drop in blood pressure, however, cannot be entirely explained by this impact, which is transient. Vasodilation that is dependent on the endothelium is also linked to elevated plasma potassium levels, which reflect food intake. The sodium-potassium adenosine triphosphatase pump is stimulated to accomplish this, and potassium channels within the pump open, promoting blood vessel relaxation (Houston, 2011).

Apart from these effects, potassium may also influence blood pressure through other possible processes, like changing baroreflex sensitivity and adjusting the sensitivity of hormones and receptors that control the sympathetic nervous system and vascular smooth muscle. Beyond potassium's natriuretic effects, these actions probably help lower blood pressure, strengthening its overall antihypertensive effect. Many studies showed that dietary potassium intake lowers the risk of stroke. Higher potassium consumption is inversely and the dose is proportionally linked with stroke, according to the Health Professionals Study, the Nurses' Health Study, and the National Health and Nutrition Examination Survey I (NHANES-I) (Bazzano et al., 2001). The favorable impact of increasing dietary (by 10 mmol/day) on risk initially documented (Khaw and Barrett-Connor, 1987). Even slight drops in blood pressure with slight increases in potassium consumption could be part of the protective mechanism behind this effect.

8.6.2 Magnesium

Magnesium (Mg) is important for maintaining blood pressure and cardiovascular health in general. Studies like those by Altura et al. (1981) have demonstrated that magnesium affects vascular tone and contractility by changing calcium (Ca) levels, even though it doesn't directly contribute to the molecular processes of muscle contraction. Magnesium's important activity in vascular function is shown by the fact that variations in magnesium concentration can alter calcium-triggered smooth muscle contraction (Altura and Altura, 1981; Turlapaty and Altura, 1980). Magnesium naturally limits the flow of calcium into heart cells by acting as a built-in calcium channel blocker. This helps maintain a regular heartbeat and supports healthy heart function (Agus et al., 1989; Iseri and French, 1984). A magnesium deficiency can increase the formation of thromboxane and vasoconstrictor prostaglandins, which can raise blood pressure, by stimulating the synthesis of aldosterone through the angiotensin II pathway (Gröber, 2019).

Magnesium also helps the vascular endothelium by regulating the release of vital molecules that support healthy vascular function, including prostacyclin, endothelin-1, and nitric oxide (Maier et al., 2004). Consequently, a magnesium deficit may increase the risk of hypertensive diseases, and magnesium plays a modulatory function in blood pressure regulation. Altura's research showed that vascular hyper-reactivity, which raises blood pressure, is caused by magnesium deficiency (Altura BM and Altura, 1991). Numerous disturbances in magnesium homeostasis have been observed in hypertension, even though serum magnesium levels are normal in hypertensive persons. Higher magnesium intake adversely affects blood pressure, according to several studies, which conveys that dietary magnesium may help to control hypertension (Joffres et al., 1987).

Apart from its involvement in vascular health, magnesium is necessary for healthy heart function. Low potassium (hypokalemia), hypomagnesemia, and other electrolyte abnormalities can cause arrhythmias in the heart. A magnesium shortage can make cardiac glycosides more prone to arrhythmogenic effects. Atrial and atrioventricular nodal refractory periods are prolonged by magnesium, which affects heart conduction and can assist control of heart rate and rhythm, especially in diseases like atrial fibrillation (DiCarlo et al., 1986). A common supportive, non-pharmacological treatment for atrial and ventricular arrhythmias is magnesium supplementation, and low magnesium levels connected to risk of atrial (Khan et al., 2013; McLean, 1994; Singh et al., 1976).

9. Future Prospects and Challenges

As the discipline of nutrition science advances, the significance of functional foods is increasingly recognized, giving possibilities for increasing health and preventing illnesses. Among the diverse bioactive ingredients in these foods, vitamins, and minerals are particularly significant, providing targeted health benefits (Barik et al., 2022; R. S. Khan et al., 2013). With increasing consumer awareness and a growing preference for natural, health-promoting products, the demand for functional foods fortified with essential micronutrients is expected to rise considerably (Baker et al., 2022; Topolska et al., 2021).The potential of such fortified foods extends beyond improving general health, and ushers in a new era of personalized nutrition, a field in which dietary advice is tailored to an individual's unique health needs (Gupta, 2020).

Despite the results achieved, several issues still persist. An important issue is the stability and bioavailability of vitamins and mineral substances in functional foods. The quality and stability of micronutrients are related to the degradation by high temperature, light and pH followed by difficult maintenance during processing, storage and digestion (Granato et al., 2020). One of the most important issues in creating functional foods is that the nutrients contained in them must be effectively absorbed and utilized by the body. This bioavailability concern is pivotal to their efficacy. Besides, the regulatory climate is quite stringent any claims of health should be backed by strong scientific evidence, hence the need for companies to present adequate proof of safety as well as established health advantages prior to such products hitting the shelves of consumers. (Vergari et al., 2010). Both manufacturers and scientists have the challenge of working within such regulatory guidelines while maintaining consumer confidence (Khan et al., 2013; Nazir et al., 2019). Secondly, consumer education is important for market acceptance since it is imperative that people understand the benefits of vitamin- and mineral-fortified functional foods and trust in their effectiveness.

In response to these challenges, research and technology are driving innovative solutions. Burke and Jacobson (2010) noted that nanotechnology and microencapsulation play a significant role in enhancing the bioavailability and stability of vitamins and minerals. These approaches have therefore been included among the selected strategies for delivering vitamins and minerals through functional foods (Granato et al., 2020). Moreover, the expanding accumulation of evidence on the health promoting effects of these foods is paving the way for their development and consumption (Ahluwalia, 2021).

In the future, fortified foods with essential vitamins and minerals are expected to play an even more central role in nutrition, filling gaps in nutrients, supporting immune systems, and building overall health. While challenges still exist, the intersection of innovative research, technological innovation, and regulatory progress is vital to realizing the full promise of these nutrient-dense foods. As the science progresses, vitamin-and mineral-enhanced functional foods will transform personalized nutrition as solutions for health needs adapted to fit the spectrum of individuals around the globe (Gupta, 2020).

10. Conclusion

During the past decade, there has been a growing emphasis on the explosion of functional foods and their broad scope of potential in many areas, as these foods, rich in functional compounds, are consumed not only for supplying the daily needs of nutrients but also to achieve a proper general sense of well-being and better quality of life, even though their functional ingredients are still relatively underexplored. Vitamins and minerals, particularly vitamin C, are essential nutrients with wide applications. Globally recognized as a nutritional enhancer, vitamin C is especially valuable when utilized as a functional antioxidant. Similarly, the discovery and development of natural vitamin E, including methylation of non-α tocopherols and related product innovations, present significant opportunities for the functional foods sector. Advances in biotechnology have also simplified the extraction of such compounds, further accelerating innovation in complex functional foods. These foods have been found to be useful in children, supporting bone, nervous system, intestinal, and immune system growth, but additional studies are necessary to determine their optimal utilization in different populations, and the concept of functional foods, initially created in Japan in the 1980s to promote health and prevent disease, has now reached all over the globe as a movement with the shifted mindset of people towards diet and well-being. The market for functional foods is expected to grow steadily as consumers increasingly recognize the health benefits of natural products. However, realizing their full potential requires overcoming significant challenges, including the instability of bioactive molecules, complexity in manufacturing scale-up, limited clinical evidence, regulatory hurdles, and consumer skepticism. Addressing these obstacles will demand collaborative efforts from scientists, industry, regulators, and the public. It is welcome news that there is a significant rise in remedies to overcome most of these challenges through technological and scientific advances to improve both stability, bioavailability, and the manufacture of bioactive metabolites, and in this new age, vitamins and minerals reveal the potential to which much-studied and well-managed nutrients can be the precursor of a healthier future. Keeping functional foods safe, sustainable, and accessible will rely upon ongoing research, open regulatory systems, and open communication among stakeholders, and with these efforts functional foods should become an integral part of modern diets.

References

Abd-Rabou, N.S., Zaghloul, A.H., Seleet, F.L. and El-Hofi, M.A. 2010. Properties of Edam cheese fortified by dietary zinc salts. *J. Am. Sci.*, *6*, 441–446.

Afsar, R.E., Kanbay, M., Ibis, A. and Afsar, B. 2021. In-depth review: Is hepcidin a marker for the heart and the kidney? *Mol. Cell. Biochem.*, *476*, 3365–3381.

Agus, Z.S., Kelepouris, E., Dukes, I. and Morad, M. 1989. Cytosolic magnesium modulates calcium channel activity in mammalian ventricular cells. *Am. J. Physiol.*, *256* Pt 1, C452–C455.

Ahad, F. and Ganie, S.A. 2010. Iodine, iodine metabolism, and iodine deficiency disorders revisited. *Indian J. Endocrinol. Metab.*, *14*, 13–17.

Ahluwalia, M.K. 2021. Nutrigenetics and nutrigenomics: A personalized approach to nutrition. *Adv. Genet.*, *108*, 277–340.

Akhtar, S., Anjum, F.M., Rehman, S.U., Sheikh, M.A. and Farzana, K. 2008. Effect of fortification on physico-chemical and microbiological stability of whole wheat flour. *Food Chem.*, *110*, 113–119.

Alsaqqa, A.H., Alkahlout, M.A. and Abu-Naser, S.S. 2022. Using deep learning to classify different types of vitamins. *Int. J. Acad. Eng. Res.*, *6*, 1–6.

Altura, B.M. and Altura, B.T. 1981. Magnesium ions and contraction of vascular smooth muscles: Relationship to some vascular diseases. *Fed. Proc.*, *40*, 2672–2679.

Altura, B.M. and Altura, B.T. 1991.. Cardiovascular risk factors and magnesium: Relationships to atherosclerosis, ischemic heart disease, and hypertension. *Magnes. Trace Elem.*, *10*, 182–192.

Ascherio, A., Rimm, E.B., Hernán, M.A., Giovannucci, E.L., Kawachi, I., Stampfer, M.J., et al., 1998. Intake of potassium, magnesium, calcium and fiber, and risk of stroke among U.S. men. *Circulation*, *98*, 1198–1204.

Ashoori, M. and Saedisomeolia, A. 2014. Riboflavin (vitamin B_2) and oxidative stress: A review. *Br. J. Nutr.*, *111*, 1985–1991.

Bag, S., Mondal, A., Majumder, A. and Banik, A. 2022. Tea and its phytochemicals: Hidden health benefits and modulation of signaling cascade by phytochemicals. *Food Chem.*, *371*, 131098.

Baker, M.T., Lu, P., Parrella, J.A. and Leggette, H.R. 2022. Consumer acceptance toward functional foods: A scoping review. *Int. J. Environ. Res. Public Health*, *19*, 1217.

Bang, H.O. and Dyerberg, J. 1972. Plasma lipids and lipoproteins in Greenlandic west coast Eskimos. *Acta Med. Scand.*, *192*, 85–94.

Ball, G.F. 2004. *Vitamins: Their Role in the Human Body*. John Wiley & Sons, Chichester.

Barik, M., Chowdhury, S. and Tewari, S. 2022. Preventive and therapeutic role of functional foods: A review. *Int. J. Food Nutr. Sci.*, *11*, 361–371.

Bates, C.J., Thurnham, D.I., Bingham, S.A., Margetts, B.M. and Nelson, M. 1997. Biochemical markers of nutrient intake. *Design Concepts in Nutritional Epidemiology*, *2*, 170–240.

Bazzano, L.A., He, J., Ogden, L.G., Loria, C., Vupputuri, S., Myers, L., et al., 2001. Dietary potassium intake and risk of stroke in U.S. men and women: National Health and Nutrition Examination Survey I epidemiologic follow-up study. *Stroke*, *32*, 1473–1480.

Beard, J.L. 2001. Iron biology in immune function, muscle metabolism, and neuronal functioning. *J. Nutr.*, *131*, 568S–579S.

Becker, L.K., Bilezikian, J.P. and Ovid Technologies, Inc. 2001. Part III: The thyroid gland. In: *Principles and Practice of Endocrinology and Metabolism*. Lippincott Williams & Wilkins, Philadelphia, PA, USA.

Belluci, M.M., Giro, G., Del Barrio, R.A.L., Pereira, R.M.R., Marcantonio Jr., E. and Orrico, S.R.P. 2011. Effects of magnesium intake deficiency on bone metabolism and bone tissue around osseointegrated implants. *Clin. Oral Implants Res.*, *22*, 716–721.

Bender, D.A. 2003. *Nutritional Biochemistry of the Vitamins* (2nd Edn.). Cambridge University Press, Cambridge, UK

Bergwitz, C. and Jüppner, H. 2010. Regulation of phosphate homeostasis by PTH, vitamin D and FGF23. *Annu. Rev. Med.*, *61*, 91–104.

Beto, J.A. 2015. The role of calcium in human aging. *Clin. Nutr. Res.*, *4*, 1–8.

Bhagavan, N.V. and Ha, C.E. 2015. Vitamin metabolism. In: *Essentials of Medical Biochemistry* (pp. 683–699). Academic Press.

Bizhanova, A. and Kopp, P. 2009. The sodium-iodide symporter NIS and pendrin in iodide homeostasis of the thyroid. *Endocrinology*, *150*, 1084–1090.

Blaner, W.S., Li, Y., Brun, P.J., Yuen, J.J., Lee, S.A. and Clugston, R.D. 2016. Vitamin A absorption, storage, and mobilization. *Biochem. Retinoid Signal. II Physiol. Vitamin A Uptake*, 95–125.

Brozmanová, J., Mániková, D., Vlčková, V. and Chovanec, M. 2010. Selenium: A double-edged sword for defense and offence in cancer. *Arch. Toxicol.*, *84*, 919–938.

Butusov, M. and Jernelöv, A. 2011. *Phosphorus: An Element That Could Have Been Called Lucifer*. Springer-Verlag, New York, NY, USA.

Carr, A.C. and Maggini, S. 2017. Vitamin C and immune function. *Nutrients*, *9*, 1211.

Carazo, A., Macáková, K., Matoušová, K., Krčmová, L.K., Protti, M. and Mladěnka, P. (2021). Vitamin A update: Forms, sources, kinetics, detection, function, deficiency, therapeutic use, and toxicity. *Nutrients*, *13*, 1703.

Cashman, K.D. and Flynn, A. 1999. Optimal nutrition: Calcium, magnesium, and phosphorus. *Proc. Nutr. Soc.*, *58*, 477–487.

Choi, J.H., Yates, Z., Veysey, M., Heo, Y.R. and Lucock, M. 2014. Contemporary issues surrounding folic acid fortification initiatives. *Prev. Nutr. Food Sci.*, *19*, 247–253.

Combs, G.F. Jr. 2005. Current evidence and research needs to support a health claim for selenium and cancer prevention. *J. Nutr.*, *135*, 343–347.

Combs, G.F. 2008. The vitamins. In: *Fundamental Aspects in Nutrition and Health* (3rd Edn.). Elsevier Academic Press, California, USA.

Da Silva, V.R. and Gregory, J.F. III. 2020. Vitamin B6. In: *Present Knowledge in Nutrition* (pp. 225–237). Elsevier, Amsterdam, Netherlands.

Davis, C.D., Tsuji, P.A. and Milner, J.A. 2012. Selenoproteins and cancer prevention. *Annu. Rev. Nutr.*, *32*, 73–95.

de Oliveira Maia, M., Batista, B.A.M., Sousa, M.P., de Souza, L.M. and Maia, C.S.C. 2020. Selenium and thyroid cancer: A systematic review. *Nutr. Cancer*, *72*, 1255–1263.

Decker, K.F. 1993, Biosynthesis and function of enzymes with covalently bound flavin. *Annu. Rev. Nutr.*, *13*, 17–41.

Delaney, K.Z. and Barke, L. 2017. More about vitamins. In: *Fundamentals of Health and Physical Activity*. Concordia University Library, Canada.

Denu, J.M. 2005. Vitamin B3 and sirtuin function. *Trends Biochem. Sci.*, *30*, 479–483.

Di Jeso, B. and Arvan, P. 2016. Thyroglobulin from molecular and cellular biology to clinical endocrinology. *Endocr. Rev.*, *37*, 2–36.

Di Matola, T., Zeppa, P., Gasperi, M. and Vitale, M. 2014. Thyroid dysfunction following a kelp-containing marketed diet. *Case Rep.*, *2014*, bcr2014206330.

DiCarlo, L.A. Jr., Morady, F., de Buitleir, M., Krol, R.B., Schurig, L. and Annesley, T.M. 1986. Effects of magnesium sulfate on cardiac conduction and refractoriness in humans. *J. Am. Coll. Cardiol.*, *7*, 1356–1362.

Duntas, L.H. 2018. The catalytic role of iodine excess in loss of homeostasis in autoimmune thyroiditis. *Curr. Opin. Endocrinol. Diabetes Obes.*, *25*, 347–352.

Engelhardt, U.H. 2020. Tea chemistry: What do and what don't we know? A micro review. *Food Res. Int.*, *132*, 109120.

Farrell, D.J. and Gibson, R.A. 1991. *The Enrichment of Eggs with Omega-3 Fatty Acids and Their Effects in Humans*. Unpublished report.

Farrell, D.J. 1998. Enrichment of hen eggs with n–3 long-chain fatty acids and evaluation of enriched eggs in humans. *Am. J. Clin. Nutr.*, *68*, 538–544.

Fattal-Valevski, A. 2011. Thiamine (vitamin B1). *J. Evid.-Based Complement. Altern. Med.*, *16*, 12–20.

Femi-Oladunni, O.A., Martínez-Ruiz, M.P., Ruiz-Palomino, P. and Muro-Rodríguez, A.I. 2023. Food values influencing consumers' decisions in a sub-Saharan African country. *Br. Food J.*, *125*, 1805–1823.

Ferrier, L.K., Caston, L.J., Leeson, S., Squires, J., Weaver, B.J., Holub, B.J., et al., 1995. α-Linolenic acid- and docosahexaenoic acid-enriched eggs from hens fed flaxseed: Influence on blood lipids and platelet phospholipid fatty acids in humans. *Am. J. Clin. Nutr.*, *62*, 81–86.

Fischer, V., Haffner-Luntzer, M., Amling, M. and Ignatius, A. 2018. Calcium and vitamin D in bone fracture healing and post-traumatic bone turnover. *Eur. Cell Mater.*, *35*, 365–385.

Frank, L.L. 2015. Thiamin in clinical practice. *J. Parenter. Enteral Nutr.*, *39*, 503–520.

Frewer, L., Scholderer, J. and Lambert, N. 2003. Consumer acceptance of functional foods: Issues for the future. *Br. Food J.*, *105*, 714–731.

Friso, S., De Santis, D., Pizzolo, F. and Udali, S. 2020. Vitamins and epigenetics. In: *Molecular Nutrition* (pp. 633–650). Academic Press, USA.

Gasperi, V., Sibilano, M., Savini, I. and Catani, M.V. 2019. Niacin in the central nervous system: An update of biological aspects and clinical applications. *Int. J. Mol. Sci.*, *20*(4), 974.

Gengatharan, A., Dykes, G.A. and Choo, W.S. 2015. Betalains: Natural plant pigments with potential application in functional foods. *LWT-Food Sci. Technol.*, *64*(2), 645–649.

Gharibzahedi, S.M.T. and Jafari, S.M. 2017. The importance of minerals in human nutrition: Bioavailability, food fortification, processing effects, and nanoencapsulation. *Trends Food Sci. Technol.*, *62*, 119–132.

Glattre, E., Nygård, J.F. and Aaseth, J. 2012. Selenium and cancer prevention: Observations and complexity. *J. Trace Elem. Med. Biol.*, *26*(2–3), 168–169.

Gorman, S., Buckley, A.G., Ling, K.M., Berry, L.J., Fear, V.S., Stick, S.M., et al., 2017. Vitamin D supplementation of initially vitamin D-deficient mice diminishes lung inflammation with limited effects on pulmonary epithelial integrity. *Physiol. Rep.*, *5*(15), e13371.

Granato, D., Barba, F. J., Bursać Kovačević, D., Lorenzo, J.M., Cruz, A. G. and Putnik, P. 2020. Functional foods: Product development, technological trends, efficacy testing, and safety. *Annu. Rev. Food Sci. Technol.*, *11*(1), 93–118.

Gröber, U. 2019. Magnesium and drugs. *Int. J. Mol. Sci.*, *20*(9), 2094.

Guilland, J.C. and Aimone-Gastin, I. 2013. Vitamin B9. *Rev. Prat. 63*(8), 1079–1081.

Gupta, E., Shyam, H., Devi, S. and Maurya, N.K. 2020. Nutrigenomics in functional foods and personalized nutrition. *Gedrag & Organisatie*, *33*(2), 670–686.

Gupta, S., Jaworska-Bieniek, K., Lubiński, J. and Jakubowska, A. 2013. Can selenium be a modifier of cancer risk in CHEK2 mutation carriers? *Mutagenesis*, *28*(6), 625–629.

Hadjimbei, E., Botsaris, G. and Chrysostomou, S. 2022. Beneficial effects of yoghurts and probiotic fermented milks and their functional food potential. *Foods*, *11*, 2691.

Halsted, C.H. 2003. Dietary supplements and functional foods: Two sides of a coin. *Am. J. Clin. Nutr.*, *77*(4), 1001S–1007S.

Hamed, I., Özogul, F., Özogul, Y. and Regenstein, J.M. 2015. Marine bioactive compounds and their health benefits: A review. *Compr. Rev. Food Sci. Food Saf.*, *14*(4), 446–465.

Hanna, M., Jaqua, E., Nguyen, V. and Clay, J.B. 2022. Vitamins: Functions and uses in medicine. *Perm. J.*, *26*(2), 89–97.

Hasler, C.M. 2002. Functional foods: Benefits, concerns, and challenges: A position paper from the American Council on Science and Health. *J. Nutr.*, *132*(12), 3772–3781.

Havaldar, R., Pilli, S.C. and Putti, B.B. 2013. Effects of magnesium on mechanical properties of human bone. *IOSR J. Pharm. Biol. Sci.*, *7*(3), 8–14.

He, F.J. and MacGregor, G.A. 2002. Effect of modest salt reduction on blood pressure: A meta-analysis of randomized trials. *J. Hum. Hypertens.*, *16*(11), 761–770.

He, F.J. and MacGregor, G.A. 2009. A comprehensive review on salt and health and current experience of worldwide salt reduction programmes. *J. Hum. Hypertens.*, *23*(6), 363–384.

Hekmat, S. and McMahon, D.J. 1997. Manufacture and quality of iron-fortified yogurt. *J. Dairy Sci.*, *80*, 3114–3122.

Hirsch, J.A. and Parrott, J. 2012. New considerations on the neuromodulatory role of thiamine. *Pharmacology*, *89*(1–2), 111–116.

Houston, M.C. 2011. The importance of potassium in managing hypertension. *Curr. Hypertens. Rep.*, *13*(4), 309–317.

Hunt, C., Chakravorty, N.K., Annan, G., Habibzadeh, N. and Schorah, C.J. (1994). The clinical effects of vitamin C supplementation in elderly hospitalised patients with acute respiratory infections. *International journal for vitamin and nutrition research, 64*(3), 212–219.

Iseri, L.T. and French, J.H. 1984. Magnesium: Nature's physiologic calcium blocker. *Am. Heart J.*, *108*(1), 188–193.

Ismail, N., Kureishy, N., Church, S.J., Scholefield, M., Unwin, R.D., Xu, J., et al., 2020. Vitamin B5 (D-pantothenic acid) localizes in myelinated structures of the rat brain: Potential role for cerebral vitamin B5 stores in local myelin homeostasis. *Biochem. Biophys. Res. Commun.*, *522*(1), 220–225.

Jee, S.H., Miller, E.R., Guallar, E., Singh, V.K., Appel, L.J. and Klag, M.J. 2002. The effect of magnesium supplementation on blood pressure: A meta-analysis of randomized clinical trials. *Am. J. Hypertens.*, *15*(8), 691–696.

Jeffery, L.E., Burke, F., Mura, M., Zheng, Y., Qureshi, O.S., Hewison, M., et al., 2009. 1,25-Dihydroxyvitamin D_3 and IL-2 combine to inhibit T cell production of inflammatory cytokines and promote development of regulatory T cells expressing CTLA-4 and FoxP3. *J. Immunol.*, *183*(9), 5458–5467.

Jiménez-Colmenero, F., Sánchez-Muniz, F.J. and Olmedilla-Alonso, B. 2010. Design and development of meat-based functional foods with walnut: Technological, nutritional, and health impact. *Food Chem.*, *123*(4), 959–967.

Joffres, M.R., Reed, D.M. and Yano, K. 1987. Relationship of magnesium intake and other dietary factors to blood pressure: The Honolulu Heart Study. *Am. J. Clin. Nutr.*, *45*(2), 469–475.

Kahraman, O. and Ustunol, Z. 2012. Effect of zinc fortification on Cheddar cheese quality. *J. Dairy Sci. 95*, 2840–2847.

Kennedy, D.O. 2016. B vitamins and the brain: Mechanisms, dose, and efficacy: A review. *Nutrients*, *8*(2), 68.

Khan, A.M., Lubitz, S.A., Sullivan, L.M., Sun, J.X., Levy, D., Vasan, R.S., et al., 2013. Low serum magnesium and the development of atrial fibrillation in the community: The Framingham Heart Study. *Circulation*, *127*(1), 33–38.

Khan, R. S., Grigor, J., Winger, R. and Win, A. 2013. Functional food product development: Opportunities and challenges for food manufacturers. *Trends Food Sci. Technol.*, *30*(1), 27–37.

Khaw, K.T. and Barrett-Connor, E. 1987. Dietary potassium and stroke-associated mortality. *N. Engl. J. Med.*, *316*(5), 235–240.

Kligman, A.M., Grove, G.L., Hirose, R. and Leyden, J.J. 1986. Topical tretinoin for photoaged skin. *J. Am. Acad. Dermatol.*, *15*(4), 836–859.

Kohrle, J. 2013. Selenium and the thyroid. *Curr. Opin. Endocrinol. Diabetes Obes*, *20*(5), 441–448.

Kombanda, K.T., Margerison, C., Booth, A. and Worsley, A. 2022. Socio-psychological factors associated with young Australian adults' consumption of energy dense and nutrient poor (EDNP) foods. *Nutrients*, *14*(4), 812.

Kozłowska, B., Sochanowicz, B., Kraj, L., Palusińska, M., Kołsut, P., Szymański, Ł., et al., 2022. Clinical and molecular aspects of iron metabolism in failing myocytes. *Life*, *12*(8), 1203.

Küçükgöz, K. and Trząskowska, M. 2022. Nondairy probiotic products: Functional foods that require more attention. *Nutrients*, *14*(4), 753.

Lakhal-Littleton, S., Wolna, M., Chung, Y.J., Christian, H.C., Heather, L.C., Brescia, M., et al., 2016. An essential cell-autonomous role for hepcidin in cardiac iron homeostasis. *eLife*, *5*, e19804.

Lanham-New, S.A., Hill, T.R., Gallagher, A.M. and Vorster, H.H. (Eds.). 2019. *Introduction to Human Nutrition.* John Wiley & Sons, UK.

Lee, J.I. and Burckart, G.J. 1998. Nuclear factor kappa B: Important transcription factor and therapeutic target. *J. Clin. Pharmaco*l., *38*(11), 981–993.

Lehman, P.A., Slattery, J.T. and Franz, T.J. 1988. Percutaneous absorption of retinoids: Influence of vehicle, light exposure, and dose. *J. Invest. Dermatol.*, *91*(1), 56–61.

Leidi, M., Dellera, F., Mariotti, M. and Maier, J.A. 2011. High magnesium inhibits human osteoblast differentiation *in vitro. Magnes. Res.*, *24*(1), 1–6.

Lešková, E., Kubíková, J., Kováčiková, E., Košická, M., Porubská, J. and Holčíková, K. 2006. Vitamin losses: Retention during heat treatment and continual changes expressed by mathematical models. *J. Food Compos. Anal.*, *19*(4), 252–276.

Lips, P. 2006. Vitamin D physiology. *Prog. Biophys. Mol. Biol.*, *92*(1), 4–8.

Liu, Y., Huang, H., Zeng, J. and Sun, C. 2013. Thyroid volume, goiter prevalence, and selenium levels in an iodine-sufficient area: A cross-sectional study. *BMC Public Health*, *13*, 1153.

Lukaski, H.C. 2004. Vitamin and mineral status: Effects on physical performance. *Nutr.*, *20*(7–8), 632–644.

Magne, D., Bluteau, G., Faucheux, C., Palmer, G., Vignes-Colombeix, C., Pilet, P., et al., 2003. Phosphate is a specific signal for ATDC5 chondrocyte maturation and apoptosis-associated mineralization: Possible implication of apoptosis in the regulation of endochondral ossification. *J. Bone Miner. Res.*, *18*, 1430–1442.

Maguire, M.E. and Cowan, J.A. (2002). Magnesium chemistry and biochemistry. *Biometals*, *15*, 203–210.

Maiani, G., Periago Castón, M.J., Catasta, G., Toti, E., Cambrodón, I.G., Bysted, A., et al., 2009. Carotenoids: Actual knowledge on food sources, intakes, stability and bioavailability, and their protective role in humans. *Mol. Nutr. Food Res.*, *53*, S194–S218.

Maier, J.A., Bernardini, D., Rayssiguier, Y. and Mazur, A. 2004. High concentrations of magnesium modulate vascular endothelial cell behaviour *in vitro*. *Biochimica et Biophysica Acta*, *1689*, 6–12.

Majewska, E., Kozłowska, M., Gruszczyńska-Sękowska, E., Kowalska, D. and Tarnowska, K. 2019. Lemongrass (Cymbopogon citratus) essential oil: Extraction, composition, bioactivity, and uses for food preservation: A review. *Polish Journal of Food and Nutrition Sciences*, *69*, 4.

Mamede, A.C., Tavares, S.D., Abrantes, A.M., Trindade, J., Maia, J.M. and Botelho, M.F. 2011. The role of vitamins in cancer: A review. *Nutrition and Cancer*, *63*, 479–494.

McLean, R.M. 1994. Magnesium and its therapeutic uses: A review. *American Journal of Medicine*, *96*, 63–76.

Mejia, L.A., Kuo, W.Y. and Beltran-Velazquez, F. 2019. Provision of micronutrients in coexisting public health programs and risk of excessive intake: Regulatory considerations. *Annals of the New York Academy of Sciences*, *1446*, 66–80.

Moore-Schiltz, L., Albert, J.M., Singer, M.E., Swain, J. and Nock, N.L. 2015. Dietary intake of calcium and magnesium, and the metabolic syndrome in the National Health and Nutrition Examination (NHANES) 2001–2010 data. *British Journal of Nutrition*, *114*, 924–935.

Mukarram, M., Choudhary, S., Khan, M.A., Poltronieri, P., Khan, M.M.A., Ali, J., et al., 2021. Lemongrass essential oil components with antimicrobial and anticancer activities. *Antioxidants*, *11*, 20.

Mullur, R., Liu, Y.Y. and Brent, G.A. 2014. Thyroid hormone regulation of metabolism. *Physiological Reviews*, *94*, 355–382.

Nazir, M., Arif, S., Khan, R.S., Nazir, W., Khalid, N. and Maqsood, S. 2019. Opportunities and challenges for functional and medicinal beverages: Current and future trends. *Trends in Food Science & Technology*, *88*, 513–526.

Nicola, J.P., Basquin, C., Portulano, C., Reyna-Neyra, A., Paroder, M. and Carrasco, N. 2009. The Na^+/I^- symporter mediates active iodide uptake in the intestine. *American Journal of Physiology-Cell Physiology*, *296*, C654–C662.

Niki, E. and Traber, M.G. 2012. A history of vitamin E. *Annals of Nutrition and Metabolism*, *61*, 207–212.

Ofoedu, C.E., Iwouno, J.O., Ofoedu, E.O., Ogueke, C.C., Igwe, V.S., Agunwah, I.M., et al., 2021. Revisiting food-sourced vitamins for consumer diet and health needs: A perspective review, from vitamin classification, metabolic functions, absorption, utilization, to balancing nutritional requirements. *PeerJ*, *9*, e11940.

Oh, C.R. and Brown, D.L. (2003). Vitamin B12 deficiency. *American Family Physician, 67*(5), 979–986.

Osendarp, S.J., Martinez, H., Garrett, G.S., Neufeld, L.M., De-Regil, L.M., Vossenaar, M., et al., 2018. Large-scale food fortification and biofortification in low- and middle-income countries: A review of programs, trends, challenges and evidence gaps. *Food Nutr. Bull.*, *39*, 315–331.

Ottaway, P.B. 1993. Stability of vitamins in food. In: *The Technology of Vitamins in Food* (pp. 90–113). Springer, Boston, MA, USA.

Ozen, A.E., del Mar Bibiloni, M., Pons, A. and Tur, J.A. 2013. Sociodemographic and lifestyle determinants of functional food consumption in an adult population of the Balearic Islands. *Annals of Nutrition and Metabolism*, *63*, 200–207.

Pastore, S.M., Gomes, P.C., Rostagno, H.S., Albino, L.F.T., Calderano, A.A., Vellasco, C.R., et al., 2012. Calcium levels and calcium: Available phosphorus ratios in diets for white egg layers from 42 to 58 weeks of age. *Rev. Bras. Zootec.* *41*, 2424–2432.

Patrick, L. (2008). Iodine: Deficiency and therapeutic considerations. *Alternative Medicine Review*, *13*, 116–127.

Plantone, D., Pardini, M. and Rinaldi, G. 2021. Riboflavin in neurological diseases: A narrative review. *Clinical Drug Investigation*, *41*, 513–527.

Portulano, C., Paroder-Belenitsky, M. and Carrasco, N. 2014. The Na^+/I^- symporter (NIS): Mechanism and medical impact. *Endocrine Reviews*, *35*, 106–149.

Precup, G., Pocol, C.B., Teleky, B.E. and Vodnar, D.C. 2022. Awareness, knowledge, and interest about prebiotics: A study among Romanian consumers. *Int. J. Environ. Res. Public Health*, *19*(3), 1208.

Rai, S., Wai, P.P., Koirala, P., Bromage, S., Nirmal, N.P., Pandiselvam, R., et al., 2023. Food product quality, environmental, and personal characteristics affecting consumer perception toward food. *Frontiers in Sustainable Food Systems*, *7*, 1222760.

Ravera, S., Reyna-Neyra, A., Ferrandino, G., Amzel, L.M. and Carrasco, N. 2017. The sodium/iodide symporter (NIS): Molecular physiology and preclinical and clinical applications. *Annual Review of Physiology*, *79*, 261–289.

Renko, K., Hofmann, P.J., Stoedter, M., Hollenbach, B., Behrends, T., Köhrle, J., et al., 2009. Down-regulation of the hepatic selenoprotein biosynthesis machinery impairs selenium metabolism during the acute-phase response in mice. *FASEB Journal*, *23*, 1758–1765.

Richardson, D.R., Lane, D.J., Becker, E.M., Huang, M.L.H., Whitnall, M., Rahmanto, Y.S. and Ponka, P. 2010. Mitochondrial iron trafficking and the integration of iron metabolism between mitochondrion and cytosol. *Proceedings of the National Academy of Sciences*, USA, *107*, 10775–10782.

Rizzo, G. and Laganà, A.S. 2020. A review of vitamin B12. *Molecular Nutrition*, 105–129.

Ross, A.C. 2010. Vitamin A. In: *Bioactive Compounds and Cancer* (pp. 335–356). Totowa, NJ: Humana Press, USA.

Said, H.M. (2011). Intestinal absorption of water-soluble vitamins in health and disease. *Biochemical Journal*, *437*, 357–372.

Salvatore, F.P., Adamashvili, N. and Conto, F. 2022. Factors affecting consumer purchasing behavior of functional food: A comparative analysis for consumer management. *Br. Food J.*, *124*, 1519–1536.

Samanta, S. 2022. Potential bioactive components and health promotional benefits of tea (*Camellia sinensis*). *Journal of the American Nutraceutical Association*, *41*, 65–93.

Sangeetha, P.T., Ramesh, M.N. and Prapulla, S.G. 2005. Recent trends in microbial production, analysis, and application of fructooligosaccharides. *Trends in Food Science & Technology*, *16*, 442–457.

Schaefer, H. 1993. Penetration and percutaneous absorption of topical retinoids: A review. *Skin Pharmacology*, *6*, 17–23.

Selhub, J. 1999. Homocysteine metabolism. *Annual Review of Nutrition*, *19*, 217–246.

Sgroi, F., Moscato, C.M. and Moscato, R. 2022. Consumer preferences for the Mediterranean diet: Results of an empirical analysis. *J. Agric. Food Res.*, *10*, 100371.

Shearer, M.J. and Newman, P. 2014. Recent trends in metabolism and cell biology of vitamin K with special reference to vitamin K cycling and MK-4 biosynthesis. *Journal of Lipid Research*, *55*, 345–362.

Shi, L.H., Balakrishnan, K., Thiagarajah, K., Mohd Ismail, N.I. and Yin, O.S. (2016). Beneficial properties of probiotics. *Tropical Life Sciences Research*, *27*, 73–90.

Shakoor, H., Feehan, J., Al Dhaheri, A.S., Ali, H.I., Platat, C., Ismail, L.C., et al., 2021. Immune-boosting role of vitamins D, C, E, zinc, selenium and omega-3 fatty acids: Could they help against COVID-19? *Maturitas*, *143*, 1–9.

Shimada, T., Mizutani, S., Muto, T., Yoneya, T., Hino, R., Takeda, S., et al., 2001. Cloning and characterization of FGF23 as a causative factor of tumor-induced osteomalacia. *Proc. Natl. Acad. Sci.*, USA, *98*, 6500–6505.

Singh, R.B., Manmohan, M.D., Dube, K.P. and Singh, V.P. 1976. Serum magnesium concentrations in atrial fibrillation. *Acta Cardiologica*, *31*, 221–226.

Sousa, R., Figueirinha, A., Batista, M.T. and Pina, M.E. (2021). Formulation effects in the antioxidant activity of extract from the leaves of cymbopogon citratus (Dc) stapf. *Molecules*, *26*(15), 4518.

Subcommittee on Upper Reference Levels of Nutrients, Standing Committee on the Scientific Evaluation of Dietary Reference Intakes, Panel on Folate, Other B Vitamins and Choline. 2000. *Dietary Reference Intakes for Thiamin, Riboflavin, Niacin, Vitamin B6, Folate, Vitamin B12, Pantothenic Acid, Biotin and Choline*. National Academy Press, Washington, DC, USA

Szymanska, R., Nowicka, B., Trela, A. and Kruk, J. 2020. Vitamin E: Structure and forms. In: V.B. Patel Ed.), *Molecular Nutrition* (pp. 67–90). Academic Press, USA.

Tan, Y. and McClements, D.J. 2021. Improving bioavailability of oil-soluble vitamins by optimizing food matrix effects: A review. *Food Chemistry*, *348*, 129148.

Teegarden, M.D. 2014. *Determination of Biologically Relevant Vitamin D Metabolites in a Mouse Model of Non-melanoma Skin Cancer*. (Master's Thesis, The Ohio State University, USA).

Thomas, D.R. 2006. Vitamins in aging, health, and longevity. *Clinical Interventions in Aging*, *1*, 81–91.

Thomas, P. R. and Earl, R. (1994). Committee on opportunities in the nutrition and food sciences. *Food and Nutrition Board, Institute of Medicine*, 98–142.

Topolska, K., Florkiewicz, A. and Filipiak-Florkiewicz, A. 2021. Functional food: Consumer motivations and expectations. *International Journal of Environmental Research and Public Health*, *18*, 5327.

Tripathi, B. and Platel, K. 2010. Finger millet (*Eleusine coracana*) flour as a vehicle for fortification with zinc. *Journal of Trace Elements in Medicine and Biology*, *24*, 46–51.

Tripathi, B. and Platel, K. 2011. Iron fortification of finger millet (*Eleusine coracana*) flour with EDTA and folic acid as co-fortificants. *Food Chemistry*, *126*, 537–542.

Turlapaty, P.D. and Altura, B.M. 1980. Magnesium deficiency produces spasms of coronary arteries: Relation to etiology of sudden-death ischemic heart disease. *Science*, *208*, 198–200.

Urala, N., Arvola, A. and Lähteenmäki, L. 2003. Strength of health-related claims and their perceived advantage. *International Journal of Food Science & Technology*, *38*, 815–826.

Van Dael, P., Davidsson, L., Aoz-Box, R.M., Fay, L.B. and Barclay, D. 2001. Selenium absorption and retention from a selenite- or selenate-fortified milk-based formula in men measured by a stable-isotope technique. *British Journal of Nutrition*, *85*, 157–163.

Verbeke, W. 2006. Functional foods: Consumer willingness to compromise on taste for health? *Food Quality and Preference*, *17*, 126–131.

V Vergari, F., Tibuzzi, A. and Basile, G. 2010. An overview of the functional food market: From marketing issues and commercial players to future demand from life in space. In: *Bio-Farms for Nutraceuticals: Functional Food and Safety Control by Biosensors* (pp. 308–321). Springer, Germany.

Wang, H., Li, L., Qin, L.L., Song, Y., Vidal-Alaball, J. and Liu, T.H. 2018. Oral vitamin B12 versus intramuscular vitamin B12 for vitamin B12 deficiency. *Cochrane Database of Systematic Reviews*, (3).

Weber, T.J., Liu, S., Indridason, O.S. and Quarles, L.D. 2003. Serum FGF23 levels in normal and disordered phosphorus homeostasis. *Journal of Bone and Mineral Research, 18*, 1227–1234.

Wendt, J., Considine, C. and Kogan, M. 2021. *Integrative Geriatric Nutrition: A Practitioner's Guide to Dietary Approaches for Older Adults*. Springer Nature, Germany.

WHO, G.S. 2018. FAO/WHO framework for the provision of scientific advice on food safety and nutrition.

Xu, W., Barrientos, T., Mao, L., Rockman, H.A., Sauve, A.A. and Andrews, N.C. 2015. Lethal cardiomyopathy in mice lacking transferrin receptor in the heart. *Cell Reports*, *13*, 533–545.

Yamazaki, Y., Okazaki, R., Shibata, M., Hasegawa, Y., Satoh, K., Tajima, T., et al., 2002. Increased circulatory level of biologically active full-length FGF-23 in patients with hypophosphatemic rickets/osteomalacia. *J. Clin. Endocrinol. Metab.*, *87*, 4957–4960.

Zbigniew, S. 2017. Role of iodine in metabolism. *Recent Patents on Endocrine, Metabolic, & Immune Drug Discovery*, *10*, 123–126.

Zempleni, J., Wijeratne, S.S.K. and Hassan, Y.I. 2009. *Biotin. Biofactors*, *35*(1), 36–46.

Zhao, Y., Wang, D. and Yang, S. 2016. Effect of organic and conventional rearing system on mineral content of pork. *Meat Science*, 118, 103–107.

Zhou, S.S., Li, D., Chen, N.N. and Zhou, Y. 2015. Vitamin paradox in obesity: Deficiency or excess? *World Journal of Diabetes*, *6*, 1158.

Zhu, K., Greenfield, H., Zhang, Q., Du, X., Ma, G., Foo, L.H., et al., 2008. Growth and bone mineral accretion during puberty in Chinese girls: A five-year longitudinal study. *J. Bone Miner. Res.*, *23*, 167–172.

CHAPTER 4

Role of Phytochemicals in Functional Foods

Jannatul Ferdushi Jany[1*]

1. Introduction

Food is any substance consisting essentially of protein, carbohydrate, fat, and other nutrients used in the body of an organism to furnish energy, provide nourishment, and assist with various metabolic functions. In the early 1900s, scientists identified essential elements, such as vitamins, and their functions in preventing various dietary deficiency diseases (Semba 2012). Nutritional research has expanded significantly over the past few decades, emphasizing both the nutrients and non-nutritive elements with their active effect on the prevention of disease and the reduction of risk (Crowe and Francis, 2013). Current research focuses on identifying physiologically active components in plant-based foods that may help reduce the risk of various chronic diseases. Beyond conventional sufficiency in energy, protein, essential fats, vitamins, and minerals, the novel idea of functional foods poses questions regarding food security and the appropriate choice of a sufficient diet.

Current dietary advice from epidemiological evidence recommends a minimum of five portions of fruits and vegetables a day. Fruits and vegetables are good sources of vitamins as well as phytochemicals, which are biologically active secondary metabolites providing color and flavor and natural toxicity to pests, and are sometimes termed phytoprotectants and nutraceuticals (Kalt, 2001; Liu, 2002). The Medical Aspects of Food and Nutrition Policy Committee, the diet and cancer working group of the UK, reported that there is good evidence that plant foods contain metabolically active components which can inhibit carcinogenesis under experimental conditions (COMA, 1998). However, ingestion of an adequate amount of phytochemicals, or by supplementing other foods, can reduce non-communicable diseases.

The term "functional food" can also refer to characteristics that have been intentionally incorporated into prominent consumable plants to lower the risk of chronic diseases beyond their basic nutritional purposes, have a traditional appearance, and be consumed as part of a regular diet (Gupta and Mishra, 2021). According to the Functional Food Market Forecast 2024-2028, the functional food industry is expected to grow by USD 127.46 billion between 2023 and 2028. Demand for these foods is rising quickly in the APAC area, especially in developing countries like China, India, and Indonesia (Technavio, 2024). Functional foods are considered a rich source of phytochemicals, which are beneficial to health, especially for the prevention or treatment of long-

[1] Department of Food Engineering and Technology, Faculty of Agricultural Engineering and Technology, Bangladesh Agricultural University, Mymensingh-2202, Bangladesh.
* Corresponding author: jany.ftri@bau.edu.bd
DOI: 10.1201/9781003659808-4

term sickness when administered in consistent, non-toxic, and defined doses. Currently, the market for health food supplements is typically one of the few where phytochemicals are used commercially. Consequently, it requires special attention due to their effectiveness in encouraging and supporting health. An important area of nutrition research is the relationship between phytochemicals and functional foods. It is crucial to extract and characterize novel natural constituents containing phytochemicals and incorporate them into functional foods to improve consumer health.

2. Functional Foods

A natural food or food component that requires extraction or addition via chemical processes or biotechnology is referred to as a functional food. "Functional foods" were defined as foods that might control biological processes and help avoid disease (Shimizu, 2003). The purpose of these functional foods is to prevent, treat, or manage chronic illnesses or issues associated with them (Martirosyan and Miller, 2018). Additional definitions of functional foods are presented in Table 1.

Table 1 Definitions of functional foods from various organizations, including the source of each definition.

Organization	*Explanation*	*References*
Functional Food Center (FFC)	"Natural or processed foods containing known or unknown biologically active compounds that provide a scientifically supported and established health benefit for the prevention, management, or treatment of chronic diseases are prescribed, appropriate, and non-toxic amounts."	(Martirosyan & Pisarski, 2017)
Food and Drug Administration (FDA)	There is no legal definition for the term "functional foods", which is now being used as a marketing term by the company.	(Martirosyan and Singh, 2015)
International Life Sciences Institute (ILSI)	"Foods that provide health benefits beyond basic nutrition due to the presence of physiologically active dietary components."	(Crowe and Francis 2013)
Institute of Food Technologists (IFT)	"Foods and food ingredients that have health benefits in addition to basic nutrition." In addition to other pharmacologically active ingredients, these products provide vital nutrients in amounts that are frequently higher than those needed for regular functioning, well-being, and production.	(MacAulay et al. 2005)
European Commission Concerted Action Group on Functional Food Science in Europe (FUSOSE), International Life Sciences Institute	"Functional" food is defined as food that has been scientifically proven to enhance one or more bodily processes and go beyond basic nutritional requirements. Such food contributes to an enhanced state of health and well-being or lowers the risk of certain conditions."	(Blades, 2000)
European Food Safety Authority (EFSA)	"A food, which beneficially affects one or more target functions in the body, beyond adequate nutritional effects, in a way that is relevant to either an improved state of health and well-being and/or reduction of risk of disease. A functional food can be a natural food or a food to which a component has been added or removed by technological or biotechnological means, and it must demonstrate its effects in amounts that can normally be expected to be consumed in the diet."	(Martirosyan and Singharaj, 2016)
Food and Nutrition Board (FNB)	"A product with potentially beneficial elements is referred to as a functional food, which includes' any altered food or food component that has the potential to provide a health benefit beyond the conventional nutrients it contains."	(Havel et al., 1994; Kruger and Mann, 2003)

2.1 Classification of the Functional Foods

Functional foods can be classified into two categories, i.e., conventional foods and modified foods.

2.1.1 Conventional Foods

Foods that are considered conventional are in their original, unaltered state and have not been modified (Ghevariya and Singh, 2023). They have healthy ingredients, including vitamins and minerals. antioxidants and good fatty acids. Conventional foods can be further divided into two categories: those derived from plants and those derived from animals. Dry fruits, fresh fruits, leafy green and non-leafy vegetables, legumes, seeds, nuts, herbs, and spices are examples of traditional plant-based foods. like buckwheat, brown rice, cereals, nuts, beans, and many more. While animal based conventional foods are milk, eggs, fish, and meat.

2.1.2 Modified Foods

Food items that have been altered or added specific elements with nutrients to support health advantages are referred to as modified foods (Ahmad et al., 2021). Probiotics, fiber, vitamins, and minerals are among the extra nutrients incorporated in these for particular health advantages. Fortified, enriched, and enhanced foods are categories of modified foods. Fortification is a technique where other nutrients are added that aren't typically present in foods. Fortification helps to balance a food's overall nutritional profile and repair or avoid frequent nutrient consumption deficiencies, replenish nutrients lost during processing, or draw in customers. Conversely, food enrichment is the act of restoring micronutrients that may be lost or destroyed during food processing. "Enhanced foods" are foods that have been modified from their original state to provide additional health or functional advantages (Siro et al., 2008). A wide range of goods, each with special enhancements, might be referred to as "Enhanced foods" under this general term.

3. Phytochemicals

Phytochemicals are "plant chemicals" found in vegetables, fruits, beans, grains, nuts, and seeds (Figure 1). The potential health advantages of phytochemicals with distinct bioactivities toward animal biochemistry and metabolism are being extensively studied. Nowadays, it is generally acknowledged that consuming a plant-based diet promotes well-being and delays the start of several illnesses in people (Soria-Florido et al., 2020). In recent years, a lot of research has been conducted on the preventive benefits of dietary phytochemicals against inflammation and oxidative stress linked to the pathogenesis of many prevalent diseases (Li et al., 2018; Wan et al., 2019). In a variety of metabolic processes, phytochemicals function as substrates, cofactors, inhibitors, absorbents/sequestrants, intracellular receptors, scavengers, enhancers of essential nutrient absorption, and elective growth factors. Although various bioactives are known to have positive effects, these effects are complex that can function singly and in combination (Yuan, 2008). Carotenoids, phenolic compounds (flavonoids, phytoestrogens, phenolic acids), phytosterols and phytostanols, tocotrienols, organosulfur compounds (allium compounds and glucosesinolates), and nondigestible carbohydrates (dietary fiber and prebiotics) are a few categories of phytochemicals that have or seem to have substantial health potentials.

3.1 Different Phytochemicals as Functional Roles

3.1.1 Phenolic Compounds

The largest class of phytochemicals, known as phenolics, are secondary metabolites that are widely distributed and have a wide range of characteristics (Table 2). They act as antioxidants that

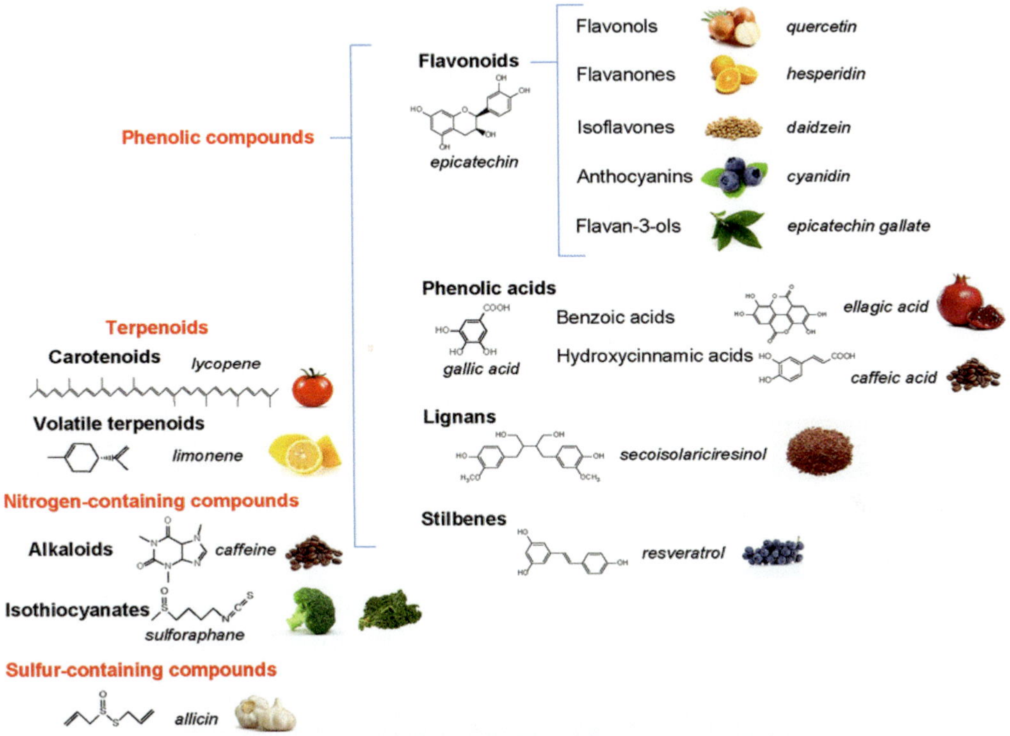

Fig. 1 Foods containing different phytochemicals (Fraga et al., 2019).

shield cells from damage caused by free radicals, control lipogenesis to control obesity, and also inflammatory hormones and adipokines implicated in the disease's processes (Martinez et al., 2022). Recent years have seen a huge increase in research and industry interest in phenolic compounds because of their potential to prevent and treat chronic illnesses in humans. Instead of depending solely on conventional food components, the growing trend of creating innovative functional food and nutraceutical products has rekindled the quest for alternate sources for extracting phenolic chemicals (Bandara and Chalamaiah, 2019).

Table 2 Categories of phenolic compounds with functional roles.

Categories	*Source*	*Functional Roles*	*Reference*
Phenolic acid	Blackberry, plum, raspberry, white vegetables like potatoes, onion, kiwi, carrots, wheat, eggplant, lettuce.	• Act as antioxidant and anti-inflammatory agents to reduce Chronic illnesses and oxidative damage. • Boost blood pressure, insulin resistance, and lipid profiles to enhance cardiovascular health. • Influence the composition of the gut microbiome and gut health	(Singh and Chhabra, 2022)
Flavonoids	Berry, blueberry, apple, onion, grape.	• Reduce the risk of cancer by neutralizing free radicals. • Prevent Alzheimer's and chronic illnesses.	(Panche et al., 2016)

Contd.

Table 2 *Contd.*

Categories	*Source*	*Functional Roles*	*Reference*
Catechins	Tea	• Reduce the incidence of diabetes, obesity, and heart disease.	(Baldi et al., 2020)
Tannins	Green coffee beans, fruits, and dates	• Bind to iron in the food and reduce its bioavailability.	(Gaffney et al., 2004)
Anthocyanins	Berries, grapes, cereal grains, roots, and tubers	• Prevent oxidative stress, cardiovascular and neurodegenerative diseases.	(Singh and Chhabra, 2022)
Coumarins	Tomatoes, grapefruits and cinnamons	• Act as antioxidants • Reduce the risk of heart disease • Reduce the development of cancer cells • Control hypoglycemia and diabetes • Suppress inflammation • Offer photoprotection	(Singh and Chhabra, 2022)
Caffeine	Coffee beans	• Mental alertness or wakefulness during fatigue or drowsiness	(Babbar, 2015)

3.1.2 Alkaloids

Alkaloids are heterocyclic compounds with alkaline nitrogen with strong pharmacological activity produced from higher plants. Alkaloids are far too common, have a complex chemical structure, and have a wide range of impacts and metabolic activities (Table 3).

Table 3 Classification of alkaloids with functional roles.

Class	*Sources*	*Functional Role*	*References*
Pyrrole and Pyrrolidine	*Erythroxylum coca, Erythroxylum truxillense*	• Pyrrole acts as an antipsychotic, β-adrenergic, antagonist, anxiolytic, anticancer antibacterial, antifungal, antiprotozoal, antimalarial agent • Act as building block in pharma ceuticals	(Moor, 1994; Evans, 1981)
Tropane	*Atropa belladonna, Dhatura stramonium, Erythroxylon coca, Schizanthus porrigens*	• Act as anticholinergics or stimulants. • Affect the central and peripheral nervous system.	(Munoz and Cortes, 1998)
Quinoline	*Cinchona officinalis, Cinchona calisaya, Almeidea rubra*	• Help in the digestion of haemoglobin in the blood stages of the malaria life cycle.	(Scherlach and Hertweck, 2006)
Indole	*Strychnos nuxvomica, Claviceps purpurea, Rauwolfia serpentine, Catharanthus roseus, Gelsemium elegans*	• Regulate several bacterial functions, particularly virulence, drug resistance, spore production, plasmid stability, and biofilm development.	(Dembitsky et al. 2005; Lin, 1996)
Steroidal	*Solanum tuberosum, Veratrum album, Whithania somniferum, Holarrhena antidysenterica*	• Lessen inflammation, or redness and swelling. • Inflammatory diseases like eczema and asthma may benefit from this.	(Dembitsky et al., 2005)
Amino alkaloid	*Ephedra gerardiana, Ephedra sinica, Colchicum autumnale*	• Show protection against inflammation, malarial activity, microorganisms, hypersensitivities, and diabetes.	(Dembitsky et al., 2005)

3.1.3 Flavonoids

Flavonoids are part of a family of secondary metabolites from plants with a polyphenolic structure. In most of the angiosperm groups, many flavonoids are easy to identify as floral pigments. Flavanones, anthocyanidins, flavones, flavanols, and isoflavones are subclasses of flavonoids (Figure 2). Each subgroup has distinct major sources. It has long been known that certain parts of plants synthesize flavonoids, which are responsible for the color and aroma of flowers. They also draw pollinators to fruits, which helps to spread the fruit. Flavonoids have several health benefits, and there is currently interest in using them for chemoprevention and disease treatment. Numerous disorders, including cancer, Alzheimer's disease (AD), atherosclerosis, and others, are linked to their numerous beneficial biochemical and antioxidant properties (Ovando et al., 2009; Lee et al., 2009). Many enzymes, including phosphoinositide 3-kinase, lipoxygenase, cyclo-oxygenase, and xanthine oxidase (XO), are also strongly inhibited by flavonoids (Walker et al., 2000). This characteristic has led to the widespread use of flavonoids in food safety, nutrition, and health. Free radical scavenging activity is one of the ways flavonoids can stop damage from free radicals. Radicals oxidize flavonoids, producing a less reactive, more stable radical in the process. The antiviral, antifungal, and antibacterial properties of flavonoids and phenolic acid esters have also been studied (Table 4). Numerous studies have shown that flavonoids can influence the metabolism of arachidonic acid by inhibiting the activities of lipo-oxygenase (LO) and cyclo-oxygenase (COX). It has been shown that flavonoids, particularly quercetin, have antidiabetic properties that promote insulin secretion and cause the regeneration of pancreatic islets (Vessal, 2003)

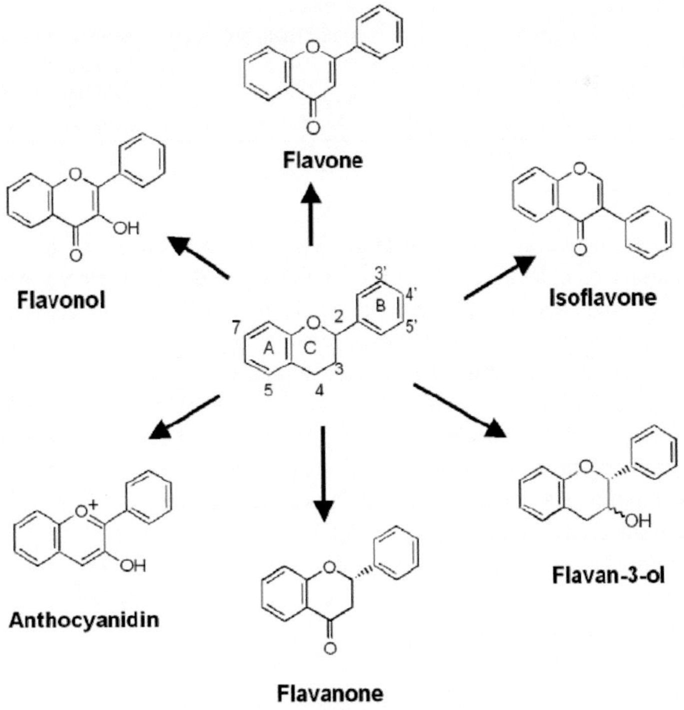

Fig. 2 Chemical structure of flavonoids (Nishiumi et al., 2011).

Flavonoids can be classified into several subgroups.

Table 4 Classification of flavonoids with Functional roles.

Class	*Subclass*	*Sources*	*Functional Role*
Flavonols	Myricetin Rutin Quercetin Morin Kaempferol	Onions, kale, lettuce, tomatoes, apples, grapes and berries, tea, red wine	• Have antioxidant potential and reduced risk of vascular disease.
Flavones	Apigenin Tangeretin Baicalein Ropifolin	Celery, parsley, red peppers, chamomile, mint	• Reduce inflammation and the risk of cardiovascular disease.
Flavanones	Fisetin Hesperetin Narigin Naringenin Taxifolin	Citrus fruit Citrus peel	• Exert interesting pharmacological effects as antioxidant, anti-inflammatory, blood lipid-lowering, and cholesterol-lowering agents.
Isoflavones	Genistin Genistein Daidzin Daidzein Glycitein	Soyabeans and leguminous plants	• Daidzein and genistein are frequently considered phyto-oestrogens due to their oestrogenic properties in specific animal models. • Genistein induces hormonal and metabolic changes in the body.
Chalcones	Phloridzin Arbutin Phloretin Chalconaringenin	Tomatoes, pears, strawberries, bearberries, and certain wheat products	• Provide various biological activities such as anti-obesity, antioxidative, anti-inflammatory, anti-atherosclerosis, antihyperglycemia, and antihypercholesterolemia.

3.1.4 Carotenoids

Carotenoid pigments, which are found in large quantities in a variety of fruits and vegetables, have been the subject of extensive research over the years due to their various applications in photobiology, photochemistry, and photomedicine. They give fruits and vegetables their yellow, orange, and red hues and are found in them as micro components (Figure 3). It is believed that carotenoids give fruits and vegetables their healthful qualities in preventing cancer, heart disease, and other chronic illnesses (Table 5). While many typical meals for humans include carotenoids, the main dietary sources are intensely pigmented fruits, juices, and vegetables, with yellow-orange fruits and vegetables supplying most of the α-carotene and β-carotene. α-carotene, β-carotene, lycopene, lutein, and cryptoxanthin make up about 90% of the carotenoids found in food and the human body. Tomatoes and tomato products, dark green vegetables that provide lutein, and orange fruits that provide cryptoxanthin, lycopene, where egg yolk contains lutein and zeaxanthin. The additional benefit of carotenoids like carotene and cryptoxanthin is their capacity to be converted to Vitamin A, which is linked to the development and prevention of disease. Carotenoids like lycopene and carotene have been shown to have antioxidant qualities in a few *in vitro*, animal, and human studies. In a dose-dependent way, carotenoid was able to inhibit the up-regulation of heme oxygenase-1 gene expression in human dermal fibroblasts (FEK4) exposed to UVA radiation (Elliott, 2005). Epidemiological research indicates a significant correlation between a lower risk of chronic diseases and increased food consumption and tissue concentrations of carotenoids (Johnson, 2002).

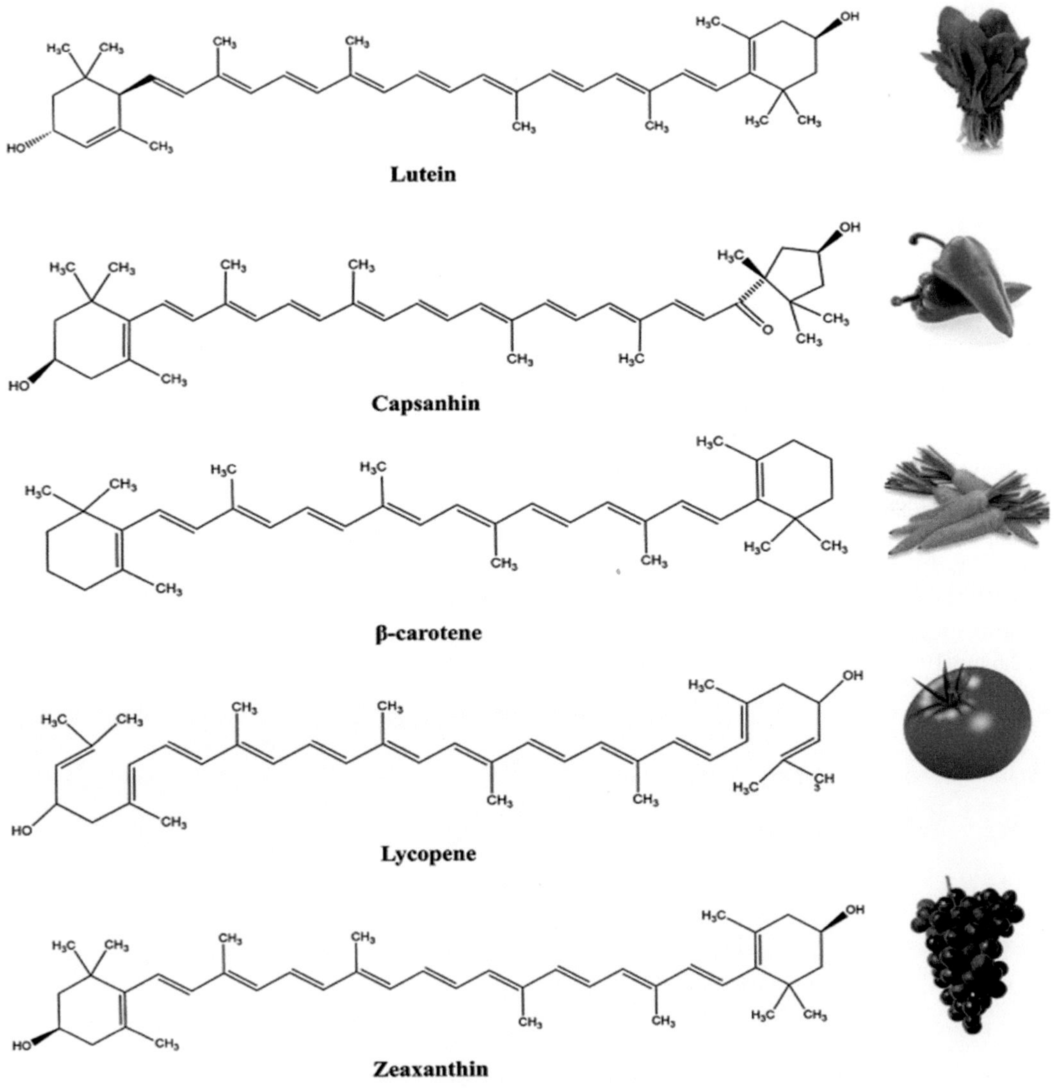

Fig. 3 Structure of carotenoids (Kolašinac et al., 2021).

Table 5 Categories and functional roles of carotenoids.

Categories	*Sources*	*Functional Roles*	*References*
β-carotene	Apricot, carrots, spinach, green broccoli, tomato	• Prevent night blindness and liver fibrosis • Act as an antioxidant	(Shaish et al., 2006; Dufossé et al., 2005)
Lycopene	Tomato, raw tomato paste, tomato juice, tomato ketchup	• Proposed for the treatment of cardiovascular diseases and prostate cancer	(Langi et al., 2018)
Lutein	Spinach, green collard beet, green broccoli, peas	• Protect from oxidative damage • Act as a filter of ultraviolet light • Lower incidence of age-related macular degeneration • Protect against the development of cancer	(Langi et al., 2018)

Contd.

Table 5 *Contd.*

Categories	*Sources*	*Functional Roles*	*References*
Astaxanthin	salmon, trout, krill, shrimp, and crayfish	• Improves the ability of protective white blood cells to surround and destroy infecting organisms • Help in diabetic retinopathy, macular degeneration, eye strain and fatigue, and seeing in fine detail.	(Langi et al., 2018)
Zeaxanthin	Egg yolk, broccoli, carrot, corn, leafy vegetables	• Protect from oxidative damage • Reduced prevalence of cataracts and age-related macular degeneration (AMD)	(Langi et al., 2018)

3.1.5 Phytosterols

Plant cell membranes naturally include a class of steroid alcohols called phytosterols and their esters, which help control the permeability and fluidity of cell membranes. The phytosterols found in fruits and vegetables are brassicasterol, campesterol, stigmasterol, and β-sitosterol (Figure 4). β-sitosterol, campesterol, and stigmasterol are the three most prevalent phytosterols. Recently, phytosterols in a few foods have been measured (Table 6). Vegetables have been shown to contain 1.1–53.7 mg/100g of edible part, with peas, cauliflower, and broccoli having the highest percentage, as well as romaine lettuce (Bacchetti et al., 2011). The phytosterol contents in fruits range between 1.6 and 32.6 mg/100g (Chen and Blumberg, 2008).

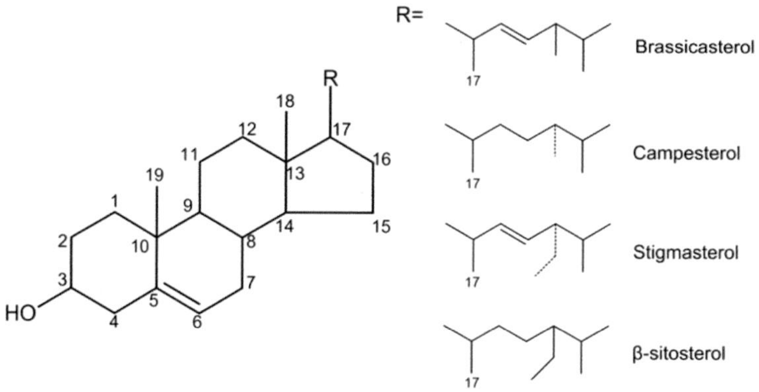

Fig. 4 Phytosterols found in fruits and vegetables (Li et al., 2022).

Table 6 Total phytosterol content in vegetables, fruits, and oils.

Vegetables	*Total Sterol Content (mg/100g)*	*Fruits*	*Total Sterol Content (mg/100g)*	*Nuts and Seeds*	*Total Sterol Content (mg/100g)*
Black olives	50	Passion fruit	44	Sesame	360
Brussels sprout	43	Orange	24	Sunflower	300
Cauliflower	40	Fig	22	Pistachios	276
Broccoli	39	Grapefruit	18	Almond	183
Green olives	35	Lemon	18	Cashew	158
Mushrooms	18	Pineapple	17	Walnut	127
Celery	17	Banana	16	Pecan	108
Carrot	16	Peach	15	Peanut	107
Onion	8	Apple	13		

3.1.6 Phytosterols as Functional Roles

The usage of foods containing phytosterols is a relatively recent trend in human nutrition. As useful components in food, phytosterols seem like a sensible and secure way to lower people's cholesterol levels. Despite having structural similarities, cholesterol and phytosterols differ in their metabolism and intestinal absorption. In the human gut, phytosterols and phytostanols are extremely poorly absorbed, and bile aggressively excretes what has been ingested. The intestinal absorption of plant sterols varies between 0.02–0.3% for their saturated counterparts and between 0.4–5% for phytosterols.

The impact of phytosterol on the absorption of fat-soluble vitamins has been the subject of conflicting reports. In addition to decreasing cholesterol absorption in males with normal cholesterol levels, compared to plant-free phytosterols (Jones et al., 2007). The capacity of phytosterols to obstruct the intestinal tract's absorption of dietary and biliary cholesterol has been linked to the hypocholesterolemic impact. Plant sterols have been shown *in vitro* to impact LDL cholesterol levels and inhibit the hyperproliferation of vascular smooth muscle cells, which contributes to the development of atherosclerosis (Awad et al., 2001).

4. Future of Functional Foods

As awareness of the link between diet and health increases, functional foods are gaining popularity. Indeed, it is anticipated that the global market for functional foods would grow to $275 billion by 2025. Researchers are creating novel functional foods that are intended to address health issues. Worldwide, government, academic, and private research institutions are actively investigating functional foods to improve health and prevent chronic diseases. Significant innovation has also been sparked by market expansion, which is noteworthy given that the food business is typically seen as having a low level of research intensity when compared to other sectors. Consumer desire for convenience and health products, along with physiological and technological breakthroughs in food processing, have greatly influenced innovation. In the US, the Food and Drug Administration controls the claims that producers of food and supplement products can make, including the nutritional content of products that fall under the category of functional foods (Ahmad et al., 2021). Nutrigenomics, a new field that examines the relationship between diet and illness development depending on a person's genetic profile, will significantly impact the future of functional foods research and development initiatives. Biotechnology will also greatly influence the future of functional foods.

5. Conclusion

Phytochemicals are the critical enhancers for the health properties of functional foods. These naturally occurring bioactive compounds found in fruits, vegetables and grains, and herbs have antioxidant, anti-inflammatory, and antibacterial activities and are useful in disease prevention and overall health promotion. Phytochemicals in functional foods enhance nutritional value and provide specific health benefits, such as reducing the incidence of diabetes, cardiovascular diseases, and cancer. The addition of phytochemicals in food systems seems to be another hopeful way for creating next-generation functional foods as consumer awareness of nutrition and health continues to rise. However, further study is necessary to optimize their bioavailability, stability, and effectiveness in various dietary matrices.

References

Ahmad, M.H.S. Safdar, S. Kousar, M. Nadeem and Z. Asghar. 2021. Functional foods and human health: An overview. *Functional Foods: Phytochemicals and Health Promoting Potential, 3*. http://dx.doi.org/10.5772/intechopen.99000.

Awad, A.B., A.J. Smith and C.S. Fink. 2001. Plant sterols regulate rat vascular smooth muscle cell growth and prostacyclin release in culture. *Prostaglandins Leukot Essent Fatty Acids, 64*, 323–330.

Babbar, N. 2015. An introduction to alkaloids and their applications in pharmaceutical chemistry. *The Pharma Innovation Journal*, *4*(10), 74–75.

Bacchetti, T., S. Masciangelo, V. Bicchiega, E. Bertoli and G. Ferretti. 2011. Phytosterols, phytostanols, and their esters: From natural to functional foods. *Mediterranean Journal of Nutrition and Metabolism*, *4*, 165–172. https://doi.org/10.1007/s12349-010-0049-0.

Baldi, A., H. Abramovič, N. Poklar Ulrih and M. Daglia. 2020. Tea Catechins. In: Xiao J., Sarker S., Asakawa Y. (Eds.), *Handbook of Dietary Phytochemicals*. Springer, Singapore. https://doi.org/10.1007/978-981-13-1745-3_19-2.

Bandara, N. and M. Chalamaiah. 2019. Bioactives from agricultural processing by-products. In: *Reference Module in Food Science*. doi: 10.1016/B978-0-08-100596-5.22408-6.

Blades, M. 2000. Functional foods or nutraceuticals. *Nutrition & Food Science*, *30*(2), 73–76. https://doi.org/10.1108/00346650010314313/FULL/PDF.

Chen, C.Y. and J.B. Blumberg. 2008. Phytochemical composition of nuts. *Asia Pacific Journal of Clinical Nutrition*, *17*, 329–332.

COMA. 1998. *Nutritional Aspects of the Development of Cancer*. Report of the Working Group on Diet and Cancer of the Committee on Medical Aspects of Food and Nutrition Committee (No. 48) HMSO, London, UK.

Crowe, K.M. and C. Francis. 2013. Position of the academy of nutrition and dietetics: Functional foods. *Journal of the Academy of Nutrition and Dietetics*, *113*(8), 1096–1103. https://doi.org/10.1016/J.JAND.2013.06.002.

Dembitsky, V.M., T. Rezanka, J. Spizek and L.O. Hanus. 2005. *Phytochemistry*, 66, 747–769. https://doi.org/10.1016/j.phytochem.2005.02.017.

Dufossé, L., P. Galaup, A. Yaron, S.M. Arad, P. Blanc, K.N.C. Murthy and G.A. Ravishankar. 2005. Microorganisms and microalgae as sources of pigments for food use: A scientific oddity or an industrial reality? *Trends in Food Science & Technology*, *16*, 389–406. https://doi.org/10.1016/j.tifs.2005.02.006.

Elliott, R. 2005. Mechanisms of genomic and non-genomic actions of carotenoids. *Biochim Biophys Acta*, *1740*, 147–154. https://doi.org/10.1016/j.bbadis.2004.12.009.

Evans, W.C. and J. Ethno. 1981. Pharmacology 3: 265. FFMA. 2024. Functional Food Market Analysis APAC, North America, Europe, South America, Middle East and Africa - US, China, Japan, Germany. 165.

Fraga, C.G., K.D. Croft, D.O. Kennedy and F.A. Tomás-Barberán. 2019. The effects of polyphenols and other bioactives on human health. *Food & Function*, *10*(2), 514–528. https://doi.org/10.1039/C8FO01997E.

Functional Food Market Analysis, Size and Forecast 2024–2028: North America (US, Canada, and Mexico), Europe (France, Germany, and UK), Middle East and Africa (UAE), APAC (China, India, Indonesia, Japan, Malaysia, South Korea, and Thailand), South America (Brazil), and Rest of World (ROW). 2024, 165. https://www.technavio.com/report/functional-food-market-industry-analysis.

Gaffney, S., Williams, V., Flynn, P., Carlino, R., Mowry, C., Dierenfeld, E., Babb, C., Fan, J. and Tramontano, W.A. 2004. Tannin/polyphenol effects on iron solubilization in vitro. *BIOS, 75*(2), 43–52. https://doi.org/10.1893/0005-3155(2004)75

Ghevariya, H. and R. Singh. 2023. *Functional Foods and Phytochemicals. Current Trends and Research in Medical Nutrition Therapy to Control Diseases* (1st Edn.). Kripa-Drishti Publications, Pune, 120–136.

Gupta, E. and P. Mishra. 2021. Functional Food with Some Health Benefits, So Called Superfood: A Review. *Current Nutrition & Food Science*, *17*(2), 144–166. https://doi.org/10.2174/22123881MTA4nMjQ8w.

Havel, R.J., J.W. Erdman, C. Garza, et al., 1994. Opportunities in the nutrition and food sciences: Research challenges and the next generation of investigators. *Journal of Nutrition*, *124*(6), 763–769. https://doi.org/10.1093/JN/124.6.763.

Johnson, E.J. 2002. The role of carotenoids in human health. *Nutr. Clin. Care*, *5*(2), 47–49.

Jones, P.J., I. Demonty, Y.M. Chan, Y. Herzog and D. Pelled. 2007. Fish-oil esters of plant sterols differ from vegetable-oil sterol esters in triglycerides lowering, carotenoid bioavailability and impact on plasminogen activator inhibitor-1 (PAI-1) concentrations in hypercholesterolemic subjects. *Lipids in Health and Disease*, *6*(1), 28. https://doi.org/10.1186/1476-511X-6-28.

Kalt, W. 2001. Health functional phytochemicals of fruit. *Horticultural Reviews*, *27*, 269–315. https://doi.org/10.1002/9780470650813.ch7.

Kolašinac, S.M., Z. Dajić-Stevanović, S.N. Kilibarda and A.Ž. Kostić. 2021. Carotenoids: New applications of "old" pigments. *Phyton*, *90*(4), 1041–1062. https://doi.org/10.32604/phyton.2021.015996.

Kruger, C.L. and S.W. Mann. 2003. Safety evaluation of functional ingredients. *Food and Chemical Toxicology*, *41*(6): 793–805. https://doi.org/10.1016/S0278 6915(03)00018-8.

Langi, P., S. Kiokias, T. Varzakas and C. Proestos. 2018. Carotenoids: From plants to food and feed industries. *Microbial Carotenoids: Methods and Protocols*, 57–71. https://doi.org/10.1007/978-1-4939-8742-9_3.

Lee, Y.K., D.Y. Yuk, J.W. Lee, S.Y. Lee, T.Y. Ha, K.W. Oh, ... and J.T. Hong. 2009. Epigallocatechin-3-gallate prevents lipopolysaccharide-induced elevation of beta-amyloid generation and memory deficiency. *Brain Research*, *1250*, 164–174. https://doi.org/10.1016/j.brainres.2008.10.012.

Li, X., Y. Guo, Y. Zhuang and Y. Qin. 2018. Nonvolatile taste components, nutritional values, bioactive compounds and antioxidant activities of three wild Chanterelle mushrooms. *International Journal of Food Science and Technology*, *53*, 1855–1864. https://doi.org/10.1111/ijfs.13769.

Li, X., Y. Xin, Y. Mo, P. Marozik, T. He and H. Guo. 2022. The bioavailability and biological activities of phytosterols as modulators of cholesterol metabolism. Molecules, 27(2), 523. https://doi.org/10.3390/molecules27020523.

Lin, L.Z., S.F. Hu and G.A. Cordell. 1996. *Phytochemistry*, *43* (3), 723–726. https://doi.org/10.1016/0031-9422(96)00280-4.

Liu, R.H. 2002. Supplement quick fix fails to deliver. *Food Technology International*, 71–72.

MacAulay, J., B. Petersen and F. Shank. 2005. *Functional Foods: Opportunities and Challenges*. Institute of Food Technologists (IFT) Expert Report.

Martinez, R.M., C. de O.R.P., de Almeida, T.P.B. Lima, M.S. Figueiredo and A.J. Teodoro. 2022. Obesity mechanisms and importance of bioactive compounds from fruits in its regulation: A narrative review. *Research, Society, and Development*, *11*(4), e11411427153. https://doi.org/10.33448/rsd-v11i4.27153.

Martirosyan, D.M. and J. Singh. 2015. A new definition of functional food by FFC: What makes a new definition unique? *Functional Foods in Health and Disease*, *5*(6): 209–223. https://doi.org/10.31989/ffhd.v5i6.18.

Martirosyan, D.M. and B. Singharaj. 2016. *Health claims and functional food: The future of functional foods under FDA and EFSA Regulation*. Research Gate.

Martirosyan, D. and K. Pisarski. 2017. Bioactive compounds: Their role in functional food and human health, classifications, and definitions. In: Danik Martirosyan and Jin-Rong Zhou (Eds.), *Bioactive Compounds and Cancer* (pp. 238–277). San Diego: Food Science Publisher.

Martirosyan, D.M. and E. Miller. 2018. Bioactive compounds: The key to functional foods. *Bioactive Compounds in Health and Disease*, *1*(3), 36–39. https://doi.org/10.31989/BCHD.V1I3.539.

Moor, J.M. 1994. *Phytochemistry*, *36*, 357. https://doi.org/10.1016/S0031-9422(00)97075-4.

Munoz, O. and S. Cortes. 1998. *Pharm. Biol.*, *36*, 162. https://doi.org/10.1076/phbi.36.3.162.6341.

Nishiumi, S., S. Miyamoto, K. Kawabata, K. Ohnishi, R. Mukai, A. Murakami and J. Terao 2011. Dietary flavonoids as cancer-preventive and therapeutic biofactors. *Front Biosci.*, *3*(3), 1332–1362. https://doi.org/10.2741/229.

Ovando, A., M.L. Pacheco-Hernández, J.A. Rodríguez and C.A. Galán-Vidal. 2009. Chemical studies of anthocyanins: A review. *Food Chemistry*, *113*, 859–871. https://doi.org/10.1016/j.foodchem.2008.09.001.

Panche, A.N., A.D. Diwan and S.R. Chandra. 2016. Flavonoids: An overview. *Journal of Nutritional Science, 5*, e47. https://doi.org/10.1017/jns.2016.41.

Scherlach, K. and C. Hertweck. 2006. *Org. Biomolecule Chemistry*, *4*, 3517–3520. https://doi.org/10.1039/B607011F.

Semba, R.D. 2012. The discovery of the vitamins. *International Journal of Vitamin and Nutrition Research*, *82*(5), 310–315. https://doi.org/10.1024/0300-9831/a000124.

Shaish, A., A. Harari, L. Hananshvili, H. Cohen, R. Bitzur, T. Luvish, ... and D. Harats. 2006. 9-cis β-carotene-rich powder of the alga Dunaliella bardawil increases plasma HDL-cholesterol in fibrate-treated patients. *Atherosclerosis*, *189*, 215–221. https://doi.org/10.1016/j.atherosclerosis.2005.12.004.

Shimizu, T. 2003. Health claims on functional foods: The Japanese regulations and an international comparison. *Nutrition Research Reviews*, *16*(2), 241–252. https://doi.org/10.1079/NRR200363.

Singh, P. and P. Chhabra. 2022. *Current Trends and Research in Medical Nutrition Therapy to Control Diseases*. Kripa Drishti Publications.

Siró, I., E. Kápolna, B. Kápolna and A. Lugasi. 2008. Functional food. Product development, marketing, and consumer acceptance-A review. *Appetite*, *51*(3), 456–467. Academic Press. https://doi.org/10.1016/j.appet.2008.05.060.

Soria-Florido, M.T., O. Castañer, C. Lassale, R. Estruch, J. Salas-Salvadó, M.A. Martínez-M., A. González, ... and M. Fitó. 2020. Dysfunctional high-density lipoproteins are associated with a greater incidence of acute coronary syndrome in a population at high cardiovascular risk: A nested case–control study. *Circulation*, *141*(6), 444–453. https://doi.org/10.1161/CIRCULATIONAHA.119.041658.

Technavio. 2024. Functional food market growth and trends in APAC countries. In Functional food market growth analysis and forecast 2024–2028.

Vessal, M., M. Hemmati and M. Vasei. 2003. Antidiabetic effects of quercetin in streptozocin induced diabetic rats. Comp. Biochem. Physiol., 135, 357–364. https://doi.org/10.1023/A:1022201608413.

Walker, E.H., Pacold, M.E., Perisic, O., Stephens, L., Hawkins, P.T., Wymann, M.P., and Williams, R.L. 2000. Structural determinants of phosphoinositide 3-kinase inhibition by wortmannin, LY294002, quercetin, myricetin, and staurosporine. *Molecular Cell, 6*(4), 909–919. https://doi.org/10.1016/S1097-2765(05)00089-4

Wan, M.L.Y., P.C. Turner, V.A. Co, M.F. Wang, K.M.A. Amiri and H. El-Nezami. 2019. Schisandrin A protects intestinal epithelial cells from deoxynivalenol-induced cytotoxicity, oxidative damage, and inflammation. *Scientific Reports*, *9*, 19173. https://doi.org/10.1038/s41598-019-55821-4.

Yuan, Y.U. 2008. Marine algal constituents. In: Barrow, C. and Shahidi, F. (Eds.), *Marine Nutraceuticals and Functional Foods* (pp. 259–296). Taylor and Francis. https://doi.org/10.1201/9781420015812.ch11.

CHAPTER 5

Probiotics and Prebiotics
Gut Health and Beyond

Ashish Kumar Singh[1], Nidhi Gossai[2], Surajbhan Sevda[3] and Meena Krishania[4*]

1. Introduction: Microbial Modulators for Human Health

Throughout evolution, humans and animals have interacted closely with an intricate and varied microbial community, including microfauna and microflora (Dubos et al., 1965). This prolonged association implies that the host is adapted to these microorganisms and depends on these microbes for essential physiological functions (Willing and Finlay, 2009). The term "indigenous microbiota" includes opportunistic pathogenic and symbiotic microbial species (Dubos et al., 1965). Also, "normal microbiota" refers explicitly to those microorganisms consistently found in approximately all individuals within a given population and play a crucial role in host physiology (Savage, 1977). The indigenous microbiota includes three components: first, the microbiota coevolved with the host species, known as autochthonous microbiota. Second, microbiota, which are consistently present in the host, are known as normal microbiota; third, pathogens are acquired from external sources but persist and sometimes lead to diseases (Dubos et al., 1965). The GI tract (gastrointestinal) of animals, as well as humans, harbours a dynamic and dense microbial ecosystem, and in this ecosystem, the microbial cells do not undergo cellular ageing, unlike human cells. However, this microbial composition may lead to the onset of concurrent ageing-related disorders (O'Toole and Jeffery, 2015). In the early 19th century, Pasteur, Koch, Escherich, and Metchnikoff pioneered this host-microbe relationship foundational concept (Falk et al., 1998). Metchnikoff emphasized the importance of microbial composition for host well-being and proposed pioneering ideas connecting systemic health with intestinal flora (Cavaillon and Legout, 2016). At the same time, Pasteur highlighted the need for microbes for normal physiological function. Escherich, specifically, emphasized the importance of gut microbiota for both digestive physiology and pathology of intestinal diseases (Falk et al., 1998). The gut microbiota comprises a mutual consortium of bacteria, viruses, and fungi. This gut microbiota colonization commences prenatally and prevails throughout life, shaping a mutualistic relationship with the host (Hillman et al., 2017; Ragonnaud and Biragyn, 2021). This heterogeneous community plays a vital role in modulating the immune system, maintaining systemic homeostasis, nutrient metabolism, gut integrity, and pathogen defence (Falk et al., 1998; Martín et al., 2014; Shao et al., 2023). The collective genome of gut microbes is a microbiome that shows a functional capacity above that of the human genome (Falk et al., 1998). The host's combined genomic and

[1,4] National Agri-Food and Biomanufacturing Institute (BRIC-NABI), Formerly CIAB, Knowledge City, Mohali 140306, India.
[2] Department of Biotechnology, Punjab University, Chandigarh 160014, India.
[3] National Institute of Technology Warrangle, Hanamkonda, Telangana 50604, India.
* Corresponding author: meena@ciab.res.in

DOI: 10.1201/9781003659808-5

metabolic potential and associated microbiota are termed the hologenome, which governs several aspects of metabolic and physiological characteristics (Gagliardi et al., 2018; Kuziel and Rakoff-Nahoum, 2022). The altered condition of gut microbiota, known as dysbiosis, is linked to several diseases, such as obesity, metabolic syndrome, inflammatory bowel disease, celiac disease, and other systemic disorders (Delgado Jiménez and Benakis, 2021; Ragonnaud and Biragyn, 2021; Ren et al., 2023). This dysbiosis is also responsible for age-related diseases like Type 2 diabetes, Alzheimer's disease, neurodegenerative disorders, and cardiovascular disease (Ragonnaud and Biragyn, 2021). The current research focuses on developing therapeutic approaches to maintain or restore microbial stability within the gut ecosystem, which is known as eubiosis (Delgado Jiménez and Benakis, 2021). This approach used probiotics, synbiotics, and prebiotics, along with targeted approaches such as faecal microbial transplantation and bacterial consortium transplantation (Gagliardi et al., 2018). The human gut microflora is primarily dominated by specifically four phyla: Proteobacteria, Firmicutes, Actinobacteria, and Bacteroidetes, underlining its crucial role as a commensal "superorganism" which comprises about 98% of the total gut population (Hillman et al., 2017; Valdes et al., 2018; Badal et al., 2020; Ragonnaud and Biragyn, 2021). This gut microflora coevolved with the host to perform various functions such as the production of short-chain fatty acids (SCFAs), sustaining epithelial barrier integrity, modulation of the immune response, and synthesizing vitamins (Martín et al., 2014; Xue et al., 2020; Ragonnaud and Biragyn, 2021; Ren et al., 2023). "This chapter explores the composition of gut microbiota and maintenance, with a particular focus on prebiotics and probiotics and their potential roles in gut health and beyond."

2. Gut Microbiome: A Hidden Organ

The GI tract of mammals and human beings contains a massive and intricate community of microorganisms, the gut microbiome, which is vital for host health (Aldars-García et al., 2021). Some studies reported that microbial cells in the gut are equal to or even more than the number of human cells in the body (O'Hara and Shanahan, 2006; Adams and Gutiérrez, 2019; Aldars-García et al., 2021). The gut microbiome is responsible for about 3.3 million non-self genes, which encode 150 times more genetic information than the human genome. So, the GI microbiota is acknowledged as a "Hidden Organ" of mammals (Varela-Trinidad et al., 2022). The diverse gut microbiota contains bacteria, viruses, fungi, and archaea, the most dominant of which are bacteria (O'Hara and Shanahan, 2006; Chattopadhyay and Mythili, 2018; Pushpanathan et al., 2019). The conserved set of gut colonizers, known as core gut microbiota and its associated genes, is present in each individual, but each person has a particular set of gut microbiota (Kinross et al., 2011; Pushpanathan et al., 2019). Despite significant individual variability, about 60% of microbial phylotypes remain stable in all individuals until two years, highlighting a constant host-microbe relationship (Kinross et al., 2011). The human gut is sterile during birth but quickly colonized by microbes instantaneously after delivery, and this colonization is influenced by diet, hygiene, medication, and delivery mode (Grönlund et al., 1999). The key function of the gut microbiota is the digestion of indigestible dietary polysaccharides, which are fermented by anaerobic bacteria in the large intestine into SCFAs that act as an energy source for the cell lining of the colon, i.e., colonocytes and enhance colonic health (Martin, 2020). The gut microbiome is critical in facilitating different host physiology, such as vitamin synthesis, epithelial barrier function, immune modulation, detoxification, digestion, and protection against various pathogens (Figure 1) (O'Hara and Shanahan, 2006; Kinross et al., 2011; Chattopadhyay and Mythili, 2018) (Figure 1). It was also reported that the gut microbiome influences the enteric nervous system (ENS) and central nervous system (CNS) and plays a crucial role in maintaining neuromodulation and homeostasis. These gut microbiota produce metabolites that control host physiology and are involved in complex signalling with the host cells that significantly affect disease and health outcomes (Sorboni et al., 2022). All these physiological processes are governed by gut microbiota genes, which often act as organs but are lacking in the human genome. Therefore, the gut microbiome is called the "neglected organ"

(O'Hara and Shanahan, 2006). The early microbiome research started in the 1680s, when Antonie van Leeuwenhoek first studied the microorganisms in faecal and oral samples, pioneering the study of microbiota studies (Leeuwenhoek A Van, 1683). Along with time, advanced sequencing techniques like 16S rRNA analysis used by researchers to identify five major bacterial phyla in the gut, viz., Proteobacteria, Actinobacteria, Verrucomicrobia, Bacteroidetes, and Firmicutes, as well as Archaea in which Firmicutes and Bacteroidetes are two dominant phyla in the gut that account for 98% of the total sequence of the human gut (Figure 1) (Kinross et al., 2011; Chattopadhyay and Mythili, 2018; Pushpanathan et al., 2019). A few minor phyla, such as Fusobacteria, Spirochaeatae, and Cyanobacteria, are also in the gut (Pushpanathan et al., 2019).

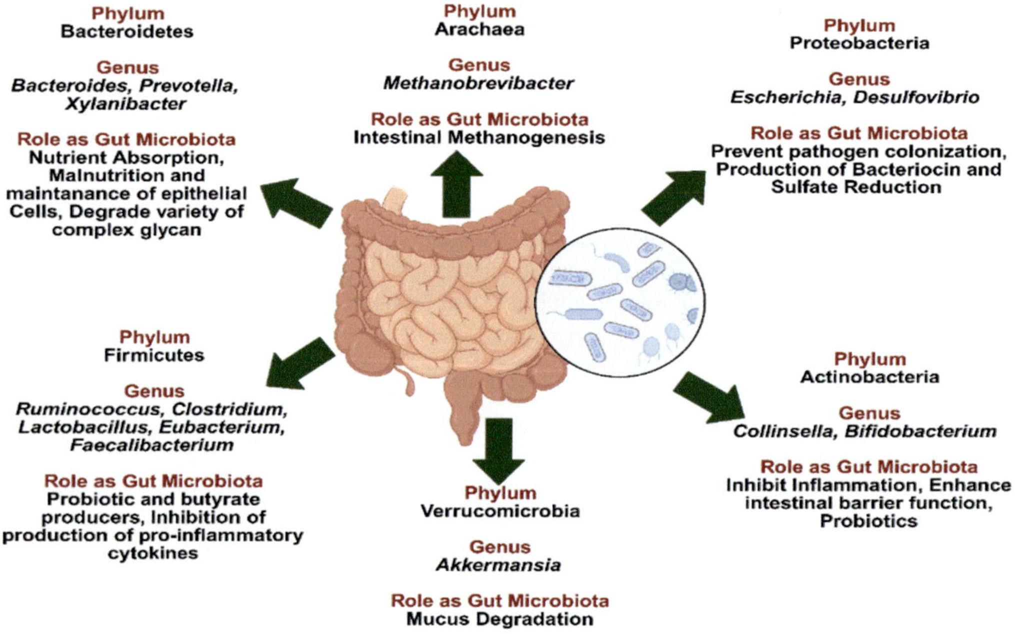

Fig. 1 Predominant bacterial genera and their key functions.

3. Evolution of Probiotics

Probiotics have historical roots originating from ancient civilizations that developed fermented foods for preservation due to their apparent health benefits. The application of live microorganisms for health benefits has a historical origin because fermented food has been widely used for preservative and medicinal purposes. Several dairy-based fermented foods, such as kumis, mastoni, kefir and yogurt, were common in different civilizations and cultures and were often used as a geroprotective (Gogineni et al., 2013). The timeline reflects the development of probiotics from their ancient origin in fermented food to their scientific endorsement in modern medicine. The study started with beer brewing in Mesopotamia and fermented milk in India, highlighting the traditional awareness of fermentation. "In the early 20th century, Nobel Prize laureate Elie Metchnikoff gave the scientific concept of probiotics to the live microorganisms that confer health benefits when consumed adequately by the host (Metchnikoff, 1908)." He proposed that the toxic compounds, such as amines and ammonia, produced by proteolytic bacteria in the large intestine lead to autotoxication, damaging both nervous and vascular systems. So, he hypothesized that gut microbiota alteration by beneficial bacteria delayed ageing and enhanced health effects (Metchnikoff, 1908). Metchnikoff also observed that fermented milk products that contain lactic acid bacteria produce lactic acid after lactose fermentation, creating an acidic environment and inhibiting the growth of proteolytic microorganisms. Consequently, he isolated the strains of *Lactobacillus* from Bulgarian fermented milk and named it *Lactobacillus bulgaricus* (Metchnikoff, 1908; Calatayud and Suárez, 2017). The first bacterial-based therapeutics, Lactobacillin, were produced in Saint Petersburg in 1912 (Suvorov, 2013). This is one of the earliest scientific studies to identify and exploit bacteria for health purposes (Tannock, 2003). However, the interest in consuming products that contain lactic acid bacteria vanished at the beginning of World War I. Still, in the 1930s, it re-emerged in the United States of America with the production of "Acidophilus milk", using *Lactobacillus acidophilus* isolated from the human gut. This acidophilus milk was supposed to help improve conditions such as diarrhoea, eczema, sprue, constipation, and colitis (R. B. Parker, 1974; Tannock, 2003). The Greek physician Hippocrates proposed the idea of food that could serve as medicine and advocated dietary interventions in health. Still, in modern times, this philosophy has evolved into the concept of "functional foods", food or ingredients offering health benefits beyond basic nutrition (Chow, 2002). Historically, the term "Probiotics" derived from the Greek words 'Pro' and 'bios' meaning "for life" (Guarner et al., 2005; Gul and Durante-Mangoni, 2024). In 1907, Metchnikoff had set the theoretical groundwork; the term "probiotics", introduced in 1965 by Lilly and Stillwell, defined it as a substance derived by microorganisms that promotes the growth of another microorganism (Lilly and Stillwell, 1965). Afterwards, Parker revised it in 1974, stating that probiotics are organisms and substances that help maintain intestinal homeostasis (R.B. Parker, 1974). After that, Fuller, in 1989, redefined probiotics as "the supplement of live microbial cells which enhanced the bowel activity of the host animal (Fuller, 1994)". Again, in 1998, Guarner highlighted that consuming many live microorganisms will enhance gut health (Guarner et al., 2005). Finally, in 20021, the official definition of probiotics was proposed by the World Health Organization (WHO) and Food and Agriculture Organization (FAO), which was widely accepted. According to FAO/WHO, "Probiotics are live microorganisms which, when administered in adequate amounts, confer a health benefit on the host" (Das et al., 2022; Singh et al., 2023). In 2013, the International Scientific Association for Probiotics and Prebiotics (ISAPP) redefined the definition of FAO/WHO. It highlighted three main characteristics of probiotic microorganisms: first, they must be microbial; second, viable; and third, they confer health benefits (Das et al., 2022). Currently, probiotics are extensively used in pharmaceuticals and functional foods, and Bifidobacteria and Lactic acid bacteria dominate. Several genera, such as *Lactococcus, Enterococcus, Streptococcus, Lactobacillus, Staphylococcus, Leuconostoc,* and *Pediococcus,* belong to Lactic acid bacteria, which initially colonized plant-based environments before inhabiting dairy-rich niches, demonstrating their evolutionary inclination towards lactose-containing habitats (Suvorov, 2013). Subsequently, in the 20th century, Eli Metchnikoff proposed the health benefits of lactic acid bacteria, a turning point in probiotic science. Further regulatory milestones include officially defining probiotics by FAO/WHO and integrating probiotics in food and nutraceutical supplements until 2021 (Figure 2) (International Probiotic Association, 2021; Plamada and Vodnar, 2021).

4. Probiotic Microorganisms

Several studies have highlighted various bacteria, including Lactic acid bacteria (LAB) such as *L. casei, L. gasseri, L. plantarum, L. brevis, L. acidophilus, L. helveticus, L. rhamnosus, L. reuteri, L. gallinarum, L. fermentum L. murinus, L. johnsonii,* etc; *Bifidobacterium* species like *B. longum, B. lactis, B. bifidum, B. infantis, B. breve, B. adolascentis, B. animalis* etc; Yeast *viz., S. carlsbergensis, S. lactis, S. boulardii,* etc. and all these have probiotic potential (Figure 3) (Michail, 2008; Calatayud and Suárez, 2017; Cremon et al., 2018; Sanders et al., 2018; Das et al., 2022). The LAB strains, such as *Enterococcus, Lactococcus, Leuconostoc, Lactobacillus,* and *Streptococcus, are* non-spore-forming, facultative anaerobes, non-motile and obtain their energy through sugar fermentation (Plaza-Diaz et al., 2019; Santacroce et al., 2019). These microbial strains are naturally present in the human gut as well as in different fermented foods such as curd, pickles, sauerkraut, and cheese.

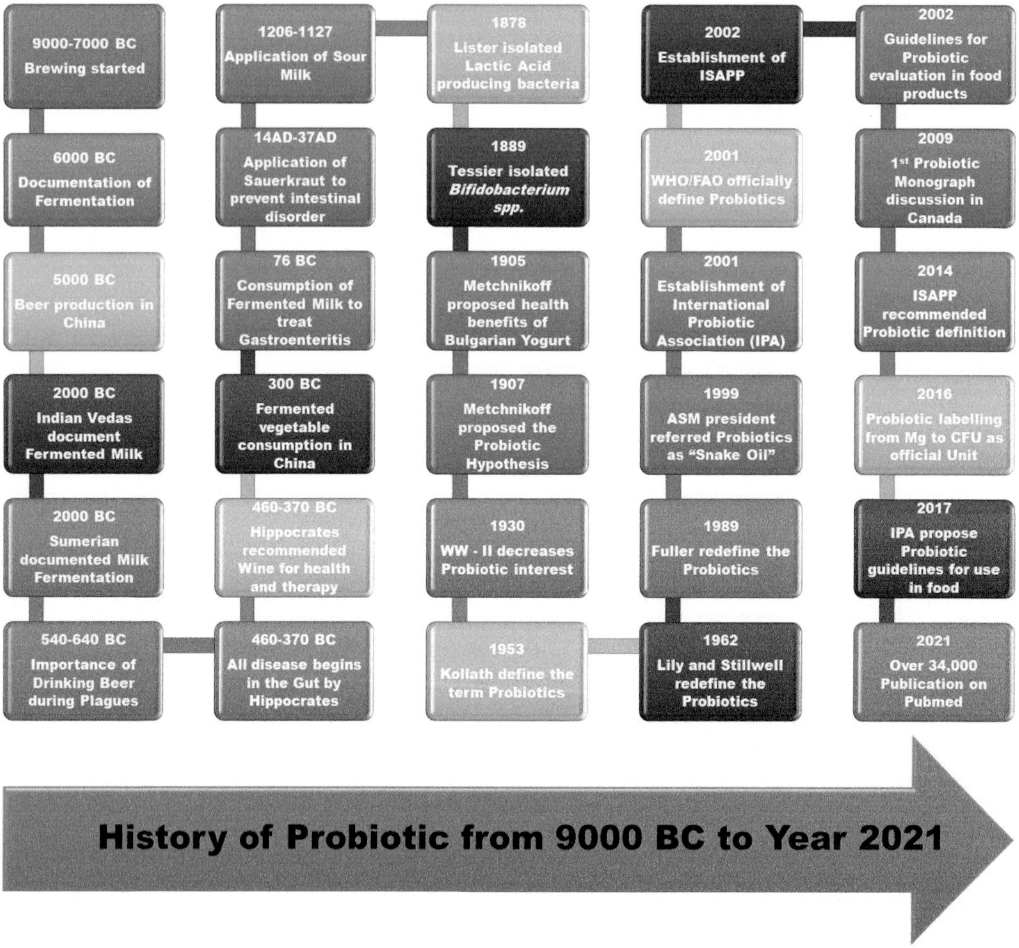

Fig. 2 Probiotic Evolution presents a historical timeline illustrating the progression from ancient fermentation practices to modern scientific advances. The ascending flowchart highlights key milestones, beginning with the early consumption of fermented foods (9000–7000 BC), followed by Metchnikoff's scientific contributions, and culminating in the regulatory recognition of probiotics by organizations such as WHO and FAO in 2021 (International Probiotic Association, 2021).

The probiotic strains, mainly *Bifidobacterium, Lactococcus,* and *Enterococcus,* were predominantly delivered as supplements or food ingredients (Gul and Durante-Mangoni, 2024). Although the probiotics are generally considered safe for consumption, safety concerns remain for a few strains, such as *Enterococcus faecalis* and *Enterococcus faecium,* that act as opportunistic pathogens and bear transferable antibiotic resistance genes (Chow, 2002; Anadón et al., 2010; Plaza-Diaz et al., 2019; Santacroce et al., 2019). Different criteria, such as resistance to digestive enzymes, human origin, safety, acid, and bile tolerance, are used for probiotic selection (Anadón et al., 2010).

5. Next-generation Probiotics

The development in culturing, genome and metagenome sequencing, and genome editing tools has uncovered a new phase of probiotic development, allowing the development of host-specific probiotics. These microorganisms are generally referred to as next-generation probiotics (NGPs) or live biotherapeutic products (LBPs) under the US FDA regulatory framework. These microorganisms are traditionally not used for health purposes but have emerging therapeutic potential. Although the

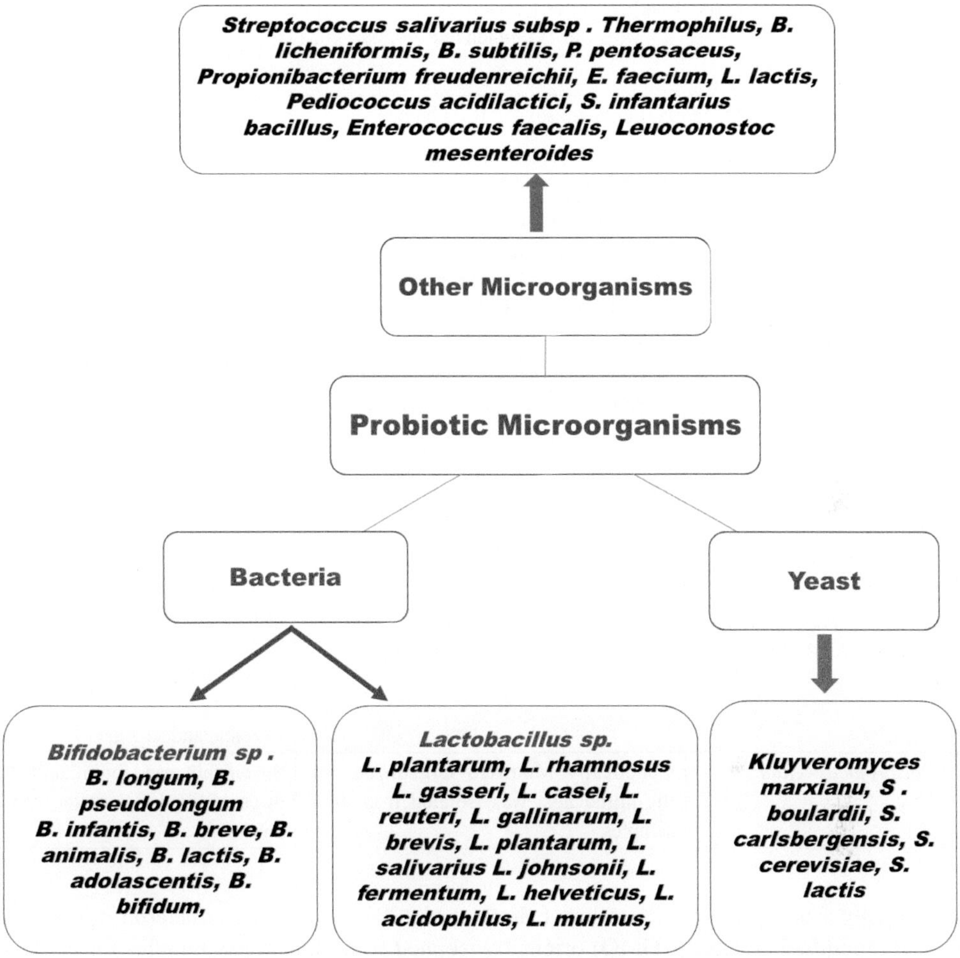

Fig. 3 Probiotic microorganisms include various species of lactic acid bacteria, Bifidobacterium, yeast, and other bacterial species (Das et al., 2022).

NGPs follow the approved WHO/FAO Probiotic definition, they differ in regulatory and clinical application. In contrast, the FDA defines LBPs as biological products and live organisms used to prevent or treat human diseases but not as vaccines (Sun et al., 2016). The advanced sequencing technique, such as high-throughput sequencing of GI microbiota, identified a broad range of microorganisms, like *Bacteroidales* and *Clostridiales*, specifically from the colon, which act as potential candidates for NGP development (Table 1) (Plamada and Vodnar, 2021). For example, *Bacteroides fragilis* strain Zy-132 have immunomodulatory properties, but it produces fragilysin that raises safety concerns because it causes colorectal cancer (Moncrief et al., 1998). Meanwhile, the *Bacteroides xylanisolvens* strain DSM 23964 is a non-toxigenic strain and an ideal candidate for NGP (Ulsemer et al., 2012). Moreover, *Bacteroides dorei* D8 can transform cholesterol to caprostanol, suggesting its potential application in managing cardiovascular disease and hypercholesterolemia (Plamada and Vodnar, 2021). Comparatively, the *Clostridium* genus is largely unexplored, but *Clostridium butyricum* strain MIYAIRI 588 are primarily used in Asia for hyperlipidemia, cancer, and infection with *Helicobacter pylori* and *Clostridium difficile* (Woo et al., 2011). Another example of a colon bacterium is *Faecalibacterium prausnitzii*, which is used for treating inflammatory bowel disease (IBD), aligning with its therapeutic application (Sokol et al., 2008). Advanced gene editing techniques are used to modify Lactococcus lactis into an anti-inflammatory agent, which

Table 1 Selected next-generation probiotics isolated from human and food sources and their therapeutic applications (Plamada and Vodnar, 2021).

Microorganism	*Type*	*Disease Target*
Bacteroides xylanisolvens DSM 23694	Natural and isolated from humans	Cancer
Bacteroides ovatus D-6	Natural and isolated from humans	Cancer
Bacteroides ovatus V975	Genetically Modified Organism (GMOs) originally isolated from human gut samples for expressing Keratinocyte Growth Factor-2 (KGF-2)	Intestinal inflammation
Bacteroides ovatus V975	Genetically Modified Organism (GMOs) originally isolated from human gut samples for expressing Transforming Growth Factor-β1 (TGF-β1)	Intestinal inflammation
Bacteroides dorei D8	Natural and isolated from humans	Heart disease
Bacteroides fragilis ZY-312	Natural and isolated from humans	Removal of Infectious agents
Clostridium butyricum MIYAIRI 588	Natural and isolated from humans	Applicable for multiple targeting like inflammation, Cancer, and infectious agents
Faecalibacterium prausnitzii	Natural and isolated from humans	Specifically for Inflammatory Bowel Disease (IBD), but also applied to Type 2 diabetes, Eczema, and Asthma
Lactococcus lactis::elafin	Genetically Modified Organisms, but the initial host was isolated from food samples.	Specifically used for inflammatory diseases like Inflammatory Bowel Disease

is traditionally a non-probiotic bacterium (Motta et al., 2012). Likewise, *Bacteroides ovatus* is a genetically modified organism (GMO) originally isolated from human gut samples for expressing Keratinocyte Growth Factor-2 (KGF-2) used in cancer and intestinal inflammation therapy (Plamada and Vodnar, 2021).

6. Health Benefits of Probiotics

Probiotic Lactic acid bacteria offer several health benefits, such as intestinal infection reduction, enhancing lactose digestion, lowering blood cholesterol, and mitigating probable cancer risk (Anadón et al., 2010; Das et al., 2022), as shown in Figure 4. The GI tract of mammals, as well as humans, serves a critical role in the digestion and absorption of nutrients from food and in protecting the body against several harmful pathogens (Shehata et al., 2022; Latif et al., 2023). As described earlier, the gut is home to trillions of microbes from different communities known as gut microbiota, and they play a crucial role in the host gut as well as overall health. The gut microbiota function and composition are affected by several external factors such as probiotics, diet, prebiotics, stress, medication, antibiotics, infection, and faecal microbiota transplantation (Shahverdi, 2016). The healthy gut balance is known as eubiosis, whereas an imbalance of this microbiota is known as dysbiosis. Dysbiosis is responsible for several diseases, such as Diarrhea, inflammatory bowel syndrome, cancer, skin infections like ulcers, allergies, cancer, and Diabetes and also affects the nervous system (Saber et al., 2017; Ranjha et al., 2021; Bodke and Jogdand, 2022). So, restoring the gut microbial composition using probiotic strains is now considered an impending therapeutic strategy for tackling these complications.

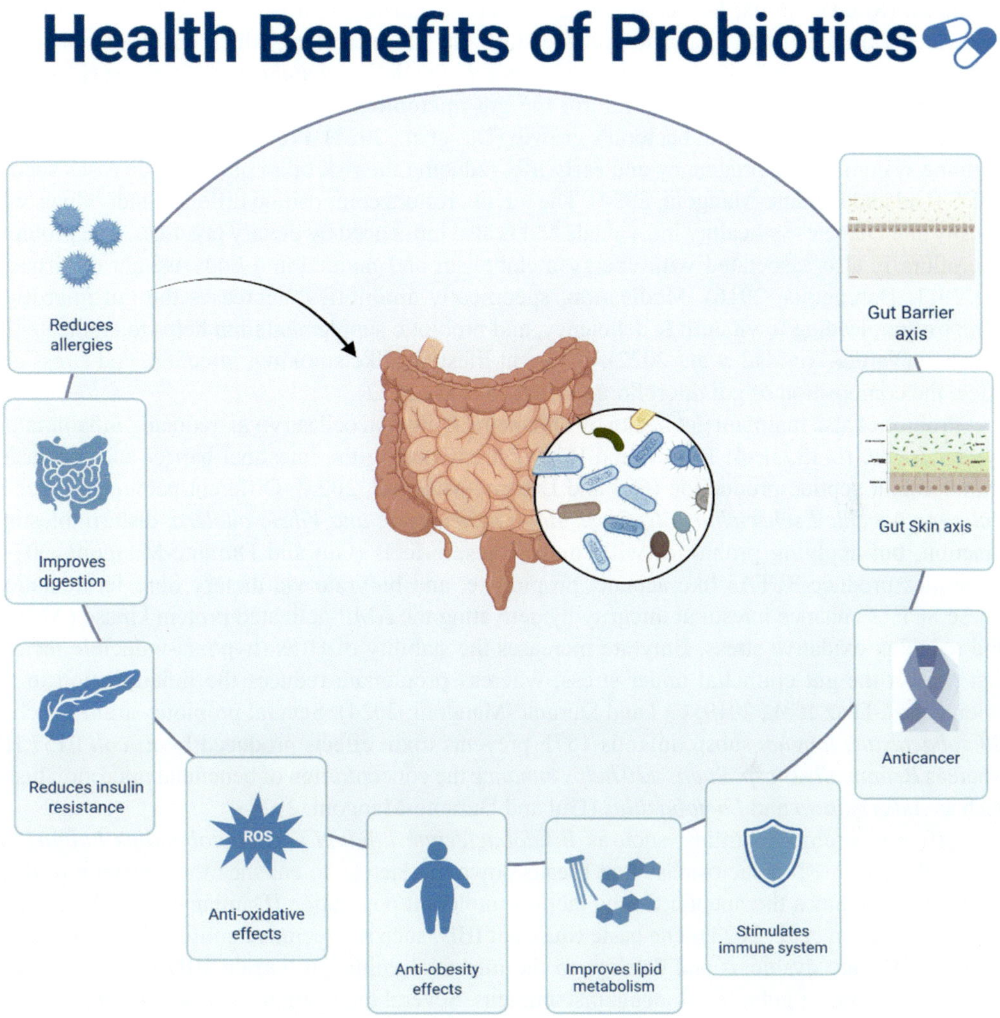

Fig. 4 The figure illustrates several health benefits of probiotics, including anti-obesity, antidiabetic, anticancer, anti-allergic, immunomodulatory effects, as well as impacts on the gut–brain axis and gut–skin axis.

6.1 Probiotics Effects on Intestinal Diseases

Probiotics therapy enhances the diversity of gut flora in the host, which helps increase enzyme production like lactase for lactose digestion, helps the intestinal epithelial barrier and facilitates immune response (Latif et al., 2023). Diarrhoea, either due to antibiotics or infection with pathogens, explicitly affects children who are immunocompromised. Different strains of probiotics manage diarrhoea by improving immune response, hydrolysing toxins, competing with nutrients and adhesion sites, producing antimicrobial peptides, preventing cell death, restoring gut lining and maintaining microbial balance (Woo et al., 2011; Suvorov, 2013; Quigley, 2019). The widely applied probiotics are used to manage diarrhoea, specifically in children infected with rotavirus, which targets the small intestine. Different probiotic strains such as *Lactobacillus casei* Shirota, *Bifidobacterium animalis*, and *Lactobacillus rhamnosus* GG have reduced the extent and severity of diarrhoea by the mechanism of improving immune response, producing antiviral compounds and blocking the binding of pathogens (Varela-Trinidad et al., 2022). The NSP1 protein of rotavirus reduces the production of interferons by degrading the different regulatory factors managed by probiotics (Michail, 2008; Das et al., 2022). Probiotic consumption is also effective in antibiotic-associated diarrhoea (AAD), as antibiotics disturb the gut microflora, which enables the attack of different harmful microbes, such as *Clostridium difficile,* to grow and cause diarrhoea (Das et al., 2022). The probiotic bacteria help reform the gut microbiota by enhancing beneficial bacteria's growth and inhibiting harmful bacteria's growth (Das et al., 2022). Probiotics also help develop the immune system during pregnancy and early life, reducing the risk of autoimmune diseases such as AIDS (Gul and Durante-Mangoni, 2024). The gut microflora composition differs in individuals with obesity and Diabetes vs healthy individuals and is also influenced by dietary practices. The probiotic microflora is also associated with energy metabolism and maintaining body weight (Kinross et al., 2011; Barengolts, 2016). Medication, specifically antibiotics, decreases the gut microflora composition, leading to vitamin K deficiency, and probiotic supplementation helps to eradicate this situation (Varela-Trinidad et al., 2022). Different lifestyles like smoking, infection, and stress also affect the composition of gut microflora (Manzoor et al., 2022).

Probiotics also maintain gut health by enhancing epithelial cell survival, reducing inflammatory cytokines like IL-13, IL-4, TNF-α, and IFN-γ, strengthening the intestinal barrier and increasing antimicrobial peptide production (Gul and Durante-Mangoni, 2024). Different pathogens, such as *Enteropathogenic Escherichia coli, Clostridium perfringens, and Vibrio cholera*, disturb intestinal function, but applying probiotics will control these effects (Gul and Durante-Mangoni, 2024). Probiotics produce SCFAs like acetate, propionate, and butyrate via dietary fibre fermentation. These SCFAs enhance intestinal integrity by activating the AMP-activated protein kinase (AMPK) and reducing oxidative stress. Butyrate increases the stability of HIFs (hypoxia-inducible factors) that protect the gut epithelial under stress, whereas propionate reduces the inflammation in the colon (Plaza-Diaz et al., 2019; Gul and Durante-Mangoni, 2024). Several probiotic strains, such as *Bifidobacterium longum* subsp. infantis 157F prevents toxin effects produced by *E. coli* 0157:H7, whereas *B. animalis subsp. Lactis 420 helps enhance* the concentration of beneficial microorganisms such as *Akkermansia* and *Lactobacillus* (Gul and Durante-Mangoni, 2024).

Different probiotic strains, such as *Bifidobacterium bifidum* and *Lactobacillus bulgaricus,* secrete the enzyme β-galactosidase that breaks down the lactose to enhance the digestion of dairy products and act as a therapeutic for the lactose-intolerant population (Damián et al., 2022; Das et al., 2022; Ibrahim et al., 2023). The basic causes of IBD, such as ulcerative colitis (UC) and Crohn's disease (CD), are dysbiosis and damage to the intestinal epithelial barrier. IBD leads to unusual immune reactions of gut microflora against antigens. Several medications, such as 5-aminosalicylic acid and corticosteroids, are used for their treatment, but probiotics offer a natural treatment of IBD through modulating immunity, increasing barrier function, conquering pathogens and decreasing inflammation (Weichselbaum, 2009; Soccol et al., 2010; Motta et al., 2012; Quigley, 2019).

During CD treatment, probiotics enhance SCFA production, balancing immune responses of Th1/Th2, alleviating cytokine production and strengthening mucosal health (Plaza-Diaz et al., 2019). The increased level of Enterobacteriaceae and decreased level of bacterial populations, such as Bifidobacteria and Lactobacilli, marked the symptoms of ulcerative colitis. Different probiotic mixtures, such as De Simone formulation and symbiotic-like combination of Synergy 1 fibre and *Bifidobacterium longum,* are used for the treatment of UC and to reduce inflammation (Woo et al., 2011; Maynard and Weinkove, 2018; Manzoor et al., 2022; Saville et al., 2025). Several butyrate-producing bacteria, such as *Faecalibacterium prausnitzii*, *Roseburia hominis, Anaerostipes caccae, Butyricicoccus pullicaecorum*, etc., also elevate the mucus secretion and strengthen the epithelial cells in case of IBD (Figure 4) (Plaza-Diaz et al., 2019; Jan et al., 2024; Saville et al., 2025).

6.2 Immunomodulatory and Anti-inflammatory Effects of Probiotics

Probiotics play a critical role in immune function by managing cell proliferation, metabolism and strengthening the gut barrier. Early colonization of probiotic bacteria, such as *Bifidobacterium* and *Bacteroides,* have a crucial role in immune development. Several factors, like antibiotics, birth methods, maternal diet, and environment, can impact microbial colonization in infancy (Mishra and Acharya, 2021; Ranjha et al., 2021; Gul and Durante-Mangoni, 2024). The immune cells of humans contain Toll-like receptors (TLR) that can differentiate between pathogenic and commensal microbes. The gut barrier function is enhanced by the consumption of probiotics that promote mucus secretion, increase tight junctions, decrease pathogen attachment, and improve epithelial survival by enhancing IgA and defensin production (Plaza-Diaz et al., 2019; Gul and Durante-Mangoni, 2024). Probiotics are nowadays supplemented in several foods like cereals, bread, sauces, yoghurt, and drinks and the daily intake of probiotics ranges from 2 to 20 grams, depending on the purpose and product (Plaza-Diaz et al., 2019; Mafe et al., 2025; Saville et al., 2025; Thakur et al., 2025). The host's immune system could also be modulated by the consumption of probiotics like fermented milk, which contains *Lacticaseibacillus casei* DN114001, increases the mucosal immunity, and facilitates the nonspecific barrier of the gut (Ibrahim et al., 2023). Also, the probiotic strain *Bifidobacterium longum* induces different cytokine responses in mononuclear blood cells that signified the strain-specific modulation (Plaza-Diaz et al., 2019; Varela-Trinidad et al., 2022). The study also reported that dietary supplementation with *Lactobacillus acidophilus* HN017, *Lacticaseibacillus rhamnosus* HN001, *Bifidobacterium lactis* HN019, *Bifidobacterium lactis* DR10, and *Lacticaseibacillus rhamnosus* DR20 enhanced both adaptive and innate immunity (Gill et al., 2000; Ibrahim et al., 2023). Further, it was suggested that probiotics stimulate immune cell maturation in the gut and integrative organs like the spleen and lymph nodes. Due to the immunomodulatory properties of probiotics, they can also reduce tumour development (Goldin, 2011; Ibrahim et al., 2023; Maciel-Fiuza et al., 2023). Probiotic strains such as *Lactobacilli* alter the immune system by increasing the activity of immune cells, such as natural killer cells, cytotoxic cells, and natural killer cells, leading to enhanced IgA production and TLR activation. These effects are due to the interaction with pattern recognition receptors on immune cells and epithelial cells, whereas reactive oxygen and lactic acid produced by these bacteria inhibit the pathogen adhesion (Ranjha et al., 2021; Das et al., 2022; Manzoor et al., 2022). The probiotic efficacy depends on the dosage, strain, host immune status, and route of administration (Martín et al., 2014; Das et al., 2022; Shao et al., 2023). The *Lactobacillus* species also shows anti-inflammatory effects by activating dendritic and T-regulatory (Treg) cells and helping treat inflammation (Figure 4) (Yahfoufi et al., 2018). It was also reported that oral administration of *Lactobacillus casei* promotes rheumatoid tolerance by enhancing Forkhead box P3 (Foxp3) expression and decreasing T-helper cells-1 (Th1) response (Yahfoufi et al., 2018; Plaza-Diaz et al., 2019; Ren et al., 2023; Un-Nisa et al., 2023).

6.3 Anti-Allergy Effects of Probiotics

Allergy is a hypersensitive immune disorder characterized as Type I hypersensitivity, and it was defined as the aggregation of immune reactions against harmless antigens (Latif et al., 2023; Mafe et al., 2025). The prevalence of allergies has risen considerably, affecting nearly 50% of Europe and North America. Environmental substances act as common allergens that trigger conditions like allergic rhinitis, eczema, angioedema, hypersensitivities, asthma, and atopic dermatitis (Figure 4) (Vyas and Ranganathan, 2012; Das et al., 2022; Latif et al., 2023; Saville et al., 2025). The gut microbiota is gradually recognized as having therapeutic potential for allergic diseases because it regulates inflammatory and immune responses (Latif et al., 2023). The allergic disorders are typically associated with immune disparity, chiefly the response of T helper-2 (Th2) cells. The Th2 cells secrete different cytokines like IL-1, IL-5 and IL-4 at the site of inflammation that recruit the basophils, eosinophils, and mast cells, which induce the B-cells to enhance the secretion of IgE (Gill et al., 2000; Varela-Trinidad et al., 2022; Latif et al., 2023). Probiotics seem to positively affect allergy via regulating antigen penetration, increasing mucosal barrier, and boosting anti-inflammatory and secretory IgA production, which decreases inflammation and allergic onset (Latif et al., 2023). Probiotics also help restore the immune balance by increasing the Th1 response while reducing the Th2 prevalence and maintaining the Th1/Th2 homeostasis (Costanzo et al., 2016).

Probiotics also modulate the dendritic cells that activate peripheral regulatory T cells (Tregs). These activated Tregs alleviate the hyperactive immune response by enhancing the production of immunosuppressive cytokines and altering the response of antibodies such as IgG and IgA (Plaza-Diaz et al., 2019; You et al., 2022). It was also reported that probiotic strains *Lactiplantibacillus plantarum* SY11 and SY12 were showing significant reduction in the production of tumour necrosis factor-α, cyclooxygenase-2, nitric oxide, nitric oxide synthase and Th-2 related cytokines. Similarly, oral administration of *Limosilactobacillus reuteri* enhances the composition of gut microflora, downregulates GATA3 expression, increases IL-10, reduces allergic diarrhoea and increases mast cell activation and IgE production (Latif et al., 2023; Gul and Durante-Mangoni, 2024).

6.4 Metabolic Health of Probiotics

6.4.1 Anti-obesity

The gut microbiota of humans plays a significant role in obesity, which affects lipid metabolism, inflammatory pathways, and energy homeostasis. Certain microorganisms, such as *Escherichia coli, Staphylococcus aureus,* and a few gut fungi, were dominant in obese people, which suggests their role in increased fat deposition and energy consumption. Conversely, probiotic bacteria such as *Lactobacillus* and *Bifidobacterium* show anti-obesity and protective effects due to their impact on metabolism and gut ecology. They maintain energy metabolism via the fermentation of dietary fibers and the production of SCFAs, which act as a signalling molecule, altering glucose and lipid metabolism. Probiotics have therapeutic potential in altering the composition of gut microflora to decrease obesity-related disorders (Figure 4). For example, *Lactobacillus reuteri* has anti-inflammatory effects by declining the TNF-α, enhancing intestinal barrier stability and reducing adiposity potential. This probiotic strain also improves bone density (Mu et al., 2018; Di Porzio et al., 2023; Mafe et al., 2025). Studies also reported the role of *Bifidobacterium pseudocatenulatum* SPM 1204, *Bifidobacterium longum* SPM 1205, and SPM 1207 in showing weight reduction and enhanced lipid profile via altering the GI microbiome (Mu et al., 2018; Ranjha et al., 2021; Di Porzio et al., 2023). Additionally, probiotics also show cholesterol metabolism. For example, *Lactobacillus plantarum* strains decrease serum cholesterol by altering the bile salts through bile salt hydrolase activity (BSH), reducing cholesterol absorption and enabling increased faecal bile acid extraction (Daliri and Lee, 2015; Chattopadhyay and Mythili, 2018; Plaza-Diaz et al., 2019). Another probiotic bacterium, *L. plantarum* DMDL 9010, also reduces serum triglycerides, cholesterol, and hepatic lipids in the *in vivo* model (Latif et al., 2023; Naeem et al., 2024). Probiotic strains also regulate lipid metabolism via gene expressions like *Lacticaseibacillus paracasei* TD3, which suppresses the gene HMG-CoA reductase (HMGCR) and CYP7A1, which are responsible for bile acid production and cholesterol synthesis, leading to a considerable reduction in cholesterol (Li et al., 2020; Latif et al., 2023). So, probiotics act as a promising representative for managing obesity by altering the gut microflora, reducing inflammation, modulating lipid metabolism, and enhancing the production of SCFAs.

6.4.2 Antidiabetic

Nowadays, a global health concern is Diabetes, which is characterized by impaired insulin secretion or mechanism of action, which leads to hyperglycaemia (Ranjha et al., 2021; Naeem et al., 2024). Type 1 diabetes (T1D) is an autoimmune disease which occurs due to damage to the pancreatic β-cell. In contrast, Type 2 diabetes (T2D) is known as insulin resistance, in which the body is unable to either effectively secrete or utilize insulin (Ranjha et al., 2021; Naeem et al., 2024). Both T1D and T2D are linked with considerable alteration in gut microbiota composition. The metagenomic study highlighted the correlation between bacterial genes, gut microorganisms and metabolic pathways of T2D (Figure 4). This study was conducted on T2D patients by altering the *Clostridium* sp. and *Lactobacillus* species and finding the association between crucial metabolic parameters like HbA1c,

HDL cholesterol, insulin, and fasting glucose (Ranjha et al., 2021; Jan et al., 2024; Silva et al., 2024). About 60% of T2D patients suffer from diabetic retinopathy, showing dominance of coagulase-negative *Staphylococcus* and Gram-positive bacteria (Chakravarty et al., 2024; Gul and Durante-Mangoni, 2024). In T2D pathophysiology, gut microbiota plays a critical role by impelling several mechanisms such as SCFA production, metabolic endotoxemia, incretin secretion, and insulin resistance, which leads to altering the immune response of the intestine (Santacroce et al., 2019; Gul and Durante-Mangoni, 2024). Probiotics also increase the antioxidant defence mechanisms by inhibiting lipid peroxidation, such as glutathione synthesis, so they reduce oxidative damage (Di Porzio et al., 2023; Gul and Durante-Mangoni, 2024). Moreover, probiotics also increase the activity of natural killer T-cells in the liver and alter the inflammation pathways, such as regulating nuclear transcription factor-κB (NF-κB) and TNF-α, hence managing the insulin activity (Plaza-Diaz et al., 2019). Probiotics also maintain better glycaemic control by increasing the bioavailability of antidiabetic drugs like gliclazide and preventing the intestinal absorption of glucose (Ranjha et al., 2021; Manzoor et al., 2022). Enhancing the gut microflora diversity has been shown to reduce inflammation, improve nutrient absorption and insulin production, and help gastrointestinal integrity, essential for maintaining glucose homeostasis (Ranjha et al., 2021; Di Porzio et al., 2023). Furthermore, probiotics also play a critical role in controlling gestational diabetes mellitus (GDM) by altering the gut microbiota composition and pregnancy metabolic environment that ultimately affects the lipid and glucose metabolism (Di Porzio et al., 2023).

6.5 Role of Probiotics in Gut-Brain Axis

The human nervous system is categorized into the peripheral nervous system (PNS), which contains ganglia and organs connecting nerves, and the second one is the CNS, which includes the spinal cord and brain (Fung, 2020; Varela-Trinidad et al., 2022). The PNS are further divided into the somatic and autonomic nervous systems, in which the autonomic system, specifically the ENS, copes with involuntary function and incorporates parasympathetic and sympathetic fibers that perform two-way communication between the gastrointestinal tract and CNS (Varela-Trinidad et al., 2022). Due to the complicated autonomous activity, the ENS is often considered the body's "second brain", and it spreads throughout the GI tract and makes two leading networks: submucosal and myenteric networks, which are responsible for mucosal immunity and gut movement (Farshbafanadi et al., 2021). So, this communication route and close association between gut microbiota and ENS is termed the "Microbiota-Gut-Brain-Axis (MGB)" (Figure 4) (Almeida et al., 2022). This MGB axis connects the endocrine, neural, and microbial systems, affecting CNS and GI function. Studies also reported the link between gut microbiota and different psychological and neurological disorders, proposing the role of microbes in brain health (Plaza-Diaz et al., 2019; Fung, 2020; Delgado Jiménez and Benakis, 2021; Hou et al., 2022). Key mediators are involved in this axis, such as tryptophan metabolism, SCFAs, and microbes-derived neurotransmitters. Several SCFAs, specifically butyrate, acetate, and propionate cross the blood-brain barrier or sometimes interact with their receptors, affecting CNS activity and neurological conditions (Plaza-Diaz et al., 2019; Li et al., 2020). Gut microflora-produced tryptophan metabolites provide the signal to peripheral organs comprising the brain and help regulate the immune response and inflammation via microglial activation (Silva et al., 2024). Moreover, gut microbes-derived neurotransmitters like catecholamines and GABA alter the thought, mood, and gut function (Varela-Trinidad et al., 2022). Several probiotic strains, such as *L. helveticus* and *Lactobacillus plantarum*, have positive effects in decreasing psychological distress and other neurological symptoms like chronic fatigue and autism symptoms (Plaza-Diaz et al., 2019; Darwesh et al., 2024). Social interactions can also affect microbial similarity among individuals, and this transmission can offer health advantages. In ancient times, microbial flora significantly impacted brain development and behaviour, whereas inflammation and stress altered the microbial composition and integrity, specifically in preterm children (Manzoor et al., 2022). Gut dysbiosis was reported in children who have autism spectrum disorder (ASD), and intake of probiotic bacteria like *Bacteroides fragilis* may relieve this ASD symptom (Plaza-Diaz et al., 2019; Fung, 2020). Additionally, maternal Diabetes and obesity have also been associated with ASD, and the consumption of probiotic bacteria like *Lactobacillus reuteri* restores the social behaviour exaggerated by maternal high-fat diet and maintains the stress-related microbial disparity to improve maternal health (Mu et al., 2018; Plaza-Diaz et al., 2019).

6.6 Anticancer Potential

A large number of *in vivo* and *in vitro* studies reported the anticancer ability of probiotics. However, the exact mechanism is still unexplored due to a lack of human clinical trials (Williams, 2010; Suvorov, 2013; Lu et al., 2021; Di Porzio et al., 2023; Naeem et al., 2024). It was also reported that bacterial strains such as *Bifidobacterium, Lactobacillus*, and *Escherichia coli* strain Nissle showed antimutagenic activity through neutralizing the mutagenic substances in the *in vitro* study (Geier et al., 2007). It was found that the consumption of probiotics inhibited cancer development by producing antimutagenic compounds, immunomodulation, and changing gut physiology and metabolism (Mishra and Acharya, 2021). Also, probiotics act as an adjuvant in cancer therapy because they maintain gut microflora and boost the innate and adaptive immune system (Ibrahim et al., 2023). The probiotics help prevent tumour formation and growth in intestinal and extra-intestinal cancers (Figure 4) (So et al., 2017; Samanta, 2022). Probiotic metabolites such as peptides, organic acids, and bacteriocins interact with the key metabolic pathways responsible for apoptosis, inflammation, angiogenesis, metastasis, and proliferation (Latif et al., 2023). Probiotic strains also compete with infectious microbes, enhance the production of SCFAs, bind with carcinogens, decrease carcinogenic bile salts, suppress proliferative genes, and boost apoptosis (Konishi et al., 2016; Latif et al., 2023). Probiotic strains like *L. reuteri* alter the Nuclear Factor kappa-light-chain-enhancer of activated B cells (NF-κB) and Mitogen-Activated Protein Kinase pathways (MAPK) that lead to the apoptosis of cancer cells (Suvorov, 2013; Latif et al., 2023; Naeem et al., 2024). Probiotic strains also show promising results in liver cancer by decreasing the tumour size and preventing the process of angiogenesis by modulating pro-inflammatory T helper 17 cells (Th17) (Lu et al., 2021; Cheng et al., 2023; Chen and Bie, 2025; Nakashima et al., 2025). Additionally, probiotic strains like *Lactobacillus* show promising results in breast cancer by suppressing the hypoxia-inducible factors (HIFs) and reducing tumour persistence (Esfandiary et al., 2016; Lu et al., 2021). Several probiotic strains, such as *L. fermentum*, *L. acidophilus*, and *L. casei* Shirota, enhance healthy colon formation and reduce malignant cells (Manzoor et al., 2022). So, the probiotic strains also show anti-cancerous activity, but human clinical trials are needed for their validation.

6.7 Probiotics role in the Gut-Skin Axis

In innate immunity, skin and mucosal barriers act as a key component, and they act as a first line of defence against several environmental pathogens. They act as a protective shield and not only protect the body from chemical, physical, and biological threats but also control the temperature, hydration, and immune responses to maintain homeostasis (Falk et al., 1998; Ding et al., 2019; Mishra and Acharya, 2021). The epidermis, the outermost layer of skin, acts as a protective barrier consisting of several layers, like keratinocytes originating from basal stem cells and differentiated and finally forming the stratum corneum (Manzoor et al., 2022). Finally, keratinocytes become corneocytes, which are flattened, keratinized cells arranged in a "brick and mortar" structure containing different immune cells like Langerhans cells to produce cytokines and antimicrobial peptides to protect against pathogens (Sanford and Gallo, 2013; Varela-Trinidad et al., 2022). Skin is also encountered with several microorganisms, including pathogenic and non-pathogenic, similar to the respiratory and gastrointestinal tract (Un-Nisa et al., 2023). The disturbance in the symbiotic association between these microbes and host immune cells leads to skin ailments like atopic dermatitis, psoriasis, and acne. So, skin and gut microbiota are disturbed in this skin complication, highlighting the association between skin and gut, known as the "gut-skin axis" (Figure 4) (Varela-Trinidad et al., 2022). In this axis, gut microflora influences skin health by producing several metabolites in circulation and immune modulation (Salem et al., 2018). Several strains of probiotics, like *Bifidobacterium* and *Lacticaseibacillus,* induce the production of Treg, which inhibits the differentiation of T cells into pro-inflammatory T helper cells, which leads to a reduction in inflammation and finally downregulates IL-5 and helps to treat Atopic Dermatitis (AD) (Parhizkar et al., 2025). It was also reported that oral supplementation of *Lactobacillus plantarum* HY7714 enhances the ceramide levels by decreasing the ceramidase mRNA expression and serine palmitoyltransferase (SPT0) expression (Gao et al., 2023). Also, oral supplementation of *Lactobacillus reuteri* enriched skin characteristics such as sebocyte production and dermal thickness. Another probiotic strain, *Lactobacillus brevis* SBC8803, increases skin hydration, minimizes water loss, and supports the systemic benefits of probiotics (Varela-Trinidad et al., 2022). It was also reported that probiotics strain Tyndallized *Lactobacillus* KCCM12625P (AL) and *Lactobacillus plantarum* HY7714 showed anti-wrinkle activity by suppressing the matrix metalloproteinase-1 (MMP-1) pathway (Gao et al., 2023). So, these findings strengthen the importance of the microbiota-gut-skin axis and propose that supplementation of probiotics is valuable for maintaining skin health and preventing dermatological complications.

7. Prebiotics: Definition, Characteristics, and History

Non-digestible food ingredients that positively impact the growth and activity of specific microorganisms in the gut, particularly the colon, are known as "Prebiotics" (Gibson and Roberfroid, 1995; Zang et al., 2024). In modern science, the definition of probiotics has evolved, and prebiotics are recognized as food components that must meet specific scientific criteria (Wen and Duffy, 2017). Prebiotic molecules should not be hydrolyzed or absorbed in the upper gastrointestinal tract and must be resistant to degradation by gastrointestinal enzymes. They should be selectively fermented by the lower gut microbiota to promote the growth and metabolic activity of beneficial microorganisms. Additionally, prebiotics must be capable of modulating the colonic microbiota toward a healthier composition by enhancing populations of Bifidobacteria and Lactobacillus while reducing pathogenic bacteria in the gut.

Most gut microflora reside in the descending colon, so prebiotic molecules are specifically focused in the lower gut. The prebiotic concept is based on enhancing the growth of beneficial bacteria naturally present in the gut instead of introducing external bacterial strains similar to probiotics. Human microbiota vary from person to person and are impacted by several internal and external variables that might change the microbiome's stability and composition (Rinninella et al., 2019). Endogenous factors include infant feeding method, weaning period, age, hormonal changes, genetics, birth gestational date, mode of delivery at birth, and health status (Penders et al., 2006). The exogenous factors include medication use, specifically antibiotics, climate, stress management level, geographical region, climate, and exercise frequency (Roper and H. Koch, 1988; Kaneko et al., 1994; Penders et al., 2006). Regular consumption of vegetables and fruits is part of a healthy diet. It carries several health benefits via its composition, such as non-digestible fibre, fructo-oligosaccharides, galacto-oligosaccharides, isomaltooligosaccharides, polyphenols, beta-glucans, and xylooligosaccharides (Table 2) (Underwood et al., 2015; Belorkar and Gupta, 2016; Hutkins et al., 2016; Carlson et al., 2018; Zendeboodi et al., 2020; Zang et al., 2024). Historically, prebiotic consumption has evolved since the early hominins of the Pleistocene and Pliocene eras, as they included fructan-rich plants in their diet, given the presence of more than 36,000 plant species (Hendry, G., 1987). The savanna-woodlands are dominated by underground plant storage organs like tubers and corns, rich in inulin-type fructans and have become vital energy sources for hominins adapted to this woodland. However, the presence of antinutritional substances limits their intake until fire arises (Hatley and Kappelman, 1980; Stahl, 1984; Laden and Wrangham, 2005; Leach et al., 2006). The widespread consumption of prebiotic-rich flora such as wild onion, camas, and agave was confirmed by archaeobotanical data collected from Holocene North America and Upper Palaeolithic

Table 2 Summarizes prebiotic molecules, including their sources, digestibility, functions, and health impacts.

Prebiotic Molecule	*Sources*	*Category*	*Digestibility and Fermentation Location*	*Function*	*Targeted Organism*	*Role*	*References*
Inulin	Chicory root, garlic, onions, and banana leeks	fructans	Non-digestible in the upper GI tract, and fermentation occurs in the Proximal Colon.	Improve calcium absorption	*Bifidobacterium*	Improve digestive health by producing acetate, butyrate and propionate, which ultimately induce apoptosis in cancer cells.	(Kolida et al., 2002)
1-Kestose, Nistose, 1-fructofuranosyl-nistose	Onions, artichokes, Jerusalem	Fructo-oligosaccharide	Fermentation occurs in the Proximal Colon and is mostly non-digestible	Non-cariogenic, Low sweetness intensity.	*Lactobacillus, Bifidobacterium, Eubacterium and some Streptococcus, Enterococcus and Bacteroides,*	It is considered high-FODMAP due to rapid fermentation; high doses can cause bloating and mild distension.	(Sabater-Molina et al., 2009)
4'-Galactosyl-lactose (4'-GL) 6'-Galactosyl-lactose (6'-GL), Allolactose	Human breast milk, legumes, and dairy products	Galacto-oligosaccharide		Reduce allergies	*Lactobacillus, Bifidobacterium*	Improve gut barrier function, mineral absorption, weight management, carcinogenesis, and allergy alleviation. Enhance Bifidobacterium growth.	(Lamsal, 2012)

Contd.

Table 2 *Contd.*

Prebiotic Molecule	*Sources*	*Category*	*Digestibility and Fermentation Location*	*Function*	*Targeted Organism*	*Role*	*References*
RS1, RS2,RS3, RS4	Whole grains, green bananas, legumes	Resistant starch	RS1 is less resistant to digestion by milling and chewing, RS2 by food processing techniques, and RS3 by food processing. RS4, chemical modification.	Dietary fibre	*Gut bacteria*	Improve glycaemic and insulinemic responses, improve bowel health, and treat cholera and chronic diarrhoea.	(Nugent, 2005)
Rhamnogalacturonan Oligosaccharides (RG-Oligosaccharides) Galacturonan Oligosaccharides (GalA-OS) Arabinogalactan Oligosaccharides (AG-OS)	Apples, citrus peels, berries, Olives, Orange, Sugar beet, Potato pulp	Pectic oligosaccharide		antioxidant activity, Anti-inflammatory effects include a Ripping agent, antibiotic therapy, immunomodulatory effects, otitis, and osteoporosis.	*Clostridia and Salmonella. Campylobacter jejuni, Bifidobacterium*	POS is used to promote bifidogenic flora. It inhibits Campylobacter jejuni, so it is used as a feedstock.	(Babbar et al., 2016)
Xylobiose (DP2), Xylotriose (DP3), Xylotetraose (DP4). Arabinoxylan-oligosaccharides (AXOS)	Bamboo shoots, oats	Xylo-oligosaccharides	XOS resists digestion in the small intestine and the colon (large intestine), primarily in the proximal colon (cecum & ascending colon). Fermentation site	Antioxidant property of XOS Immune stimulation, Lipidaemic effects,	*Bifidobacterium adolescentis, Bacteroides thetaiotaomicron, Lactobacillus brevis, Roseburia spp*	Improve metabolic health, selective growth stimulation of gut microflora, short-chain fatty acid production, and XOS on the reduction of serum triglycerides and cholesterol levels.	(Samanta et al., 2015)

Contd.

Table 2 *Contd.*

Prebiotic Molecule	*Sources*	*Category*	*Digestibility and Fermentation Location*	*Function*	*Targeted Organism*	*Role*	*References*
2'-Fucosyllactose (2'-FL), Lacto-N-tetraose (LNT), Lacto-N-neotetraose (LNnT), 3'-Sialyllactose (3'-SL), 6'-Sialyllactose (6'-SL)	Human Milk	Human milk oligosaccharides	HMOs escape breakdown in the small intestine,	The anti-adhesive effect, glycome-modifying effect, HMO as Selectin-Ligand analogues, supplements for infant formula.	*Bifidobacterium infantis (specialized HMO utilizer), Bifidobacterium bifidum, Bacteroides fragilis (some strains)*	HMOs mimic gut receptors, blocking the adhesion of E. coli, Salmonella, and Campylobacter Necrotizing Enterocolitis (NEC). Prevention of Brain Development modulating dendritic cells reduces excessive inflammation.	(Coppa et al., 2004)
Zymosan, Laminarin, Pachyman, Curdlan, Lentinan, Glomerellan, Grifolan, Crestin	Oats barley mushrooms	Beta-glucans	Minor digestion may occur via brush-border enzymes like maltase-glucoamylase, but most beta-glucans remain intact.	Cholesterol-lowering, SCFA production, Anticancer Potential, increases mucin production (MUC2) and tight junction proteins.	*Bifidobacterium, Lactobacillus, Bacteroides*	Selectively promoting the growth of Bifidobacterium, Lactobacillus	(Lam and Cheung, 2013)
Feruloylated AXOS (FAXOS),	Rye barley, Cereal grains (wheat, rye, barley, corn bran), Flaxseed	Arabinoxylan oligosaccharides	Proximal colon	Ferulic acid (antioxidant) has a strong bifidogenic effect	*Bifidobacterium, Lactobacillus, Bacteroides, Bifidobacterium adolescentis, Bifidobacterium longum*	Microbiota modulation, production of short-chain fatty acids and propionic acid in the gut. Antiobetic and antidiabetic effect	(Grootaert et al., 2007)

Contd.

Table 2 *Contd.*

Prebiotic Molecule	*Sources*	*Category*	*Digestibility and Fermentation Location*	*Function*	*Targeted Organism*	*Role*	*References*
Quercetin, Catechins, Anthocyanins, Chlorogenic acid, Ellagic acid, Secoisolariciresinol diglucoside (SDG) Resveratrol	onions, apples, berries, green tea, cocoa, purple sweet potatoes, oats, walnuts	polyphenols	Not directly fermented but metabolized by gut microbiota	Plant secondary metabolites	*Bifidobacteriaceae and Lactobacillaceae, Akkermansia muciniphila, Bacteroides, F. prausnitzii, Lactobacillus rhamnosus, and decrease Coprococcus, Ruminococcus, Clostridiales, and Oscillospira*	have potential antioxidants as well as antimicrobial activity. Also, beneficial alterations were observed in the Firmicutes and Bacteroidetes groups (after intervention)	(Plamada and Vodnar, 2021)
Isomaltose, isomaltotriose, panose,	Fermented foods (soy sauce, sake), honey, Corn syrup, IMO supplements	Isomalto-oligosaccharides	In the small intestine of the GI tract, they are partially resistant, whereas mostly fermented in the colon.	Low glycemic impact, improved dental health,	*Bifidobacterium, Lactobacillus*	Used as a feed supplement,	(Wu et al., 2017)

Europe approximately 9,000 years ago (Leach et al., 2006). Finally, prebiotics came into existence scientifically in the mid-20th century when Gyorgy first gave the bifidogenic activity of N-acetyl-glucosamine, and Petuel reported lactulose as a bifidus factor in 1957 (Tymczyszyn et al., 2014). In the subsequent decades, Japanese researchers highlighted the importance of oligosaccharides as bifidogenic agents. Afterwards, Gibson and Roberfroid formally gave the scientific definition of "Prebiotics" in 1995, and modern definitions of probiotics were given by Wen and Duffy in 2017 (Gibson and Roberfroid, 1995; Wen and Duffy, 2017) (Figure 5).

8. Prebiotics and Their Health Impact

8.1 Isomalto-oligosaccharide

Isomalto-oligosaccharides (IMO) are identified as a potential prebiotic molecule that contains α-(1→6) glycosidic bonds between α-d-glucopyranose units (Saburi et al., 2006; Ketabi et al., 2011; Nguyen and Haltrich, 2013). IMO, they have prebiotic potential and are used for treating hyperlipidemia and chronic constipation and maintaining haemodialysis (Nguyen and Haltrich, 2013) (Table 2). Several studies have demonstrated the positive health benefits of IMO on rats (Sung et al., 2004), broilers (Rehman et al., 2009), and pigs (Li et al., 2010). Among the commercialized synthesized prebiotic molecules, IMO are synthesized by enzymatic conversion of starch (Villéger et al., 2022). Widely used as oligosaccharides as food ingredients or additives, IMO has attained great interest due to its stability, availability and low-cost production (Nguyen and Haltrich, 2013). IMO, those are non-digestible oligosaccharides that confer health benefits to the host. They selectively promote the growth or activity of non-pathogenic bacteria while restricting certain pathogenic bacteria in the colon (Gibson and Roberfroid, 1995; Ziemer and Gibson, 1998; Roberfroid, 1999; Rycroft et al., 2001). For the industrial production of IMOS, maltose and maltodextrins were used as a substrate via using *Aspergillus* sp. produced transglucosidase (EC 2.4.1.24) and produced IMOS with a low degree of polymerization (DP) in the range of IMO02 to IMO06 (Roper and H. Koch, 1988). The short-chain IMOS have several limitations, such as being partially digested and absorbed in the large intestine. Long-chain IMOS are reported to be more effective than short-chain, as they enhance the gut microflora like *Bifidobacteria* and *Lactobacillus* (Kaneko et al., 1994). It was also noted that the application of IMOS with the probiotic bacteria *B. subtilis* CU1 as a symbiotic product enhanced the proliferation and fitness of gut microflora (Villéger et al., 2022). Several foods, such as sake, miso, soy, starchy foods, and sauce act as natural sources of IMOS (Mohanty et al., 2018) (Figure 6).

8.2 Arabinoxylanoligo Saccharides

Arabinoxylans (AX) are primary non-starch polysaccharides in the cell wall of cereal grains such as rice, wheat, corn, oats, barley, sorghum, and rye, which act as a component of dietary fibres (Hespell and Cotta, 1995; Pushpanathan et al., 2019; Hernández-Pinto et al., 2024). Structurally, AX is made up of a linear chain of β-(1,4)-xylopyranosyl units, which are substituted by α-L-arabinofuranosyl moieties at O-3 or O-2 position (Bangar et al., 2025). Some intestinal bacteria with AX-degrading enzymes degrade the AX in the colon and produce the arabinoxylan oligosaccharides (AXOS). The AX-derived AXOS acts as an emerging food ingredient as a prebiotic that enhances the proliferation of several health-promoting microflora in the gut (Hernández-Pinto et al., 2024). Like other prebiotic molecules, the AX is a non-digestible fibre that stimulates the *Lactobacillus* and *Bifidobacteria* in the large intestine and has prebiotic potential. There are several health benefits of the AXs, such as preventing the dysbiosis impacts of gut microbiota, inhibiting the proliferation of pathogens, and alleviating gastrointestinal disorders, which have been reported (Table 2) (Hernández-Pinto et al., 2024). The structure of the AXs depends on the type of cereal used for extraction, the methodologies used for their extraction and the recovery method (Schupfer et al., 2021). The symbiotic potential of

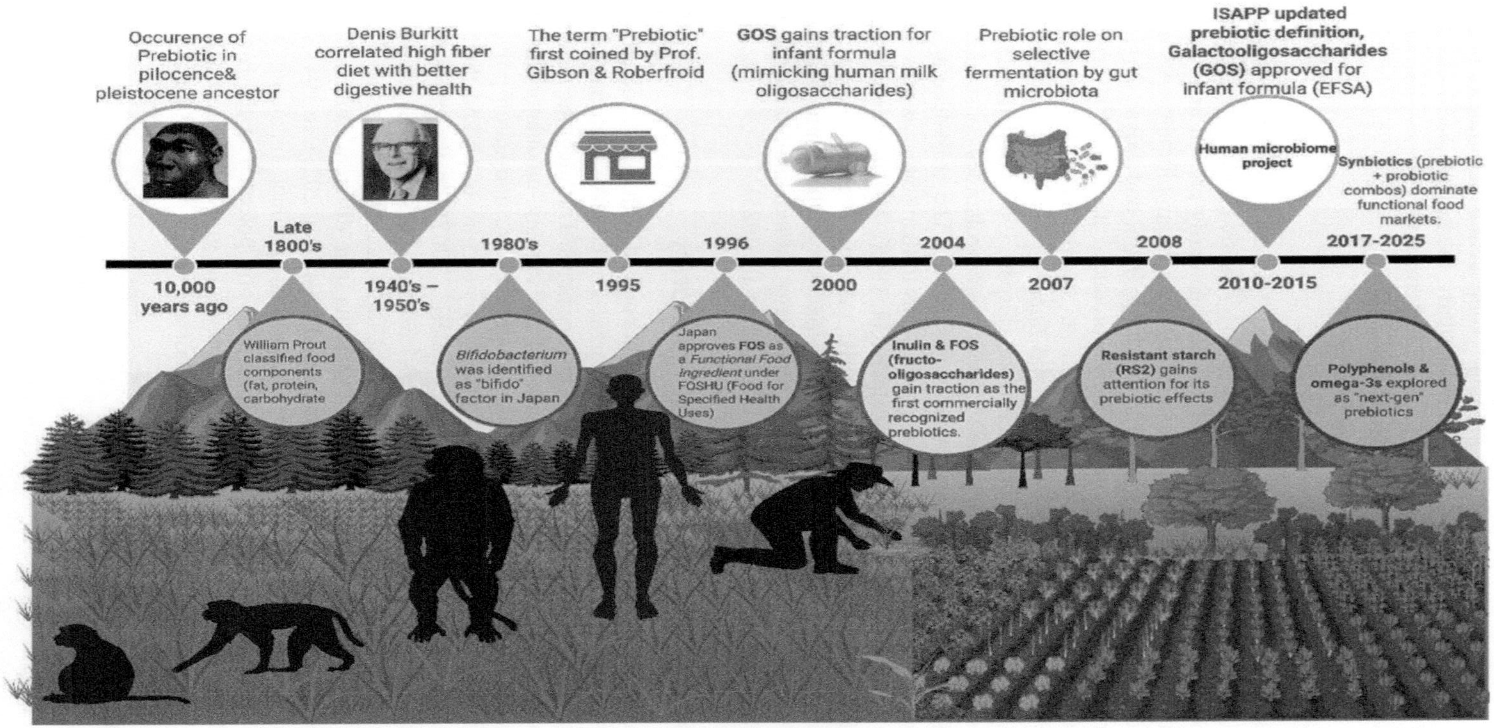

Fig. 5 A brief history of the discovery of prebiotic molecules and their development for health benefits. The figure highlights the evolutionary and scientific progress in prebiotic research, from the consumption of prebiotic molecules during the Pleistocene to modern advancements, including the approval of galactooligosaccharides (GOS) in infant formula for healthy consumption (created with BioRender.com).

AXs with the *Lactobacillus reuteri* ATCC23272 showed an anti-inflammatory effect and modulated the microbiome-brain axis (Hernández-Pinto et al., 2024; Bangar et al., 2025). The fermentation of AX by the gut microflora showed a slight rise in the propionic acid concentration, which showed cholesterol-reducing properties (Amrein et al., 2003). Similarly, the human faecal samples showed a high concentration of butyrate, propionate and acetate after consuming the AX-enriched food for three weeks (Figure 6) (Lopez et al., 1999; Grasten et al., 2003).

8.3 Xylooligosaccharides

Xylooligosaccharides (XOS) are oligomers originating from forestry as well as agricultural waste composed of 2–7 xylose units interconnected by β (1 → 4) xylosidic bonds and having a prebiotic activity (Patel and Goyal, 2011; Ali et al., 2025). XOS are naturally present and extracted from vegetables, honey, fruits, milk, bamboo, and extracted from several agricultural wastes like brewery spent grains, straw, corn cob, almond shell, and barley hulls (Patel and Goyal, 2011; Ali et al., 2025). From the hydrolysis of xylan-rich material, XOS is produced at an industrial scale (Vázquez et al., 2005). Several gut microflora, such as, *Fibrobacter succinogenes, Ruminococcus albus, Ruminococcus flavefaciens, Prevotella ruminicola*, and *Eubacterium ruminantium* secreted the xylan-degrading enzymes (Dehority, 1966; Whitehead and Hespell, 1990; Hespell and Cotta, 1995). The partial digestion of xylan leads to the production of oligosaccharides consumed by the gut bacteria, and this type of cooperation is known as cross-feeding. The cross-feeding association between xylan-degrading *Butyrivibrio fibrisolvens* and the XOS-utilizing *Selenomonas ruminantium* is a suitable example of cross-feeding (Cotta and Zeltwanger, 1995). The XOS inhibits pathogenic bacteria's growth and prevents intestinal disorders like IBD, gastritis, diarrhea, and constipation. As the pharmaceutical ingredient, this XOS controls encephalopathy, hyperlipidemia, and constipation, as well as XOS has antimicrobial, anti-inflammatory, antiallergy, immunomodulatory, and selective cytotoxicity (Table 2) (Figure 6) (Patel and Goyal, 2011; Ali et al., 2025).

8.4 Long-chain ß-glucans as Potential Prebiotics

β-glucans are polysaccharides of long D-glucose chains connected with β-glycosidic linkage. The long-chain β-glucans are abundant in nature, and different sources like bacteria, yeasts, algae, mushrooms, oats, and barley are rich sources of these dietary fibres. This prebiotic molecule shows anti-inflammatory, anti-ageing, anti-tumour, anti-microbial, and immuno-modulatory activity (Table 2) (Singh and Bhardwaj, 2023; Sarkar et al., 2025). β-glucans are non-starch polysaccharides containing glucose molecules in a long chain or branched structure form, and their branching patterns depend on their origin (Lazaridou and Biliaderis, 2007). In the case of microbial β-glucans, β-D-glucopyranose units are linked with β-1,3-bonds, whereas side chains form the β-1,6 linkage (Ahmad et al., 2012). β-glucans are critical for a healthy diet, so the US Food and Drug Administration recommends consuming 3g of β-glucans from different sources like oats and barley, which help reduce heart-related diseases (Singh and Bhardwaj, 2023). In 1941, "Zymosan" was the first discovery of a pharmaceutical insoluble yeast crude product containing 50% glucans and polysaccharides (Pillemer and Ecker, 1941). The prebiotic potential of the β-glucans was first reported in the 1990s and in the early 2000s, enhancing the growth of intestinal microflora, specifically *Bifidobacteria* and *Lactobacilli* (Figure 6) (Jaskari et al., 1998; Kontula et al., 1998).

8.5 Human Milk Oligosaccharide

Human milk oligosaccharides are the key component of breast milk, an essential nutritional source for infants for the first few months. Human milk comprises about 7% of the carbohydrates, of which 80% are lactose and the other 20% are oligosaccharides (Sun et al., 2024; Urashima et al., 2025). The ISAPP (International Scientific Association of Probiotics and Prebiotics) 2008 recommended

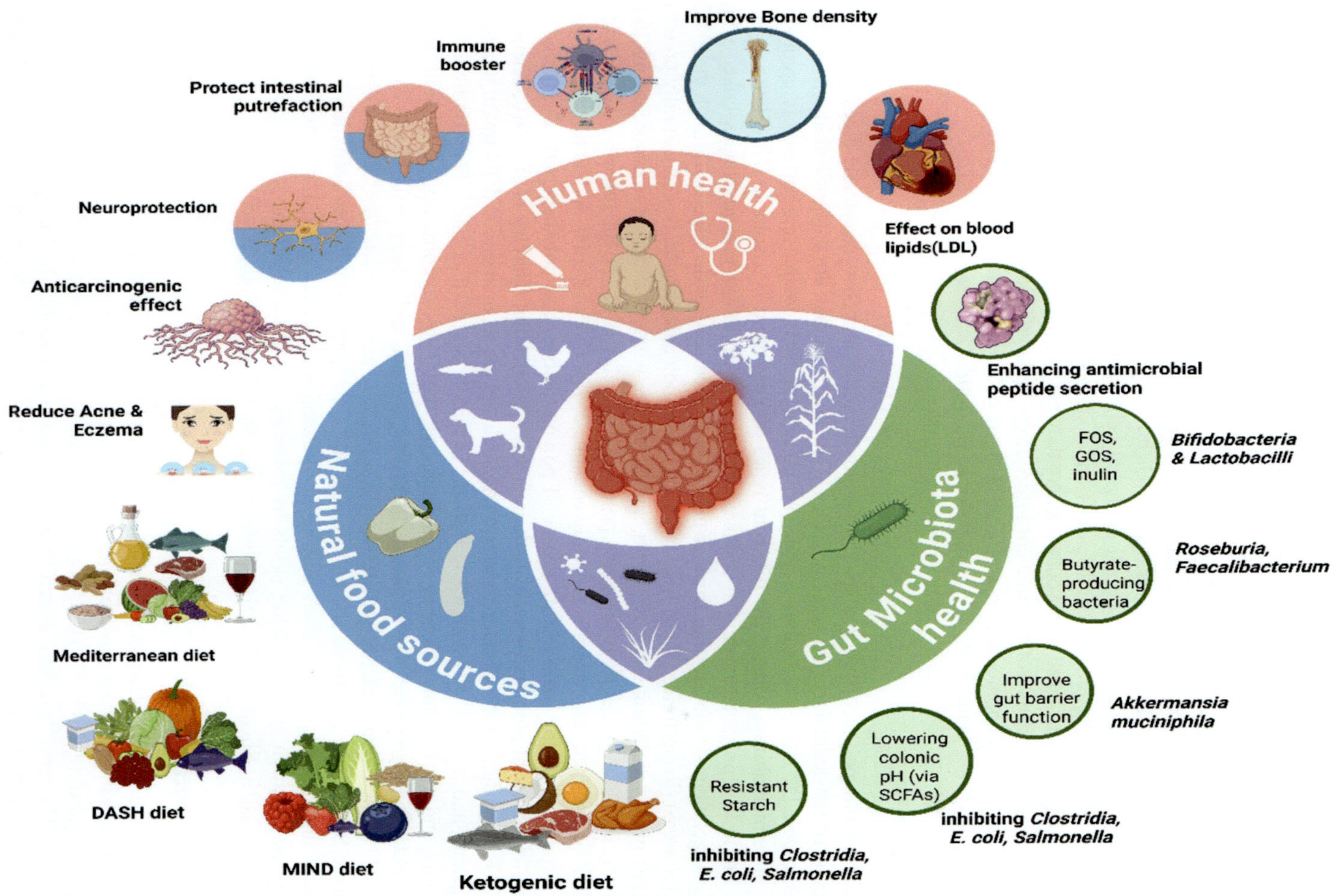

Fig. 6 The correlation of gut microbiota, food sources, and prebiotic molecules such as resistant starch, SCFAs, FOS, GOS, inulin, and butyrate, along with their health impacts, including anticarcinogenic, neuroprotective, cardioprotective, and immunomodulatory effects (created with BioRender.com).

that HMOs be infants' most significant prebiotic molecules (Figure 6) (Adams and Gutiérrez, 2019). Because of their resistance to gastrointestinal digestion, HMOs ferment in the colon and stimulate the growth of beneficial microorganisms to develop a healthy gut ecosystem (Sun et al., 2024). They also protect from pathogens by acting as analogue receptors of epithelial cells, blocking their adhesion, colonization, and invasion (Asakuma et al., 2011). HMOs are complex sugar molecules, a building block of several carbohydrate types. The HMOs' core comprises lactose and several sugar molecules like galactose, fucose, sialic acid, glucose, and N-acetylglucosamine. The content of HMOs changes over time after birth. It is about 17.7 grams per litre in colostrum in the first five days and 13.3 grams per litre in 6 to 14 days. Further, 11.3 grams in mature milk, i.e., 15 days to 3 months (Urashima et al., 2025). It was reported that HMOs have glycome-modifying effects, i.e., they can modulate the glycocalyx of the intestinal epithelial cells so that any pathogenic bacteria can not bind to the intestinal cells and prevent diseases (Angeloni et al., 2005). Like other oligosaccharides, the HMOs are also unabsorbed in the colon, where they act as prebiotic molecules that enhance the growth of beneficial gut microflora, prevent infection by pathogenic bacteria, strengthen the colon barrier, act as an immunomodulator, and protect from necrotizing enterocolitis (NEC) (Table 2) (Holman et al., 1989; Uauy et al., 1991; Costalos et al., 2008; Urashima et al., 2025).

8.6 Inulin

The carbohydrate polymer of more than 10 monomeric units that resists enzymatic digestion in the GI tract is acknowledged as "Dietary fibre (DF)". The Dietary fibres are categorized into insoluble DF and soluble DF based on solubility, which are completely fermentable and partially fermentable based on fermentability. Among the different types of DF, inulin is a soluble DF (Sheng et al., 2023). Inulin is a polysaccharide that consists of a long chain of fructose molecules linked with β-2,1 linkage, whereas the terminal fructose moiety is connected with the glucose through the α-1,2 bond (Figure 6) (Roberfroid, 1999). Like other prebiotic molecules, inulin is also naturally present in various plants, including Chicory, dahlia, and artichoke, whereas different microorganisms like *Streptococcus* and *Aspergillus* species also produce it (Barclay et al., 2010; Teferra, 2021; Karimi et al., 2025). The plant-based inulin generally contains 2–200 fructose units, and its functionality is highly dependent on the chain length, degree of polymerization, and polydispersity. Compared to plant inulin, bacterial inulin shows a high degree of polymerization (Barclay et al., 2010; Sheng et al., 2023). According to the US Department of Agriculture, the recommended daily inulin intake is 25-36 grams, and in 2003, it was characterized as "Generally regarded as safe". Their optimal dose is 5 grams, and the maximum dose is 15-20 grams per day. The overconsumption of inulin leads to gastrointestinal symptoms like vomiting, flatulence, nausea, and finally, IBD (Table 2) (Sheng et al., 2023). The colon bacteria, specifically *Bifidobacteria* and *Lactobacillus,* ferment the inulin to produce SCFAs like propionic and butyric acid and show tumour-inhibitory activities, lowering the risk of high triacylglycerol levels and blood lipogenesis, lowering the risk of atherosclerosis, relieving constipation and increasing stools by lowering the pH level of the gut (Figure 6) (Al-Sheraji et al., 2013; Shoaib et al., 2016; Sheng et al., 2023; Karimi et al., 2025).

8.7 Resistant Starch

Carbohydrates are a key dietary component mainly used for energy supply, and their gastrointestinal digestion affects the body's glucose balance. Based on the digestion of starch, it can be categorised into slowly digestible starch (SDS), rapidly digestible starch (RDS) and resistant starch (RS) (Baptista et al., 2024). The term "resistant starch" was first used and described by Englyst et al. (1982) and defined as "a small fraction of starch that was resistant to hydrolysis by exhaustive treatment with amylase and pullulanase *in vitro*" (Englyst et al., 1996). The RDS and SDS are digested in the small intestine, whereas resistance starch resists digestion, and on that resists digestion, it is classified into five types: RS1–RS5. Unlike the SDS and RDS, the resistant starch reaches the colon without digestion, where it acts as a fermentation substrate for gut microflora and, after fermentation, produces SCFAs (Figure 6) (Tian and Sun, 2020; Bojarczuk et al., 2022). The RS is naturally present in several starchy foods, such as unripe bananas, legumes, cooled cooked starches, raw potatoes, and whole grains (Kembabazi et al., 2025). Chemically, starch is a polymer of glucose molecules linked by α-1,4 and α-1,6 linkages. Based on the linkage structure, starch is categorized into amylose and amylopectin. Amylose is a linear alpha 1-4 and consists of 15-20% starch and amylopectin, with both linkage and branched molecules being the principal constituents of starch (Topping and Clifton, 2001). Resistant starch is a low-calorie food that plays a beneficial role in controlling blood glucose levels and decreases several diseases like colon cancer, metabolic disorders and heart disease (Table 2). After fermentation in the colon, RS produces SCFAs that provide energy for colonocytes, managing healthy gut lining, as well as lowering the blood glucose level (Figure 6) (Tian and Sun, 2020; Bojarczuk et al., 2022; Kembabazi et al., 2025).

8.8 Galactooligosaccharides (GOS)

Galactooligosaccharides are lactose-derived molecules comprising 2–8 sugar units, mainly the galactose moiety and terminal glucose molecules (Torres et al., 2010). GOS is primarily produced by lactose syrup via a transglycosylation reaction mediated by the lactase enzyme. The enzyme is known for its hydrolytic and transgalactosylation, causing GOS to have a different saccharide composition based on the enzyme reaction kinetics, structural diversity and degree of polymerisation occurring in GOS (Figure 6) (Gosling et al., 2010). The structure of the GOS provides differences in their chemical properties and applications. Commercially available food-grade GOS are a mixture of oligosaccharides, unreacted lactose, and glucose to some extent, and their physicochemical properties depend on the mixture composition (Sako et al., 1999). Like other prebiotic molecules, the GOS also promotes a healthy gut by enhancing the growth of beneficial bacteria, inhibiting the growth of pathogenic bacteria by inhibiting their adhesion and acting as an immunomodulatory (Table 2). The GOS also prevent gut dysbiosis, ulcerative colitis, and neurological conditions and is anti-allergic and helps in mineral absorption (Figure 6) (Macfarlane et al., 2008; Torres et al., 2010; Olivares-Tenorio et al., 2022; Ignatova et al., 2025).

8.9 Lactulose

Lactulose is a synthetic sugar composed of galactose and fructose through the isomerization of lactose. Lactulose was identified in 1929, but its bifidogenic effects were studied and reported in 1957 when Petuely found that after giving lactulose to infants, the *Bifidobacteria* composition was enhanced and coined the term "bifidogenic factor". Later, this prebiotic molecule was proposed for treating constipation due to its laxative properties (Karakan et al., 2021). It is a 4-O-β-D-galactopyranosyl-D-fructofuranose synthetic disaccharide molecule that contains mainly fructose and glucose linked with the β-1,4-glycosidic bond. It is sweeter than lactose and resistant to the gastrointestinal enzymes of mammals and humans (Figure 6) (Olano and Corzo, 2009). The gut microflora ferments the lactulose into acetic, lactic, and formic acid, along with CO_2 production. These acids extract fluid into the intestinal bowel and soften the stool; hence, this prebiotic molecule is used as a laxative substance (Table 2). In the gut, lactulose enhances the growth of several gut microflora such as *Bifidobacteria longum, Bifidobacteria adolescentis, Bifidobacteria bifidum, Lactobacillus casei, Lactobacillus bulgaricus,* etc. (Figure 6) (Singh et al., 2021; Wang et al., 2022; Vicic et al., 2024).

9. Conclusion

The gastrointestinal tract of humans is dominated by extremely specialized microflora that significantly affect the physiology and metabolism of the host. Advanced studies on gut microbiome research have revealed the profound role of prebiotics and probiotics in enhancing gut health and well-being. The probiotic strains alter the intestinal ecosystem, increase immune response, improve the gut's barrier integrity, and play a crucial role in neurological and metabolic health. They showed promising effects on managing different health effects, viz., metabolic diseases, IBD, allergies, and cancer. Simultaneously, prebiotic molecules enhance the beneficial microbial community in the gut, increasing the production of short-chain fatty acids, aiding immunomodulation, mineral absorption, and energy metabolism. Several prebiotic molecules, like human milk oligosaccharides, GOS, FOS, resistance starch, inulin, and beta-glucans, act as critical dietary substances nurturing the balanced gut microbiota. The emerging studies on the microbiota-gut-brain axis, as well as the microbiota-gut-skin axis, underscore the significance of probiotics and prebiotics in gut health beyond digestion. However, challenges continue, such as the need for strain-specific validation in a particular disease, and primarily, comprehensive human clinical trials are needed urgently. Advancements in science-targeted approaches, such as next-generation probiotics and microbiome therapeutics, embrace promising potential. A comprehensive understanding of host-microbe interaction and specific microbiota-based therapies will further revolutionize the therapeutic and protective strategies.

References

Adams, C.A. and Gutiérrez, B. 2019. Human Milk Oligosaccharides: The First Prebiotics. *Acta Scientific Nutritional Health*, *3*, 172–175.

Ahmad, A., Munir, B., Abrar, M., Bashir, S., Adnan, M.V. and Tabassum, M. 2012. Perspective of β-Glucan as functional ingredient for food industry. *Journal of Nutrition and Food Sciences*, *2*, 100–133.

Al-Sheraji, S.H., Ismail, A., Manap, M.Y., Mustafa, S., Yusof, R.M. and Hassan, F.A. 2013. Prebiotics as functional foods: A review. *Journal of Functional Foods*, *5*, 1542–1553.

Aldars-García, L., Chaparro, M. and Gisbert, J.P. 2021. Systematic review: The gut microbiome and its potential clinical application in inflammatory bowel disease. *Microorganisms*, *9*, 977.

Ali, K., Niaz, N., Waseem, M., Ashraf, W., Hussain, M., Khalid, M.U., et al., 2025. Xylooligosaccharides: A comprehensive review of production, purification, characterization, and quantification. *Food Research International*, *201*, 115631. doi: 10.1016/j.foodres.2024.115631.

Almeida, P.P., Tavares-Gomes, A.L. and Stockler-Pinto, M.B. 2022. Relaxing the "second brain": Nutrients and bioactive compounds as a therapeutic and preventive strategy to alleviate oxidative stress in the enteric nervous system. *Nutrition Reviews*, *80*, 2206–2224.

Amrein, T.M., Gränicher, P., Arrigoni, E. and Amadò, R. 2003. *In vitro* digestibility and colonic fermentation of aleurone from wheat. *LWT - Food Science and Technology*, *36*, 451–460.

Anadón, A., Martínez-Larrañaga, M.R., Caballero, V. and Castellano, V. 2010. Assessment of Prebiotics and Probiotics: An Overview. In: V.R.P. Ronald Ross Watson (Ed.), *Bioactive Foods in Promoting Health*, (pp. 19–41). (Academic Press). doi: 10.1016/B978-0-12-374938-3.00002-5.

Angeloni, S., Ridet, J.L., Kusy, N., Gao, H., Crevoisier, F., Guinchard, S., et al., 2005. Glycoprofiling with micro-arrays of glycoconjugates and lectins. *Glycobiology*, *15*, 31–41.

Asakuma, S., Hatakeyama, E., Urashima, T., Yoshida, E., Katayama, T., Yamamoto, K., et al., 2011. Physiology of consumption of human milk oligosaccharides by infant gut-associated bifidobacteria. *Journal of Biological Chemistry*, *286*, 34583–34592.

Babbar, N., Dejonghe, W., Gatti, M., Sforza, S. and Elst, K. 2016. Pectic oligosaccharides from agricultural by-products: Production, characterization, and health benefits. *Critical Reviews in Biotechnology*, *36*, 594–606.

Badal, V.D., Vaccariello, E.D., Murray, E.R., Yu, K.E., Knight, R., Jeste, D.V., et al., 2020. The Gut Microbiome, Aging, and Longevity: A Systematic Review. *Nutrients*, *12*, 3759.

Bangar, S.P., Gumber, S., Whiteside, W.S. and Phimolsiripol, Y. 2025. Arabinoxylan-based films and coatings for fresh produce: A review of emerging trends in food packaging. *International Journal of Biological Macromolecules*, *310*, 143097. doi: 10.1016/j.ijbiomac.2025.143097.

Baptista, N.T., Dessalles, R., Illner, A.K., Ville, P., Ribet, L., Anton, P.M., et al., 2024. Harnessing the power of resistant starch: A narrative review of its health impact and processing challenges. *Frontiers in Nutrition*, *11*. doi: 10.3389/fnut.2024.1369950.

Barclay, T., Ginic-Markovic, M., Cooper, P. and Petrovsky, N. 2010. Inulin: A versatile polysaccharide with multiple pharmaceutical and food chemical uses. *Journal of Excipients and Food Chemicals*, *1*, 27–50.

Barengolts, E. 2016. Gut microbiota, prebiotics, probiotics, and synbiotics in management of obesity and prediabetes: Review of randomized controlled trials. *Endocrine Practice*, *22*, 1224–1234. doi: 10.4158/EP151157.RA.

Belorkar, S.A. and Gupta, A.K. 2016. Oligosaccharides: A boon from nature's desk. *AMB Express*, *6*, 82. doi: 10.1186/s13568-016-0253-5.

Bodke, H. and Jogdand, S. 2022. Role of Probiotics in Human Health. *Cureus*, *14*, 1–8. doi: 10.7759/cureus.31313.

Bojarczuk, A., Skąpska, S., Mousavi Khaneghah, A. and Marszałek, K. 2022. Health benefits of resistant starch: A review of the literature. *Journal of Functional Foods*, *93*, 105094. doi: 10.1016/j.jff.2022.105094.

Calatayud, G.A. and Suárez, J.E. 2017. A new contribution to the history of probiotics. *Beneficial Microbes*, *8*, 323–325. doi: 10.3920/BM2017.x002.

Carlson, J.L., Erickson, J.M., Lloyd, B.B. and Slavin, J.L. 2018. Health effects and sources of prebiotic dietary fiber. *Current Developments in Nutrition*, *2*, 1–8. doi: 10.1093/cdn/nzy005.

Cavaillon, J.M. and Legout, S. 2016. Centenary of the death of Elie Metchnikoff: A visionary and an outstanding team leader. *Microbes and Infection*, *18*, 577–594. doi: 10.1016/j.micinf.2016.05.008.

Chakravarty, K., Gaur, S., Kumar, R., Jha, N.K. and Gupta, P.K. 2024. Exploring the Multifaceted Therapeutic Potential of Probiotics: A Review of Current Insights and Applications. *Probiotics and Antimicrobial Proteins*, 341–363. doi: 10.1007/s12602-024-10328-x.

Chattopadhyay, A. and Mythili, S. 2018. The journey of gut microbiome: An introduction and its influence on metabolic disorders. *Frontiers in Biology*, *13*, 327–341. doi: 10.1007/s11515-018-1490-6.

Chen, M. and Bie, L. 2025. Intratumoral microbiota for hepatocellular carcinoma : from preclinical mechanisms to clinical cancer treatment. *Cancer Cell International*, *25*, 152. doi: 10.1186/s12935-025-03745-7.

Cheng, L., Shi, J., Peng, H., Tong, R., Hu, Y. and Yu, D. 2023. Probiotics and liver fibrosis: An evidence-based review of the latest research. *Journal of Functional Foods*, *109*, 105773. doi: 10.1016/j.jff.2023.105773.

Chow, J.M. 2002. Probiotics and prebiotics: A brief overview. *Journal of Renal Nutrition*, *12*, 76–86. doi: 10.1053/jren.2002.31759.

Coppa, G.V, Bruni, S., Morelli, L., Soldi, S. and Gabrielli, O. 2004. The first prebiotics in humans: Human milk oligosaccharides. *J. Clin. Gastroenterol.*, *38*, S80–S83.

Costalos, C., Kapiki, A., Apostolou, M. and Papathoma, E. 2008. The effect of a prebiotic supplemented formula on growth and stool microbiology of term infants. *Dev., Early Hum.*, *84*, 45–49.

Costanzo, M. Di, Amoroso, A. and Canani, R.B. 2016. Gut microbiota as a target for food allergy. *J. Pediatr. Gastroenterol. Nutr.*, *63*, S48–S13. doi: 10.1097/01. mpg.0000489618.97688.b8.

Cotta, M.A. and Zeltwanger, R.L. 1995. Degradation and utilization of xylan by the ruminal bacteria Butyrivibrio fibrisolvens and Selenomonas ruminantium. *Applied Environmental Microbiology*, *61*, 4396–4402.

Cremon, C., Barbaro, M.R., Ventura, M. and Barbara, G. 2018. Pre- and probiotic overview. *Current Opinion in Pharmacology*, *43*, 87–92. doi: 10.1016/j.coph.2018.08.010.

Daliri, E.B.M. and Lee, B.H. 2015. New perspectives on probiotics in health and disease. *Food Science and Human Wellness*, *4*, 56–65. doi: 10.1016/j.fshw.2015.06.002.

Damián, M.R., Cortes-Perez, N.G., Quintana, E.T., Ortiz-Moreno, A., Noguez, C.G., Cruceño-Casarrubias, C.E., et al., 2022. Functional Foods, Nutraceuticals and Probiotics: A Focus on Human Health. *Microorganisms*, *10*, 1–13. doi: 10.3390/microorganisms10051065.

Darwesh, M.-A.K., Bakr, W., Omar, T.E.I. and Azzam, M.A. E.-K. N.F. 2024. Unraveling the relative abundance of psychobiotic bacteria in children with Autism Spectrum Disorder. *Scientific Reports*, *14*, 24321.

Das, T.K., Pradhan, S., Chakrabarti, S., Mondal, K.C. and Ghosh, K. 2022. Current status of probiotic and related health benefits. *Applied Food Research*, *2*, 100185. doi: 10.1016/j.afres.2022.100185.

Dehority, B.A. 1966. Characterization of several bovine rumen bacteria isolated with a xylan medium. *Journal of Bacteriology*, *91*, 1724–1729.

Delgado Jiménez, R. and Benakis, C. 2021. The Gut Ecosystem: A Critical Player in Stroke. *NeuroMolecular Medicine*, *23*, 236–241. doi: 10.1007/s12017-020-08633-z.

Di Porzio, A., Barrella, V., Gatto, C., Cigliano, L., Spagnuolo, M.S., Crescenzo, R., et al., 2023. Protective effect of probiotic Limosilactobacillus reuteri DSM17938 against western diet-induced obesity and associated metabolic alterations. *Journal of Functional Foods*, *109*, 105805. doi: 10.1016/j.jff.2023.105805.

Ding, R. Xue, Goh, W.R., Wu, R. na, Yue, X. Qing, Luo, X., Khine, W. W T., et al., 2019. Revisit gut microbiota and its impact on human health and disease. *Journal of Food and Drug Analysis*, *27*, 623–631. doi: 10.1016/j.jfda.2018.12.012.

Dubos, R., Schaedler, R.W., Costello, R. and Hoet, P. (1965). Indigenous, Normal, and Autochthonous Flora of the Gastrointestinal. *The Journal of Experimental Medicine*, *122*, 67–76. doi: 10.1084/jem.122.1.67.

Englyst, H.N., Kingman, S.M., Hudson, G.J. and Cummings, J.H. 1996. Measurement of resistant starch *in vitro* and *in vivo*. *British Journal of Nutrition*, *75*, 749–755.

Esfandiary, A., Abedin-Do, Z.T.-E. A., Mirfakhraie, R., Shirzad, M. and Motevaseli, S.G.-F.E. 2016. Lactobacilli modulate hypoxia-inducible factor (HIF)-1 regulatory pathway in triple negative breast cancer cell line. *Cell J.*, *18*, 237–244. doi: 10.22074/cellj.2016.4319.

Falk, P.G., Hooper, L.V., Midtvedt, T. and Gordon, J.I. 1998. Creating and Maintaining the Gastrointestinal Ecosystem: What We Know and Need to Know from Gnotobiology. *Microbiology and Molecular Biology Reviews*, *62*, 1157–1170. doi: 10.1128/mmbr.62.4.1157-1170.1998.

Farshbafnadi, M., Agah, E. and Rezaei, N. 2021. The second brain: The connection between gut microbiota composition and multiple sclerosis. *J. Neuroimmunol.*, *360*, 577700.

Fuller, R. 1994. Probiotics : An Overview. In: S.A.W. Gibson (Ed.), *Human Health* (pp. 1–2). London: Springer Series in Applied Biology. Springer. doi: https://doi.org/10.1007/978-1-4471-3443-5_4.

Fung, T.C. 2020. The microbiota-immune axis as a central mediator of gut-brain communication. *Neurobiology of Disease*, *136*, 104714. doi: 10.1016/j.nbd.2019.104714.

Gagliardi, A., Totino, V., Cacciotti, F., Iebba, V., Neroni, B., Bonfiglio, G., et al., 2018. Rebuilding the gut microbiota ecosystem. *International Journal of Environmental Research and Public Health*, *15*, 1679. doi: 10.3390/ijerph15081679.

Gao, T., Wang, X., Li, Y. and Ren, F. 2023. The Role of Probiotics in Skin Health and Related Gut–Skin Axis: A Review. *Nutrients*, *15*, 3123. doi: 10.3390/nu15143123.

Geier, M.S., Butler, R.N., Giffard, P.M. and Howarth, G.S. 2007. Lactobacillus fermentum B.R11, a potential new probiotic, alleviates symptoms of colitis induced by dextran sulfate sodium (DSS) in rats. *Int. J. Food Microbiol.*, *114*, 267–274.

Gibson, G.R. and Roberfroid, M.B. 1995. Dietary Modulation of the Human Colonic Microbiota: Introducing the Concept of Prebiotics. *The Journal of Nutrition*, *125*, 1401–1412.

Gill, H.S., Rutherfurd, K.J., Prasad, J. and Gopal, P.K. 2000. Enhancement of natural and acquired immunity by Lactobacillus rhamnosus (HN001), Lactobacillus acidophilus (HN017) and Bifido-bacterium lactis. *The British Journal of Nutrition*, *83*, 167–176.

Gogineni, V.K., Morrow, L.E., Gregory, P.J. and Malesker, M.A. 2013. Probiotics: History and Evolution. *Journal of Infectious Diseases & Preventive Medicine*, *01*, 1–7. doi: 10.4172/2329-8731.1000107.

Goldin, B.R. 2011. Probiotics and Health : From History to Future. In: W. Kneifel and S. Salminen (Eds.), *Probiotics and Health Claims* (pp. 1–16). New Jersey: Blackwell Publishing Ltd.

Gosling, A., Stevens, G.W., Barber, A.R., Kentish, S.E. and Gras, S.L. 2010. Recent advances refining galactooligosaccharide production from lactose. *Food Chemistry*, *121*, 307–318.

Grasten, S., Liukkonen, K.H., Chrevatidis, A., ElNezami, H., Poutanen, K. and Mykkanen, H. 2003. Effects of wheat pentosan and inulin on the metabolic activity of fecal microbiota and on bowel function in healthy humans. *Nutrition Research*, *23*, 1503–1514.

Grönlund, M.M., Lehtonen, O.P., Eerola, E. and Kero, P. (1999). Fecal microflora in healthy infants born by different methods of delivery: Permanent changes in intestinal flora after cesarean delivery. *J. Pediatr. Gastroenterol. Nutr.*, *28*, 19–25.

Grootaert, C., Delcour, J.A., Courtin, C.M., Broekaert, W.F., Verstraete, W. and Wiele, T. Van de 2007. Microbial metabolism and prebiotic potency of arabinoxylan oligosaccharides in the human intestine. *Trends in Food Science & Technology*, *18*, 64–71.

Guarner, F., Perdigon, G., Corthier, G., Salminen, S., Koletzko, B. and Morelli, L. 2005. Should yoghurt cultures be considered probiotic? *British Journal of Nutrition*, *93*, 783–786.

Gul, S. and Durante-Mangoni, E. 2024. Unraveling the Puzzle: Health Benefits of Probiotics: A Comprehensive Review. *Journal of Clinical Medicine*, *13*, 1436. doi: 10.3390/jcm13051436.

Hendry, G.A.F. 1987. The ecological significance of fructan in a contemporary flora. *New Phytol.*, *106*, 201–216.

Hatley, T. and Kappelman, J. (1980). Bears, pigs, and plio- pleistocene hominids: A case for the exploitation of belowground food resources. *Hum. Ecol.*, *8*, 371–387.

Hernández-Pinto, F.J., Miranda-Medina, J.D., Natera-Maldonado, A., Vara-Aldama, Ó., Ortueta-Cabranes, M.P., Vázquez del Mercado-Pardiño, J.A., et al., 2024. Arabinoxylans: A review on protocols for their recovery, functionalities, and roles in food formulations. *International Journal of Biological Macromolecules*, *259*, 129309. doi: 10.1016/j.ijbiomac.2024.129309.

Hespell, R.B. and Cotta, M.A. (1995). Degradation and utilization by Butyrivibrio fibrisolvens H17c of xylans with different chemical and physical properties. *Applied Environmental Microbiology*, *61*, 3042–3050.

Hillman, E.T., Lu, H., Yao, T. and Nakatsu, C.H. 2017. Microbial ecology along the gastrointestinal tract. *Microbes and Environments*, *32*, 300–313. doi: 10.1264/jsme2.ME17017.

Holman, R.C., Stehr-Green, J.K. and Zelasky, M.T. 1989. Necrotizing enterocolitis mortality in the United States, 1979–85. *American Journal of Public Health*, *79*, 987–989.

Hou, K., Wu, Z.X., Chen, X.Y., Wang, J.Q., Zhang, D., Xiao, C., et al., 2022. Microbiota in health and diseases. *Signal Transduction and Targeted Therapy*, *7*, 135. doi: 10.1038/s41392-022-00974-4.

Hutkins, R.W., Krumbeck, J.A., Bindels, L.B., Cani, P.D., Fahey, G., Goh, Y.J., et al., 2016. Prebiotics: Why definitions matter. *Current Opinion in Biotechnology*, *37*, 1–7. doi: 10.1016/j.copbio.2015.09.001.

Ibrahim, S.A., Yeboah, P.J., Ayivi, R.D., Eddin, A.S., Wijemanna, N.D., Paidari, S., et al., 2023. A review and comparative perspective on health benefits of probiotic and fermented foods. *International Journal of Food Science and Technology*, *58*, 4948–4964. doi: 10.1111/ijfs.16619.

Ignatova, I., Arsov, A., Petrova, P. and Petrov, K. 2025. Prebiotic Effects of α- and β-Galactooligosaccharides: The Structure-Function Relation. *Molecules*, *30*, 803. doi: 10.3390/molecules30040803.

International Probiotic Association 2021. *Highlights from the History of Probiotics*. Available at: www.internationalprobiotics.org.

Jan, T., Negi, R., Sharma, B., Kumar, S., Singh, S., Rai, A.K., et al., 2024. Next generation probiotics for human health: An emerging perspective. *Heliyon*, *10*, e35980. doi: 10.1016/j.heliyon.2024.e35980.

Jaskari, J., Kontula, P., Siitonen, A., Jousimies-Somer, H., Sandholm, T.M. and Poutaten, K. 1998. Oat β-glucan and xylan hydrolysates as selective substrates for Bifidobacterium and Lactobacillus strains. *Applied Microbiology and Biotechnology*, *49*, 175–181.

Kaneko, T., Kohmoto, T., Kikuchi, H., Shiota, M., Iino, H. and Mitsuoka, T. 1994. Effects of Isomaltooligosaccharides with Different Degrees of Polymerization on Human Fecal Bifidobacteria. *Bioscience, Biotechnology, and Biochemistry*, *58*, 2288–2290.

Karakan, T., Tuohy, K.M. and Janssen-van Solingen, G. 2021. Low-Dose Lactulose as a Prebiotic for Improved Gut Health and Enhanced Mineral Absorption. *Frontiers in Nutrition*, *8*, 1–15. doi: 10.3389/fnut.2021.672925.

Karimi, I., Ghowsi, M., Mohammed, L.J., Haidari, Z., Nazari, K. and Schiöth, H.B. 2025. Inulin as a Biopolymer: Chemical Structure, Anticancer Effects, Nutraceutical Potential, and Industrial Applications: A Comprehensive Review. *Polymers*, *17*, 412. doi: 10.3390/polym17030412.

Kembabazi, S., Mutambuka, M., Shukri, R., Anwar, F. and Zawawi, N. 2025. Unlocking the potential of resistant starches from underutilized tropical fruits as substrates for fermentation into short-chain fatty acids. *Journal of Functional Foods*, *124*, 106630. doi: 10.1016/j.jff.2024.106630.

Ketabi, A., Dieleman, L.A. and Gaenzle, M.G. (2011). Influence of isomalto- oligosaccharides on intestinal microbiota in rats. *J. Appl. Microbiol.*, *110*, 1297–1306.

Kinross, J.M., Darzi, A.W. and Nicholson, J.K. 2011. The medical importance of the human microbiome. *Genome Medicine*, *3*, 1–12. Available at: http://genomemedicine.com/content/3/3/14.

Kolida, S., Tuohy, K. and Gibson, G.R. (2002). Prebiotic effects of inulin and oligofructose. *British Journal of Nutrition*, *87*, S193–S197.

Konishi, H., Fujiya, M., Tanaka, H., Ueno, N., Moriichi, K., Sasajima, J., et al., 2016. Probiotic-derived ferrichrome inhibits colon cancer progression via JNK-mediated apoptosis. *Nat. Commun.*, *7*, 12365.

Kontula, P., Wright, A. von and Mattila-Sandholm, T. 1998. Oat bran beta-gluco- and xylo-oligosaccharides as fermentative substrates for lactic acid bacteria. *Int. J. Food Microbiol.*, *8*, 163–169.

Kuziel, G.A. and Rakoff-Nahoum, S. 2022. The gut microbiome. *Current Biology*, *32*, R257–R264. doi: 10.1016/j.cub.2022.02.023.

Laden, G. and Wrangham, R. 2005. The rise of the hominids as an adap tive shift in fallback foods: Plant underground storage organs (USOs) and australpith origins. *J. Hum. Evol.*, *49*, 482–498.

Lam, K.L. and Cheung, P.C.K. 2013. Non-digestible long chain beta-glucans as novel prebiotics. *Bioactive Carbohydrates and Dietary Fibre*, *2*, 45–64.

Lamsal, B.P. 2012. Production, health aspects, and potential food uses of dairy prebiotic galactooligosaccharides. *Journal of the Science of Food and Agriculture*, *92*, 2020–2028.

Latif, A., Shehzad, A., Niazi, S., Zahid, A., Ashraf, W., Iqbal, M.W., et al., 2023. Probiotics: Mechanism of action, health benefits, and their application in food industries. *Frontiers in Microbiology*, *14*, 1216674. doi: 10.3389/fmicb.2023.1216674.

Lazaridou, A. and Biliaderis, C.G. (2007). Molecular aspects of cereal β-glucan functionality: Physical properties, technological applications, and physiological effects. *Journal of Cereal Science*, *46*, 101–118.

Leach, J.D., Gibson, G.R. and Loo, J. Van. 2006. Human Evolution, Nutritional Ecology, and Prebiotics in Ancient Diet. *Bioscience and Microflora*, *25*, 1–8. doi: 10.12938/bifidus.25.1.

Leeuwenhoek A Van. 1683. An abstract of a Letter from Antonie van Leeuwenhoek about Animals in the Scruff of the Teeth. *Philosophical Transactions of the Royal Society of London*, *14*, 568–574.

Li, M., Liu, W.L., Zhao, G.Y., Yu, L.M., Wu, Y.H., Wu, Y.F., et al., 2010. Effect of different oligosaccharides on immunity and production performance of piglets. *Anim. Husb. Feed Sci.*, *Z2*, 9–11.

Li, X., Liu, L., Cao, Z., Li, W., Li, H., Lu, C., et al., 2020. Gut microbiota as an "invisible organ" that modulates the function of drugs. *Biomedicine and Pharmacotherapy*, *121*, 109653. doi: 10.1016/j.biopha.2019.109653.

Lilly, D.M. and Stillwell, R.H. 1965. Probiotics: Growth-promoting factors produced by microorganisms. *Science*, *147*, 747–748.

Lopez, H.W., Levrat, M.A., Guy, C., Messager, A., Demigne, C. and Remesy, C. 1999. Effects of soluble corn bran arabinoxylans on cecal digestion, lipid metabolism, and mineral balance (Ca, Mg) in rats. *Journal of Nutritional Biochemistry*, *10*, 500–509.

Lu, K., Dong, S., Wu, X., Jin, R. and Chen, H. 2021. Probiotics in Cancer. *Frontiers in Oncology*, *11*, 638148. doi: 10.3389/fonc.2021.638148.

Macfarlane, G.T., H, S. and Macfarlane, S. (2008). Bacterial metabolism and health-related effects of galacto-oligosaccharides and other prebiotics. *J. Appl. Microbiol.*, *104*, 305–344.

Maciel-Fiuza, M.F., Muller, G.C., Campos, D.M.S., do Socorro Silva Costa, P., Peruzzo, J., Bonamigo, R.R., et al., 2023. Role of gut microbiota in infectious and inflammatory diseases. *Frontiers in Microbiology*, *14*, 1098386. doi: 10.3389/fmicb.2023.1098386.

Mafe, A.N., Edo, G.I., Majeed, O.S., Gaaz, T.S., Akpoghelie, P.O., Isoje, E.F., et al., 2025. A review on probiotics and dietary bioactives: Insights on metabolic well-being, gut microbiota, and inflammatory responses. *Food Chemistry Advances*, *6*, 100919. doi: 10.1016/j.focha.2025.100919.

Manzoor, S., Wani, S.M., Ahmad Mir, S. and Rizwan, D. 2022. Role of probiotics and prebiotics in mitigation of different diseases. *Nutrition*, *96*, 111602. doi: 10.1016/j.nut.2022.111602.

Martin, C. 2020. Gut check. *Current Biology : CB*, *30*, R1099–R1104. doi: 10.1016/j.cub.2020.09.003.

Martín, R., Miquel, S., Ulmer, J., Langella, P. and Bermúdez-Humarán, L.G. 2014. Gut ecosystem: How microbes help us. *Beneficial Microbes*, *5*, 219–233. doi: 10.3920/BM2013.0057.

Maynard, C. and Weinkove, D. 2018. The gut microbiota and ageing. *Subcellular Biochemistry*, *90*, 351–371. doi: 10.1007/978-981-13-2835-0_12.

Metchnikoff, E. 1908. The prolongation of the life. Optimistic study. In: P.C. Mitchell (Ed.). English translation. (1–376). G.P. Putnam's Sons. New York and London.

Michail, S. 2008. Probiotics; Past, Present and Future Perspectives. *Current Pediatric Reviews*, *4*, 96–102. doi: 10.2174/157339608784462034.

Mishra, S. and Acharya, S. 2021. A Brief Overview on Probiotics: The Health Friendly Microbes. *Biomedical and Pharmacology Journal*, *14*, 1869–1880. doi: 10.13005/bpj/2285.

Mohanty, D., Misra, S., Mohapatra, S. and Sahu, P.S. 2018. Prebiotics and synbiotics: Recent concepts in nutrition. *Food Bioscience*, *26*, 152–160. doi: 10.1016/j.fbio.2018.10.008.

Moncrief, J.S., Duncan, A.J., Wright, R.L., Barroso, L.A. and Wilkins, T.D. 1998. Molecular characterization of the fragilysin pathogenicity islet of enterotoxigenic Bacteroides fragilis. *Infection and Immunity*, *66*, 1735–1739. doi: 10.1128/iai.66.4.1735-1739.1998.

Motta, J.P., Bermúdez-Humarán, L.G., Deraison, C., Martin, L., Rolland, C., Rousset, P., et al., 2012. Food-grade bacteria expressing elafin protect against inflammation and restore colon homeostasis. *Science Translational Medicine*, *4*, 158ra144. doi: 10.1126/scitranslmed.3004212.

Mu, Q., Tavella, V.J. and Luo, X.M. 2018. Role of Lactobacillus reuteri in human health and diseases. *Frontiers in Microbiology*, *9*, 1–17. doi: 10.3389/fmicb.2018.00757.

Naeem, H., Hassan, H.U., Shahbaz, M., Imran, M., Memon, A.G., Hasnain, A., et al., 2024. Role of Probiotics against Human Cancers, Inflammatory Diseases, and Other Complex Malignancies. *Journal of Food Biochemistry*, *2024*, 1–23. doi: 10.1155/2024/6632209.

Nakashima, M., Fukumoto, A. and Matsuda, S. 2025. Beneficial Probiotics with New Cancer Therapies for Improved Treatment of Hepatocellular Carcinoma. *Diseases*, *2025,* 13, 111.

Nguyen, T.H. and Haltrich, D. 2013. Microbial production of prebiotic oligosaccharides. In: B. Mcneil, D. Archer, I. Giavasis and L. Harvey (Eds.), *Microbial Production of Food Ingredients, Enzymes, and Nutraceuticals* (pp. 494–530). Austria: Woodhead Publishing Limited. doi: 10.1533/9780857093547.2.494.

Nugent, A.P. 2005. Health properties of resistant starch. *Nutrition Bulletin, 30*, 27–54.

O'Hara, A.M. and Shanahan, F. 2006. The gut flora as a forgotten organ. *EMBO Reports*, *7*, 688–693. doi: 10.1038/sj.embor.7400731.

O'Toole, P.W. and Jeffery, I.B. 2015. Gut microbiota and aging. *Science*, *350*, 1214–1215. doi: 10.1126/science.aac8469.

Olano, A. and Corzo, N. 2009. Lactulose as a food ingredient. *J. Sci. Food Agric.*, *89*, 1987–1990.

Olivares-Tenorio, M.-L., Orrego, D., Klotz-Ceberio, B.-F., Palanca, C. and Tortajada-Serra, M. 2022. Galactooligosaccharides Food technological applications prebiotic health. *JSFA Reports*, *2*, 578–590.

Parhizkar, E., Vosough, P., Baneshi, M., Keshavarzi, A., Lohrasbi, P., Taghizadeh, S., et al., 2025. Probiotics and gut microbiota modulation: Implications for skin health and disease management. *Archives of Microbiology*, *207*, 1–22. doi: 10.1007/s00203-025-04267-6.

Parker, R.B. 1974. Probiotics, the other half of the antibiotic story. *Animal Nutrition and Health*, *29*, 4–8.

Patel, S. and Goyal, A. 2011. Functional oligosaccharides: Production, properties, and applications. *World Journal of Microbiology and Biotechnology*, *27*, 1119–1128. doi: 10.1007/s11274-010-0558-5.

Penders, J., Thijs, C., Vink, C., Stelma, F.F., Snijders, B., Kummeling, I., et al., 2006. Factors Influencing the Composition of the Intestinal Microbiota in Early Infancy. *Pediatrics*, *118*, 511–521.

Pillemer, L. and Ecker, E.E. 1941. Anticomplementary Factor in Fresh Yeast. *Journal of Biological Chemistry*, *137*, 139–142.

Plamada, D. and Vodnar, D.C. 2021. Polyphenols—Gut Microbiota Interrelationship: A Transition to a New Generation of Prebiotics. *Nutrients, 14*, 137.

Plaza-Diaz, J., Ruiz-Ojeda, F.J., Gil-Campos, M. and Gil, A. 2019. Mechanisms of Action of Probiotics. *Advances in Nutrition, 10*, S49–S66. doi: 10.1093/advances/nmy063.

Pushpanathan, P., Mathew, G., Selvarajan, S., Seshadri, K. and Srikanth, P. 2019. Gut microbiota and its mysteries. *Indian Journal of Medical Microbiology, 37*, 268–277. doi: 10.4103/ijmm.IJMM_19_373.

Quigley, E.M.M. 2019. Prebiotics and Probiotics in Digestive Health. *Clinical Gastroenterology and Hepatology, 17*, 333–344. doi: 10.1016/j.cgh.2018.09.028.

Ragonnaud, E. and Biragyn, A. 2021. Gut microbiota as the key controllers of "healthy" aging of elderly people. *Immunity and Ageing, 18*, 1–11. doi: 10.1186/s12979-020-00213-w.

Ranjha, M.M.A.N., Shafique, B., Batool, M., Kowalczewski, P.Ł., Shehzad, Q., Usman, M., et al., 2021. Nutritional and health potential of probiotics: A review. *Applied Sciences (Switzerland), 11*, 11204. doi: 10.3390/app112311204.

Rehman, H., Vahjen, W., Kohl-Parisini, A., Ijaz, A. and Zentek, J. 2009. Influence of fermentable carbohydrates on the intestinal bacteria and enteropathogens in broilers. *Worlds Poult. Sci. J., 65*, 75–90.

Ren, J., Li, H., Zeng, G., Pang, B., Wang, Q. and Wei, J. 2023. Gut microbiome-mediated mechanisms in aging-related diseases: Are probiotics ready for prime time? *Frontiers in Pharmacology, 14*, 1–17. doi: 10.3389/fphar.2023.1178596.

Rinninella, E., Raoul, P., Cintoni, M., Franceschi, F., Miggiano, G.A.D., Gasbarrini, A., et al., 2019. What is the Healthy Gut Microbiota Composition? A Changing Ecosystem across Age, Environment, Diet, and Diseases. *Microorganisms, 7*, 14. doi: 10.3390/microorganisms7010014.

Roberfroid, M.B. (1999). Concepts in functional foods: The case of inulin and oligofructose. *The Journal of Nutrition, 129*, 1396–1401.

Roper, H. and Koch. H. 1988. New carbohydrate-derivatives for biotechnical and chemical processes. *Starch, 40*, 453-459.

Rycroft, C.E., Jones, M.R., Gibson, G.R. and Rastall, R.A. 2001. A comparative *in vitro* evaluation of the fermentation properties of prebiotic oligosaccharides. *J. Appl. Microbiol., 91*, 878–887.

Sabater-Molina, M., Larqué, E., Torrella, F. and Zamora, S. 2009. Dietary fructooligosaccharides and potential benefits on health. *Journal of Physiology and Biochemistry, 65*, 315–328.

Saber, A., Alipour, B., Faghfoori, Z. and Khosroushahi, A.Y. 2017. Cellular and molecular effects of yeast probiotics on cancer. *Crit. Rev. Microbiol., 43*, 96–115. doi: 10.1080/1040841X.2016.1179622.

Saburi, W., Mori, H., Saito, S., Okuyama, M. and Kimura, A. 2006. Structural elements in dextran glucosidase responsible for high specificity to long chain substrate. *Biochim. Biophys. Acta, 1764*, 688–698.

Sako, T., Matsumoto, K. and Tanaka, R. 1999. Recent progress on research and applications of non-digestible galacto-oligosaccharides. *Int. Dairy J., 9*, 69–80.

Salem, I., Ramser, A., Isham, N. and Ghannoum, M.A. 2018. The Gut Microbiome as a Major Regulator of the Gut-Skin Axis. *Front. Microbiol., 9*, 1459.

Samanta, A.K., Jayapal, N., Jayaram, C., Roy, S., Kolte, A.P., Senani, S., et al., 2015. Xylooligosaccharides as prebiotics from agricultural by-products: Production and applications. *Bioactive Carbohydrates and Dietary Fibre, 5*, 62–71.

Samanta, S. 2022. Potential impacts of prebiotics and probiotics on cancer prevention. *Anti Cancer Agents Med. Chem., 22*, 605–628. doi: doi: 10.2174/1871520621999201210220442.

Sanders, M.E., Benson, A., Lebeer, S., Merenstein, D.J. and Klaenhammer, T.R. 2018. Shared mechanisms among probiotic taxa: Implications for general probiotic claims. *Current Opinion in Biotechnology, 49*, 207–216. doi: 10.1016/j.copbio.2017.09.007.

Sanford, J.A. and Gallo, R.L. 2013. Functions of the skin microbiota in health and disease. *Semin. Immunol., 25*, 370–377.

Santacroce, L., Charitos, I.A. and Bottalico, L. 2019. A successful history: Probiotics and their potential as antimicrobials. *Expert Review of Anti-Infective Therapy, 17*, 635–645. doi: 10.1080/14787210.2019.1645597.

Sarkar, N., Mahajan, A.A., Pathak, S., Seth, P., Chowdhury, A., Ghose, I., et al., 2025. Beta-Glucans in Biotechnology : A Holistic Review with a Special Focus on Yeast. *Bioengineering, 12*, 365. doi: https://doi.org/10.3390/bioengineering12040365.

Savage, D.C. 1977. Microbial ecology of the gastrointestinal tract. *Annual Review of Microbiology, 31*, 107–133. doi: 10.1146/annurev.mi.31.100177.000543.

Saville, S.H., Younes, J. A., Paraskevakos, G. and Venema, K. 2025. The prebiotic landscape: History, health and physiological benefits, and regulatory challenges, an IPA perspective part 1. *Beneficial Microbes 16*, 1–33. doi: 10.1163/18762891-bja00056.

Schupfer, E., Pak, S.C., Wang, S., Micalos, P.S., Jeffries, T., Ooi, S.L., et al., 2021. The effects and benefits of arabinoxylans on human gut microbiota: A narrative review. *Food Bioscience, October*, 101267.

Shahverdi, E. 2016. Probiotics and gastrointestinal diseases. *Int. J. Dig. Dis., 2*, 1–2. doi: 10.4172/2472-1891.100022.

Shao, T., Hsu, R., Rafizadeh, D.L., Wang, L., Bowlus, C.L., Kumar, N., et al., 2023. The gut ecosystem and immune tolerance. *Journal of Autoimmunity, 141*, 103114. doi: 10.1016/j.jaut.2023.103114.

(2022). Shehata, A.A., Yalçın, S., Latorre, J.D., Basiouni, S., Attia, Y.A., Abd el-Wahab, A., et al., (2022). Probiotics, Prebiotics, and Phytogenic Substances for Optimizing Gut Health in Poultry. *Microorganisms*, *10*, 395. doi: 10.3390/microorganisms10020395.

Sheng, W., Ji, G. and Zhang, L. 2023. Immunomodulatory effects of inulin and its intestinal metabolites. *Frontiers in Immunology*, *14*, 1–16. doi: 10.3389/fimmu.2023.1224092.

Shoaib, M., Shehzad, A., Omar, M., Rakha, A., Raza, H., Sharif, H.R., et al., 2016. Inulin: Properties, health benefits, and food applications. *Carbohydrate Polymer*, *147*, 444–454.

Silva, T.F. da, Glória, R. de A., Americo, M.F., Freitas, A. dos S., de Jesus, L.C.L., Barroso, F.A.L., et al., 2024. Unlocking the Potential of Probiotics: A Comprehensive Review on Research, Production, and Regulation of Probiotics. *Probiotics and Antimicrobial Proteins*, *16*, 1687–1723. doi: 10.1007/s12602-024-10247-x.

Singh, A.K., Upadhyay, S.K. and Singh, S.P. 2023. Microbial Bioreactor: An Introduction,. In: S.K. Upadhyay and S.P. Singh (Eds.), *Microbial Bioreactor for Industrial Molecules* (pp. 1–17). Wiley. 1–17.

Singh, R.P. and Bhardwaj, A. 2023. B-Glucans: A Potential Source for Maintaining Gut Microbiota and the Immune System. *Frontiers in Nutrition*, *10*, 1–22. doi: 10.3389/fnut.2023.1143682.

Singh, S., Kola, P., Kaur, D., Singla, G., Mishra, V., Panesar, P.S., et al., 2021. Therapeutic Potential of Nutraceuticals and Dietary Supplements in the Prevention of Viral Diseases: A Review. *Frontiers in Nutrition*, *8*, 1–16. doi: 10.3389/fnut.2021.679312.

So, S.S., L.Wan, M. and El-Nezami, H. 2017. Probiotics-mediated suppression of cancer. *Curr. Opin. Oncol.*, *29*, 62–72.

Soccol, C.R., Vandenberghe, L.P. de S., Spier, M.R., Medeiros, A.B.P., Yamaguishi, C.T., De Dea Lindner, J., et al., 2010. The potential of probiotics: A review. *Food Technology and Biotechnology*, *48*, 413–434.

Sokol, H., Pigneur, B., Watterlot, L., Lakhdari, O., Bermudez-Humaran, L.G., Gratadoux, J.-J., et al., 2008. Faecalibacterium prausnitzii is an anti-inflammatory commensal bacterium identified by gut microbiota analysis of Crohn disease patients. *PNAS*, *105*, 16731–16736 SEE. doi: 10.1002/path.1711620408.

Sorboni, S.G., Moghaddam, H.S., Jafarzadeh-Esfehani, R. and Soleimanpour, S. 2022. A Comprehensive Review on the Role of the Gut Microbiome in Human Neurological Disorders. *Clinical Microbiology Reviews*, *35*, e00338-20. doi: 10.1128/CMR.00338-20.

Stahl, A.B. 1984. Hominid dietary selection before fire. *Curr. Anthropol.*, *25*, 151–157.

Sun, W., Tao, L., Qian, C., Xue, P.P., Du, S.S. and Tao, Y.N. 2024. Human milk oligosaccharides: Bridging the gap in intestinal microbiota between mothers and infants. *Frontiers in Cellular and Infection Microbiology*, *14*, 1–19. doi: 10.3389/fcimb.2024.1386421.

Sun, X., Fiala, J.L. and Lowery, D. 2016. Patent watch: Modulating the human microbiome with live biotherapeutic products: intellectual property landscape. *Nat. Rev. Drug Discov.*, *15*, 224–225.

Sung, H.Y., Jeong, H.J. and Choi, Y. S. 2004. Effects of fructans and isomalto-oligosaccharide on large bowel mass and plasma and fecal immunoglobulin A in rat. *Nutr. Sci.*, *7*, 196–200.

Suvorov, A. 2013. Gut Microbiota, Probiotics, and Human Health. *Bioscience of Microbiota, Food and Health*, *32*, 81–91. doi: 10.3389/fphar.2013.00011.

Tannock, G.W. 2003. Probiotics: Time for a dose of realism. *Current Issues in Intestinal Microbiology*, *4*, 33–42.

Teferra, T.F. (2021). Possible actions of inulin as prebiotic polysaccharide: A review. *Food Frontiers*, *2*, 407–416. doi: 10.1002/fft2.92.

Thakur, A., Sharma, P., Sharma, A.J., Sharma, N., Singh, S., Chamoli, N., et al., 2025. Probiotics: Microbes for human health and beyond. *Food Materials Research*, *5*, e001. doi: 10.48130/fmr-0024-0029.

Tian, S. and Sun, Y. 2020. Influencing factor of resistant starch formation and application in cereal products: A review. *International Journal of Biological Macromolecules*, *149*, 424–431. doi: 10.1016/j.ijbiomac.2020.01.264.

Topping, D.L. and Clifton, P.M. 2001. Short-chain fatty acids and human colonic function: Roles of resistant starch and nonstarch polysaccharides. *Physiological Reviews*, *81*, 1031–1064.

Torres, D.P.M., Gonçalves, M. do P.F., Teixeira, J.A. and Rodrigues, L.R. 2010. Galacto-Oligosaccharides: Production, properties, applications, and significance as prebiotics. *Comprehensive Reviews in Food Science and Food Safety*, *9*, 438–454. doi: 10.1111/j.1541-4337.2010.00119.x.

Tymczyszyn, E.E., Santos, M.I., Costa, M.C., Illanes, A. and Gómez-Zavaglia, A. 2014. History, synthesis, properties, applications, and regulatory issues of prebiotic oligosaccharides. *Research Signpost*, *37661*, 127–154. Available at: https://www.trnres.com/ebook/uploads/gilcontent/T_1401355846gil 5.pdf.

Uauy, R.D., Fanaroff, A.A., Korones, S.B., Phillips, E.A., Phillips, J.B. and Wright, L.L. 1991. Necrotizing enterocolitis in very low birth weight infants: Biodemographic and clinical correlates. *The Journal of Pediatrics*, *119*, 630–638.

Ulsemer, P., Toutounian, K., Schmidt, J., Karsten, U. and Goletz, S. 2012. Preliminary safety evaluation of a new Bacteroides xylanisolvens isolate. *Applied and Environmental Microbiology*, *78*, 528–535. doi: 10.1128/AEM.06641-11.

Un-Nisa, A., Khan, A., Zakria, M., Siraj, S., Ullah, S., Tipu, M.K., et al., 2023. Updates on the Role of Probiotics against Different Health Issues: Focus on Lactobacillus. *International Journal of Molecular Sciences*, *24*, 142. doi: 10.3390/ijms24010142.

Underwood, M.A., Gaerlan, S.C., Leoz, M.L.A. De, Dimapasoc, L.M., Kalanetra, K.M., Lemay, D., et al., 2015. Human milk oligosaccharides in premature infants: Absorption, excretion, and influence on the intestinal microbiota. *Pediatr. Res.*, *78*, 670–677.

Urashima, T., Ajisaka, K., Ujihara, T. and Nakazaki, E. 2025. Recent advances in the science of human milk oligosaccharides. *BBA Advances*, *7*, 100136. doi: 10.1016/j.bbadva.2024.100136.

Valdes, A.M., Walter, J., Segal, E. and Spector, T.D. 2018. Role of the gut microbiota in nutrition and health. *BMJ* (*Online*), *361*, 36–44. doi: 10.1136/bmj.k2179.

Varela-Trinidad, G.U., Domínguez-Díaz, C., Solórzano-Castanedo, K., Íñiguez-Gutiérrez, L., Hernández-Flores, T. de J. and Fafutis-Morris, M. 2022. . Probiotics: Protecting Our Health from the Gut. *Microorganisms*, *10*, 1–43. doi: 10.3390/microorganisms10071428.

Vázquez, M.J., Garrote, G., Alonso, J.L., Domínguez, H. and Parajó, J.C. 2005. Refining of autohydrolysis liquors for manufacturing xylooligosaccharides: Evaluation of operational strategies. *Bioresource Technology*, *96*, 889–896. doi: 10.1016/j.biortech.2004.08.013.

Vicic, V., Pandel Mikuš, R. and Ferjančič, B. 2024. Review of history and mechanisms of action of lactulose (4-O-β-D-Galactopyranosyl-β-D-fructofuranose): Present and future applications in food. *Journal of Food Science and Technology*, *61*, 2036–2045. doi: 10.1007/s13197-024-05997-z.

Villéger, R., Pinault, E., Vuillier-Devillers, K., Grenier, K., Landolt, C., Ropartz, D., et al., 2022. Prebiotic Isomaltooligosaccharide Provides an Advantageous Fitness to the Probiotic Bacillus subtilis CU1. *Applied Sciences* (*Switzerland*), *12*, 6404. doi: 10.3390/app12136404.

Vyas, U. and Ranganathan, N. 2012. Probiotics, prebiotics, and synbiotics: Gut and beyond. *Gastroenterology Research and Practice*, *2012*, 1–16. doi: 10.1155/2012/872716.

Wang, M., Wang, L., Lyu, X., Hua, X., Goddard, J.M. and Yang, R. 2022. Lactulose production from lactose isomerization by chemo-catalysts and enzymes: Current status and future perspectives. *Biotechnology Advances*, *60*, 108021. doi: 10.1016/j.biotechadv.2022.108021.

Weichselbaum, E. 2009. Probiotics and health: A review of the evidence. *Nutrition Bulletin*, *34*, 340–373. doi: 10.1111/j.1467-3010.2009.01782.x.

Wen, L. and Duffy, A. 2017. Factors Influencing the Gut Microbiota, Inflammation, and Type 2 Diabetes. *The Journal of Nutrition*, *147*, 1468S–1475S.

Whitehead, T.R. and Hespell, R.B. 1990. The genes for three xylan-degrading activities from Bacteroides ovatus are clustered in a 3.8-kilobase region. *Journal of Bacteriology*, *172*, 2408–2412. doi: 10.1128/jb.172.5.2408-2412.1990.

Williams, N.T. 2010. Probiotics. *American Journal of Health-System Pharmacy*, *67*, 449–458. doi: 10.2146/ajhp090168.

Willing, B.P. and Finlay, B.B. 2009. Gut Microbiology: Fitting into the Intestinal Neighbourhood. *Current Biology*, *19*, R457–R459. doi: 10.1016/j.cub.2009.04.025.

Woo, T.D.H., Oka, K., Takahashi, M., Hojo, F., Osaki, A., Hanawa, T., et al., 2011. Inhibition of the cytotoxic effect of Clostridium difficile *in vitro* by Clostridium butyricum MIYAIRI 588 strain. *Journal of Medical Microbiology*, *60,* 1617–1625. doi: 10.1099/jmm.0.033423-0.

Wu, Y., Pan, L., Shang, Q.H., Ma, X.K., Long, S.F., Xu, Y.T., et al., 2017. Effects of isomalto-oligosaccharides as potential prebiotics on performance, immune function, and gut microbiota in weaned pigs. *Animal Feed Science and Technology*, *230*, 126–135.

Xue, J., Ajuwon, K. M. and Fang, R. 2020. Mechanistic insight into the gut microbiome and its interaction with host immunity and inflammation. *Animal Nutrition*, *6*, 421–428. doi: 10.1016/j.aninu.2020.05.007.

Yahfoufi, N., Mallet, J.F., Graham, E. and Matar, C. 2018. Role of probiotics and prebiotics in immunomodulation. *Current Opinion in Food Science*, *20*, 82–91. doi: 10.1016/j.cofs.2018.04.006.

You, S., Ma, Y., Yan, B., Pei, W., Wu, Q., Ding, C., et al., 2022. The promotion mechanism of prebiotics for probiotics: A review. *Frontiers in Nutrition*, *9*, 1000517. doi: 10.3389/fnut.2022.1000517.

Zang, T., Han, L., Lu, Z., Tan, L., Liang, D., Shen, X., et al., 2024. The History and Prediction of Prebiotics and Postbiotics: A Patent Analysis. *Nutrients*, *16*, 380. doi: https:// doi.org/10.3390/nu16030380.

Zendeboodi, F., Khorshidian, N., Mortazavian, A.M. and da Cruz, A.G. 2020. Probiotic: Conceptualization from a new approach. *Current Opinion in Food Science*, *32*, 103–123. doi: 10.1016/j.cofs.2020.03.009.

Ziemer, C.J. and Gibson, G.R. (1998). An overview of probiotics, prebiotics, and synbiotics in the functional food concept: Perspectives and future strategies. *Int. Dairy J.*, *8*, 473–479.

CHAPTER 6

Impact of Functional Foods on Immune System Support

Meena Tippana[1], Pisal Ketali Prakash[2] and Priya Prasad[3*]

1. Introduction

A food can be termed "functional" if it has been adequately proven to positively affect any number of target functions in the body, in addition to adequate nutritional effects ("Scientific Concepts of Functional Foods in Europe Consensus Document," 1999). According to the Institute of Medicine, USA, functional food is any modified food or food ingredient that can give a health benefit beyond the traditional nutrients it provides (Trumbo et al., 2002). Functional foods are an emerging field in nutrition science and public health, providing benefits beyond simple sustenance. These are foods that, through naturally occurring or added bioactive compounds-such as dietary fibre, probiotics, antioxidants, or phytochemicals, can have a favourable impact on health, including the prevention or treatment of chronic diseases. Functional foods, as opposed to supplements, are ingested as part of the regular diet and are widely recognized for supporting physiological functioning and improving overall well-being. ("Scientific Concepts of Functional Foods in Europe Consensus Document," 1999). The notion of functional foods originated in Japan in the 1980s with the introduction of "Foods for Specified Health Uses" (FOSHU), and it has since attracted global interest as dietary habits and health challenges change (Iwatani and Yamamoto, 2019). The rising burden of noncommunicable illnesses has fuelled interest in preventive nutrition, in which functional foods play an important role. Yogurt is supplemented with probiotics to support digestive health, oats containing beta-glucan to lower cholesterol, and fortified foods such as milk with additional vitamin D to encourage bone health (Morvaridzadeh et al., 2021). As scientific understanding of food components expands, functional foods are being incorporated into global dietary standards and public health programmes, highlighting their value in promoting long-term health and lowering illness risk (Trumbo et al., 2002). These foods also play a major role in controlling the immune system.

The immune system consists of an integrated network of specialized cells, tissues, and organs responsible for protecting the body against pathogens and foreign agents. Its major function is to recognize and eradicate dangerous agents such as bacteria, fungi, viruses, and parasites while separating them from the body's normal healthy cells to prevent autoimmunity (Miller, 2008). In addition to host defence, the immune system is involved in wound healing, tissue regeneration, tumour surveillance, and overall body homeostasis (Belokhvostova et al., 2018). The immune

[1-3] Department of Biotechnology, National Institute of Technology Warangal, Warangal, Telangana-506004, India.
* Corresponding author: priyap@nitw.ac.in

DOI: 10.1201/9781003659808-6

system has two primary components: innate immunity and adaptive immunity. The innate immunity is the initial line of defence, delivering a rapid, nonspecific reaction to pathogens. It consists of physical and chemical barriers, including the skin, mucosal surfaces, and antimicrobial peptides, as well as cellular components like macrophages, dendritic cells, neutrophils, and natural killer (NK) cells. Engagement of pattern recognition receptors with pathogen-associated molecular patterns (PAMPs) activates early inflammatory signalling pathways (Wang et al., 2024). In addition to the innate response, adaptive immunity is extremely specialized and capable of creating immunological memory. B-Lymphocytes, which create antigen-specific antibodies, and T cells, which include helper T cells (CD4+), cytotoxic T cells (CD8+), and regulatory T cells, are responsible for adaptive immune responses. Helper T cells use cytokine signalling to coordinate immune responses, cytotoxic T cells eliminate infected or aberrant cells, whereas regulatory T cells maintain immunological tolerance (Sun et al., 2023). Adaptive immunity is slower to respond during initial exposure to a pathogen, but upon subsequent exposures, memory B and T cells facilitate a more rapid and robust immune reaction that forms the basis of vaccination strategies (Quast and Tarlinton, 2021). The synchronization of innate and adaptive immunity is critical for an efficient immune response. Antigen-presenting cells (APCs), notably dendritic cells, bridge the gap between the two systems by collecting pathogen antigens and delivering them to naive T lymphocytes in lymphoid organs (Banchereau and Steinman, 1998). Small signalling molecules known as cytokines and chemokines facilitate communication among immune cells, regulating their growth, differentiation, and migration to infection sites. A fully functioning immune system maintains a careful balance of activation and control. Dysregulation can lead to immunodeficiency (e.g., HIV/AIDS) or autoimmune diseases, in which the immune system destroys self-tissues (Hoang Nguyen et al., 2025). Hypersensitivity reactions, including allergies and anaphylaxis, occur when the immune system overreacts to harmless environmental antigens (Boasso et al., 2009). With advances in immunology, the immune system is increasingly recognized as a key component not only in defending against infectious diseases but also in the treatment of cancer, aging, chronic inflammation, and emerging pathogens. Immunotherapy advances, such as immune checkpoint inhibitors and CAR T-cell therapy, demonstrate the therapeutic promise of modifying immune responses for disease treatment (Mellman et al., 2011).

Functional foods have recently acquired popularity as a critical tool for maintaining immunological health, particularly in light of rising global health issues such as infectious illnesses, chronic inflammation, and lifestyle-induced immune disorders. Functional foods are those that provide health advantages beyond basic nutrition because they contain bioactive components such as antioxidants, vitamins, minerals, probiotics, and essential fatty acids (Worqlul, 2024). These items are key to dietary strategies for modifying immune responses and increasing disease resistance. Nutrition plays a crucial role in regulating immune function. Dietary micronutrients, including vitamins A, C, D, E, B6, B12, folate, as well as minerals such as zinc, selenium, and iron, are essential for immune cell development, activity, and regulation. These nutritional deficiencies can impair immunological responses, increase infection susceptibility, and lengthen recovery duration. Functional foods high in these nutrients, either naturally or fortified, can help restore immunological balance and boost both innate and adaptive immunity (Pecora et al., 2020). The impact of probiotics on immunity is one of the most extensively investigated areas of functional food research. Probiotics, derived from fermented foods like curd, yogurt, kimchi, and kefir, aid in maintaining a healthy gut microbiota, which is associated with immunological homeostasis. The gastrointestinal tract includes more than 70% of the body's immune cells; therefore, gut health is critical for avoiding systemic inflammation and immunological dysfunction. Probiotics have been demonstrated to activate natural killer cells, macrophages, and dendritic cells while controlling cytokine production (Lomax and Calder, 2009). In addition to probiotics, prebiotics (non-digestible fibres) found in foods such as chicory root, garlic, onions, and bananas, act as substrates for good gut flora. Their fermentation in the colon produces short-chain fatty acids (SCFAs), which have immunomodulatory effects and help to maintain the gut barrier (Victoria Obayomi et al., 2024). Polyphenols, notably flavonoids present in berries, tea, cocoa, and citrus fruits, have been shown to have immuno-supportive and anti-inflammatory properties via antioxidant activity and modulation of signalling pathways that control immune cell behaviour. These chemicals can regulate cytokine production, reduce oxidative stress, and improve the function of T and B cells (Rudrapal et al., 2022). Omega-3 fatty acids, particularly docosahexaenoic acid (DHA) and eicosapentaenoic acid (EPA), contribute to immune regulation through their anti-inflammatory and immunomodulatory properties. They also contribute to inflammation resolution by producing specific pro-resolving lipid mediators and regulating T cell activity and cytokine expression (Calder, 2013). During the COVID-19 pandemic, public interest in functional foods increased, indicating a broader cultural movement towards preventive health. Functional foods are widely regarded not only as nutritional supplements but also as essential components of immunological resilience at various life stages. Continued research and innovation in this area are critical for establishing tailored dietary interventions that lower illness risk and improve overall health. Research and innovation in this area are critical for establishing tailored dietary interventions that lower illness risk and improve overall health (Farzana et al., 2022).

2. Fundamentals of the Immune System

Understanding the immune system is critical for understanding how the body protects itself against infections and maintains homeostasis. This section discusses the fundamentals of immunity, important immune cells, immunomodulation, and the impact of nutrition on immune function.

2.1 Components of Innate and Adaptive Immunity

Innate and adaptive immunity represent two integrated arms of the immune system with specialized roles in host protection.

2.1.1 Innate Immunity

Innate immunity is the body's first line of defense, generating quick and nonspecific responses to invading pathogens. It includes physical barriers such as the skin, mucosal membranes, and cilia, which prevent pathogen invasion (Anaya et al., 2013). Chemical barriers, such as lysozymes in saliva and the acidic environment of the stomach, further limit bacterial development (Matwiejczyk et al., 2025). Innate immunity's cellular components include macrophages, dendritic cells, neutrophils, eosinophils, basophils, and NK cells, each of which plays a unique role in pathogen detection and elimination. These cells detect PAMPs using pattern recognition receptors, including Toll-like receptors (TLRs), to initiate inflammatory responses (Chaplin, 2010). The complement system, a group of plasma proteins, further aids by opsonizing pathogens, forming membrane attack complexes, and recruiting immune cells (Li and Wu, 2021).

2.1.2 Adaptive Immunity

Adaptive immunity evolves and produces a particular response to infections. B lymphocytes differentiate into plasma cells, which release antibodies against specific antigens. T lymphocytes include CD4+ helper T cells that coordinate immune responses and CD8+ cytotoxic T cells that eliminate infected cells (Chi et al., 2024). Adaptive immunity is characterized by the development of immunological memory, allowing the host to mount faster and more robust secondary responses following re-exposure to previously encountered antigens (Deo et al., 2024; Lebien et al., 2008). The innate and adaptive systems are highly intertwined. Dendritic cells APCs, connecting innate sensing to adaptive activation by delivering processed antigens to naive T cells (Banchereau and Steinman, 1998).

2.2 Key Immune Cells and Their Function

The immune system includes a diverse range of specialized cells, each with unique roles in pathogen recognition, elimination, and immune regulation.

2.2.1 Innate Immune Cells

Macrophages are essential phagocytic cells that engulf pathogens and present antigens to T cells, while neutrophils act as the first responders, performing phagocytosis and releasing reactive oxygen species (ROS) to neutralize infectious agents. As key components of innate immunity, dendritic cells initiate adaptive immune responses through antigen presentation, whereas NK cells contribute to early immune surveillance by targeting virus-infected and tumour cells (Marshall et al., 2018).

2.2.2 Adaptive Immune Cells

B lymphocytes differentiate into antibody-secreting plasma cells and long-lived memory B cells that contribute to sustained immune protection. T lymphocytes are broadly classified into helper T cells ($CD4^{+}$) and cytotoxic T cells ($CD8^{+}$) (Luckheeram et al., 2012). $CD4^{+}$ helper T cells further differentiate into distinct functional subsets, including Th1 cells that support cell-mediated immunity, Th2 cells that promote humoral immune responses, Th17 cells that drive inflammatory pathways, and regulatory T cells (Tregs) that restrain excessive immune activation and maintain immune tolerance (Kan et al., 2024). Cytotoxic T cells release perforins and granzymes to eliminate infected or abnormal cells. Together, B and T lymphocytes coordinate antigen-specific immune defence and long-term memory formation (Gagliani and Huber, 2017).

2.2.3 Other Important Cells

Eosinophils help fight parasite infections and mediate allergy responses. γδ T cells and innate lymphoid cells produce cytokines and respond to stress signals, bridging innate and adaptive immunity. Each immune cell type helps to maintain a delicate balance of immune activation and regulation, ensuring efficient defense while avoiding tissue harm (de Oliveira Henriques and Penido, 2012; Morita et al., 2016).

Each cell type contributes to a coordinated immune response by balancing activation and control to protect the host while minimizing damage.

2.3 Immunomodulation and Its Significance

Immunomodulation is the change of the immune response by internal or external stimuli, which can either enhance or reduce immunological function. It comprises immunomodulation, immunostimulation, and immunosuppression (Strzelec et al., 2023). Immunomodulation involves the alteration of immune responses through either enhancement or suppression. Immunostimulation enhances immune function and is beneficial in infections or cancer. Conversely, immunosuppression reduces immune activity, which is necessary in autoimmune disorders and during organ transplantation (Sharma et al., 2024). Immunotolerance maintains immune balance by preventing reactions to self-antigens or harmless environmental agents (Han et al., 2025).

Mechanisms of immunomodulation involve cytokine signalling, antigen presentation, receptor-ligand interactions, and immune cell differentiation. Pharmacological agents such as corticosteroids and monoclonal antibodies are well-known immunomodulators (Strzelec et al., 2023). Additionally, lifestyle factors, including diet, stress, sleep, and exercise, influence immune regulation. Clinically, immunosuppressive drugs such as methotrexate are used to control autoimmune disorders, while immunostimulatory therapies like checkpoint inhibitors enhance the immune response against cancer (Ren et al., 2024; Temizoz et al., 2016). Natural immunomodulators such as probiotics, polyphenols, and medicinal herbs are increasingly recognized for maintaining immune balance without severe side effects (Alhazmi et al., 2021).

2.4 Relationship between Nutrition and Immune Function

Nutrition profoundly influences immune function. Micronutrients include vitamins A, D, C, E, as well as trace elements such as iron, zinc, and selenium. play vital roles in immune cell development, antioxidant defence, and cytokine regulation (Calder, 2020). Macronutrients, including proteins and essential fatty acids, are equally important, providing the necessary substrates for immunoglobulin synthesis and regulating inflammation (Basak et al., 2022). A balanced diet also supports a healthy gut microbiota, which maintains immune homeostasis through prebiotics and probiotics (Yoo et al., 2020). Conversely, obesity disrupts immune function by promoting chronic inflammation through excessive cytokine production (Pugliese et al., 2022). Functional foods enriched with bioactive compounds such as flavonoids, omega-3 fatty acids, and probiotics exhibit immunomodulatory effects, offering a preventive approach against chronic inflammatory and immune-related disorders (Damián et al., 2022).

3. Categories of Immune-boosting Functional Foods

Categories of immune-boosting functional foods include probiotics and fermented foods, bioactive compounds, essential vitamins and minerals, omega-3 fatty acids, and dietary fibres. These categories enhance immune defence through modulation of gut health, attenuation of inflammatory responses, and support of immune cell function. The primary components of each group are summarized in Table 1, in addition to their main immunological benefits.

3.1 Probiotics and Fermented Foods

Probiotics are live bacteria that, when taken in sufficient quantities, provide health advantages, primarily by modifying gut microbiota and improving immune responses. *Lactobacillus*, *Enterococcus*, *Pediococcus*, *Bifidobacterium*, and *Saccharomyces boulardii* are common bacteria used in the dairy and bakery industries. These organisms interact with intestinal epithelial and immunological cells, which improves mucosal barrier function and stimulates innate and adaptive immunity (Latif et al., 2023). Probiotics increase secretory IgA synthesis, regulate dendritic cell activity, and influence T-regulatory cell populations, all of which promote an anti-inflammatory milieu (Parmar et al., 2025). Furthermore, fermented foods contain metabolites such as SCFAs, which aid in gut health and immunological homeostasis (Fusco et al., 2023). Clinical and epidemiological studies indicate that probiotic supplementation can reduce both the incidence and severity of respiratory infections, particularly among vulnerable populations such as children and older adults (Darbandi et al., 2021).

3.2 Bioactive Compounds and Phytochemicals

Bioactive compounds and phytochemicals are naturally occurring plant-based components that improve immunological function. These include flavonoids, polyphenols, alkaloids, terpenoids, and glucosinolates, which are plentiful in fruits, vegetables, tea, spices, and herbs. Flavonoids, including quercetin from onions and apples and catechins abundant in green tea, exhibit potent antioxidant, anti-inflammatory, and immunomodulatory properties. These bioactive compounds neutralize reactive oxygen species and modulate key signalling pathways, such as NF-κB and MAPK, thereby influencing cytokine production and immune cell activation (Jomova et al., 2025). Curcumin, an active ingredient in turmeric, has been extensively investigated for its capacity to modify immunological responses by improving macrophage activity, increasing antibody production, and suppressing pro-inflammatory cytokines such as TNF-α and IL-6 (Sharifi-Rad et al., 2020). Other polyphenols, such as resveratrol and sulforaphane, have shown promise in lowering oxidative stress and improving cellular immunity. These chemicals can alter gene expression in immune cells and induce apoptosis in infected or malignant cells (Mileo and Miccadei, 2016).

3.3 Essential Vitamins and Minerals

Micronutrients, such as vitamins and minerals, are essential for immune system formation, maintenance, and function. Deficiencies in these nutrients can cause immunological dysfunction, increased susceptibility to infections, and negative effects. Vitamin C protects the epithelial barrier, improves the function of phagocytes and lymphocytes, and helps to reduce oxidative stress. It is necessary for neutrophil migration and microbial destruction (Carr and Maggini, 2017). Vitamin D influences both innate and adaptive immunity by modulating antimicrobial peptide production and reducing proinflammatory cytokines. Low vitamin D levels are linked to an increased risk of respiratory infections and autoimmune illnesses (Ghaseminejad-Raeini et al., 2023). Vitamin A is essential for maintaining mucosal surfaces, as well as supporting lymphocyte function and antibody responses (Shastak and Pelletier, 2023). Adequate zinc and selenium intake is required for optimal immune competence, as deficiencies in these micronutrients adversely affect T-cell function, antibody production, and cellular immune responses (Stefanache et al., 2023). Sufficient intake of key vitamins and minerals appears to be important for maintaining immune competence, especially under conditions of infection or immune compromise.

3.4 Omega-3 Fatty Acids and Others

Omega-3 fatty acids, particularly EPA and DHA, are polyunsaturated lipids with potent immunomodulatory and anti-inflammatory properties. Abundant in fatty fish such as salmon and sardines, as well as in flaxseeds and walnuts, these fatty acids regulate immune cell membrane fluidity and modulate eicosanoid synthesis (Banaszak et al., 2024). These fatty acids serve as precursors for specialized pro-resolving mediators (SPMs) such as resolvins and protectins, which resolve inflammation and restore tissue homeostasis. They reduce the production of inflammatory cytokines, such as IL-1β, IL-6, and TNF-α, while increasing macrophage phagocytic activity (Julliard et al., 2022; Spite et al., 2014). Omega-3 supplementation has been found to lessen the occurrence and severity of inflammatory and autoimmune disorders like rheumatoid arthritis, asthma, and inflammatory bowel disease (Simopoulos, 2002). Furthermore, omega-3s improve B-cell activity and antibody production, hence promoting adaptive immunity. Other immunomodulatory substances include beta-glucans (found in oats and mushrooms), arginine (an amino acid required for lymphocyte proliferation), and glutamine, which fuels rapidly dividing immune cells and promotes gut immunity. Together, omega-3 fatty acids and these other bioactives are critical dietary components for immunological function and inflammation-related diseases (Basak and Gokhale, 2022; Murphy et al., 2020).

3.5 Dietary Fibre and Prebiotics

Dietary fibre and prebiotics are indigestible food components that have a substantial impact on gut function and immunity. Dietary fibre broadly comprises plant-derived carbohydrates resistant to digestion by human enzymes, but prebiotics are a specific subset of fibres that preferentially encourage the growth and metabolic activity of beneficial gut microorganisms, including *Bifidobacteria* and *Lactobacilli* (Holscher, 2017). Foods containing prebiotics include garlic, onions, bananas, chicory root, and whole grains. Gut bacteria digest these substances, producing SCFAs such as acetate, propionate, and butyrate, which are essential for gut integrity and immune response regulation (Kumar et al., 2025). SCFAs have immunomodulatory effects by affecting the development and activity of regulatory T cells (Tregs), lowering pro-inflammatory cytokine production, and improving epithelial barrier function. They also provide energy to colonocytes and help to protect the colon from infections by reducing its pH (Trompette et al., 2014). Furthermore, a fibre-rich diet promotes microbial diversity, which is linked to lower systemic inflammation and better immune surveillance. Clinical studies have demonstrated that fibre-rich diets can lower the risk of chronic inflammatory illnesses and respiratory infections, as well as alter vaccination responses (Makki et al., 2018). Thus, dietary fibre and prebiotics are critical components of immune-supportive nutrition, since they strengthen the gut-immune axis and protect against both infectious and noncommunicable diseases.

Table 1 Categories of immune-boosting functional foods, their key bioactive components, and the corresponding immunological benefits.

Category	*Key Components*	*Primary Immune Benefits*
Probiotics and Fermented Foods	*Lactobacillus*, *Bifidobacterium*, kefir, yogurt, kimchi, sauerkraut	Enhance gut microbiota, improve gut barrier function, stimulate innate and adaptive immunity
Bioactive Compounds and Phytochemicals	Polyphenols, flavonoids, carotenoids, curcumin, allicin	Anti-inflammatory, antioxidant, and modulate cytokine production
Bioactive Compounds and Phytochemicals	Polyphenols, flavonoids, carotenoids, curcumin, allicin	Anti-inflammatory, antioxidant, and modulate cytokine production
Essential Vitamins and Minerals	Vitamins A, D, C, E, iron, zinc, and selenium	Support immune cell function, antioxidant defence, and improve resistance to infection
Omega-3 Fatty Acids and Marine Foods	EPA, DHA (from fish oil), seaweed polysaccharides	Reduce chronic inflammation, support immune resolution pathways
Dietary Fibre and Prebiotics	Inulin, oligosaccharides, resistant starch (from bananas, garlic, onions)	Promote beneficial gut bacteria, regulate immune signalling via the gut-immune axis

4. Specific Functional Foods and Their Immunomodulatory Effects

Functional foods are elements of food that provide health advantages beyond basic nutritional requirements, primarily by supporting or enhancing physiological activities such as immunological responses. Functional foods include yogurt and fermented dairy products, fruits and vegetables, mushrooms, herbs and spices, different teas, and marine products (Figure 1). Among their various health-promoting properties, the immunomodulatory potential of functional foods has drawn significant scientific attention. Many of these foods are rich in bioactive compounds, such as probiotics, antioxidants, polyphenols, and essential fatty acids, that help modulate immune activity, enhance the body's defence against pathogens, and reduce inflammation. Understanding the types and immunological impacts of functional foods is crucial for designing effective dietary strategies for immune health and disease prevention.

Figure 1 illustrates various types of functional foods, including fermented dairy products, fruits and vegetables, herbs and spices, seafood, mushrooms, and polyphenol-rich teas. Each category consists of bioactive compounds that support immune function through antioxidant, anti-inflammatory, or gut-modulating effects. These foods play a vital role in enhancing immunity and promoting overall health.

4.1 Yoghurt and Fermented Dairy Products

Yogurt and other fermented dairy products contain significant levels of probiotics, which are living bacteria that can provide health benefits when consumed in adequate amounts. These foods, particularly those containing *Lactobacillus* and *Bifidobacterium* strains, are important for regulating gut microbiota and improving immunological function (Marco et al., 2017). Yoghurt's immune-supportive properties stem from its potential to influence both innate and adaptive immunity. Probiotic strains in yoghurt increase the activity of macrophages, NK cells, and dendritic cells, which improves pathogen identification and phagocytosis. These microorganisms reduce pro-

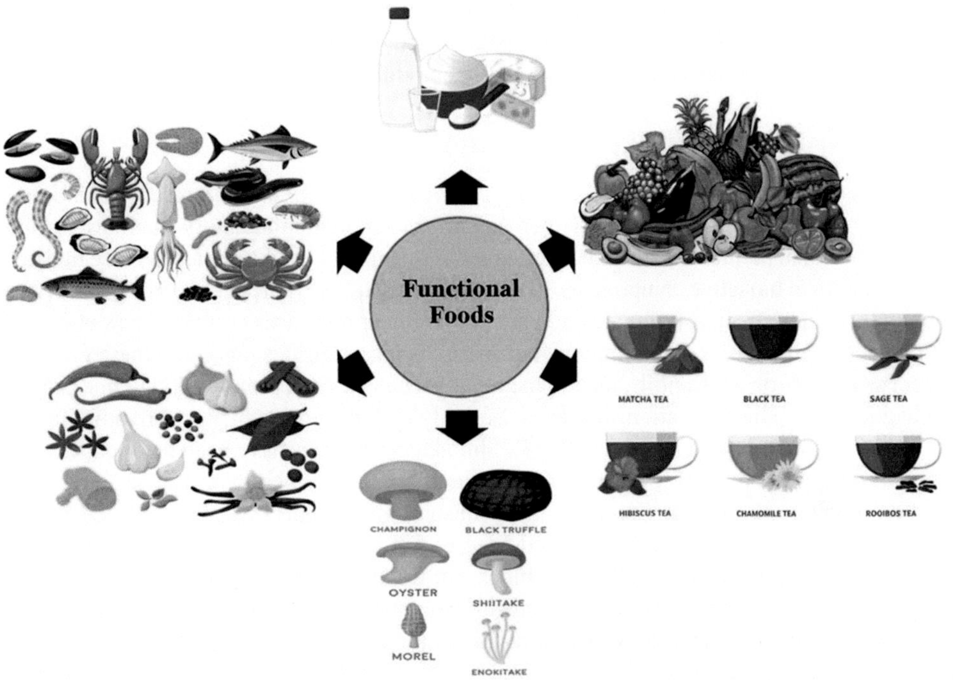

Fig. 1 Different types of functional foods.

inflammatory mediators like TNF-α and increase anti-inflammatory cytokines like IL-10 (Gao et al., 2021). Fermented dairy products have been linked to enhanced secretory immunoglobulin A (sIgA), a key mediator of mucosal immunity, as well as improved gut barrier integrity, which reduces systemic inflammation and prevents pathogen translocation (Chen et al., 2020). Furthermore, regular yogurt eating has been related to a lower risk of gastrointestinal and respiratory illnesses, particularly in susceptible groups like children and the elderly (Fernandez and Marette, 2017). In addition to probiotics, yogurt contains critical minerals like calcium, vitamin D, and bioactive peptides, all of which help to enhance immune function. Together, these ingredients make fermented dairy an essential component of functional diets designed to enhance immunity (Hadjimbei et al., 2022).

4.2 Fruits and Vegetables Rich in Antioxidants

Fruits and vegetables constitute key components of immune-supportive diets, as they are rich sources of antioxidants, vitamins, minerals, and diverse phytochemicals. These bioactive compounds protect immune cells from oxidative stress, modulate cellular signalling pathways, and enhance the host's capacity to respond to infections and inflammatory challenges (Carr and Maggini, 2017). Fruits and vegetables include important antioxidants such as vitamin C, vitamin E, beta-carotene, flavonoids, and polyphenols. Vitamin C, found in citrus fruits, strawberries, and bell peppers, improve neutrophil function, protect the epithelial barrier, and aids in the formation of collagen and antimicrobial substances. Similarly, vitamin E, found in green leafy vegetables and avocados, functions as a lipid-soluble antioxidant, protecting cell membranes and promoting T-cell-mediated immune responses (Pisoschi et al., 2022). Carotenoids such as beta-carotene (found in carrots and sweet potatoes) are transformed into vitamin A, which is essential for mucosal surface maintenance and immune cell development (Pisoschi et al., 2022). Flavonoids and polyphenols found in berries, apples, grapes, and cruciferous vegetables can reduce inflammation by regulating signalling pathways like NF-κB and MAPK, as well as encouraging good gut flora (Saleh et al., 2021). The diversity and synergy of these phytochemicals and nutrients help to build a stronger, more resilient immune system. High fruit and vegetable consumption has been related to a decreased frequency and severity of respiratory infections, less systemic inflammation, and a better reaction to vaccines. Thus, fruits and vegetables are not only food providers, but also potent immunomodulators (Hassanizadeh et al., 2024).

4.3 Mushrooms (Shiitake, Reishi, and Maitake)

Medicinal mushrooms like shiitake (*Lentinula edodes*), reishi (*Ganoderma lucidum*), and maitake (*Grifola frondosa*) are well-known for their significant immunomodulatory qualities, which are based on traditional medicine and are becoming more scientifically substantiated. These mushrooms contain a variety of bioactive components, including beta-glucans, triterpenoids, lectins, and phenolic compounds, which affect both innate and adaptive immune responses (Balakrishnan et al., 2023). Beta-glucans are the best-studied immune-enhancing polysaccharides present in these mushrooms. They act as PAMPs on neutrophils, macrophages, and dendritic cells promoting cytokine production and phagocytosis. This activation improves the body's ability to identify and remove infections more effectively (Varnosfaderani et al., 2024). Shiitake mushrooms have been reported to stimulate the proliferation and activity of T lymphocytes and NK cells, thereby enhancing antiviral and anticancer immune responses. Eating shiitake mushrooms regularly has been proven in clinical investigations to increase secretory IgA while decreasing inflammatory markers (Dai et al., 2015). Triterpenoids found in Reishi mushrooms block pro-inflammatory mediators such as TNF-α and IL-6, providing immunomodulatory and anti-inflammatory actions. In addition, they aid in the restoration of immunological homeostasis in autoimmune disorders (Oke et al., 2022). Maitake mushrooms have been researched for their ability to stimulate dendritic cell maturation and NK cell activity, implying a role in vaccination adjuvancy and tumour immunity. Because of their complex bioactivity, these mushrooms are increasingly being employed as functional foods or nutraceuticals to boost immunological resilience, particularly in aged populations or those with chronic immune suppression (Kodama et al., 2005).

4.4 Herbs and Spices

Herbs and spices have long been regarded in traditional medical systems for their medicinal properties, notably their ability to improve immunological function. Common culinary herbs like turmeric, ginger, garlic, oregano, and basil, as well as spices like black pepper, cinnamon, and cloves, include bioactive components with antioxidant, anti-inflammatory, and immunomodulatory properties (El-Sayed and Youssef, 2019). Turmeric, a South Asian culinary staple, contains curcumin, a polyphenolic molecule that has been shown to influence numerous immunological responses. Curcumin suppresses the NF-κB pathway, reducing pro-inflammatory cytokines such IL-1, IL-6, and TNF-α. It also improves antibody response and macrophage function (Aggarwal and Harikumar, 2009). Ginger contains chemicals such as gingerols and shogaols, which have been demonstrated to reduce inflammation while increasing T and NK cell activity (Yücel et al., 2022). Garlic (*Allium sativum*) is another powerful immunomodulator, with its sulfur-containing component allicin stimulating phagocytosis, increasing lymphocyte proliferation, and improving cytokine secretion (Arreola et al., 2015). Oregano and thyme, which include carvacrol and thymol, have anti-inflammatory and antibacterial properties that help regulate immune responses and gastrointestinal health. These herbs and spices also serve as prebiotics, promoting the establishment of healthy gut flora, which in turn influences systemic immunological activity (Gholami-Ahangaran et al., 2022). Regular use of these nutrients has been linked to lower rates of infection, enhanced gut barrier function, and reduced chronic inflammation.

4.5 Green Tea and Other Polyphenol-Rich Beverages

Green tea (*Camellia sinensis*) and other polyphenol-rich beverages, including black tea, matcha, and various herbal infusions, have been recognized for their immunomodulatory properties. Green tea is notably high in catechins, a type of flavonoid with potent antioxidant and anti-inflammatory activities (Gholami-Ahangaran et al., 2022). The most prevalent and active catechin is epigallocatechin gallate (EGCG), which has a considerable effect on immune cell signalling and function. EGCG affects immune responses by decreasing pro-inflammatory cytokines, including IL-6, TNF-α, and IL-1β, and increasing anti-inflammatory mediators like IL-10. It also affects the activity of dendritic cells, T cells, and regulatory T cells (Tregs), which helps to maintain balanced immunity. EGCG lowers the NF-κB pathway and minimizes oxidative stress, protecting immune cells during inflammation or infection (Yang et al., 1998). Green tea polyphenols also have antiviral and antibacterial properties. EGCG has been demonstrated to suppress viral entry and replication in diseases such as influenza, HIV, and SARS-CoV-2, which could protect during viral outbreaks. Furthermore, green tea improves gut microbiota composition by encouraging the growth of good bacteria such as *Lactobacillus* and *Bifidobacterium*, which indirectly aids systemic immune function (S. Wang et al., 2021a; Y. Q. Wang et al., 2021). Other beverages, such as black tea, rooibos, and hibiscus tea, contain similar polyphenolic components, such as theaflavins and anthocyanins, which influence oxidative and inflammatory responses (Cleverdon et al., 2018). Regular intake of green tea and kindred beverages may lower the risk of chronic diseases and improve immunological resilience, making them an important addition to immune-supportive diets.

4.6 Fish and Marine-Derived Compounds

Fish and other marine-derived foods contain distinctive bioactive compounds that contribute significantly to immune regulation. Fatty fish, such as salmon, mackerel, sardines, and tuna, are particularly rich in omega-3 polyunsaturated fatty acids (PUFAs), notably EPA and DHA. These important fatty acids are well-known regulators of inflammation and immunological function (Awuchi et al., 2022). Omega-3 PUFAs impact immunological responses by incorporating into cell membranes, altering membrane fluidity and receptor signalling. They produce specialized SPMs such as resolvins, protectins, and maresins, which actively resolve inflammation without reducing host defence. These chemicals suppress pro-inflammatory cytokines (e.g., IL-6, TNF-α) and increase phagocytic activity in macrophages, resulting in effective pathogen clearance (Kar et al., 2023). In addition to fatty acids, marine sources contain bioactive peptides, vitamin D, and selenium, all of which are necessary for immunological function. Vitamin D, which is naturally abundant in fatty fish and fish liver oils, promotes innate and adaptive immunity by increasing the pathogen-fighting abilities of macrophages and monocytes by modifying T cell responses. Polysaccharides produced from marine sources, such as seaweed and shellfish, are likewise immunomodulatory (Thirumdas et al., 2021). Fucoidan, a sulphated polysaccharide found in brown algae, has anti-inflammatory and antiviral properties, promoting NK cell activity and increasing antibody responses (Phull and Kim, 2017). Regular consumption of fish and marine-derived nutrients is linked to a decreased frequency of inflammatory illnesses, a lower risk of respiratory infections, and better vaccine efficiency. These effects highlight marine foods as essential components of functional diets designed to boost immunological health (Mendivil, 2021).

5. Mechanism of Action of Functional Foods

Genetics, age, lifestyle, and environment affect immune function and vulnerability to disease. Certain food compounds can regulate cellular processes, influencing disease development. Consuming specific functional foods may enhance immune cell activity, potentially protecting against cancer and infections. These functional foods have different key components that act on the immune system and affect its activity (Table 2). Few foods show immunostimulatory effects by modulating the signaling pathways (Figure 2), preventing cancer development and protecting against pathogens.

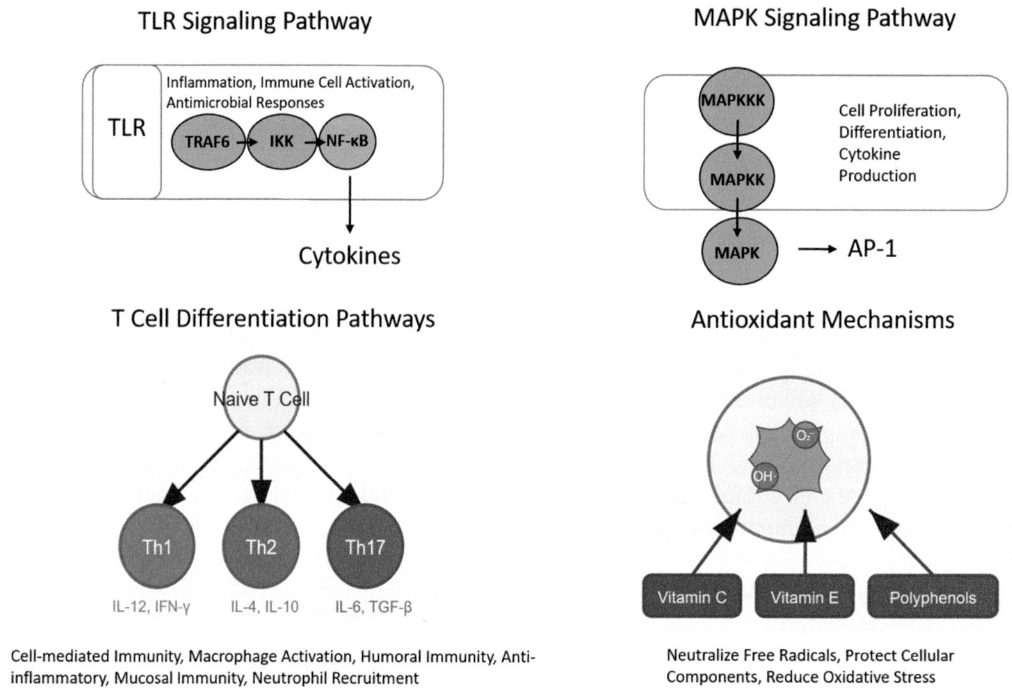

Fig. 2 This figure shows the key signalling pathways in the immune system.

The **TLR signalling pathway** shows the activation sequence from TLR receptors through TRAF6 and IKK to NF-κB, leading to pro-inflammatory cytokine production. Ginseng and mushroom compounds activate this pathway, while green tea polyphenols inhibit it. The **MAPK signalling pathway** illustrates the MAPKKK→MAPKK→MAPK cascade, leading to AP-1 activation and subsequent cell proliferation and cytokine production regulation. Red ginseng activates this pathway via ERK and JNK. T-Cell differentiation depicts naive T cell differentiation into Th1 (cell-mediated immunity), Th2 (humoral immunity), and Th17 (mucosal immunity) subsets. Different natural compounds selectively promote specific T cell populations. Antioxidant mechanisms shows how antioxidants like Vitamin C, Vitamin E, and polyphenols neutralise ROS (O_2^- and OH·), protecting cells from oxidative damage and indirectly supporting immune function.

5.1 White Ginseng

White ginseng and its compounds (ginsenosides and polysaccharides) show immunostimulant effects through different mechanisms (Lee et al., 2010; Shibata, 2001). The ginseng extract improves macrophage function by increasing the phagocytic activity, stimulating NO production, and upregulating iNOS expression (Jang and Shin, 2010; Lim et al., 2016; Um et al., 2020). Macrophages exhibit increased expression of IL-6, IL-1β, IL-1α, IL-23, and TNF-α when treated with white ginseng extract (Kang and Min, 2012; Um et al., 2020). These immunomodulatory effects are mediated by activating the MKK4-JNK-c-Jun pathway for immune cell recruitment and cytokine expression, and the MAPK/JNK pathway for overall immunomodulation. Additionally, white ginseng enhances phagocytosis activity of macrophage by activating NF-κB, PI3K/AKT, and MAPK signalling pathways through TLR2/4-dependent fashion (Lim et al., 2016; Shen et al., 2017; Yang et al., 2018). White ginseng oligopeptides boost innate immune responses, including macrophage phagocytosis and NK cell activity, while also promoting dendritic cell maturation and proinflammatory cytokine production in PBMCs (Peripheral Blood Mononuclear Cells) (He et al., 2017; Larsen et al., 2004). Ginsan, a white ginseng polysaccharide, induces CD86 and Class II MHC production in dendritic cells, for cytokine production and T cell activation, CD4+ T cells proliferation (Kim et al., 2009). Through increased cytokine production, ginseng strengthens immunity by enhancing NK cell function, antibody production (IgA, IgM, IgG), and T-helper responses. Specific ginsenosides like Rg1 promote TH2 responses by boosting CD4+ T cells and IL-4 levels.

5.2 Red Ginseng

Red ginseng increases white blood cells (WBCs), macrophages, natural killer (NK) cells production, and the size of the thymus and spleen (Shin et al., 2020). It stimulates macrophage activation via the JNK, ERK, AP-1, and NF-κB pathways, and induces cytokines (IL-1, TNF-α, and IL-6 (Byeon et al., 2012). Red ginseng extract in humans proliferates CD3+ T lymphocytes through the enhanced CD25 and CD69 expression (Kim et al., 2016). Red ginseng ginsenosides (Rg1, Rd, Re) stimulate TH1 and TH2 lymphocyte growth and increase ovalbumin-specific IgG in mouse models (Song et al., 2010; Su et al., 2012; Sun et al., 2008; Yang et al., 2007). Rg1 and Re activate NF-κB and inflammatory cytokines in normal mice but not in mice without TLR-4, indicating that these immune-enhancing and adjuvant effects rely on TLR-4 signalling pathways (Bremner and Heinrich, 2010; Su et al., 2012).

5.3 Mushrooms

Agaricus blazei (A. blazei)

A. blazei improves the production and proliferation of cytokines and immune cells, respectively, while also increasing IFN-γ production and IP-10 mRNA expression in mouse spleen cells. It elevates CD69 expression, indicating increased T-cell infiltration in tumor cells, boosts macrophage and NK cell activity (Lin et al., 2012; Takimoto et al., 2008, 2004). The extract of *A. blazei* reduces ROS activity, lipid peroxidation, and the severity of *Plasmodium berghei*-induced cerebral malaria, indicating the potential of the mushroom in therapeutics and disease management (Val et al., 2015).

Ganoderma lucidum (G. lucidum, reishi)

G. lucidum enhances macrophage secretion of TNF-α and IL-6 by upregulating signaling pathways (ERK1/2, MAPK, p38, and JNK). It also increases IFN-γ, IL-6, and IL-2 concentration through NF-κB upregulation and it increases CD56+ cells. The NK cell activity is improved (34.5% ± 11.8% vs. 26.6% ± 8.3%) compared to baseline (Gao et al., 2003; Guo et al., 2009; Wang et al., 2007).

Grifola frondosa (maitake D)

A β-glucan-rich polysaccharide extract from *Grifola frondosa*, exerts its immunomodulatory and antitumor effects through multiple pathways. It promotes a TH1-dominant immune response by macrophages and T cells activation, elevating TH1 cytokines, and suppressing TH2 cytokine production. IFN-γ and IL-12 expression is markedly increased in APCs (Antigen-Presenting Cells), resulting in augmented functionality of NK cells and macrophages (Harada et al., 2003; Inoue et al., 2002; Kodama et al., 2002). The fraction stimulates dendritic cell maturation via the dectin-1 pathway, increasing IFN-γ-expressing T cells (CD4+ and CD8+). The β-glucans interact with various receptors on immune cells, including dectin-1, TLR, and CR3, triggering cytokine production and activation (Chan et al., 2009; Kim et al., 2011).

5.4 Bacillus subtilis chungkookjang (Poly-γ-Glutamate, γ-PGA)

Poly-γ-glutamate, a biomaterial produced by *Bacillus subtilis chungkookjang* interacts with TLR4 receptors on macrophages derived from bone marrow, triggering the production of pro-inflammatory cytokines, including IL-1β, TNF-α, IL-6, and NLRP3. γ-PGA also enhances Natural Killer cells function by stimulating the synthesis of IFN-γ, increasing NK cell numbers, and enhancing cytotoxicity. Furthermore, it induces the secretion of various cytokines, specifically TNF-α, IL-12, and IFN-γ, by active NK dendritic cells (Ahn et al., 2018; Kim et al., 2007; Lee et al., 2014).

5.5 Chlorella

Chlorella, a single-celled green alga extensively utilized as a functional food and nutraceutical because of its abundant quantity of proteins, dietary fibres, vitamins, minerals, and other bioactive items. Among the Chlorella species, *Chlorella vulgaris* and *Chlorella pyrenoidosa* are the most researched for their application as nutritional supplements (Bax et al., 2021; Rani et al., 2018).

Chlorella vulgaris

Chlorella vulgaris enhances immunity through multiple mechanisms: it restores cytokine production (IL-2, TNF-α, IFN-γ, IL-12) in immunosuppressed mice, boosts NK cell activity, increases lymphocyte proliferation, and improves macrophage phagocytosis (Cheng et al., 2017). Human clinical trials confirm these effects, showing elevated NK cell function and increased cytokine levels after supplementation (Kwak et al., 2012). *C. vulgaris* demonstrates significant antibacterial effects against *Listeria monocytogenes* and *E. coli* by reducing bacterial loads while enhancing immune cell activity (Takashi Hasegawa et al., 1989). Its mechanism depends heavily on TLR2 signalling for its activity on the immune system (Hasegawa et al., 2002). Additionally, it increases antibody production (IgG and IgM) in animal models, with fresh liquid formulations showing superior efficacy to dried powder supplements (An et al., 2016).

Chlorella pyrenoidosa

C. pyrenoidosa boosts immunity by activating macrophages and enhancing antibody responses. Its hot water extracts increase macrophage phagocytosis, proliferation, and nitric oxide production while inducing HLA expression in human macrophages (Hsu et al., 2010). Mechanistically, it upregulates inflammatory cytokines (TNF-α, IL-1β) and costimulatory molecules (CD80, CD86) through TLR4 signaling and NF-κB activation (Hsu et al., 2010; Pugh et al., 2001). Clinical studies show it increases breast milk IgA in pregnant women, potentially protecting nursing infants (Nakano et al., 2007). In animal models, *C. pyrenoidosa* enhances IgM and IgG production, while human trials demonstrated improved influenza antibody responses specifically in middle-aged adults (50–55 years), though not across all age groups (Halperin et al., 2003).

5.6 Lactobacillus plantarum

L. plantarum exerts its immunomodulatory effects through multiple mechanisms targeting the innate and adaptive immune systems. In the innate immune system, *L. plantarum* and its exopolysaccharides (EPSs) enhance macrophage phagocytic activity, including iNOS (inducible nitric oxide synthase) expression, and NO production (Adams, 2010; Jang and Shin, 2010; Ren et al., 2019). They also increase TLR2 mRNA expression and stimulate pro-inflammatory cytokine (TNF-α and IL-6) production via NF-κB and MAPK signaling pathways (Jang and Shin, 2010; Jeong et al., 2019; Lee et al., 2019; Wang et al., 2018). L. plantarum upregulates various cytokines (TNF-α, IL-20, p40, IL-10, p70, and IL-12) to modulate dendritic cells' function through NF-κB and MAPK pathways (Rigaux et al., 2009). In the adaptive immune system, *L. plantarum* promotes TH1 and TH17 responses, increases IFN-γ production, and enhances IgA production (Kim et al., 2022; Lee et al., 2015).

5.7 Other Probiotics

Probiotics are living microorganisms, and when administered in an appropriate dose, they provide health benefits to the host" (FAO/WHO, 2001). The regular consumption of *Bifidobacterium lactis* HN109 and *Lactobacillus rhamnosus* HN001 increases tumoral activity mediated by NK cells in murine splenic leucocytes. Ogawa et al. (2006) proved that consuming *L. casei subsp. casei* JCM 1134 with dextran increases the NK cell activity in mononuclear cells of the spleen from BALB/c mice (Ogawa et al., 2006). In addition, this bacteria activated the production of IL-12 in peripheral blood mononuclear cells (PBMCs) and IL-15 expression in the IEC for promoting intestinal intraepithelial NK cells from the Caco-2 cell line. It appears that probiotic strains (*L. rhamnosus GG, B. lactis Bb12*) can induce B cells to produce IgA. Probiotics (*L. paracasei spp. and Lactobacillus johnsonii La1*) can also affect many immune cells like enterocytes, DC, Th1, Th2, and Treg, to regulate the immune response promoting pro-inflammation. By stimulating macrophages, phagocytosis, and NO production, *Lactobacillus helveticus LH-2* and its F5 (a peptide fraction) increases the synthesis of cytokines (Santiago-López et al., 2015a). The active bacterial components of *Bifidobacterium breve C50* can pass through intestinal epithelial cells and regulate cytokine production. Probiotics alter the intestinal microbiota by enhancing the growth of beneficial microbes. Consuming kefir or kefiran by animal models elevated useful bacteria count, mainly Lactobacillus and Bifidobacterium, and suppressed harmful microbes like *Clostridium perfringens*. Specific Lactobacillus strains from kefir adhere to intestinal cells and prevent the *Salmonella typhimurium* and *Escherichia coli* O157 adhesion. Additionally, Lactobacillus shields the cells from extracellular components released from *Bacillus cereus* and shiga toxin type II from *E. coli* O157 (García-Burgos et al., 2020).

5.8 Fruits and Vegetables

The mechanism of action of fruits and vegetables in supporting immune function involves multiple synergistic pathways. Antioxidants like vitamin C, E, polyphenols, flavonoids, lycopene, carotenoids, and glutathione neutralize free radicals, protecting cells from oxidative damage. Flavonoids exhibit anti-inflammatory effects by inhibiting pro-inflammatory genes and cytokines. They modulate intracellular signalling cascades, influencing cell function and gene expression at lower concentrations than those required for direct antioxidant effects (Baidya and Sethy, 2020). STAT-1, NF-κB, and AP-1 signalling is inhibited in the active macrophages and endothelial cells by kaempferol, a flavonol from broccoli and tea, minimizes the activity of inflammation-related genes including iNOS and COX-2. Another flavonol, quercetin from leafy vegetables, broccoli, onions, apples, and grapes, suppresses TNF-induced NF-κB recruitment to pro-inflammatory gene promoters and inhibits pro-inflammatory cytokines expression in mast cells, is a potent antioxidant and anti-inflammatory, t and suppressing agent. Vitamin C enhances antibody production, promotes lymphocyte proliferation, and supports various immune responses (Jideani et al., 2021a). Vitamin E protects against lipid peroxidation, reacts with reactive species, and aids T-cell differentiation. Specific nutrients stimulate the immune system through increased production of infection-fighting cells, T-cells, white blood cells, and NK cells, while enhancing antibody production. Anti-inflammatory compounds like bromelain, ginger, and curcumin reduce inflammation, which weakens the immune response. Essential micronutrients such as zinc, selenium, potassium, iron, vitamin A, and B vitamins support various immune functions and overall health. Carotenoids act as vitamin A precursors, suppress inflammatory responses, and increase lymphocyte populations. Selenium improves both innate and acquired immunity while reducing inflammation. Zinc, essential for numerous proteins and enzymes, regulates key genes and supports antibody-mediated and cell-mediated immunity. Additionally, fruits and vegetables help maintain metabolic health by regulating blood sugar levels, reducing heart disease risk factors, and supporting healthy blood pressure, indirectly benefiting immune function. This complex array of bioactive compounds and nutrients enhances the body's immune defences through multiple complementary mechanisms (Jideani et al., 2021b; Stephen et al., 2023).

5.9 Herbs and Spices

Herbs and spices have been used in traditional medicine for centuries. Turmeric shows positive immune enhancement and anti-inflammatory effects on phospholipases, COX-2, thromboxane, prostaglandins, NO, lipoxygenase, TNF, and IL-12. Curcumin has been shown to suppress inflammatory cascades, resulting in inhibition of macrophage, dendritic cell, and T cell subset activation, maturation, and cytokine production like TNF-α, IL-1β, and the activation of NF-κB in human monocyte-derived cells. The active compound in black pepper (piperine) reduces pro-inflammatory cytokines prostaglandin E_2, IL-1β, TNF-α, and IL-6. It also inhibits macrophage inflammation and treats allergic inflammation through major T helper cell groups. Quercetin, a pepper component, improves immunity through antiviral and antibacterial properties. Oregano shows dose-dependent effects, with high and lower doses potentially having pro-inflammatory and anti-inflammatory properties, respectively. It increases lysozyme and bactericidal activity in fish and increases specific immunoglobulins in poultry (Garnier and Shahidi, 2021). Tulsi (*Ocimum sanctum*) exhibits antimicrobial activity and increases T-helper cells and NK cells against viral infections. Ajwain (Trachyspermum ammi) has thymol, which shows antifungal, antimicrobial, and anti-aggregatory effects. The oil content of *T. ammi* seeds ranges from 2%–4.4%. Thymol, an important component, which is used to treat bronchial, gastrointestinal, and appetite issues. In humans, the oil has fungicidal, anti-aggregatory, and antimicrobial properties. Asafoetida produces natural antiviral compounds effective against the influenza A virus (H1N1). Garlic's active ingredients, alliinase and allicin, effectively treat fungal and parasitic infections and confer antiviral activity against various viruses by stimulating and protecting immune cells. Peppermint (*Mentha piperita*) suppresses inflammatory mediators and enhances innate immunity, potentially counteracting respiratory syncytial virus infections. MPE suppresses production of nitric oxide (NO), tumour necrosis factor (TNF)-α, interleukin-6 (IL-6), and prostaglandin E2 (PGE2) in macrophages. Ginger exhibits a vast anti-inflammatory response by reducing monocyte chemoattractant protein-1, iNOS, CRP, COX-2, and NF-κB subunit p65 (inducible nitric oxide synthase). Fresh ginger inhibits viral attachment and stimulates mucosal cells for interferon-beta secretion, mostly contributing as a defence to viral infection and internalization. Ginger improves humoral immunity, and volatile oil influences cell-mediated response and nonspecific T-lymphocyte proliferation. At the same time, specific compounds like gingerols and zingerone prevent viral replication and cell entry (Shrivastava, 2020). Cardamom aqueous extract was found to have an immunomodulatory role and enhance Natural Killer cell activity of the initial innate immune response against infections. Cinnamon bark extracts improved immune systems' humoral, cellular, and innate responses in immune-compromised mice. Capsicum extracts increased serum immunoglobulin concentrations (IgG and IgM). Fenugreek exhibits a distinguished high phagocytic index by the reticulo-endothelial system (Sudhakar Johnson and Narayana, 2021).

5.10 Green Tea and Other Polyphenols

Green tea polyphenols exert immunomodulatory effects through multiple mechanisms, particularly epicatechin-3-gallate (ECG) and epigallocatechin-3-gallate (EGCG). They modulate TLR signalling by inhibiting IKKβ and NF-κB activation, affecting inflammatory responses. These compounds influence T cell function and differentiation by altering the Th1/Th2 balance, inhibiting Th1-like immunity, and promoting anti-inflammatory patterns. Green tea polyphenols (GTPs) reduce STAT1 phosphorylation, suppress iNOS and NF-κB upregulation, and inhibit TNF-α and IFNγ, primarily blocking Th1 immunity. They increase anti-inflammatory IL-4, promoting Th2 cell activity and shifting the IFNγ/IL-4 ratio toward anti-inflammatory responses. While GTPs enhance immune defense by promoting T cell proliferation, increasing CD4+ T cells, improving CD4+/CD8+ ratios, and enhancing lymphocyte transformation, EGCG specifically inhibits T cell division dose-dependently, affecting CD4+ T cells more than CD8+ T cells, and suppresses T cell mitogen concanavalin A (Con A)-stimulated splenocyte proliferation at physiological concentrations. They also suppress antigen-presenting and dendritic cells' maturation and function by MAPK and NF-κB pathways inhibition. GTPs impact neutrophils and macrophages by inhibiting elastase secretion, reducing neutrophil transmigration, and attenuating NF-κB activation. They stimulate B-cell proliferation and antibody production, enhancing humoral immunity. These compounds exhibit anti-inflammatory properties by reducing inflammatory mediator expression and suppressing NLRP3 inflammasome activation by blocking mitochondrial DNA synthesis. Their antioxidative mechanisms help combat oxidative stress, which is closely associated with impaired immunity. GTPs have a unique aromatic polycyclic hydrocarbon structure that reduces oxidation of cellular components and oxidative free radicals. Polyphenols in green tea can decrease body fat, cholesterol, and hepatic triacylglycerols (S. Wang et al., 2021b). They also influence gut microbiota composition, potentially contributing to their immunomodulatory effects. Fermented green tea extract recovers gut microbiota composition along with the ratios of Firmicutes/Bacteroidetes and Bacteroides/Prevotella linked to the development of obesity and insulin resistance (Sun et al., 2022). Cyanidin 3-O-glucoside, which is frequently present in grapes, suppressed the cyclo-oxygenase enzymes COX-1 and COX-2 similar to non-steroidal drugs such as aspirin and ibuprofen. Grape seed-derived proanthocyanidin lowered oxidative DNA damage in lymphocytes (Schütz et al., 2010). Cocoa extract rich in polyphenols inhibits M1-type inflammatory responses and promotes the alternative M2 activation pathway in macrophages. By lowering the expression of inducible nitric oxide synthase (iNOS) and cyclooxygenase-2 (COX-2), pomegranate peel polyphenols (PPPs) and their constituents such as punicalagin (PC) and ellagic acid (EA) decrease pro-inflammatory cytokines and downregulate inflammatory mediators such as NO and prostaglandin E2 (PGE2). Additionally, they suppress TLR4 protein levels and inhibit LPS-induced intracellular ROS (Shakoor et al., 2021).

5.11 Fish and Marine-Derived Compounds

Marine-derived compounds exhibit diverse immunomodulatory mechanisms. Protein-based mechanisms involve hemocyanins interacting with immune cells to enhance host responses, lectins controlling immune function through pathogen recognition and cell activation, and taurine protecting cells from oxidative stress while modulating the immune system via NF-κB and PPAR-γ activation. Antimicrobial peptides (including callinectin, clavanins, crustins, defensins, myticins, mytilins, and others) disrupt microbial membranes, stimulate immune responses, suppress inflammation in sepsis, destroy biofilms, and modulate cytokine levels. Protein hydrolysates from sources like *Chlorella, Ecklonia, Porphyra,* and *Cyclina sinensis* enhance hemopoiesis, increase leukocyte counts, stimulate macrophage activity, modulate cytokine production, and activate signaling pathways like NF-κB and MAPK. General immunomodulatory mechanisms include macrophage activation and phagocytosis stimulation, cytokine modulation, NK cell stimulation, lymphocyte proliferation and differentiation, and mucosal immunity enhancement. These diverse mechanisms contribute to the potential therapeutic applications of marine-derived compounds in immunomodulation (Kang et al., 2019). E-JUR-94013 is a natural lipoprotein supplement derived from Atlantic horse mackerel (*Trachurus trachurus*) harvested from Galicia's Atlantic coast. This lipofishin, processed through mechanical transformation and lyophilization to preserve its biological properties, demonstrates significant immune-enhancing and anti-inflammatory effects. Research shows it stimulates antibody production, enhances phagocytosis, modifies cytokine production, and improves acquired immunity. Spinacin, an amino acid (4,5,6,7-tetrahydroimidazo [4,5-c] pyridine-6-carboxylic acid) from the component analysis, may be responsible for the compound's immunomodulatory qualities (Lombardi et al., 2018). Su (2011) identified pelteobagrin from the skin mucus of yellow catfish with the amino acid sequence GKLNLFLSRLEILKLFVGAL. Pelteobagrin demonstrated broad-spectrum efficacy against a variety of bacteria with no hemolytic activity (Su, 2011). According to Huang et al. (2007) tilapia TH2–3 is comparable to the 26-amino acid peptide JF2 (Japanese flounder (*Paralichthys olivaceus*), tilapia TH2–2 is the same as a 19 amino acid Japanese flounder (JF1), and tilapia TH1–5 is comparable to sea bream (*Chrysophrys major*) hepcidin regulate iron homeostasis (critical for immune function), possess direct antimicrobial properties, and modulate inflammatory responses (Khora, 2013). Fish oil's n-3 PUFAs (EPA and DHA) work through multiple immune pathways by restructuring cell membranes and lipid rafts, affecting cellular signaling, reducing inflammation by activating GPR120 and inhibiting NF-κB. These oils produce specialized pro-resolving mediators (resolvins, protectins, maresins) that terminate inflammatory processes, regulate T cell responses by suppressing CD4+ T cell polarization, enhance B cell development and antibody production, and promote beneficial gut bacteria while reducing intestinal inflammation (Ellulu et al., 2015; Parolini, 2019).

Table 2 Immunomodulatory mechanism of major functional foods.

Functional Food	*Key Bioactive Compounds*	*Primary Immunomodulatory Mechanisms*
White Ginseng	Ginsenosides, Polysaccharides	Improves macrophage function, phagocytic activity, stimulates NO production, upregulates iNOS expression, and promotes proinflammatory cytokine expression (IL-1β, IL-6, IL-1α, TNF-α, IL-23) through MKK4-JNK-c-Jun and MAPK/JNK pathways
Red Ginseng	Ginsenosides (Rd, Re, Rg1)	Increases production of WBCs, macrophages, NK cells; stimulates macrophage activation via ERK, JNK, NF-Kb, and AP-1 pathways; proliferates CD3+ T lymphocytes; stimulates TH1 and TH2 lymphocyte growth; increases ovalbumin-specific IgG
A. blazei	Polysaccharides	Improves cytokine production, increases IFN-γ production, elevates CD69 expression, boosts macrophage and NK cell activity, reduces ROS activity and lipid peroxidation
G. lucidum (Reishi)	Polysaccharides	Enhances macrophage secretion of TNF-α and IL-6 via MAPK signalling pathways; increases IFN-γ, IL-6, and IL-2 through NF-κB upregulation; improves NK cell activity
G. frondosa (Maitake D)	β-glucan polysaccharides	Promotes TH1-dominant immune response, activates macrophages and T cells, increases TH1 cytokines, stimulates dendritic cell maturation via the dectin-1 pathway
Poly-γ-Glutamate (γ-PGA)	Biomaterial from *B. subtilis*	Interacts with TLR4 receptors, triggers proinflammatory cytokine expression, enhances NK cell function and cytotoxicity, and induces cytokine secretion from activated NK dendritic cells
Chlorella vulgaris	Proteins, fibers, vitamins, minerals	Restores cytokine production, boosts NK cell activity, increases lymphocyte proliferation, improves macrophage phagocytosis, enhances antibody production via TLR2 signalling
Chlorella pyrenoidosa	Bioactive compounds	Activates macrophages, enhances antibody responses, increases phagocytosis and NO production, upregulates inflammatory cytokines through TLR4 signalling and NF-κB activation
Lactobacillus plantarum	Exopolysaccharides	Enhances macrophage phagocytic activity, increases TLR2 mRNA expression, stimulates pro-inflammatory cytokine production via NF-κB and MAPK pathways, promotes TH1 and TH17 responses
Other Probiotics	Live microorganisms	Increases NK cell-mediated tumoral activity, enhances IL-12 production, stimulates IgA production by B cells, modulates immune response, and changes intestinal microbiota composition

Contd.

Table 2 *Contd.*

Functional Food	*Key Bioactive Compounds*	*Primary Immunomodulatory Mechanisms*
Fruits and Vegetables	Antioxidants, vitamins, and flavonoids	Neutralizes free radicals, exhibits anti-inflammatory effects, modulates intracellular signalling, enhances antibody production, stimulates immune cell production, and reduces chronic inflammation
Herbs and Spices	Curcumin, piperine, thymol	Suppresses inflammatory cascades, reduces pro-inflammatory cytokines, exhibits antimicrobial and antiviral properties, enhances innate immunity, and improves humoral immune response
Green Tea	Polyphenols (EGCG, ECG)	Modulates TLR signalling, influences T cell function and differentiation, reduces STAT1 phosphorylation, suppresses inflammatory mediators, and exhibits antioxidative mechanisms
Fish/Marine Compounds	Proteins, peptides, n-3 PUFAs	Enhances host responses, protects from oxidative stress, disrupts microbial membranes, stimulates immune responses, restructures cell membranes, and reduces inflammation

6. Research Evidence and Clinical Studies

White ginseng extract has demonstrated significant immunomodulatory effects on various immune cells. In RAW 264.7 murine macrophages, the extract enhanced phagocytic activity and dose-dependently boosted NO generation and iNOS expression. JNK inhibitor pretreatment markedly suppressed the production of immunomodulatory factors such as NO, IL-6, and TNF-α in white ginseng extract-stimulated RAW264.7 cells. Subsequently, Um et al. (2020) demonstrated that increased macrophage phagocytic activity occurred via TLR2/4-mediated stimulation of MAPK, NF-κB, and PI3K/AKT signalling pathways (Um et al., 2020). In BALB/c mice, oral administration of oligopeptides from white ginseng bolstered innate immunity, as evidenced by increased phagocytic capacity of macrophages and NK cell activity. Additionally, ginsan, a component of white ginseng, induced expression of CD86 and class II MHC *in vitro* in both murine BMDCs and human monocyte-derived DCs (Kim et al., 2022).

In mouse models, red ginseng extract has demonstrated significant antiviral effects against influenza viruses. When administered to mice infected with the H1N1 virus, the extract increased NKp46 expression on NK cells and enhanced IFN-γ production. Similarly, in mice infected with the A/PR8 influenza virus, red ginseng extract boosted serum IgA and all IgG subtype levels. The extract's efficacy extended to reducing accumulation of viral plaque and expression of lytic genes induced by H1N1 virus, ultimately improving mouse survival rates. (Yoo et al., 2012) further confirmed the extract's antiviral properties against both H1N1 and H3N2 influenza viruses in mice. Notably, red ginseng extract administered through oral route enhanced cross-protection against antigenically distinct strains of these viruses, highlighting its potential as a broad-spectrum antiviral agent (Kim et al., 2022).

In clinical applications, oral administration of γ-PGA from *Bacillus* has demonstrated increased histological remission of cervical intraepithelial neoplasia (CIN 1) and enhanced human papillomavirus (HPV) clearance compared to control subjects. However, the sustainability of its effects on NK cell activity appears time-dependent, potentially diminishing after pausing the consumption. Beyond cancer models, γ-PGA also exhibits antiviral properties, protecting against H1N1 influenza A virus infection through infection inhibition and increasing specific activity by cytotoxic T cells against influenza, ultimately improving survival rates in infected mice (Kim et al., 2022).

An 8-week long, placebo-controlled clinical trial in healthy humans demonstrated that *C. vulgaris* extract supplements enhanced immune function by increasing natural killer cell activity and boosting serum levels of specific cytokines (especially IFN-γ and IL-1β) (Kim et al., 2022).

A Japanese study found that healthy subjects who consumed *L. casei* Shirota-fermented milk for 3 weeks experienced increased NK cell activity, which could help prevent disease in people with initially low NK cell function. In a trail of placebo-controlled crossover with 10 volunteers, NK cell activity was similar in both groups before treatment but significantly higher in those consuming the fermented milk than in those consuming the placebo, demonstrating that *L. casei* Shirota has meaningful effects on immune function (Santiago-López et al., 2015b).

Treatment with EGCG in a transgenic zebrafish system reduced neutrophil activity and downregulated IL-1β, TNF-α, and their related signalling mechanisms. In animal models of lupus nephritis (mice) and contrast-induced nephropathy (rats), EGCG when given orally or intravenously lowered renal NLRP3 expression at both mRNA and protein levels. Furthermore, tea polyphenols in BALB/c mice promoted humoral immunity by increasing splenic antibody-secreting cells, and significantly boosting dose-dependent IgM and IgG immune responses to BSA and SRBCs antigens. In male DBA/1 mice, GTPs significantly reduce the incidence of arthritis (33–50%) and minimize inflammatory mediator expression in arthritic joints, including COX2, IFN-γ, and tumour necrosis factor. Dietary GTPs exhibit significant immunomodulatory potential in oxidatively stressed piglets, by promoting T lymphocyte proliferation and activation, elevating the CD4+/CD8+ ratio, attenuating pro-inflammatory IL-1, reducing serum IFN-γ concentration, enhancing anti-inflammatory cytokine IL-4 levels, immune shift from Th1 to Th2, and alleviating growth depression. When combined with *Lactobacillus plantarum*, green tea powder enhances the diversity of intestinal bacteria, increases the proliferation of Lactobacillus in the intestine, and reduces inflammation brought on by a high-fat diet in C57BL/6J mice (S. Wang et al., 2021b).

E-JUR-94013 supplementation (750 mg/day for 6 months) demonstrated significant immunomodulatory effects across three studies. In 56 subjects, it increased leukocyte counts, particularly neutrophils and eosinophils, while raising immunoglobulins A, G, and M levels and slightly decreasing allergy-related IgE. A more extensive study with 205 immunocompromised patients showed normalized white cell counts, increasing them in immunodeficient patients and reducing them in those with elevated counts. A third study with 1,500 random subjects confirmed this regulatory effect on various leukocyte populations, with responses influenced by genetic variations in immune-related genes and a reduced inflammatory C-reactive protein to normal levels in many cases (Lombardi et al., 2018).

7. Applications

Functional foods give health benefits more than essential nutrition through biologically active compounds influencing physiological functions. These foods contain naturally occurring compounds or are fortified with ingredients that confer specific benefits to health if taken as part of a regular diet (Dillard and Bruce German, 2000; Martirosyan and Singh, 2015).

7.1 Cardiovascular Health

Omega-3 fatty acid-rich foods such as fatty fish, flaxseeds, and walnuts show abilities for reducing inflammation, improving lipid profiles, and lower blood pressure (Mozaffarian and Wu, 2018). Plant sterols and stanols found in cereals effectively reduce LDL (low density lipid) cholesterol levels by inhibiting cholesterol absorption (Santhakumar et al., 2018). Polyphenol-rich foods like berries, dark chocolate, and green tea have shown cardioprotective effects through antioxidant activity, elevated functions of endothelial cells, and poor platelet aggregation (Del Bo et al., 2019). Soluble fibre in oats, barley, and psyllium binds to bile acids, promoting their excretion and forcing the body to use cholesterol to produce more bile acids, lowering serum cholesterol levels (Li and Komarek, 2017). Fermented foods containing bioactive peptides can inhibit angiotensin-converting enzyme (ACE), contributing to blood pressure control (Weaver, 2014). Allicin and related organosulfur compounds in garlic have demonstrated antihypertensive, cholesterol-lowering, and antithrombotic properties (Gupta and Prakash, 2014). Anthocyanins in berries and red/purple fruits improve endothelial function and vascular elasticity while reducing oxidative stress (Del Bo et al., 2019). Soy proteins and isoflavones have been shown to lower LDL cholesterol and triglycerides while increasing HDL cholesterol levels (Mozaffarian and Wu, 2018).

7.2 Diabetes Management

Foods with a low glycemic index and high fibre content (oats, barley, legumes) help regulate blood glucose levels and improve insulin sensitivity (Li and Komarek, 2017). Cinnamon contains bioactive compounds that enhance insulin signaling and glucose uptake (Gupta and Prakash, 2014). Fermented dairy products containing probiotics can improve glycemic control through modulation of gut microbiota, which influences glucose metabolism and inflammation (Conlon and Bird, 2015). Alpha-lipoic acid in spinach, broccoli, and organ meats enhances glucose uptake and improves insulin sensitivity (Mirmiran, 2014). Bitter melon (Momordica charantia) contains compounds that mimic insulin action and activate AMP-activated protein kinase pathways (Mirmiran, 2014). Chromium-rich foods like broccoli, whole grains, and nuts enhance insulin receptor signaling (Weaver, 2014). Berberine-containing plants demonstrate hypoglycemic effects through multiple mechanisms, including AMPK activation and intestinal disaccharidase inhibition (Mirmiran, 2014). Polyphenols in tea, coffee, and cocoa improve glucose tolerance through reduced carbohydrate digestion and absorption, enhanced insulin secretion, and decreased hepatic glucose output (Del Bo et al., 2019).

7.3 Cancer Prevention

Cruciferous vegetables (broccoli, cauliflower, kale) contain glucosinolates and isothiocyanates that induce detoxification enzymes and inhibit proliferation of cancer cells (Gupta and Prakash, 2014). Curcumin from turmeric demonstrates anti-inflammatory and antioxidant properties that may inhibit cancer development and progression (Shahidi and Ambigaipalan, 2015). Resveratrol from grapes and berries has shown potential in modulating multiple cellular pathways involved in cancer development, potentially inhibiting initiation, promotion, and progression stages (Del Bo et al., 2019). Lycopene in tomatoes and watermelon protects against prostate cancer through antioxidant activity and modulation of cell cycle progression (Dillard and Bruce German, 2000). Catechins in green tea exhibit anticarcinogenic activity via. cell cycle arrest, apoptosis, and tumour invasion and angiogenesis inhibition (Shahidi and Ambigaipalan, 2015). Allium vegetables (garlic, onions) contain organosulfur compounds that inhibit carcinogen activation and detoxification (Gupta and Prakash, 2014). Selenium-rich foods like Brazil nuts and seafood support antioxidant enzyme systems that protect against DNA damage (Weaver, 2014). Dietary fiber reduces colorectal cancer risk through decreased transit time, dilution of carcinogens, and production of beneficial short-chain fatty acids like butyrate (Li and Komarek, 2017).

7.4 Digestive Health

Probiotics in fermented foods (yogurt, kefir, sauerkraut) deliver bacteria to enhance gut barrier function, alter immune responses, and compete with pathogens (Conlon and Bird, 2015). Prebiotics from foods like onions, garlic, and bananas provide substrates to bacteria for promoting growth and metabolism (Gibson et al., 2017). Fermented foods contain bioactive peptides that enhance intestinal barrier integrity by upregulating tight junction proteins (Conlon and Bird, 2015). Polyphenols in fruits, vegetables, and tea modulate gut microbiota composition, promoting beneficial bacteria growth (Del Bo et al., 2019). Ginger contains compounds that enhance gastric emptying and intestinal transit while reducing inflammation in the digestive tract (Gupta and Prakash, 2014). Digestive enzymes in pineapple (bromelain) and papaya (papain) aid protein digestion and reduce digestive discomfort (Weaver, 2014). Glutamine-rich foods like bone broth support intestinal cell proliferation and maintenance of gut barrier function (Conlon and Bird, 2015).

7.5 Cognitive Function

Fatty fish containing DHA support neuron membrane integrity and function (Kris-Etherton et al., 2021). Flavonoid-rich foods like berries and cocoa improve cerebral blood flow and neural signalling, potentially enhancing learning and brain power (Del Bo et al., 2019). Curcumin from turmeric has demonstrated neuroprotective properties through anti-inflammatory and antioxidant mechanisms, with potential applications in neurodegenerative disease prevention (Shahidi and Ambigaipalan, 2015). Choline-containing foods like eggs and liver support acetylcholine synthesis, a neurotransmitter critical for memory and cognition (Weaver, 2014). Vitamin E from nuts and seeds protects neurons from oxidative damage and may slow cognitive decline (Dillard and Bruce German, 2000). Caffeine in coffee and tea enhances alertness, attention, and mental processing through adenosine receptor antagonism (Del Bo et al., 2019). L-theanine in tea promotes relaxation while maintaining alertness, creating a "calm attention" (Shahidi and Ambigaipalan, 2015). Resveratrol enhances cerebral blood flow and potentially activates sirtuin pathways that protect against neurodegenerative processes (Del Bo et al., 2019).

7.6 Bone Health

Isoflavones in soy foods act as selective estrogen receptor modulators, potentially reducing bone loss in postmenopausal women (Mozaffarian and Wu, 2018). Prebiotic fibres enhance calcium absorption through bacterial fermentation and production of short-chain fatty acids that lower intestinal pH (Gibson et al., 2017). Silicon in whole grains, green beans, and some mineral waters contributes to collagen synthesis and bone mineralization (Weaver, 2014). Omega-3 fatty acids reduce bone resorption by decreasing inflammatory cytokine production (Kris-Etherton et al., 2021). Dried plums (prunes) contain boron, vitamin K, and polyphenols that suppress bone resorption and enhance bone formation (Del Bo et al., 2019). Magnesium from green leafy vegetables and nuts activates vitamin D, essential for calcium absorption (Weaver, 2014). Carotenoids in fruits and vegetables reduce the level of oxidative stress that can contribute to bone loss (Shahidi and Ambigaipalan, 2015). Vitamin K2-rich fermented foods like natto direct calcium to bones rather than arteries, enhancing bone mineralization (Weaver, 2014).

8. Factors Affecting the Efficacy of Functional Foods

The health benefits derived from functional foods are not solely determined by the presence of bioactive compounds, but rather by a complex interplay of multiple factors that influence their bioavailability, metabolism, and physiological impact. While functional foods hold significant promise for disease prevention and health promotion, their efficacy varies considerably among individuals and under different conditions. The factors influencing functional food efficacy can be broadly categorised into food-related factors (such as dosage, processing, and timing of consumption), individual biological factors (including genetics, age, and health status), lifestyle, and environmental factors (including cultivation conditions and storage). Understanding these determinants is required to optimise functional food benefits and developing personalized nutrition strategies for maximum health benefits.

8.1 Dosage and Concentration

The concentration of bioactive compounds in functional foods is critical to their efficacy. Research indicates that many functional ingredients demonstrate dose-dependent effects, with minimum thresholds required for physiological impact. For example, studies on omega-3 fatty acids show that beneficial cardiovascular effects typically require specific dosage levels that may vary based on the targeted health outcome (Mozaffarian and Wu, 2018). The American Heart Association recommends at least 250–500 mg/day of EPA and DHA for cardiovascular protection in healthy individuals, while therapeutic benefits for reducing triglycerides may require 2–4 g/day (Mozaffarian and Wu, 2018). Similarly, plant sterols demonstrate cholesterol-lowering effects only when consumed at doses of 1.5–3 g/day (Ras et al., 2014). Research on polyphenols suggests that minimum effective doses vary considerably depending on the specific compound and health outcome targeted, with some studies indicating that 400–600 mg/day of flavonoids may be necessary for cardiovascular benefits (Hooper et al., 2008). The challenge lies in achieving these therapeutic doses through dietary sources alone, which often necessitates fortification or supplementation strategies (Hasler, 2002).

8.2 Individual Variability and Genetic Factors

Genetic polymorphisms can significantly affect the response of individuals to functional foods. Research in nutrigenomics has revealed that genetic variations affect the metabolism of bioactive compounds and subsequent physiological responses. For instance, variations in genes encoding for enzymes of folate metabolism affect how individuals respond to folate-fortified foods (Zetterberg et al., 2002). The methylenetetrahydrofolate reductase (MTHFR) gene variants significantly affect folate metabolism, with the C677T polymorphism reducing enzyme activity by 30–70%, thereby influencing individual requirements for folate-fortified foods (Zetterberg et al., 2002). Polymorphisms in genes encoding for fatty acid desaturases (FADS1 and FADS2) affect the conversion of plant-based omega-3 precursors to EPA and DHA, explaining why some individuals benefit more from marine-based omega-3 sources (Glaser et al., 2010). The apolipoprotein E (APOE) genotype influences cholesterol metabolism and modulates responses to dietary fat interventions, including those involving functional foods (Ordovas, 2006). Additionally, variations in taste receptor genes affect food preferences and consumption patterns, indirectly influencing functional food efficacy (Garcia-Bailo et al., 2009). These genetic factors underscore the need for personalized nutrition approaches when recommending functional foods.

8.3 Gut Microbiota Composition

The composition of an individual's gut microbiome is important in metabolizing many biologically active compounds in functional foods, particularly polyphenols and dietary fibers. The different microbial compositions can lead to varying production of beneficial metabolites from the same functional food components (Duda-Chodak et al., 2015). Polyphenols, such as ellagitannins found in pomegranates and berries, are converted by gut bacteria into urolithins, which possess anti-inflammatory and antioxidant properties (Tomás-Barberán et al., 2017). However, only 30–40% of individuals harbor bacteria capable of this conversion, creating "responders" and "non-responders" (Duda-Chodak et al., 2015). Similarly, the metabolism of isoflavones from soy products depends on specific bacterial species that convert daidzein to equol, a more bioactive metabolite; only 30-50% of Western populations are equol producers (Setchell and Clerici, 2010). The composition of the gut microbiota varies widely between individuals, contributing to different responses to the same functional food. For instance, Dietary fibers are fermented by gut bacteria into short-chain fatty acids (SCFAs) like butyrate, acetate, and propionate, but the quantity and ratio of these metabolites vary based on microbial composition (Macfarlane and Macfarlane, 2012). The ratio of Firmicutes to Bacteroidetes, along with the presence of specific bacterial genera, significantly influences the metabolic outcomes of functional food consumption (Turnbaugh et al., 2009).

8.4 Processing and Storage Conditions

Food processing techniques profoundly impact the bioavailability and stability of bioactive compounds. Thermal processing, for example, can either enhance or reduce the bioavailability of specific compounds. Cooking tomatoes increases lycopene bioavailability by 2–3 fold, but decreases the vitamin C content by 10–90% (Dewanto et al., 2002; Howard et al., 1999). Fermentation processes can enhance the bioavailability of minerals by reducing phytate content and increase the production of peptides that are biologically active and vitamins (Anal, 2019). High-pressure processing preserves more heat-sensitive compounds compared to traditional thermal processing while achieving microbial safety (Barba et al., 2012). Similarly, storage conditions and time can lead to degradation of unstable bioactive compounds (Howard et al., 1999). Anthocyanins in berries degrade rapidly when exposed to light, oxygen, and elevated temperatures (Kalt, 2005). Frozen storage generally preserves nutrients better than extended refrigerated storage, though blanching before freezing can cause initial losses (Rickman et al., 2007). Packaging technologies, including modified atmosphere packaging and light-protective materials, help maintain bioactive compound integrity during storage (Caleb et al., 2013).

8.5 Timing of Consumption

The timing of functional food consumption relative to meals and other dietary components can affect its efficacy. For instance, fat-soluble vitamins (A, D, E, K) and carotenoids require dietary fat for optimal absorption. Consuming carotenoid-rich vegetables with dietary fats (3–5 g) significantly enhances their bioavailability by 4–5 fold compared to consumption without fats (Brown et al., 2004; Unlu et al., 2005). The presence of other food components can either enhance or inhibit bioactive compound absorption; for example, iron absorption is promoted by vitamin C, while phytates and polyphenols inhibit it (Teucher et al., 2004). Consuming probiotics with meals or shortly before meals improves bacterial survival through the acidic stomach environment compared to consumption on an empty stomach (Tompkins et al., 2011). The glycemic response to functional foods containing soluble fibers is most beneficial when consumed before or with high-glycemic meals (Weickert and Pfeiffer, 2008). Circadian rhythms also influence nutrient metabolism, with some evidence suggesting that timing of antioxidant consumption relative to exercise can affect their efficacy in reducing oxidative stress (Myburgh, 2014).

8.6 Health Status and Age

An individual's health status and age can significantly influence how functional foods affect their body. Certain health conditions may alter the absorption, metabolism, or response to bioactive compounds. Studies have shown that inflammation, oxidative stress, and compromised gut integrity, often associated with ageing or disease states, can modify the response of the body to functional ingredients (Rémond et al., 2015). The bioavailability of vitamin B12, calcium, and iron typically decreases with age due to these physiological changes (Andrès et al., 2005). Inflammatory conditions and oxidative stress, common in chronic diseases, can modify the metabolism and efficacy of bioactive compounds; for instance, individuals with metabolic syndrome may show enhanced benefits from anti-inflammatory functional foods compared to healthy individuals (Esposito et al., 2004). Compromised intestinal barrier function, often seen in conditions like inflammatory bowel disease or during aging, affects the absorption of bioactive compounds and may alter immune responses to functional food components (Camilleri, 2019). Medications commonly used by older people or people with severe conditions can interact with functional food components, affecting their absorption or metabolism (Boullata and Hudson, 2012). Some functional foods may be more effective as preventive measures in healthy populations rather than therapeutic interventions in diseased states, though this varies by compound and condition (Bellisle et al., 1998).

8.7 Exercise and Physical Activity

Physical activity levels and exercise patterns modulate how the body responds to functional food components, particularly those targeting metabolism or inflammation. The anti-inflammatory effects of omega-3 fatty acids are enhanced when combined with regular exercise, with combined interventions showing greater reductions in inflammatory markers like CRPs (C-reactive proteins) and IL-6 compared to either intervention alone (Tartibian et al., 2010; Żebrowska et al., 2015). Exercise-induced oxidative stress can be mitigated by antioxidant-rich functional foods, though the timing and dose require careful consideration to avoid blunting beneficial exercise adaptations (Powers et al., 2020). Similarly, the glucose-regulating effects of certain functional foods containing starch or soluble fibre are enhanced in physically active individuals due to improved insulin sensitivity (Malin and Braun, 2013). Protein-rich functional foods and those containing branched-chain amino acids show enhanced effects on muscle protein synthesis when consumed in proximity to resistance exercise (Moore et al., 2015). Athletes and highly active individuals may have different requirements and responses to functional foods compared to sedentary populations, including increased oxidative stress and altered energy metabolism that affect bioactive compound utilization (Williams et al., 2006).

8.8 Stress and Psychological Factors

Psychological stress can alter gut function, immune responses, and metabolic processes, affecting how functional foods are absorbed and utilized. Chronic stress activates the hypothalamic-pituitary-adrenal (HPA) axis, resulting in elevated cortisol levels that impair gut barrier function, reduce digestive enzyme secretion, and modify gut motility (Karl et al., 2018; Konturek et al., 2011). Stress-induced increases in intestinal permeability can reduce the absorption efficiency of bioactive compounds while potentially increasing exposure to pro-inflammatory luminal contents (Vanuytsel et al., 2014). The composition and diversity of gut microbiota are negatively affected by chronic stress, which can impair the metabolism of functional food components requiring microbial transformation (Foster et al., 2017). Research also shows that stress hormones can alter metabolic responses to food components, including changes in glucose metabolism, lipid oxidation, and inflammatory mediator production (Kiecolt-Glaser, 2010; Kiecolt-Glaser et al., 2015). Studies indicate that individuals under chronic stress may show reduced benefits from anti-inflammatory functional foods due to sustained activation of inflammatory pathways (Steptoe et al., 2007). Conversely, some functional foods, particularly those containing omega-3 fatty acids, probiotics, or adaptogens, may help modulate stress responses and support psychological well-being (Sarris et al., 2015).

8.9 Environmental Factors

Environmental conditions during food production, including climate, soil conditions, and altitude, affect the concentration of bioactive compounds in functional foods. Studies show that grapes grown at higher altitudes or under mild water stress produce wines with higher polyphenol content, increased anthocyanin and resveratrol content due to greater UV exposure and temperature fluctuations (Castellarin et al., 2007; Mateus et al., 2004). Water stress during grape cultivation can increase polyphenol concentrations by 20–50%, though severe drought may reduce overall yield and quality (Roby et al., 2004). Soil mineral composition directly affects the mineral content and secondary metabolite production in plants; selenium-rich soils produce crops with higher selenium content, important for functional food development (Lyons et al., 2007). UV-B radiation significantly influences the production of protective compounds like flavonoids and phenolic acids in plants, with controlled UV exposure sometimes used to enhance functional properties (Jansen et al., 2008). Similarly, variations in sunlight exposure can significantly impact mushrooms' vitamin D content (Urbain et al., 2011). Mushrooms exposed to UV light post-harvest can increase vitamin D2 content from negligible amounts to therapeutically relevant levels (100–200% of daily requirements) within hours (Cardwell et al., 2018; Urbain et al., 2011). Organic versus conventional farming practices can influence phytochemical profiles, with some studies showing 20–40% higher polyphenol content in organically grown produce, possibly due to increased plant stress responses in the absence of synthetic pesticides (Barański et al., 2014).

8.10 Duration of Consumption

The time frame over which functional foods are consumed is crucial for achieving beneficial effects. Many functional ingredients require consistent, long-term consumption to demonstrate measurable health benefits. Research on plant sterols indicates that cholesterol-lowering effects require regular consumption for at least 2–3 weeks before reaching maximum efficacy (Demonty et al., 2009). Similarly, studies on probiotics show that their effects on gut microbiota composition are often temporary, with introduced bacterial strains declining to baseline levels within 1–4 weeks after cessation of consumption requiring continued consumption for sustained benefits (Derrien and van Hylckama Vlieg, 2015; Zmora et al., 2018). Omega-3 fatty acid incorporation into cell membranes occurs gradually over 8–12 weeks, with cardiovascular and anti-inflammatory benefits emerging after this incorporation period (Flock et al., 2013). Flavonoid-rich functional foods demonstrate antioxidant effects acutely (within hours), but sustained cardiovascular benefits require chronic consumption for at least 4–8 weeks (Hooper et al., 2008). Isoflavones from soy products show progressive effects on menopausal symptoms and bone health markers over 6–12 months of consistent consumption (Messina, 2014). The concept of "nutritional programming" suggests that early and consistent exposure to certain functional food components may provide long-term health benefits extending beyond the period of active consumption, though this requires further investigation (Koletzko et al., 2012). Intermittent or inconsistent consumption patterns generally fail to produce significant health outcomes, emphasizing the importance of incorporating functional foods as regular dietary components rather than occasional additions (Granadolorencio et al., 2007).

9. Conclusion and Future Perspective

The effect of functional foods on the immune system can be further improved by developing advanced delivery systems and formulating new combinations of bioactive compounds to increase the immunomodulatory effect. Interactions between the components of functional foods and the immune system can be studied using AI and Computational studies. Exploring the under-researched sources (deep seas) and biotechnological methods yield new compounds that enhance the immune system. More robust human trials with standardized immune biomarkers will provide more substantial evidence for health claims and establish transparent dose-response relationships. The advancement of immune-supporting functional foods requires collaboration across disciplines—from immunology and nutrition to microbiology, genetics, and biotechnology. This cross-disciplinary integration, combined with cutting-edge technologies, promises to maximize the potential of functional foods as immune enhancers. Such progress could fundamentally shift healthcare paradigms from reactive treatment toward proactive prevention. As research clarifies the intricate connections linking dietary choices, gut microbiota, and immune responses, functional foods are positioned to become essential components in lifelong health maintenance and disease prevention strategies.

References

Adams, C.A. 2010. The probiotic paradox: Live and dead cells are biological response modifiers. *Nutrition Research Reviews*, *23*, 37–46. https://doi.org/10.1017/S0954422410000090.

Aggarwal, B.B. and Harikumar, K.B. 2009. Potential therapeutic effects of curcumin, the anti-inflammatory agent, against neurodegenerative, cardiovascular, pulmonary, metabolic, autoimmune, and neoplastic diseases. *The International Journal of Biochemistry & Cell Biology*, *41*(1), 40–59. https://doi.org/10.1016/j.biocel.2008.06.010.

Ahn, H., Kang, S.G., Yoon, S. Il, Kim, P.H., Kim, D. and Lee, G.S. 2018. Poly-gamma-glutamic acid from Bacillus subtilis upregulates pro-inflammatory cytokines while inhibiting NLRP3, NLRC4, and AIM2 inflammasome activation. *Cellular and Molecular Immunology*, *15*, 111–119. https://doi.org/10.1038/cmi.2016.13.

Alhazmi, H.A., Najmi, A., Javed, S.A., Sultana, S., Al Bratty, M., Makeen, H.A., Meraya, A.M., Ahsan, W., Mohan, S., Taha, M.M.E, and Khalid, A. 2021. Medicinal Plants and Isolated Molecules Demonstrating Immunomodulation Activity as Potential Alternative Therapies for Viral Diseases Including COVID-19. May 13; *Front Immunol.*, *12*, 637553. doi: 10.3389/fimmu.2021.637553.

An, B.K., Kim, K.E., Jeon, J.Y. and Lee, K.W. 2016. Effect of dried Chlorella vulgaris and Chlorella growth factor on growth performance, meat qualities, and humoral immune responses in broiler chickens. Jun 14; *Springerplus*, *5*(1), 718. doi: 10.1186/s40064-016-2373-4.

Anal, A.K. 2019. Quality Ingredients and Safety Concerns for Traditional Fermented Foods and Beverages from Asia: A Review. *Fermentation*, *5*(1), 8. https://doi.org/10.3390/fermentation5010008.

Anaya, J.-Manuel., Shoenfeld, Yehuda., Rojas-Villarraga, Adriana., Levy, R.A. and Cervera, Ricard. 2013. *Autoimmunity: From Bench to Bedside*. Center for Autoimmune Diseases Research, School of Medicine and Health Sciences, El Rosario University.

Andrès, E., Affenberger, S., Vinzio, S., Kurtz, J.-E., Noel, E., Kaltenbach, G., Maloisel, F., Schlienger, J.-L., and Blicklé, J.-F. 2005. Food-cobalamin malabsorption in elderly patients: Clinical manifestations and treatment. *The American Journal of Medicine*, *118*(10), 1154–1159. https://doi.org/10.1016/j.amjmed.2005.02.026.

Arreola, R., Quintero-Fabián, S., López-Roa, R.I., Flores-Gutiérrez, E.O., Reyes-Grajeda, J.P., Carrera-Quintanar, L. and Ortuño-Sahagún, D. 2015. Immunomodulation and anti-inflammatory effects of garlic compounds. *J. Immunol Res.*, *2015*, 401630. doi: 10.1155/2015/401630. Epub 2015 Apr 19.

Awuchi, C.G., Chukwu, C.N., Iyiola, A.O., Noreen, S., Morya, S., Adeleye, A.O., Twinomuhwezi, H., Leicht, K.,, Mitaki, N.B. and Okpala, C.O.R. 2022. Bioactive Compounds and Therapeutics from Fish: Revisiting Their Suitability in Functional Foods to Enhance Human Wellbeing. Aug 5; *Biomed. Res. Int.*, *2022*, 3661866. doi: 10.1155/2022/3661866.

Baidya, B.K. and Sethy, P. 2020. Importance of Fruits and Vegetables in Boosting our Immune System amid the COVID191. *July*, 50–55.

Balakrishnan, K., Nagarajan, S., Balakrishnan, G., Manickam, M. and Dharumadurai, D. 2023. Chapter 15 - Symbiotic microbial interactions in medicinal mushroom: An insight of phenotypic and genotypic characterization methods. In: Dharumadurai, D. (Ed.), *Microbial Symbionts* (pp. 277–294). Academic Press. https://doi.org/https://doi.org/10.1016/B978-0-323-99334-0.00044-X.

Banaszak, M., Dobrzyńska, M., Kawka, A., Górna, I., Woźniak, D., Przysławski, J. and Drzymała-Czyż S. 2024. Role of Omega-3 fatty acids eicosapentaenoic (EPA) and docosahexaenoic (DHA) as modulatory and anti-inflammatory agents in noncommunicable diet-related diseases - Reports from the last 10 years. Oct; *Clin. Nutr. ESPEN*, *63*, 240–258. doi: 10.1016/j.clnesp.2024.06.053.

Banchereau, J. and Steinman, R.M. 1998. Dendritic cells and the control of immunity. Mar 19, *Nature*, *392*(6673), 245–252. doi: 10.1038/32588.

Barański, M., Średnicka-Tober, D., Volakakis, N., Seal, C., Sanderson, R., Stewart, G.B., Benbrook, C., Biavati, B., Markellou, E., Giotis, C., Gromadzka-Ostrowska, J., Rembiałkowska, E., Skwarło-Sońta, K., Tahvonen, R., Janovská, D., Niggli, U., Nicot, P. and Leifert, C. 2014. Higher antioxidant and lower cadmium concentrations and lower incidence of pesticide residues in organically grown crops: A systematic literature review and meta-analyses. *British Journal of Nutrition*, *112*(5), 794–811. https://doi.org/10.1017/S0007114514001366.

Barba, F.J., Esteve, M.J. and Frígola, A. 2012. High Pressure Treatment Effect on Physicochemical and Nutritional Properties of Fluid Foods During Storage: A Review. *Comprehensive Reviews in Food Science and Food Safety*, *11*(3), 307–322. https://doi.org/10.1111/j.1541-4337.2012.00185.x.

Basak, S., Banerjee, A., Pathak, S. and Duttaroy, A.K. 2022. Dietary Fats and the Gut Microbiota: Their impacts on lipid-induced metabolic syndrome. *Journal of Functional Foods*, *91*, 105026. https://doi.org/10.1016/j.jff.2022.105026.

Basak, S. and Gokhale J. 2022. Immunity boosting nutraceuticals: Current trends and challenges. Mar; *J. Food Biochem.*, *46*(3), e13902. doi: 10.1111/jfbc.13902.

Bax, C.E., Chakka, S., Concha, J.S.S., Zeidi, M. and Werth, V.P. 2021. The effects of immunostimulatory herbal supplements on autoimmune skin diseases. *Journal of the American Academy of Dermatology*, *84*, 1051–1058. https://doi.org/10.1016/j.jaad.2020.06.037.

Bellisle, F., Blundell, J.E., Dye, L., Fantino, M., Fern, E., Fletcher, R.J., Lambed, J., Roberfroid, M., Specter, S., Westenhöfer, J. and Westerterp-Plantenga, M.S. 1998. Functional food science and behaviour and psychological functions. *British Journal of Nutrition*, *80*(S1), S173–S193. https://doi.org/10.1079/BJN19980109.

Belokhvostova, D., Berzanskyte, I., Cujba, A.M., Jowett, G., Marshall, L., Prueller, J. and Watt, F.M. 2018. Homeostasis, regeneration, and tumour formation in the mammalian epidermis. *International Journal of Developmental Biology*, *62*, 571–582. https://doi.org/10.1387/ijdb.170341fw.

Boasso, A., Shearer, G.M., Chougnet, C. 2009. Immune dysregulation in human immunodeficiency virus infection: Know it, fix it, prevent it? *Journal of Internal Medicine,* 78–96. https://doi.org/10.1111/j.1365-2796.2008.02043.x.

Boullata, J.I. and Hudson, L.M. (2012). Drug–Nutrient Interactions: A Broad View with Implications for Practice. *Journal of the Academy of Nutrition and Dietetics*, *112*(4), 506–517. https://doi.org/10.1016/j.jada.2011.09.002.

Bremner, P. and Heinrich, M. 2010. Natural products as targeted modulators of the nuclear factor: κ B pathway . *Journal of Pharmacy and Pharmacology*, *54*, 453–472. https://doi.org/10.1211/0022357021778637.

Brown, M.J., Ferruzzi, M.G., Nguyen, M.L., Cooper, D.A., Eldridge, A.L., Schwartz, S.J. and White, W.S. 2004. Carotenoid bioavailability is higher from salads ingested with full-fat than with fat-reduced salad dressings as measured with electrochemical detection. *American Journal of Clinical Nutrition*, *80*, 396–403. https://doi.org/10.1093/ajcn/80.2.396.

Byeon, S.E., Lee, J., Kim, J.H., Yang, W.S., Kwak, Y.S., Kim, S.Y., Choung, E.S., Rhee, M.H. and Cho, J.Y. 2012. Molecular mechanism of macrophage activation by red ginseng acidic polysaccharide from Korean red ginseng. *Mediators of Inflammation*, *2012*, 7–9. https://doi.org/10.1155/2012/732860.

Calder, P.C. 2013. Omega-3 polyunsaturated fatty acids and inflammatory processes: Nutrition or pharmacology? *Br. J. Clin. Pharmacol.*, *75*, 645–662. https://doi.org/10.1111/j.1365-2125.2012.04374.x.

Calder, P.C. 2020. Nutrition, immunity, and COVID-19. *BMJ Nutr. Prev. Health*, *3*, 74–92. https://doi.org/10.1136/bmjnph-2020-000085.

Caleb, O.J., Mahajan, P.V., Al-Said, F.A.-J. and Opara, U.L. (2013). Modified Atmosphere Packaging Technology of Fresh and Fresh-cut Produce and the Microbial Consequences: A Review. *Food and Bioprocess Technology*, *6*(2), 303–329. https://doi.org/10.1007/s11947-012-0932-4.

Camilleri, M. (2019). Leaky gut: Mechanisms, measurement, and clinical implications in humans. *Gut*, *68*(8), 1516–1526. https://doi.org/10.1136/gutjnl-2019-318427.

Cardwell, G., Bornman, J.F., James, A.P. and Black, L.J. (2018). A Review of Mushrooms as a Potential Source of Dietary Vitamin D. *Nutrients*, *10*(10), 1498. https://doi.org/10.3390/nu10101498.

Carr, A.C. and Maggini S. 2017. Vitamin C and Immune Function. Nov 13; *Nutrients*, , *9*(11), 1211. doi: 10.3390/nu9111211.

Castellarin, S.D., Matthews, M.A., Di Gaspero, G. and Gambetta, G.A. 2007. Water deficits accelerate ripening and induce changes in gene expression regulating flavonoid biosynthesis in grape berries. *Planta*, *227*, 101–112. https://doi.org/10.1007/s00425-007-0598-8.

Chan, G.C.F., Chan, W.K. and Sze, D.M.Y. 2009. The effects of beta-glucan on human immune and cancer cells. *Journal of Hematology & Oncology*, *2*, 25. https://doi.org/10.1186/1756-8722-2-25.

Chaplin, D.D. 2010. Overview of the immune response. *Journal of Allergy and Clinical Immunology*, *125*, S3–S23. https://doi.org/10.1016/j.jaci.2009.12.980.

Chen, K., Magri, G., Grasset, E.K. and Cerutti A. 2020. Rethinking mucosal antibody responses: IgM, IgG, and IgD join IgA. Jul; *Nat Rev Immunol.*, *20*(7), 427–441. doi: 10.1038/s41577-019-0261-1.

Cheng, D., Wan, Z., Zhang, X., Li, J., Li, H. and Wang C. 2017. Dietary Chlorella vulgaris Ameliorates Altered Immunomodulatory Functions in Cyclophosphamide-Induced Immunosuppressive Mice. Jul 6: *Nutrients*, *9*(7), 708. doi: 10.3390/nu9070708. Erratum in: 2024 Aug 27; *Nutrients*, *16*(17), 2867. doi: 10.3390/nu16172867.

Chi, H., Pepper, M. and Thomas, P.G., 2024. Principles and therapeutic applications of adaptive immunity. *Cell*, *187*, 2052–2078. https://doi.org/10.1016/j.cell.2024.03.037.

Cleverdon, R., Elhalaby, Y., McAlpine, M.D., Gittings, W. and Ward, W.E. 2018. Total polyphenol content and antioxidant capacity of tea bags: Comparison of black, green, red rooibos, chamomile, and peppermint over different steep times. *Beverages*, *4*(1), 15. https://doi.org/10.3390/beverages4010015.

Conlon, M.A. and Bird, A.R. 2015. The impact of diet and lifestyle on gut microbiota and human health. *Nutrients*, *7*, 17–44. https://doi.org/10.3390/nu7010017.

Dai, X., Stanilka, J.M., Rowe, C.A., Esteves, E.A., Nieves, C., Spaiser, S.J., Christman, M.C., Langkamp-Henken, B. and Percival, S.S. 2015. Consuming Lentinula edodes (Shiitake) Mushrooms Daily Improves Human Immunity: A Randomized Dietary Intervention in Healthy Young Adults. *J. Am. Coll. Nutr.*, *34*, 478–487. https://doi.org/10.1080/07315724.2014.950391.

Damián, M.R., Cortes-Perez, N.G., Quintana, E.T., Ortiz-Moreno, A., Noguez, C.G., Cruceño-Casarrubias, C.E., Pardo, M.E.S. and Bermúdez-Humarán, L.G., 2022. Functional Foods, Nutraceuticals and Probiotics: A Focus on Human Health. *Microorganisms*, *10*(5), 1065 https://doi.org/10.3390/microorganisms10051065.

Darbandi, A., Asadi, A., Ghanavati, R., Afifirad, R., Darb Emamie, A., Kakanj, M. and Talebi, M., 2021. The effect of probiotics on respiratory tract infection with special emphasis on COVID-19: Systemic review 2010–20. *International Journal of Infectious Diseases*. https://doi.org/10.1016/j.ijid.2021.02.011.

de Oliveira Henriques, M.D.G.M. and Penido, C., 2012. γδ T lymphocytes coordinate eosinophil influx during allergic responses. 3 Dec;,Front. Pharmacol., *3*, 200. https://doi.org/10.3389/fphar.2012.00200.

Del Bo, C., Bernardi, S., Marino, M., Porrini, M., Tucci, M., Guglielmetti, S., Cherubini, A., Carrieri, B., Kirkup, B., Kroon, P., Zamora-ros, R., Liberona, N.H., Andres-lacueva, C. and Riso, P. 2019. Systematic Review on Polyphenol Intake and Health Outcomes : Is there Sufficient Evidence to Define a Health-Promoting Polyphenol-Rich Dietary Pattern? *Nutrients*, *11*, 1355.

Demonty, I., Ras, R.T., Van Der Knaap, H.C.M., Duchateau, G.S.M.J.E., Meijer, L., Zock, P.L., Geleijnse, J.M. and Trautwein, E.A. 2009. Continuous dose-response relationship of the LDL-cholesterol-lowering effect of phytosterol intake. *Journal of Nutrition, 139*, 271–284. https://doi.org/10.3945/jn.108.095125.

Deo, A.S., Shrijana, Sruthika, S.U., Karun, S., Bisaria, K. and Sarkar, K., 2024. Participation of T cells in generating immune protection against cancers. Oct; *Pathol. Res. Pract.*, *262*, 155534. https://doi.org/10.1016/j.prp.2024.155534.

Derrien, M. and van Hylckama Vlieg, J.E.T. (2015). Fate, activity, and impact of ingested bacteria within the human gut microbiota. *Trends in Microbiology*, *23*(6), 354–366. https://doi.org/10.1016/j.tim.2015.03.002.

Dewanto, V., Wu, X., Adom, K.K. and Liu, R.H. (2002). Thermal Processing Enhances the Nutritional Value of Tomatoes by Increasing Total Antioxidant Activity. *Journal of Agricultural and Food Chemistry*, *50*(10), 3010–3014. https://doi.org/10.1021/jf0115589.

Dillard, C.J. and Bruce German, J. 2000. Phytochemicals: Nutraceuticals and human health. *Journal of the Science of Food and Agriculture*, *80*, 1744–1756. https://doi.org/10.1002/1097-0010(20000915)80:12<1744::AID-JSFA725>3.0.CO;2-W.

Duda-Chodak, A., Tarko, T., Satora, P. and Sroka, P. 2015. Interaction of dietary compounds, especially polyphenols, with the intestinal microbiota: A review. *European Journal of Nutrition*, *54*, 325–341. https://doi.org/10.1007/s00394-015-0852-y.

Ellulu, M.S., Khaza'ai, H., Abed, Y., Rahmat, A., Ismail, P. and Ranneh, Y. 2015. Role of fish oil in human health and possible mechanism to reduce the inflammation. *Inflammopharmacology*, *23*, 79–89. https://doi.org/10.1007/s10787-015-0228-1.

El-Sayed, S.M. and Youssef, A.M. 2019. Potential application of herbs and spices and their effects in functional dairy products. Jun 28; *Heliyon*, *5*(6), e01989. https://doi.org/10.1016/j.heliyon.2019.e01989

Esposito, K., Marfella, R., Ciotola, M., Di Palo, C., Giugliano, F., Giugliano, G., D'Armiento, M., D'Andrea, F. and Giugliano, D. 2004. Effect of a Mediterranean-Style Diet on Endothelial Dysfunction and Markers of Vascular Inflammation in the Metabolic Syndrome. *JAMA*, *292*(12), 1440. https://doi.org/10.1001/jama.292.12.1440.

FAO/WHO 2001. Health and nutritional properties of probiotics in food including powder milk with live lactic acid bacteria. Cordoba, Argentina, October 1–4

Farzana, M., Shahriar, S., Jeba, F.R., Tabassum, T., Araf, Y., Ullah, M.A., Tasnim, J., Chakraborty, A., Naima, T.A., Marma, K.K.S., Rahaman, T.I. and Hosen, M.J. 2022. Functional food: Complementary to fight against COVID-19. *Beni- Suef Univ. J. Basic Appl. Sci.*, *11*(1), 33. https://doi.org/10.1186/s43088-022-00217-z.

Fernandez, M.A. and Marette, A. 2017. Potential health benefits of combining yogurt and fruits based on their probiotic and prebiotic properties. *Advances in Nutrition*, *8*, 155S–164S. https://doi.org/10.3945/an.115.011114.

Flock, M.R., Skulas-Ray, A.C., Harris, W.S., Etherton, T.D., Fleming, J.A. and Kris-Etherton, P.M. 2013. Determinants of Erythrocyte Omega-3 Fatty Acid Content in Response to Fish Oil Supplementation: A Dose–Response Randomized Controlled Trial. *Journal of the American Heart Association*, *2*(6). https://doi.org/10.1161/JAHA.113.000513.

Foster, J.A., Rinaman, L. and Cryan, J.F. 2017. Stress & the gut-brain axis: Regulation by the microbiome. *Neurobiology of Stress*, *7*, 124–136. https://doi.org/10.1016/j.ynstr.2017.03.001.

Fusco, W., Lorenzo, M.B., Cintoni, M., Porcari, S., Rinninella, E., Kaitsas, F., Lener, E., Mele, M.C., Gasbarrini, A., Collado, M.C., Cammarota, G. and Ianiro, G. 2023. Short-Chain Fatty-Acid-Producing Bacteria: Key Components of the Human Gut Microbiota. May 6; *Nutrients*, *15*(9), 2211. https://doi.org/10.3390/nu15092211.

Gagliani, N. and Huber, S. 2017. Basic Aspects of T Helper Cell Differentiation. In: *Methods in Molecular Biology*, *1514*, 19–30. Clifton, N.J., https://doi.org/10.1007/978-1-4939-6548-9_2.

Gao, D., Liu, Z., Liu, F., Chen, L., Wang, W., Ma, J., Xu, C. and Jiang, Z., Hou, J. 2021. Study of the immunoregulatory effect of Lactobacillus rhamnosus 1.0320 in immunosuppressed mice. *J. Funct. Foods*, *79*, 104423. https://doi.org/10.1016/j.jff.2021.104423.

Gao, Y., Zhou, S., Jiang, W., Huang, M. and Dai, X. 2003. Effects of Ganopoly® (a Ganoderma lucidum polysaccharide extract) on the immune functions in advanced-stage cancer patients. *Immunological Investigations*, *32*, 201–215. https://doi.org/10.1081/IMM-120022979.

García-Burgos, M., Moreno-Fernández, J., Alférez, M.J.M., Díaz-Castro, J. and López-Aliaga, I. 2020. New perspectives in fermented dairy products and their health relevance. *Journal of Functional Foods*, *72*, 104059. https://doi.org/10.1016/j.jff.2020.104059.

Garcia-Bailo, B., Toguri, C., Eny, K.M. and El-Sohemy, A. 2009. Genetic Variation in Taste and Its Influence on Food Selection. *OMICS: A Journal of Integrative Biology*, *13*(1), 69–80. https://doi.org/10.1089/omi.2008.0031.

Garnier, A. and Shahidi, F. 2021. Spices and herbs as immune enhancers and anti-inflammatory agents: A review. *Journal of Food Bioactives*, *14*, 20–52. https://doi.org/10.31665/jfb.2021.14266.

Ghaseminejad-Raeini, A., Ghaderi, A., Sharafi, A., Nematollahi-Sani, B., Moossavi, M., Derakhshani, A. and Sarab, G.A. 2023. Immunomodulatory actions of vitamin D in various immune-related disorders: A comprehensive review. Jul 14; *Front. Immunol.*, *14*, 950465. https://doi.org/10.3389/fimmu.2023.950465.

Gholami-Ahangaran, M., Ahmadi-Dastgerdi, A., Azizi, S., Basiratpour, A., Zokaei, M. and Derakhshan, M. 2022. Thymol and carvacrol supplementation in poultry health and performance. Jan; *Vet Med Sci.*, *8*(1), 267–288. https://doi.org/10.1002/vms3.663.

Gibson, G.R., Hutkins, R., Sanders, M.E., Prescott, S.L., Reimer, R.A., Salminen, S.J., Scott, K., Stanton, C., Swanson, K.S., Cani, P.D., Verbeke, K. and Reid, G. 2017. Expert consensus document: The International Scientific Association for Probiotics and Prebiotics (ISAPP) consensus statement on the definition and scope of prebiotics. *Nature Reviews Gastroenterology and Hepatology*, *14,* 491–502. https://doi.org/10.1038/nrgastro.2017.75.

Glaser, C., Heinrich, J. and Koletzko, B. 2010. Role of FADS1 and FADS2 polymorphisms in polyunsaturated fatty acid metabolism. *Metabolism*, *59*(7), 993–999. https://doi.org/10.1016/j.metabol.2009.10.022.

Granadolorencio, F., Olmedillaalonso, B., Herrerobarbudo, C., Blanconavarro, J., Perezsacristan, B. and Blazquezgarcia, S. 2007. *In vitro* bioaccessibility of carotenoids and tocopherols from fruits and vegetables. *Food Chemistry*, *102*(3), 641–648. https://doi.org/10.1016/j.foodchem.2006.05.043.

Guo, L., Xie, J., Ruan, Y., Zhou, L., Zhu, H., Yun, X., Jiang, Y., Lü, L., Chen, K., Min, Z., Wen, Y. and Gu, J. 2009. Characterization and immunostimulatory activity of a polysaccharide from the spores of Ganoderma lucidum. *International Immunopharmacology*, *9*, 1175–1182. https://doi.org/10.1016/j.intimp.2009.06.005.

Gupta, C. and Prakash, D. 2014. Phytonutrients as therapeutic agents. *Journal of Complementary and Integrative Medicine*, *11*, 151–169. https://doi.org/10.1515/jcim-2013-0021.

Hadjimbei, E., Botsaris, G. and Chrysostomou, S. 2022. Beneficial Effects of Yoghurts and Probiotic Fermented Milks and Their Functional Food Potential. Sep 3; *Foods*, *11*(17), 2691. https://doi.org/10.3390/foods11172691.

Halperin, S.A., Smith, B., Nolan, C., Shay, J. and Kralovec, J. 2003. Safety and immunoenhancing effect of a Chlorella-derived dietary supplement in healthy adults undergoing influenza vaccination: Randomized double-blind, placebo-controlled trial. *CMAJ*, *169*, 111–117.

Han, L., Wu, T., Zhang, Q., Qi, A. and Zhou, X. 2025. Immune Tolerance Regulation is Critical to Immune Homeostasis. *J. Immunol. Res.* https://doi.org/10.1155/jimr/5006201.

Harada, N., Kodama, N. and Nanba, H. 2003. Relationship between dendritic cells and the D-fraction-induced Th-1 dominant response in BALB/c tumor-bearing mice. *Cancer Letters*, *192*, 181–187. https://doi.org/10.1016/S0304-3835(02)00716-4.

Hasegawa, T., Matsuguchi, T., Noda, K., Tanaka, K., Kumamoto, S., Shoyama, Y. and Yoshikai, Y. 2002. Toll-like receptor 2 is at least partly involved in the antitumor activity of glycoprotein from Chlorella vulgaris. *International Immunopharmacology*, *2*, 579–589. https://doi.org/10.1016/S1567-5769(02)00002-4.

Hasler, C.M. 2002. Functional Foods: Benefits, Concerns, and Challenges: A Position Paper from the American Council on Science and Health. *The Journal of Nutrition*, *132*(12), 3772–3781. https://doi.org/10.1093/jn/132.12.3772.

Hassanizadeh, S., Mohammadi, V., Nadjarzadeh, A., Mirzaei, M. and Hosseinzadeh, M. 2024. The Association of Fruits and Vegetables Consumption and COVID-19: Results from Yazd Health Study and TAMYZ Study. *Curr. Nutr. Food Sci.*, *21*, 256–263. https://doi.org/10.2174/0115734013282462240319043756.

He, L.X., Ren, J.W., Liu, R., Chen, Q.H., Zhao, J., Wu, X., Zhang, Z.F., Wang, J.B., Pettinato, G. and Li, Y. 2017. Ginseng (Panax ginseng Meyer) oligopeptides regulate innate and adaptive immune responses in mice via increased macrophage phagocytosis capacity, NK cell activity and Th cells secretion. *Food and Function*, *8*, 3523–3532. https://doi.org/10.1039/c7fo00957g.

Hoang Nguyen, K.H., Le, N.V., Nguyen, P.H., Nguyen, H.H.T., Hoang, D.M. and Huynh, C.D. 2025. Human immune system: Exploring diversity across individuals and populations. *Heliyon*, *11,* e41836. https://doi.org/10.1016/j.heliyon.2025.e41836.

Holscher, H.D. 2017. Dietary fiber and prebiotics and the gastrointestinal microbiota. Mar 4; *Gut Microbes*, *8*(2), 172–184. https://doi.org/10.1080/19490976.2017.1290756.

Hooper, L., Kroon, P.A., Rimm, E.B., Cohn, J.S., Harvey, I., Le Cornu, K.A., Ryder, J.J., Hall, W.L. and Cassidy, A. 2008. Flavonoids, flavonoid-rich foods, and cardiovascular risk: A meta-analysis of randomized controlled trials. *The American Journal of Clinical Nutrition*, *88*(1), 38–50. https://doi.org/10.1093/ajcn/88.1.38.

Howard, L.A., Wong, A.D., Perry, A.K. and Klein, B.P. 1999. β-Carotene and Ascorbic Acid Retention in Fresh and Processed Vegetables. *Journal of Food Science*, *64*(5), 929–936. https://doi.org/10.1111/j.1365-2621.1999.tb15943.x.

Hsu, H.Y., Jeyashoke, N., Yeh, C.H., Song, Y.J., Hua, K.F. and Chao, L.K. 2010. Immunostimulatory bioactivity of algal polysaccharides from Chlorella pyrenoidosa activates macrophages via toll-like receptor 4. *Journal of Agricultural and Food Chemistry*, 58, 927–936. https://doi.org/10.1021/jf902952z.

Huang, P.H., Chen, J.Y. and Kuo, C.M. 2007. Three different hepcidins from tilapia, Oreochromis mossambicus: Analysis of their expressions and biological functions. *Molecular Immunology*, *44*, 1922–1934. https://doi.org/10.1016/j.molimm.2006.09.031.

Inoue, A., Kodama, N. and Nanba, H. 2002. Effect of maitake (Grifola frondosa) D-Fraction on the control of the T lymph node Th-1/Th-2 proportion. *Biological and Pharmaceutical Bulletin*, *25*, 536–540. https://doi.org/10.1248/bpb.25.536.

Iwatani, S. and Yamamoto, N. 2019. Functional food products in Japan: A review. *Food Science and Human Wellness*, *8*, 96–101. https://doi.org/10.1016/j.fshw.2019.03.011.

Jang, H. Il. and Shin, H.M. 2010. Wild panax ginseng (Panax ginseng C.A. Meyer) protects against methotrexate-induced cell regression by enhancing the immune response in RAW 264.7 macrophages. *American Journal of Chinese Medicine*, *38*, 949–960. https://doi.org/10.1142/S0192415X10008378.

Jansen, M.A.K., Hectors, K., O'Brien, N.M., Guisez, Y. and Potters, G. 2008. Plant stress and human health: Do human consumers benefit from UV-B acclimated crops? *Plant Science*, *175*(4), 449–458. https://doi.org/10.1016/j.plantsci.2008.04.010.

Jeong, M., Kim, J.H., Yang, H., Kang, S.D., Song, S., Lee, D., Lee, J.S., Yoon Park, J.H., Byun, S. and Lee, K.W. 2019. Heat-Killed lactobacillus plantarum KCTC 13314BP enhances phagocytic activity and immunomodulatory effects via activation of MAPK and STAT3 pathways. *Journal of Microbiology and Biotechnology*, *29*, 1248–1254. https://doi.org/10.4014/JMB.1905.05066.

Jideani, A.I.O., Silungwe, H., Takalani, T., Omolola, A.O., Udeh, H.O. and Anyasi, T.A. 2021a. Antioxidant-rich natural fruit and vegetable products and human health. *International Journal of Food Properties*, *24*, 41–67. https://doi.org/10.1080/10942912.2020.1866597.

Jideani, A.I.O., Silungwe, H., Takalani, T., Omolola, A.O., Udeh, H.O. and Anyasi, T.A. 2021b. Antioxidant-rich natural fruit and vegetable products and human health. *International Journal of Food Properties*, *24,* 41–67. https://doi.org/10.1080/10942912.2020.1866597.

Jomova, K., Alomar, S.Y., Valko, R., Liska, J., Nepovimova, E., Kuca, K. and Valko, M. 2025. Flavonoids and their role in oxidative stress, inflammation, and human diseases. May 25; *Chem. Biol. Interact.*, *413*, 111489. https://doi.org/10.1016/j.cbi.2025.111489.

Julliard, W.A., Myo, Y.P.A., Perelas, A., Jackson, P.D., Thatcher, T.H. and Sime, P.J. 2022. Specialized pro-resolving mediators as modulators of immune responses. Jan; *Semin. Immunol.*, *59*, 101605. https://doi.org/10.1016/j.smim.2022.101605

Kalt, W. 2005. Effects of Production and Processing Factors on Major Fruit and Vegetable Antioxidants. *Journal of Food Science*, *70*(1), R11–R19. https://doi.org/10.1111/j.1365-2621.2005.tb09053.x.

Kan, A.K.C., Tang, W.T. and Li, P.H. 2024. Helper T cell subsets: Development, function, and clinical role in hypersensitivity reactions in the modern perspective. *Heliyon*, *10*, e30553. https://doi.org/10.1016/j.heliyon.2024.e30553.

Kang, H.K., Lee, H.H., Seo, C.H. and Park, Y. 2019. Antimicrobial and immunomodulatory properties and applications of marine-derived proteins and peptides. Jun 12; *Marine Drugs*, *17*(6), 350. https://doi.org/10.3390/md17060350.

Kang, S. and Min, H. 2012. Ginseng, the "immunity boost": The effects of panax ginseng on immune system. *Journal of Ginseng Research*, *36*, 354–368. https://doi.org/10.5142/jgr.2012.36.4.354.

Kar, A., Ghosh, P., Patra, P., Chini, D.S., Nath, A.K., Saha, J.K. and Chandra Patra, B. 2023. Omega-3 fatty acids mediated Cellular signaling and its regulation in Human Health. *Clinical Nutrition Open Science*, *52*, 72–86. https://doi.org/10.1016/j.nutos.2023.10.004.

Karl, P.J., Hatch, A.M., Arcidiacono, S.M., Pearce, S.C., Pantoja-Feliciano, I.G., Doherty, L.A. and Soares, J.W. 2018. Effects of psychological, environmental, and physical stressors on the gut microbiota. *Frontiers in Microbiology*, *9*(SEP), 1–32. https://doi.org/10.3389/fmicb.2018.02013.

Khora, S.S. 2013. Marine fish-derived bioactive peptides and proteins for human therapeutics. *International Journal of Pharmacy and Pharmaceutical Sciences*, *5*, 31–37.

Kiecolt-Glaser, J.K. 2010. Stress, food, and inflammation: psychoneuroimmunology and nutrition at the cutting edge. *Psychosomatic Medicine*, 72(4), 365–369. https://doi.org/10.1097/PSY.0b013e3181dbf489.

Kiecolt-Glaser, J.K., Habash, D.L., Fagundes, C.P. andridge, R., Peng, J., Malarkey, W.B. and Belury, M.A. 2015. Daily Stressors, Past Depression, and Metabolic Responses to High-Fat Meals: A Novel Path to Obesity. *Biological Psychiatry*, *77*(7), 653–660. https://doi.org/10.1016/j.biopsych.2014.05.018.

Kim, H., Jang, M., Kim, Y., Choi, J., Jeon, J., Kim, J., Hwang, Y. Il, Kang, J.S., Lee, W.J. 2016. Red ginseng and Vitamin C increase immune cell activity and decrease lung inflammation induced by influenza A virus/H1N1 infection. *Journal of Pharmacy and Pharmacology*, *68*, 406–420. https://doi.org/10.1111/jphp.12529.

Kim, H.S., Hong, J.T., Kim, Y. and Han, S.-B. 2011. Stimulatory Effect of β-glucans on Immune Cells. *Immune Network*, *11*(4), 191–195. https://doi.org/10.4110/in.2011.11.4.191.

Kim, J.H., Kim, D.H., Jo, S., Cho, M.J., Cho, Y.R., Lee, Y.J. and Byun, S. 2022. Immunomodulatory functional foods and their molecular mechanisms. *Experimental and Molecular Medicine*, *54*, 1–11. https://doi.org/10.1038/s12276-022-00724-0.

Kim, M.H., Byon, Y.Y., Ko, E.J., Song, J.Y., Yun, Y.S., Shin, T. and Joo, H.G. 2009. Immunomodulatory activity of ginsan, a polysaccharide of Panax ginseng, on dendritic cells. *Korean Journal of Physiology and Pharmacology*, *13*, 169–173. https://doi.org/10.4196/kjpp.2009.13.3.169.

Kim, T.W., Lee, T.Y., Bae, H.C., Hahm, J.H., Kim, Y.H., Park, C., Kang, T.H., Kim, C.J., Sung, M.H. and Poo, H. 2007. Oral Administration of High Molecular Mass Poly-γ-Glutamate Induces NK Cell-Mediated Antitumor Immunity. *The Journal of Immunology*, *179*, 775–780. https://doi.org/10.4049/jimmunol.179.2.775.

Kodama, N., Asakawa, A., Inui, A., Masuda, Y., Nanba, H. 2005. Enhancement of cytotoxicity of NK cells by D-Fraction, a polysaccharide from Grifola frondosa. *Oncol. Rep.*, *13*, 497–502. https://doi.org/10.3892/or.13.3.497.

Kodama, N., Komuta, K., Sakai, N. and Nanba, H. 2002. Effects of D-Fraction, a polysaccharide from Grifola frondosa on tumor growth involve activation of NK cells. *Biological and Pharmaceutical Bulletin*, *25*, 1647–1650. https://doi.org/10.1248/bpb.25.1647.

Koletzko, B., Brands, B., Poston, L., Godfrey, K. and Demmelmair, H. 2012. Early nutrition programming of long-term health. *Proceedings of the Nutrition Society*, *71*(3), 371–378. https://doi.org/10.1017/S0029665112000596.

Konturek, P.C., Brzozowski, T. and Konturek, S.J. 2011. Stress and the gut: Pathophysiology, clinical consequences, diagnostic approach, and treatment options. *Journal of Physiology and Pharmacology: An Official Journal of the Polish Physiological Society*, *62*(6), 591–599. http://www.ncbi.nlm.nih.gov/pubmed/22314561.

Kris-Etherton, P.M., Petersen, K.S., Hibbeln, J.R., Hurley, D., Kolick, V., Peoples, S., Rodriguez, N. and Woodward-Lopez, G. 2021. Nutrition and behavioral health disorders: Depression and anxiety. *Nutrition Reviews*, *79*, 247–260. https://doi.org/10.1093/nutrit/nuaa025.

Kumar, S., Mukherjee, R., Gaur, P., Leal, É., Lyu, X., Ahmad, S., Puri, P., Chang, C.M., Raj, V.S. and Pandey, R.P. 2025. Unveiling roles of beneficial gut bacteria and optimal diets for health. Feb 18; *Front Microbiol.*, *16*, 1527755. https://doi.org/10.3389/fmicb.2025.1527755.

Kwak, J.H., Baek, S.H., Woo, Y., Han, J.K., Kim, B.G., Kim, O.Y. and Lee, J.H. 2012. Beneficial immunostimulatory effect of short-term Chlorella supplementation: Enhancement of Natural Killer cell activity and early inflammatory response (Randomized, double-blinded, placebo-controlled trial). *Nutrition Journal*, 11, 1–8. https://doi.org/10.1186/1475-2891-11-53.

Larsen, M.W., Moser, C., Høiby, N., Song, Z., Kharazmi, A. 2004. Ginseng modulates the immune response by induction of interleukin-12 production. *Apmis*, *112*, 369–373. https://doi.org/10.1111/j.1600-0463.2004.apm1120607.x.

Latif, A., Shehzad, A., Niazi, S., Zahid, A., Ashraf, W., Iqbal, M.W., Rehman, A., Riaz, T., Aadil, R.M., Khan, I.M., Özogul, F., Rocha, J.M., Esatbeyoglu, T. and Korma, S.A. 2023. Probiotics: Mechanism of action, health benefits, and their application in food industries. Aug 17; *Front Microbiol.*, *14*, 1216674. https://doi.org/10.3389/fmicb.2023.1216674.

Lebien, T.W., Thomas, * Tedder, F. 2008. B lymphocytes: How they develop and function. Sep 1; *Blood*, *112*(5), 1570–1580. doi: 10.1182/blood-2008-02-078071.

Lee, H.A., Kim, H., Lee, K.W., Park, K.Y. 2015. Dead Lactobacillus plantarum stimulates and skews immune responses toward T helper 1 and 17 polarizations in RAW 264.7 cells and mouse splenocytes. *Journal of Microbiology and Biotechnology*, *26*, 469–476. https://doi.org/10.4014/jmb.1511.11001.

Lee, J., Jung, I., Choi, J.W., Lee, C.W., Cho, S., Choi, T.G., Sohn, M. and Park, Y. Il. 2019. Micronized and heat-treated lactobacillus plantarum LM1004 stimulates host immune responses via the TLR-2/MAPK/NF-κB signalling pathway *in vitro* and *in vivo*. *Journal of Microbiology and Biotechnology*, *29*, 704–712. https://doi.org/10.4014/jmb.1812.12059.

Lee, S.W., Park, H.J., Park, S.H., Kim, N., Hong, S. 2014. Immunomodulatory effect of poly-γ-glutamic acid derived from Bacillus subtilis on natural killer dendritic cells. *Biochemical and Biophysical Research Communications*, *443*, 413–421. https://doi.org/10.1016/j.bbrc.2013.11.097.

Lee, S.Y., Kim, Y.K., Park, N., Kim, C.S., Lee, C.Y. and Park, S.U. 2010. Chemical constituents and biological activities of the berry of Panax ginseng. *Journal of Medicinal Plants Research*, *4*, 349–353.

Li, D. and Wu, M., 2021. Pattern recognition receptors in health and diseases. Aug 4; *Signal Transduct Target Ther.*, *6*(1), 291. https://doi.org/10.1038/s41392-021-00687-0.

Li, Y.O. and Komarek, A.R. 2017. Dietary fibre basics: Health, nutrition, analysis, and applications. *Food Quality and Safety*, *1*, 47–59. https://doi.org/10.1093/fqs/fyx007.

Lim, T.G., Jang, M., Cho, C.W., Hong, H. Do, Kim, K.T., Lee, S.Y., Jung, S.K. and Rhee, Y.K. 2016. White ginseng extract induces immunomodulatory effects via the MKK4-JNK pathway. *Food Science and Biotechnology*, *25*, 1737–1744. https://doi.org/10.1007/s10068-016-0265-6.

Lin, J.G., Fan, M.J., Tang, N.Y., Yang, J.S., Hsia, T.C., Lin, J.J., Lai, K.C., Sai-Chuen Wu, R., Ma, C.Y., Wood, W.G. and Chung, J.G. 2012. An extract of agaricus blazei Murill administered orally promotes immune responses in murine leukemia BALB/c mice *in vivo*. *Integrative Cancer Therapies*, *11*, 29–36. https://doi.org/10.1177/1534735411400314.

Lomax, A.R. and Calder, P.C. 2009. Prebiotics, immune function, infection, and inflammation: A review of the evidence. *British Journal of Nutrition*, *101*(5), 633–658. https://doi.org/10.1017/S0007114508055608.

Lombardi, V.R.M., Corzo, L., Carrera, I. and Cacabelos, R. 2018. The Search for Biomarine-derived Compounds with Immunomodulatory Activity. *Journal of Exploratory Research in Pharmacology*, *3*, 30–41. https://doi.org/10.14218/jerp.2018.00006.

Luckheeram, R.V., Zhou, R., Verma, A.D., Xia, B., 2012. CD4 + T Cells: Differentiation and Functions. *Clin. Dev. Immunol.*, *2012*, 1–12. https://doi.org/10.1155/2012/925135.

Lyons, M.P., Papazyan, T.T. and Surai, P.F. 2007. Selenium in Food Chain and Animal Nutrition: Lessons from Nature: Review. *Asian-Australasian Journal of Animal Sciences*, *20*(7), 1135–1155. https://doi.org/10.5713/ajas.2007.1135.

Macfarlane, G.T. and Macfarlane, S. 2012. Bacteria, Colonic Fermentation, and Gastrointestinal Health. *Journal of AOACI International, 95*(1), 50–60. https://doi.org/10.5740/jaoacint.SGE_Macfarlane.

Makki, K., Deehan, E.C., Walter, J. and Bäckhed, F., 2018. The Impact of Dietary Fiber on Gut Microbiota in Host Health and Disease. Jun 13; *Cell Host Microbe*, *23*(6), 705–715. https://doi.org/10.1016/j.chom.2018.05.012.

Malin, S.K. and Braun, B. (2013). Effect of metformin on substrate utilization after exercise training in adults with impaired glucose tolerance. *Applied Physiology, Nutrition, and Metabolism*, *38*(4), 427–430. https://doi.org/10.1139/apnm-2012-0433.

Marco, M.L., Heeney, D., Binda, S., Cifelli, C.J., Cotter, P.D., Foligné, B., Gänzle, M., Kort, R., Pasin, G., Pihlanto, A., Smid, E.J. and Hutkins, R. 2017. Health benefits of fermented foods: microbiota and beyond. *Curr. Opin. Biotechnol.*, *44*, 94–102. https://doi.org/10.1016/j.copbio.2016.11.010

Marshall, J.S., Warrington, R., Watson, W. and Kim, H.L. 2018. An introduction to immunology and immunopathology. Sep 12; *Allergy, Asthma and Clinical Immunology*, *14*(Suppl 2), 49. https://doi.org/10.1186/s13223-018-0278-1.

Martirosyan, D.M. and Singh, J. 2015. A new definition of functional food by FFC: What makes a new definition unique? *Functional Foods in Health and Disease*, *5*, 209–223. https://doi.org/10.31989/ffhd.v5i6.183.

Mateus, N., Carvalho, E., Luís, C. and de Freitas, V. 2004. Influence of the tannin structure on the disruption effect of carbohydrates on protein–tannin aggregates. *Analytica Chimica Acta*, *513*(1), 135–140. https://doi.org/10.1016/j.aca.2003.08.072.

Matwiejczyk, M., Zambrowicz, A. and Besman, M. 2025. Lysozyme Monomer, Dimer, and Oligomers: A Review with a Focus on Immunological Potential. *Int. J. Pept. Res. Ther.*, *31*, 79. https://doi.org/10.1007/s10989-025-10742-x.

Mellman, I., Coukos, G. and Dranoff, G., 2011. Cancer immunotherapy comes of age. *Nature*, *480*, 480–489. https://doi.org/10.1038/nature10673.

Mendivil, C.O. 2021. Dietary Fish, Fish Nutrients, and Immune Function: A Review. Jan 20; *Front. Nutr.*, *7*, 617652. doi: 10.3389/fnut.2020.617652. Erratum in: 21 May; *Front Nutr.*, *8*, 693773. doi: 10.3389/fnut.2021.693773.

Messina, M. 2014. Soy foods, isoflavones, and the health of postmenopausal women. *The American Journal of Clinical Nutrition*, *100*, 423S–430S. https://doi.org/10.3945/ajcn.113.071464.

Mileo, A.M. and Miccadei, S. 2016. Polyphenols as Modulator of Oxidative Stress in Cancer Disease: New Therapeutic Strategies. *Oxid. Med. Cell Longev.*, *2016*, 6475624. https://doi.org/10.1155/2016/6475624.

Miller, E. 2008. Chapter 74 - Introduction. In: Morgan, R.V (Ed.), *Handbook of Small Animal Practice* (5th Edn.), pp. 737–739. W.B. Saunders, Saint Louis. https://doi.org/https://doi.org/10.1016/B978-1-4160-3949-5.50078-9.

Mirmiran, P. 2014. Functional foods-based diet as a novel dietary approach for management of Type 2 diabetes and its complications: A review. *World Journal of Diabetes*, *5*(3), 267–281. https://doi.org/10.4239/wjd.v5.i3.267.

Moore, D.R., Churchward-Venne, T.A., Witard, O., Breen, L., Burd, N.A., Tipton, K.D. and Phillips, S.M. 2015. Protein Ingestion to Stimulate Myofibrillar Protein Synthesis Requires Greater Relative Protein Intakes in Healthy Older Versus Younger Men. *The Journals of Gerontology: Series A*, *70*(1), 57–62. https://doi.org/10.1093/gerona/glu103.

Morita, H., Moro, K. and Koyasu, S. 2016. Innate lymphoid cells in allergic and nonallergic inflammation. *Journal of Allergy and Clinical Immunology*, *138*(5), 1253–1264. https://doi.org/10.1016/j.jaci.2016.09.011.

Morvaridzadeh, M., Nachvak, S.M., Mohammadi, R., Moradi, S., Mostafai, R., Pizarro, A.B. and Abdollahzad H. 2021. Probiotic Yogurt Fortified with Vitamin D Can Improve Glycemic Status in Non-Alcoholic Fatty Liver Disease Patients: A Randomized Clinical Trial. Jan 22; *Clin. Nutr. Res.*, *10*(1), 36–47. doi: 10.7762/cnr.2021.10.1.36.

Mozaffarian, D. and Wu, J.H.Y. 2018. Flavonoids, Dairy Foods, and Cardiovascular and Metabolic Health: A Review of Emerging Biologic Pathways. Circulation Research, *122*, 369–384. https://doi.org/10.1161/CIRCRESAHA.117.309008.

Murphy, E.J., Rezoagli, E., Major, I., Rowan, N.J. and Laffey, J.G. 2020. β-glucan metabolic and immunomodulatory properties and potential for clinical application. Dec 10; *Journal of Fungi*, *6*(4), 356. https://doi.org/10.3390/jof6040356.

Myburgh, K.H. 2014. Polyphenol Supplementation: Benefits for Exercise Performance or Oxidative Stress? *Sports Medicine*, *44*(S1), 57–70. https://doi.org/10.1007/s40279-014-0151-4.

Nakano, S., Takekoshi, H. and Nakano, M. 2007. Chlorella (Chlorella pyrenoidosa) supplementation decreases dioxin and increases immunoglobulin A concentrations in breast milk. *Journal of Medicinal Food*, *10*, 134–142. https://doi.org/10.1089/jmf.2006.023.

Ogawa, T., Asai, Y., Tamai, R., Makimura, Y., Sakamoto, H., Hashikawa, S. and Yasuda, K. 2006. Natural killer cell activities of synbiotic Lactobacillus casei ssp. casei in conjunction with dextran. *Clinical and Experimental Immunology*, *143*, 103–109. https://doi.org/10.1111/j.1365-2249.2005.02975.x.

Oke, M.A., Afolabi, F.J., Oyeleke, O.O., Kilani, T.A., Adeosun, A.R., Olanbiwoninu, A.A. and Adebayo, E.A. 2022. Ganoderma lucidum: Unutilized natural medicine and promising future solution to emerging diseases in Africa. Aug 22; *Front Pharmacol.*, *13*, 952027. https://doi.org/10.3389/fphar.2022.952027.

Ordovas, J.M. 2006. Genetic interactions with diet influence the risk of cardiovascular disease. *The American Journal of Clinical Nutrition*, *83*(2), 443S–446S. https://doi.org/10.1093/ajcn/83.2.443S.

Parmar, B., Arya, R. and Patel, G. 2025. Harnessing Probiotics: Emerging Strategies for Health and Disease Management. *Indian J. Microbiol.*, https://doi.org/10.1007/s12088-025-01478-9.

Parolini, C. 2019. Effects of fish n-3 PUFAs on intestinal microbiota and immune system. *Marine Drugs*, *17*(6), 374. https://doi.org/10.3390/md17060374.

Pecora, F., Persico, F., Argentiero, A., Neglia, C. and Esposito, S. 2020. The role of micronutrients in support of the immune response against viral infections. *Nutrients*, *12*(10), 3198. https://doi.org/10.3390/nu12103198.

Phull, A.R. and Kim, S.J. 2017. Fucoidan as bio-functional molecule: Insights into the anti-inflammatory potential and associated molecular mechanisms. *J. Funct. Foods*, *38*, 415–426. https://doi.org/10.1016/j.jff.2017.09.051.

Pisoschi, A.M., Pop, A., Iordache, F., Stanca, L., Geicu, O.I., Bilteanu, L. and Serban, A.I. 2022. Antioxidant, anti-inflammatory, and immunomodulatory roles of vitamins in COVID-19 therapy. *Eur. J. Med. Chem.*, *232*, 114175. https://doi.org/10.1016/j.ejmech.2022.114175.

Powers, S.K., Deminice, R., Ozdemir, M., Yoshihara, T., Bomkamp, M.P. and Hyatt, H. 2020. Exercise-induced oxidative stress: Friend or foe? *Journal of Sport and Health Science*, *9*(5), 415–425. https://doi.org/10.1016/j.jshs.2020.04.001.

Pugh, N., Ross, S.A., Elsohly, H.N., Elsohly, M.A. and Pasco, D.S. 2001. In the present study we have identified robust macrophage stimulating activity in the crude extracts of. *Planta Med.*, *67*, 737–742.

Pugliese, G., Liccardi, A., Graziadio, C., Barrea, L., Muscogiuri, G. and Colao, A. 2022. Obesity and infectious diseases: Pathophysiology and epidemiology of a double pandemic condition. *Int. J. Obes.*, *46*(3), 449–465. https://doi.org/10.1038/s41366-021-01035-6.

Quast, I. and Tarlinton, D. 2021. B cell memory: Understanding COVID-19. *Immunity*, *54*, 205–210. https://doi.org/https://doi.org/10.1016/j.immuni.2021.01.014.

Rani, K., Sandal, N. and Sahoo, P.K. 2018. A comprehensive review on chlorella-its composition, health benefits, market, and regulatory scenario. ~ 584 ~ *The Pharma Innovation Journal*, *7*, 584–589.

Ras, R.T., Geleijnse, J.M. and Trautwein, E.A. 2014. LDL-cholesterol-lowering effect of plant sterols and stanols across different dose ranges: A meta-analysis of randomised controlled studies. *British Journal of Nutrition*, *112*(2), 214–219. https://doi.org/10.1017/S0007114514000750.

Rémond, D., Shahar, D.R., Gille, D., Pinto, P., Kachal, J., Peyron, M.-A., Dos Santos, C.N., Walther, B., Bordoni, A., Dupont, D., Tomás-Cobos, L. and Vergères, G. 2015. Understanding the gastrointestinal tract of the elderly to develop dietary solutions that prevent malnutrition. *Oncotarget*, *6*(16), 13858–13898. https://doi.org/10.18632/oncotarget.4030.

Ren, D., Wang, D., Liu, H., Shen, M., Yu, H. 2019. Two strains of probiotic Lactobacillus enhance immune response and promote naive T cell polarization to Th1. *Food and Agricultural Immunology*, *30*, 281–295. https://doi.org/10.1080/09540105.2019.1579785.

Ren, D., Xiong, S., Ren, Y., Yang, X., Zhao, X., Jin, J., Xu, M., Liang, T., Guo, L. and Weng, L. 2024. Advances in therapeutic cancer vaccines: Harnessing immune adjuvants for enhanced efficacy and future perspectives. *Comput. Struct. Biotechnol. J.*, *23*, 1833–1843. https://doi.org/https://doi.org/10.1016/j.csbj.2024.04.054.

Rickman, J.C., Barrett, D.M. and Bruhn, C.M. 2007. Nutritional comparison of fresh, frozen and canned fruits and vegetables. Part 1. Vitamins C and B and phenolic compounds. *Journal of the Science of Food and Agriculture*, *87*(6), 930–944. https://doi.org/10.1002/jsfa.2825.

Rigaux, P., Daniel, C., Hisbergues, M., Muraille, E., Hols, P., Pot, B., Pestel, J. and Jacquet, A. 2009. Immunomodulatory properties of Lactobacillus plantarum and its use as a recombinant vaccine against mite allergy. *Allergy: European Journal of Allergy and Clinical Immunology*, *64*, 406–414. https://doi.org/10.1111/j.1398-9995.2008.01825.x.

Roby, G., Harbertson, J.F., Adams, D.A. and Matthews, M.A. 2004. Berry size and vine water deficits as factors in winegrape composition: Anthocyanins and tannins. *Australian Journal of Grape and Wine Research*, *10*(2), 100–107. https://doi.org/10.1111/j.1755-0238.2004.tb00012.x.

Rudrapal, M., Khairnar, S.J., Khan, J., Dukhyil, A. Bin, Ansari, M.A., Alomary, M.N., Alshabrmi, F.M., Palai, S., Deb, P.K. and Devi, R. 2022. Dietary Polyphenols and Their Role in Oxidative Stress-Induced Human Diseases: Insights Into Protective Effects, Antioxidant Potentials and Mechanism(s) of Action. *Front. Pharmacol.*, *13*, 806470. https://doi.org/10.3389/fphar.2022.806470.

Saleh, H.A., Yousef, M.H. and Abdelnaser, A. 2021. The Anti-Inflammatory Properties of Phytochemicals and Their Effects on Epigenetic Mechanisms Involved in TLR4/NF-κB-Mediated Inflammation. *Front. Immunol.*, *12*, 606069. https://doi.org/10.3389/fimmu.2021.606069.

Santhakumar, A.B., Battino, M. and Alvarez-Suarez, J.M. 2018. Dietary polyphenols: Structures, bioavailability, and protective effects against atherosclerosis. *Food and Chemical Toxicology*, *113*, 49–65. https://doi.org/10.1016/j.fct.2018.01.022.

Santiago-López, L., Hernández-Mendoza, A., Garcia, H.S., Mata-Haro, V., Vallejo-Cordoba, B. and González-Córdova, A.F. 2015. The effects of consuming probiotic-fermented milk on the immune system: A review of scientific evidence. *International Journal of Dairy Technology*, *68*, 153–165. https://doi.org/10.1111/1471-0307.12202.

Sarris, J., Logan, A.C., Akbaraly, T.N., Amminger, G.P., Balanzá-Martínez, V., Freeman, M.P., Hibbeln, J., Matsuoka, Y., Mischoulon, D., Mizoue, T., Nanri, A., Nishi, D., Ramsey, D., Rucklidge, J.J., Sanchez-Villegas, A., Scholey, A., Su, K.-P. and Jacka, F.N. 2015. Nutritional medicine as mainstream in psychiatry. *The Lancet Psychiatry*, *2*(3), 271–274. https://doi.org/10.1016/S2215-0366(14)00051-0.

Schütz, K., Sab, M., De With, A., Graubaum, H.J. and Grünwald, J. 2010. Immune-modulating efficacy of a polyphenol-rich beverage on symptoms associated with the common cold: A double-blind, randomised, placebo-controlled, multicentric clinical study. *British Journal of Nutrition*, *104*, 1156–1164. https://doi.org/10.1017/S0007114510002047.

Scientific Concepts of Functional Foods in Europe Consensus Document, 1999. *British Journal of Nutrition*, *81*, S1–S27. https://doi.org/10.1017/S0007114599000471.

Setchell, K.D.R. and Clerici, C. 2010. Equol: History, Chemistry, and Formation. *The Journal of Nutrition*, *140*(7), 1355S–1362S. https://doi.org/10.3945/jn.109.119776.

Shahidi, F. and Ambigaipalan, P. 2015. Phenolics and polyphenolics in foods, beverages, and spices: Antioxidant activity and health effects –A review. *Journal of Functional Foods*, *18*, 820–897. https://doi.org/10.1016/j.jff.2015.06.018.

Shakoor, H., Feehan, J., Apostolopoulos, V., Platat, C., Dhaheri, A.S. Al, Ali, H.I., Ismail, L.C., Bosevski, M. and Stojanovska, L. 2021. Immunomodulatory effects of dietary polyphenols. *Nutrients*, *13*, 1–18. https://doi.org/10.3390/nu13030728.

Sharifi-Rad, J., Rayess, Y. El, Rizk, A.A., Sadaka, C., Zgheib, R., Zam, W., Sestito, S., Rapposelli, S., Neffe-Skocińska, K., Zielińska, D., Salehi, B., Setzer, W.N., Dosoky, N.S., Taheri, Y., El Beyrouthy, M., Martorell, M., Ostrander, E.A., Suleria, H.A.R., Cho, W.C., Maroyi, A. and Martins, N. 2020. Turmeric and Its Major Compound Curcumin on Health: Bioactive Effects and Safety Profiles for Food, Pharmaceutical, Biotechnological, and Medicinal Applications. Sep 15; *Front. Pharmacol.*, *11*, 01021. https://doi.org/10.3389/fphar.2020.01021.

Sharma, Y., Arora, M. and Bala, K. 2024. The potential of immunomodulators in shaping the future of healthcare. *Discover Medicine*, *1*, 37. https://doi.org/10.1007/s44337-024-00029-3.

Shastak, Y. and Pelletier, W. 2023. The role of vitamin A in non-ruminant immunology. *Frontiers in Animal Science*, *4*. https://doi.org/10.3389/fanim.2023.1197802.

Shen, T., Wang, G., You, L., Zhang, L., Ren, H., Hu, W., Qiang, Q., Wang, X., Ji, L., Gu, Z. and Zhao, X. 2017. Polysaccharide from wheat bran induces cytokine expression via the toll-like receptor 4-mediated p38 MAPK signaling pathway and prevents cyclophosphamide-induced immunosuppression in mice. *Food and Nutrition Research*, *61*(1), 1344523. https://doi.org/10.1080/16546628.2017.1344523.

Shibata, S. 2001. Chemistry and cancer preventing activities of ginseng saponins and some related triterpenoid compounds. *Journal of Korean Medical Science*, *16* (Suppl), 28–37. https://doi.org/10.3346/jkms.2001.16.S.S28.

Shin, K.K., Yi, Y.S., Kim, J.K., Kim, H., Hossain, M.A., Kim, J.H. and Cho, J.Y. 2020. Korean red ginseng plays an anti-aging role by modulating expression of aging-related genes and immune cell subsets. *Molecules*, *25*(7), 1492. https://doi.org/10.3390/molecules25071492.

Shrivastava, R. 2020. Immunity boosters: Solutions from nature: Herbs and spices. *Journal of Renal Nutrition and Metabolism*, *6*(2), 35–37. https://doi.org/10.4103/jrnm.jrnm_20_20.

Simopoulos, A.P. 2002. The importance of the ratio of omega-6/omega-3 essential fatty acids. *Biomed. Pharmacother.*, Oct; *56*(8), 365-379. doi: 10.1016/s0753-3322(02)00253-6.

Song, X., Chen, J., Sakwiwatkul, K., Li, R. and Hu, S. 2010. Enhancement of immune responses to influenza vaccine (H3N2) by ginsenoside Re. *International Immunopharmacology*, *10*, 351–356. https://doi.org/10.1016/j.intimp.2009.12.009.

Spite, M., Clària, J., Serhan, C.N. 2014. Resolvins, specialized proresolving lipid mediators, and their potential roles in metabolic diseases. *Cell Metab.*, *19*(1), 21–36. https://doi.org/10.1016/j.cmet.2013.10.006.

Stefanache, A., Lungu, I.I., Butnariu, I.A., Calin, G., Gutu, C., Marcu, C., Grierosu, C., Bogdan Goroftei, E.R., Duceac, L.D., Dabija, M.G., Popa, F. and Damir, D. 2023. Understanding How Minerals Contribute to Optimal Immune Function. *J. Immunol. Res.*, Nov 1; *2023*, 3355733. https://doi.org/10.1155/2023/3355733.

Stephen, J., Manoharan, D. and Radhakrishnan, M. 2023. Immune boosting functional components of natural foods and its health benefits. *Food Production, Processing, and Nutrition*, *5*, 61. https://doi.org/10.1186/s43014-023-00178-5.

Steptoe, A., Hamer, M. and Chida, Y. 2007. The effects of acute psychological stress on circulating inflammatory factors in humans: A review and meta-analysis. *Brain, Behavior, and Immunity*, *21*(7), 901–912. https://doi.org/10.1016/j.bbi.2007.03.011.

Strzelec, M., Detka, J., Mieszczak, P., Sobocińska, M.K. and Majka, M. 2023. Immunomodulation: A general review of the current state-of-the-art and new therapeutic strategies for targeting the immune system. *Front Immunol.*, *14*. https://doi.org/10.3389/fimmu.2023.1127704.

Su, F., Yuan, L., Zhang, L. and Hu, S. 2012. Ginsenosides Rg1 and Re act as adjuvant via TLR4 signaling pathway. *Vaccine*, *30*, 4106–4112. https://doi.org/10.1016/j.vaccine.2012.03.052.

Su, Y. 2011. Isolation and identification of pelteobagrin, a novel antimicrobial peptide from the skin mucus of yellow catfish (*Pelteobagrus fulvidraco*). *Comparative Biochemistry and Physiology - B Biochemistry and Molecular Biology*, *158*, 149–154. https://doi.org/10.1016/j.cbpb.2010.11.002.

Sudhakar Johnson, T. and Narayana, D.B.A. 2021. Role of spices in offering natural immunity to fight various diseases. *Pharmacognosy Journal*, *13*, 600–613. https://doi.org/10.5530/pj.2021.13.76.

Sun, J., Dong, S., Li, J. and Zhao, H. 2022. A comprehensive review on the effects of green tea and its components on the immune function. *Food Science and Human Wellness*, *11*, 1143–1155. https://doi.org/10.1016/j.fshw.2022.04.008.

CHAPTER 7

Functional Foods in Cancer Prevention and Management

Garima Singh[1*] and Ankush Sharma[2]

1. Introduction

The World Health Organization (WHO) estimated that over 10 million fatalities occurred in 2020 as a result of diverse cancer types, rendering it one of the primary causes of death globally (Sung et al., 2021). Despite the substantial enhancement in survival rates with conventional cancer treatments like chemotherapy, radiotherapy, and targeted therapies, these modalities frequently result in severe side effects, multidrug resistance, and considerable financial strain, necessitating the exploration of effective supplementary techniques. Doyon and Labrecque (2008) underscored the significance of functional foods in illness prevention and management, especially in cancer, characterizing them as foods that encompass bioactive substances with health-enhancing properties. Increasing data indicates that bioactive chemicals in functional foods—such as carotenoids, flavonoids, isothiocyanates, probiotics, omega-3 fatty acids, and dietary fiber—exhibit substantial anti-cancer potential (Konstantinidi and Koutelidakis, 2019). These chemicals affect several cancer-related systems, such as decreasing oxidative stress, modulating inflammatory pathways, inducing apoptosis, inhibiting angiogenesis, and enhancing immunological function (Taylor and Greenwald, 2005).

This chapter seeks to examine the scientific foundation of functional nutrients in cancer treatment. This document offers an extensive analysis of functional foods, encompassing their types and bioactive constituents, alongside their modes of action and effectiveness as evidenced by *in vitro*, animal, and human clinical research. Functional foods, initially conceived in Japan during the 1980s, are widely characterized as foods that offer health advantages beyond fundamental nutrition (Doyon and Labrecque, 2008). These foods possess physiologically active constituents that can induce physiological responses potentially diminishing the risk of chronic diseases, including cardiovascular disease, diabetes, and cancer (Lobo et al., 2010). Functional foods can naturally exist in fruits, vegetables, nuts, and whole grains, or be integrated into manufactured products like fortified cereals, probiotic-enhanced foods, and omega-3-enriched items. Specific bioactive molecules can be separated and concentrated to produce nutraceuticals, like lycopene supplements. Functional diets play a role in cancer therapy by influencing molecular pathways related to tumor initiation and progression, so augmenting conventional treatments and potentially decreasing cancer incidence (Kopustinskiene et al., 2020).

[1] School of Biosciences and Technology. Galgotias University, Greater Noida, Uttar Pradesh-203201, India.
[2] Department of Physiotherapy. G.D. Goenka University, Gurugram, Haryana-122103, India.
* Corresponding author: s.garima.o@gmail.com
DOI: 10.1201/9781003659808-7

Cancer treatment is complex and contingent upon tumor kind, stage, and molecular attributes. Standard treatment modalities encompass surgery, optimal for localized tumors but less advantageous in metastatic disease; radiotherapy, employing ionizing radiation to eradicate cancer cells while potentially harming adjacent healthy tissues; chemotherapy, aimed at rapidly proliferating cells but frequently causing neuropathy, gastrointestinal toxicity, and immunosuppression; targeted therapy, which employs agents to inhibit specific molecular pathways, such as HER2 in breast cancer; and immunotherapy, which amplifies the body's immune response via monoclonal antibodies or immune checkpoint inhibitors. Notwithstanding progress, therapeutic resistance and significant adverse effects remain substantial obstacles (Lei et al., 2023). Functional foods and their bioactive constituents are being investigated as adjunctive measures to potentially augment therapy efficacy, diminish toxicity, and boost patient well-being (Patra et al., 2021).

Nutraceuticals sourced from functional foods have garnered significant attention for their capacity to influence many cancer pathways with low harm (Giovannucci, 1999). Epidemiological studies consistently indicate that diets abundant in fruits, vegetables, and marine-derived nutrients correlate with a diminished incidence of cancer (Konstantinidi and Koutelidakis, 2019). Moreover, both preclinical and clinical studies suggest that substances including carotenoids, flavonoids, and omega-3 fatty acids may augment the efficacy of chemotherapy and radiotherapy while mitigating their detrimental effects (Kopustinskiene et al., 2020). Consequently, incorporating nutraceuticals and functional nutrients into cancer treatment signifies a viable approach for enhancing therapeutic results. By utilizing the intrinsic therapeutic properties of these food elements, researchers and clinicians can formulate novel strategies that enhance existing cancer therapies and promote holistic patient care.

2. Types of Functional Food

Functional foods encompass a diverse array of bioactive substances that have been thoroughly investigated for their possible involvement in cancer prevention and treatment (Samtiya et al., 2021). These chemicals, inherently found in fermented foods, marine goods, fruits, and vegetables, have anticancer effects, including immune control, inflammatory modulation, angiogenesis suppression, and antioxidant activity (Samtiya et al., 2021). This section delineates the principal functional foods and their bioactive constituents recognized for their therapeutic efficacy in oncology.

2.1 Astaxanthin

Astaxanthin is a carotenoid pigment present in microalgae, crustaceans including salmon, shrimp, and krill, as well as specific fungi (Ambati et al., 2014). Its significance in cancer prevention and treatment is chiefly ascribed to its potent antioxidant and anti-inflammatory effects (Ambati et al., 2014). Astaxanthin mitigates oxidative stress by neutralizing reactive oxygen species and promotes apoptosis in cancer cells through the activation of caspase-dependent pathways while downregulating anti-apoptotic proteins such as Bcl-2 (Fakhri et al., 2018). It also mitigates inflammation-induced cancer progression by regulating NF-κB signaling (Fakhri et al., 2018). *In vitro* studies provide evidence of its capacity to inhibit the proliferation of breast, liver, and colon cancer cells (Guerin et al., 2003). Animal studies also demonstrate decreased tumor growth subsequent to astaxanthin treatment (Jyonouchi et al., 2000).

2.2 Lycopene

Lycopene, a red carotenoid prevalent in tomatoes, watermelon, papaya, and pink citrus, is acknowledged for its antioxidant and chemopreventive attributes (Story et al., 2010). It mitigates oxidative stress by neutralizing free radicals, modulates the cell cycle by the induction of G1-phase arrest, and downregulates IGF-1, a principal promoter of cancer cell proliferation (Agarwal and

Rao, 2000). Epidemiological studies indicate a negative correlation between lycopene consumption and the incidence of prostate, lung, and stomach cancers (Giovannucci et al., 2002). Laboratory research validates its ability to suppress the proliferation of breast and liver cancer cells (Singh et al., 2022). Clinical research indicates that lycopene supplementation may inhibit tumor progression and decrease PSA levels in patients with prostate cancer (Kucuk et al., 2002).

2.3 Lutein

Lutein, a xanthophyll carotenoid present in green leafy vegetables including broccoli, kale, and spinach, offers protection against inflammation and oxidative stress (Singh et al., 2024a). It diminishes lipid peroxidation and DNA damage, decreases inflammatory cytokines including IL-6 and TNF-α, and induces apoptosis through the activation of caspase-3 while concurrently reducing Akt signaling (Chen et al., 2012). Research suggests that elevated dietary lutein consumption is associated with reduced risks of breast and colon cancers (Mares-Perlman et al., 2002). Research involving animals substantiates its efficacy in diminishing tumor proliferation and metastasis (Suzuki et al., 2014).

2.4 Alpha-Carotene

Alpha-carotene, present in sweet potatoes, pumpkin, and carrots, is a provitamin A carotenoid exhibiting superior antioxidant capabilities compared to beta-carotene (Nishino et al., 2009). It mitigates oxidative damage and inhibits cancer cell proliferation by modulating retinoic acid signaling pathways (Le Marchand, 2002). Population studies demonstrate a reduced incidence of lung and liver malignancies in those with elevated alpha-carotene consumption (Tanaka et al., 2012). *In vitro* studies indicate that alpha-carotene suppresses the proliferation of human liver cancer cells more efficiently than beta-carotene (Murakoshi et al., 2005).

2.5 Beta-Carotene

Beta-carotene, prevalent in carrots, apricots, and spinach, serves as a precursor to vitamin A and has been extensively researched for its potential in cancer prevention (Krinsky et al., 2005). It mitigates reactive oxygen species (ROS), thereby diminishing DNA damage, and facilitates the differentiation of cancer cells, thereby reducing their malignancy (Bendich, 1994). Certain studies link beta-carotene consumption to diminished cancer risk (Lee et al., 1999). Nonetheless, high-dose supplementation in smokers has been demonstrated to elevate lung cancer incidence (ATBC Cancer Prevention Study Group, 1994).

2.6 Cryptoxanthin

Cryptoxanthin, found in oranges, tangerines, papayas, and red peppers, is a xanthophyll carotenoid exhibiting antioxidant and anti-inflammatory activities, serving as a precursor to vitamin A (Rao and Rao, 2007). It alleviates oxidative stress, induces apoptosis through the regulation of Bax and Bcl-2 levels, and suppresses cell proliferation by altering cyclins and CDKs (Nishino et al., 2009). Epidemiological studies associate elevated cryptoxanthin consumption with reduced incidences of cervical, colorectal, and lung malignancies (Tanaka et al., 2012). Animal studies further corroborate its capacity to impede the establishment of liver and lung tumors (Nishino et al., 2005).

2.7 Fucoxanthin

Fucoxanthin, a carotenoid prevalent in brown seaweeds like wakame and kombu, demonstrates anti-inflammatory, anticancer, and anti-obesity effects (Miyashita et al., 2020). It triggers apoptosis by caspase activation, produces cell cycle arrest by modulating cyclin D1 and p21/p27, and inhibits angiogenesis by diminishing VEGF production (Li, Q. Wang, et al., 2019). *In vitro* investigations demonstrate inhibitory effects on liver, breast, and prostate carcinoma cells (D'Orazio et al., 2012). Animal models exhibit decreased tumor growth and metastasis in colorectal cancer (Kim et al., 2010).

2.8 Isothiocyanates

Isothiocyanates (ITCs), originating from cruciferous vegetables, encompass sulfur-containing chemicals such as sulforaphane, which exhibit potent anticancer properties (Jiang et al., 2018). They regulate epigenetic processes including DNA methylation and histone modification, augment detoxification via phase II enzymes such as GST, and impede angiogenesis and metastasis by diminishing VEGF and MMP activity (Lee et al., 2013). Epidemiological studies indicate a reduced cancer risk in communities that consume cruciferous vegetables (Verhoeven et al., 1996). Laboratory research demonstrates that sulforaphane inhibits the proliferation of breast, lung, and prostate cancer cells (Clarke et al., 2008).

2.9 Probiotics, Prebiotics, and Synbiotics

Probiotics, prebiotics, and synbiotics play a crucial role in cancer prevention and therapeutic assistance (O'Callaghan and van Sinderen, 2016). Probiotics augment gut microbiota diversity, decompose carcinogens such as nitrosamines, and bolster immune responses by activating dendritic cells and NK cells (Gopalakrishnan et al., 2018). Research using animals indicates a decreased colorectal tumor burden subsequent to probiotic administration (Zhu et al., 2014). Clinical trials indicate enhanced gastrointestinal function and decreased chemotherapy-induced diarrhea in cancer patients (Ishikawa et al., 2011).

2.10 Phytoestrogens

Phytoestrogens are plant-derived polyphenols that exhibit structural similarities to estrogen, occurring in soy, flaxseeds, legumes, and whole grains (Adlercreutz, 2007). They function as selective estrogen receptor modulators, promote apoptosis by caspase activation, and impede angiogenesis by downregulating VEGF production (Wu A, 2015). Epidemiological research associates elevated soy consumption with a diminished risk of breast and prostate cancer (Messina, 2016). Laboratory research indicates that genistein suppresses the multiplication of cancer cells (Lazarevia et al., 2012). Clinical evidence indicates that soy isoflavones may diminish recurrence rates in breast cancer survivors (Shu et al., 2009).

2.11 Fiber

Dietary fiber derived from whole grains, legumes, fruits, and vegetables is essential for cancer prevention and gastrointestinal health (Aune et al., 2011). Fiber promotes advantageous gut microbiota that generate short-chain fatty acids with anticancer properties, sequesters carcinogens for elimination, and modulates hormone levels, hence diminishing the risks associated with hormone-dependent cancers (Lattimer and Haub, 2010). Meta-analyses validate a decrease in colorectal cancer risk linked to elevated fiber consumption (Aune et al., 2011). Research with animals indicates that short-chain fatty acids, particularly butyrate, suppress the development of colon cancer cells (Canani et al., 2012).

2.12 Omega-3 Fatty Acids

Omega-3 polyunsaturated fatty acids from fish, flaxseeds, and walnuts exhibit significant anti-inflammatory and anticancer properties (Jump, 2011). EPA and DHA diminish inflammatory cytokines, obstruct cancer cell signaling by modifying lipid raft composition, and facilitate apoptosis and autophagy via mitochondrial malfunction (West et al., 2020). Epidemiological studies link increased omega-3 consumption to a diminished risk of breast, prostate, and colorectal cancers (Chen et al., 2019). Laboratory investigations indicate the suppression of tumor growth in lung and liver malignancies (Larsson et al., 2004). Clinical data demonstrate enhanced chemotherapeutic efficacy with omega-3 supplementation (Theinel et al., 2023).

2.13 Flavonoids

Flavonoids, present in fruits, vegetables, tea, and cocoa, function as antioxidants and modulate many cellular pathways associated with cancer growth (Ross and Kasum, 2002). They neutralize ROS, inhibit cyclin-dependent kinases to modulate the cell cycle, and trigger apoptosis by inhibiting NF-κB and upregulating pro-apoptotic proteins (Ganai and Farooqi, 2015). Population studies indicate that elevated flavonoid consumption reduces the incidence of lung, breast, and colorectal cancers (Wang et al., 2016). Laboratory research demonstrates that green tea catechins impede the proliferation of prostate and liver cancer cells (Chen et al., 2019). Clinical investigations indicate enhanced cancer prognosis and diminished recurrence rates associated with flavonoid-rich diets (Thyagarajan et al., 2020).

3. Mechanisms of Action

Functional nutrients demonstrate anticancer properties via multiple interrelated biological pathways, encompassing inflammation suppression, apoptosis induction, angiogenesis inhibition, oxidative stress reduction, and immune response modulation. Collectively, these strategies inhibit tumor advancement and reduce cancer progression. Figure 1 depicts the primary methods by which functional foods aid in cancer prevention, encompassing antioxidant, anti-inflammatory, apoptotic, anti-angiogenic, and immune-modulatory pathways.

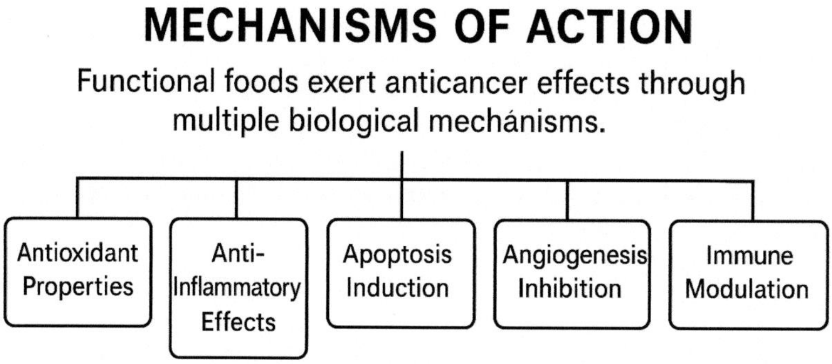

Fig. 1 Functional Foods and their mechanism of actions on cancer prevention.

3.1 Antioxidant Properties

Oxidative stress significantly contributes to cancer development as free radicals and ROS trigger lipid peroxidation, DNA damage, and protein oxidation, potentially resulting in genetic alterations, cell proliferation, and tumor start (Reuter et al., 2010). Functional foods encompass bioactive ingredients that mitigate ROS and bolster the body's antioxidant defense mechanisms. Carotenoids, including astaxanthin, lycopene, and lutein, mitigate singlet oxygen and lipid peroxides, thus averting oxidative DNA damage (Milani et al., 2017). Flavonoids like quercetin and catechins mitigate oxidative stress by enhancing the activity of antioxidant enzymes, specifically superoxide dismutase (SOD) and glutathione peroxidase (GSH-Px) (Kumar and Pandey, 2013). Omega-3 fatty acids mitigate oxidative stress by diminishing lipid peroxidation and enhancing antioxidant enzyme activity (Shaikh et al., 2016). *In vitro* studies indicate that carotenoids inhibit ROS production in lung and liver cancer cells, whilst animal research reveals that flavonoid supplementation mitigates oxidative damage and tumor development (Milani et al., 2017; Kumar and Pandey, 2013). Clinical study indicates that diets high in antioxidants correlate with a reduced incidence of colorectal and breast malignancies (Liu et al., 2018). Figure 2 presents a summary of the antioxidant pathways associated with functional meals.

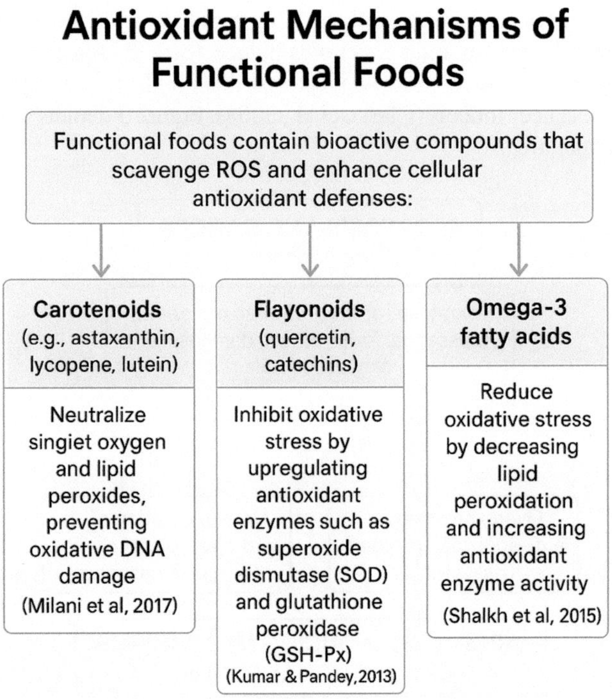

Fig. 2 Antioxidant Mechanisms of Functional Foods.

3.2 Anti-Inflammatory Effects

Chronic inflammation substantially facilitates cancer progression by causing DNA damage, enhancing immune evasion, and modifying the tumor microenvironment (Greten and Grivennikov, 2019). Pro-inflammatory cytokines, including TNF-α, IL-6, and NF-κB, promote cell proliferation and suppress apoptosis (Visht and Singh, 2023). Functional foods influence anti-inflammatory effects by regulating these molecular pathways. Polyphenols, especially flavonoids and curcumin, suppress NF-κB signaling and diminish the synthesis of inflammatory cytokines (Bankson et al., 1993). Isothiocyanates, including sulforaphane, inhibit COX-2 and prostaglandins (Kensler et al., 2013). Omega-3 fatty acids reduce pro-inflammatory eicosanoids and facilitate anti-inflammatory responses (Calder, 2013). Research with animals demonstrates that omega-3 supplementation diminishes inflammatory markers and tumor proliferation (West et al., 2020), whereas clinical investigations reveal that curcumin reduces inflammatory cytokine levels in individuals with colorectal cancer (Gupta et al., 2013).

3.3 Apoptosis Induction

Apoptosis, or programmed cell death, is crucial for the removal of damaged or cancerous cells. Cancer cells frequently circumvent apoptosis by inhibiting pro-apoptotic proteins including Bax, p53, and caspases, while simultaneously overexpressing anti-apoptotic proteins such as Bcl-2 and Mcl-1 (Elmore, 2007). Functional meals can initiate apoptosis via both internal (mitochondrial) and extrinsic (death receptor) routes. Flavonoids like quercetin and apigenin stimulate p53 and caspase-3, leading to mitochondrial instability and death (Ganai and Farooqi, 2015). Carotenoids, such as fucoxanthin and lycopene, regulate the Bax/Bcl-2 ratio and promote cytochrome-c release (D'Orazio et al., 2012). Isothiocyanates, including sulforaphane, augment TRAIL receptor expression, thereby elevating cancer cell vulnerability to apoptosis (Clarke et al., 2008). *In vitro* studies demonstrate that flavonoids and carotenoids induce death in breast and prostate cancer cells (Tezerji et al., 2022), but animal research suggest that diets high in sulforaphane facilitate tumor regression in colorectal cancer models (Clarke et al., 2008). Figure 3 depicts the apoptosis-inducing mechanisms of functional meals.

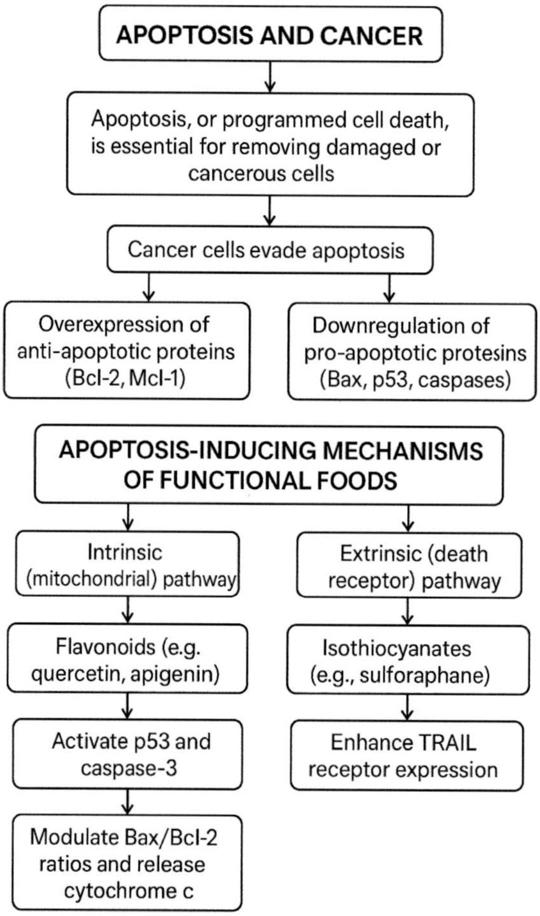

Fig. 3 Apoptosis-Inducing Mechanisms of Functional Foods.

3.4 Angiogenesis Inhibition

Angiogenesis facilitates tumor development and spread by providing oxygen and nutrients via newly formed blood vessels. Vascular endothelial growth factor (VEGF) is the principal mediator of angiogenesis in neoplasms (Folkman, 2007). Specific functional nutrients disrupt angiogenic pathways to inhibit tumor progression. Carotenoids, including lycopene and fucoxanthin, suppress VEGF expression and impede endothelial cell migration (Chen et al., 2012). Flavonoids, such as epigallocatechin gallate and genistein, inhibit matrix metalloproteinases (MMPs), therefore obstructing angiogenesis (Chen et al., 2004). Omega-3 fatty acids suppress pro-angiogenic pathways and diminish tumor vascularization (Calder, 2013). *In vitro* investigations demonstrate that flavonoids decrease VEGF levels in lung cancer cells, whilst animal studies validate that fucoxanthin diminishes tumor vascularization and metastasis (Chen et al., 2004; Chen et al., 2012).

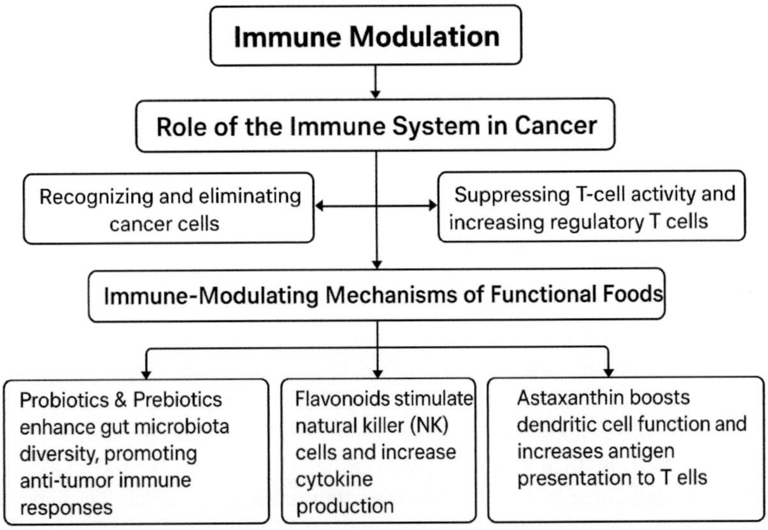

Fig. 4 Immune-Modulating Mechanisms of Functional Foods.

3.5 Immune Modulation

The immune system is essential for identifying and destroying cancer cells; yet, cancers frequently inhibit T-cell function and augment regulatory T-cell populations to escape immune detection (Dunn et al., 2004). Functional nutrients enhance immune responses and rejuvenate anticancer activity. Probiotics and prebiotics optimize gut microbiota composition, hence enhancing immune surveillance (Gopalakrishnan et al., 2018). Flavonoids enhance the activity of natural killer (NK) cells and elevate cytokine synthesis (O'Callaghan & van Sinderen, 2016). Astaxanthin augments dendritic cell functionality and facilitates antigen presentation, hence reinforcing T-cell responses (Fakhri et al., 2018). Animal studies indicate that probiotic supplementation enhances T-cell-mediated immunity (Hashemi et al., 2023), whereas human trials reveal that flavonoid-rich diets increase immune function in cancer patients (Gopalakrishnan et al., 2018). Figure 4 delineates the immune-modulating mechanisms of functional meals.

4. Efficacy and Clinical Evidence

The anticancer efficacy of functional meals has been thoroughly examined via *in vitro* investigations, animal tests, human clinical trials, case studies, and meta-analyses. These study methodologies collectively furnish essential evidence concerning the efficacy, safety, and fundamental mechanisms of nutraceuticals in cancer prevention and treatment.

4.1 *In Vitro* Studies

In vitro investigations utilizing cancer cell lines clarify the biochemical processes by which functional meals demonstrate anticancer properties. Studies indicate that carotenoids, including astaxanthin, lycopene, and fucoxanthin, impede the growth of prostate, lung, and colon cancer cells via downregulating cyclin-dependent kinases (Chen et al., 2012). They also promote apoptosis in breast cancer cells by elevating Bax/Bcl-2 ratios and activating caspases (D'Orazio et al., 2012). Flavonoids, including quercetin and epigallocatechin gallate, have demonstrated the ability to diminish tumor cell viability in lung and pancreatic cancers by inhibiting the PI3K/Akt and NF-κB signaling pathways (Tezerji S. et al., 2022), while concurrently averting oxidative DNA damage via free radical scavenging (Wilms et al., 2005). Moreover, isothiocyanates like sulforaphane impede the proliferation of colorectal cancer cells by promoting autophagy and death (Clarke et al., 2008). Nonetheless, *in vitro* studies are constrained as they fail to entirely emulate the intricacies of the tumor microenvironment, necessitating validation by animal and human research.

4.2 Animal Studies

Animal studies provide critical insights into the bioavailability, metabolism, and systemic effects of functional foods in cancer prevention and treatment. Carotenoids have demonstrated the ability to diminish tumor volume in mice lung cancer models by inhibiting angiogenesis (Chen et al., 2012). Diets high in flavonoids impede the progression of colorectal cancer in mice by altering gut microbiota and diminishing inflammation (Gopalakrishnan et al., 2018). Omega-3 fatty acids impede the advancement of liver cancer in hepatocellular carcinoma models by diminishing the production of inflammatory cytokines (West et al., 2020). Sulforaphane supplementation reduces the occurrence of colon tumors in mice by altering Wnt signaling pathways (Clarke et al., 2008). Animal studies, while their significance, possess limitations stemming from metabolic disparities between rodents and humans, potentially impacting the applicability of findings to clinical populations (Singh and Kochar, 2010). Consequently, results from animal studies require validation through extensive human trials.

4.3 Human Clinical Trials

Human clinical trials yield direct and clinically pertinent evidence concerning the efficacy and safety of functional foods in cancer patients and at-risk populations. A phase II clinical research indicated that green tea catechins (EGCG), a category of flavonoids, markedly decreased prostate-specific antigen (PSA) levels in individuals with prostate cancer (Chen et al., 2004). Isoflavones from soy, especially in cultures with a high habitual intake of soy, have been linked to a decreased recurrence of breast cancer, as evidenced by epidemiological studies conducted on Asian women (Shu et al., 2009). Moreover, genistein supplementation in randomized clinical trials has demonstrated a reduction in prostate cancer cell growth (Kumar et al., 2015). Omega-3 fatty acids, especially DHA, have been shown to enhance chemotherapy efficacy in breast cancer patients (Theinel et al., 2023). Curcumin administration in colorectal cancer patients has been demonstrated to reduce tumor-associated inflammatory markers (Gupta et al., 2013). Nonetheless, human studies encounter constraints like dose variability, uneven absorption, and the necessity for prolonged follow-up to assess enduring effects.

4.4 Case Studies

Case studies offer personalized narratives regarding the prospective advantages of nutraceuticals in actual clinical environments. A breast cancer patient adhering to a high-soy diet showed substantial decreases in estrogen receptor-positive tumor markers (Messina, 2016). A colon cancer patient using a high-fiber, plant-based diet had increased gut microbial diversity and improved immune function

(O'Keefe, 2016). Nevertheless, case studies are incapable of establishing causality due to the lack of control groups, and the highly customized characteristics of these reports restrict their applicability to wider populations.

4.5 Meta-Analyses and Systematic Reviews

Meta-analyses and systematic reviews offer robust evidence by integrating results from several studies. A systematic evaluation of 19 research determined that elevated dietary fiber consumption decreases the risk of colorectal cancer by roughly 22% (Aune et al., 2011). A meta-analysis of 35 research revealed that increased soy consumption correlates with a 26% decrease in breast cancer risk (Shu et al., 2009). Studies investigating omega-3 fatty acids have demonstrated associations between higher consumption and reduced risks of liver and breast malignancies (Chen et al., 2004). Notwithstanding their advantages, meta-analyses may be constrained by heterogeneity in study design, variability in demographic characteristics, and possible publication bias.

5. Conclusion

Functional foods and nutraceuticals present interesting supplementary approaches in cancer prevention and treatment. Their bioactive compounds—carotenoids, flavonoids, omega-3 fatty acids, probiotics, and isothiocyanates—demonstrate significant anticancer properties via mechanisms including antioxidant activity, inflammation reduction, apoptosis induction, angiogenesis inhibition, and immune response modulation. Despite significant potential demonstrated in both preclinical and clinical investigations, an additional study is essential to confirm their therapeutic efficacy, establish uniform dosages, and elucidate their interactions with conventional cancer therapies.

Recent data underscores numerous significant discoveries. Studies indicate that carotenoids, flavonoids, and omega-3 fatty acids significantly diminish oxidative stress, consequently averting DNA damage and cancer (Reuter et al., 2010; Milani et al., 2017). Polyphenols and isothiocyanates are recognized for their ability to suppress inflammatory pathways, including NF-κB and COX-2, hence mitigating cancer-promoting inflammation (Singh et al., 2019). Numerous bioactive substances activate caspase-dependent pathways that induce apoptosis, and several have demonstrated the ability to downregulate VEGF production, thereby inhibiting angiogenesis and restricting tumor blood supply (Chen et al., 2012). Probiotics and flavonoids enhance immune function, augmenting the body's capacity to identify and eradicate tumor cells (Gopalakrishnan et al., 2018). *In vitro* and animal model evidence indicates significant anticancer effects, but human studies reveal moderate advantages, especially for carotenoids, polyphenols, and omega-3 fatty acids. Meta-analyses findings further substantiate the correlation between elevated consumption of functional foods and diminished cancer risk.

The ramifications for cancer treatment are substantial. Consistent intake of fiber, flavonoids, and omega-3 fatty acids correlates with a diminished risk of colon, breast, and prostate cancers (Aune et al., 2011; Shu et al., 2009). Nutraceuticals may augment the efficacy of chemotherapy while mitigating inflammation, oxidative stress, and adverse effects associated with treatment (Theinel et al., 2023; Gupta et al., 2013). Modulating gut microbiota with probiotics and prebiotics demonstrates potential in enhancing immune responses, thereby improving outcomes for patients receiving immunotherapy (Gopalakrishnan et al., 2018). These functional meals should be regarded as adjuncts rather than substitutes for traditional cancer therapies, as their independent therapeutic efficacy has not been definitively proven.

Notwithstanding these encouraging results, some constraints remain. Numerous functional food ingredients, like curcumin and lycopene, exhibit limited bioavailability, requiring enhanced formulations to optimize their absorption and efficacy (Clarke et al., 2008). Moreover, human clinical trials frequently exhibit non-standardized dosages, hence hindering the formulation of definitive guidelines for clinical use (Chen et al., 2004). Concerns exist regarding possible interactions between nutraceuticals and chemotherapeutic or hormonal therapies, especially in hormone-sensitive malignancies (Messina, 2016). Future research must emphasize extensive randomized controlled trials to evaluate long-term efficacy, investigate personalized nutrition aligned with individual genetic profiles, analyze the synergistic effects of integrating functional foods with conventional therapies, and create advanced formulations that enhance bioavailability and therapeutic results.

Healthcare workers are urged to advocate for diets abundant in fruits, vegetables, whole grains, and other functional foods as components of a cancer-preventive lifestyle. Omega-3 supplementation may aid chemotherapy patients by mitigating inflammation and avoiding cachexia (West et al., 2020). Healthcare providers must also observe potential interactions between nutraceuticals and standard medicines. Public health initiatives could gain from integrating nutraceutical therapies and advocating for plant-based dietary patterns to diminish cancer risk at the population level.

Functional foods and nutraceuticals constitute a burgeoning and promising domain within oncology and dietary therapies. Despite existing research endorsing its anticancer potential, additional efforts are required to refine formulations, standardize dosages, and validate clinical efficacy through rigorously conducted trials. Ongoing investigation into individualized nutrition, combination medicines, and novel delivery technologies will be crucial to fully exploit the therapeutic potential of functional foods in cancer treatment.

References

Adlercreutz, H. 2007. Lignans and human health. *Crit. Rev. Clin. Lab. Sci.*, *44*, 483–525.

Agarwal, S. and Rao, A.V. 2000. Tomato lycopene and its role in human health and chronic diseases. *CMAJ*, *163*, 739–744.

Ambati, R.R., Phang, S. Ravi and Aswathanarayana, R.G. 2014. Astaxanthin: Sources, extraction, stability, biological activities, and commercial applications. *Mar. Drugs*, *12*, 128–152.

Ansari, N., Singh, G., Singh, R. and Sheetal. 2025. Innovative herbal tea formulation using *Holarrhena antidysenterica*, *Emblica officinalis*, and Stevia. *J. Neonatal Surg.*, *14*, 381–389.

Aune, D., Chan, D.S., Greenwood, D.C. et al., 2011. Dietary fiber and colorectal cancer risk: A systematic review. *BMJ*, *343*, d6617.

Bala, S., Singh, G. and Kaur, M. 2024a. Mindfulness of functional foods in cancer prevention. *Rev. Electron. Vet. 25*, 1181–1187.

Bankson, D.D., Kestin, M. and Rifai, N. 1993. Role of free radicals in cancer and atherosclerosis. *Clin. Lab. Med.*, *13*, 463–480.

Bendich, A. 1994. Recent advances in clinical research involving carotenoids. *Pure Appl. Chem.*, *66*, 1017–1024.

Blumberg, J. and Block, G. 1994. The ATBC Cancer Prevention Study in Finland. *Nutr. Rev. 52*, 242–245.

Calder, P.C. 2013. Omega-3 PUFAs and inflammatory processes. *Br. J. Clin. Pharmacol.*, *75*, 645–662.

Canani, R. Berni, R. Nocerino, Terrin, G. et al., 2012. Effect of *Lactobacillus GG* on tolerance acquisition. *J. Allergy Clin. Immunol.*, *129*, 580–582.

Chen, D., Daniel, K.G., Kuhn, D.J. et al., 2004. Green tea polyphenols in cancer prevention. *Front. Biosci. 9*, 2618–2631.

Chen, M.L., Lin, Y.H., Yang, C.M. and Hu, M.L. 2012. Lycopene inhibits angiogenesis. *Mol. Nutr. Food Res.*, *56*, 889–899.

Chakravarty, R. and Dash, A. 2014. Nanomaterial-based adsorbents: the prospect of developing new generation radionuclide generators to meet future research and clinical demands. *J. Radioanal. Nucl. Chem.*, *299*, 741–757.

Clarke, J.D., Dashwood, R.H. and Ho, E. 2008. Multi-targeted prevention of cancer by sulforaphane. *Cancer Lett.*, *269*, 291–304.

D'Orazio, N., Gammone, M.A., Gemello, E. et al., 2012. Fucoxanthin: A treasure from the sea. *Mar. Drugs*, 10, 604–616.

Doyon, M. and Labrecque, J. 2008. Functional foods: A conceptual definition. *Br. Food J. 110*, 1133–1149.

Dunn, G.P., Old, L.J. and Schreiber, R.D. 2004. Immunobiology of cancer immunosurveillance. *Immunity*, *21*, 137–148.

Elmore, S. 2007. Apoptosis: Programmed cell death. *Toxicol. Pathol.*, *35*, 495–516.

Fakhri, S., Abbaszadeh, F., Dargahi, L. and Jorjani, M. 2018. Biological activities of astaxanthin. *Pharmacol. Res.*, *136*, 1–20.

Folkman, J. 2007. Angiogenesis and drug discovery. *Nat. Rev. Drug Discov.*, *6*, 273–286.

Ganai, A.A. and Farooqi, H. 2015. Bioactivity of genistein. *Biomed. Pharmacother.*, *76*, 30–38.

Gaziano, J.M., Sesso, H.D., Christen, W.G. et al., 2012. Multivitamins and cancer prevention in men. *JAMA*, *308*, 1871–1880.

Giovannucci, E. 2002. Tomatoes, lycopene, and prostate cancer. *Exp. Biol. Med.*, *227*: 852–859.

Gopalakrishnan, V., Spencer, C.N. and Nezi, L. 2018. Microbiome modulates immunotherapy response. *Science*, *359*, 97–103.

Greten, F.R. and Grivennikov, S.I. 2019. Inflammation and cancer. *ImmunityI, 51*, 27–41.

Guerin, M., Huntley, M.E. and Olaizola, M. 2003. Haematococcus astaxanthin applications. *Trends Biotechnol.*, *21*, 210–216.

Gupta, S.C., Patchva, S. and Aggarwal, B.B. 2013. Therapeutic roles of curcumin. *AAPS J.*, *15*, 195–218.

Hashemi, B., Abdollahi, M., Abbaspour-Aghdam, S. et al., 2023. Probiotics and immune responses. *Biomed. Pharmacother.*, *159*, 114195.

Hollman, P.H. and Katan, M.B. 1999. Dietary flavonoids. *Food Chem. Toxicol.*, *37*, 937–944.

Ishikawa, H., Matsumoto, S., Ohashi, Y. et al., 2011. Probiotic and GOS effects in ulcerative colitis. *Digestion*, *84*, 128–133.

Ji, X., Liu, K., Li, Q. et al., 2022. Apigenin and genistein in prostate cancer. *Front. Pharmacol.*, *13*, 851589.

Jiang, X., Liu, Y., Ma, L. et al., 2018. Chemopreventive activity of sulforaphane. *Drug Des. Devel. Ther.*, *12*, 2905–2913.

Jump, D.B. 2011. Regulation of hepatic lipid metabolism. *Annu. Rev. Nutr.*, *31*, 235–258.

Jyonouchi, H., Sun, S., Abiru, T. et al., 2000. Dietary nucleotides and immune responses. *Nutrition*, *16*, 442–446.

Kamal, M.M., Akter, S., Lin, C.N. and Nazzal, S. 2020. Mechanisms of sulforaphane. *Arch. Pharm. Res.*, *43*, 371–384.

Kapała, A., Szlendak, M. and Motacka, E. 2022. Anti-cancer activity of lycopene. *Nutrients*, *14*, 5152.

Kensler, T.W., Egner, P.A., Agyeman, A.S. et al., 2013. Keap1-Nrf2 signaling and sulforaphane. *Top. Curr. Chem.*, *329*, 163–177.

Khorshidian, N., M.Y. Asli, Hosseini, H. et al., 2016. Anticarcinogenic effects of lactic acid bacteria. *Iranian J. of Cancer Prevention*, *9*(50). doi:10.17795/ijcp-7920.

Kim, N.H. and Kang, P.M. 2010. Apoptosis in cardiovascular disease. *Korean Circ. J.*, *40*, 299–305.

Konstantinidi, M. and Koutelidakis, A.E. 2019. Functional foods and obesity. *Medicines*, *6*, 94.

Kopustinskiene, D.M., V. Jakstas, A. Savickas and Bernatoniene, J. 2020. Flavonoids as anticancer agents. *Nutrients*, *12*, 457.

Krinsky, N.I. and Johnson, E.J. 2005. Carotenoid actions. *Mol. Aspects Med.*, *26*, 459–516.

Kucuk, O., F.H. Sarkar, Djuric, Z. et al., 2002. Lycopene supplementation and prostate cancer. *Exp. Biol. Med.*, *227*, 881–885.

Kumar, S. and Pandey, A.K. 2013. Chemistry and biological activities of flavonoids. *Sci. World J.*, *2013*, 162750.

Larsson, M.C., A.I. Domingos, Jones, W.D. et al., 2004. Odorant receptor Or83b in Drosophila. *Neuron*, *43*, 703–714.

Lattimer, J.M. and Haub. M.D. 2010. Dietary fiber and metabolic health. *Nutrients*, *2*, 1266–1289.

Lazarevic, B., C. Hammarström, Yang, J. et al., 2012. Genistein intervention in prostate cancer. *Br. J. Nutr.*, *108*, 2138–2147.

Le Marchand, L. 2002. Cancer-preventive effects of flavonoids. *Biomed. Pharmacother.*, *56*, 296–301.

Lee, Y.R., E.M. Noh, Han, J.H. et al., 2013. Sulforaphane suppresses MMP-9. *BMB Rep.*, *46*, 201–206.

Lei, Z.N., Q. Tian, Q.X. Teng, et al., 2023. Targeting resistance mechanisms in cancer. *MedComm.*, *4*, e265.

Li, Q., Wang, Y., Jia, W., Deng, H., Li, G. and Deng, W. et al., 2020. Low-dose anti-angiogenic therapy sensitizes breast cancer to PD-1 blockade. *Clin. Cancer Res.*, *26*, 1712–1724.

Liu, Z., Ren, Z., Zhang, J., Chuang, C.C., Kandaswamy, E., Zhou, T. et al., 2018. Role of ROS and nutritional antioxidants in human diseases. *Front. Physiol.*, *9*, 360203.

Lobo, V., Patil, A., Phatak, A. and Chandra, N. 2010. Free radicals, antioxidants, and functional foods: Impact on human health. *Pharmacogn. Rev.*, *4*, 118–126.

Maeda, H., Tsukui, T., Sashima, T., Hosokawa, M. and Miyashita. K. 2008. Seaweed carotenoid fucoxanthin as a multi-functional nutrient. *Asia Pac. J. Clin. Nutr.*, *17*(S1), 196–199.

Mares-Perlman, J.A., Millen, A.E., Ficek, T.L. and Hankinson, S.E. 2002. Evidence supporting lutein and zeaxanthin in delaying chronic disease. *J. Nutr.*, *132*, 518S–524S.

Messina, M. 2016. Soy and health update. Evaluation of clinical and epidemiologic literature. *Nutrients*, *8*, 754.

Milani, C., Duranti, S., Bottacini, F., Casey, E., Turroni, F., Mahony, J. et al., 2017. First microbial colonizers of the infant gut: Composition and health implications. *Microbiol. Mol. Biol. Rev.*, *81*, 10–1128.

Miyashita, K., Lutz, J., Hudgins, L.C., Toib, D., Ashraf, A.P., Song, W. et al., 2020. Chylomicronemia from GPIHBP1 autoantibodies. *J. Lipid Res.*, *61*, 1365–1376.

*Moeller, S.M., Jacques, P.F. and Blumberg, J.B. 2000. Dietary xanthophylls in cataract and macular degeneration. *J. Am. Coll. Nutr.*, *19*, 522S–527S.

Murakoshi, M., Nishino, H., Satomi, Y., Takayasu, J., Hasegawa, T., Tokuda, H. et al., 1992. Alpha-carotene suppresses carcinogenesis more effectively than beta-carotene. *Cancer Res.*, *52*, 6583–6587.

Nishino, H., Murakoshi, M., Mou, X.Y., Wada, S., Masuda, M., Ohsaka, Y. et al., 2005. Cancer prevention by phytochemicals. *Oncology*, *69*(S1), 38–40.

Nishino, H., Fujita, N., Noda, T. et al., 2009. ER subdomain and autophagosome formation. *Nat. Cell Biol.*, *11*, 1433–1437.

O'Callaghan, A. and van Sinderen, D. 2016. Bifidobacteria as members of the human gut microbiota. *Front. Microbiol.* 7: 925.

O'Keefe, S.J. 2016. Diet, microorganisms, metabolites, and colon cancer. *Nat. Rev. Gastroenterol. Hepatol.*, *13*, 691–706.

Paiva, S.A. and Russell, R.M. 1999. Carotenoids as antioxidants. *J. Am. Coll. Nutr.*, *18*, 426–433.

Pap, R., Pandur, E., Jánosa, G., Sipos, K., Nagy, T., A. Agócs, et al., 2022. Lutein decreases inflammation, oxidative stress and lipid peroxidation. *Antioxidants*, *11*, 2269.

Patra, S., Pradhan, B., Nayak, R., Behera, C., Das, S., Patra, S.K. et al., 2021. Dietary polyphenols in chemoprevention. *Phytomedicine*, *90*, 153554.

Rao, A.V. and Rao, L.G. 2007. Carotenoids and human health. *Pharmacol. Res.*, *55*, 207–216.

Reuter, S., Gupta, S.C., Chaturvedi, M.M. and Aggarwal, B.B. 2010. Oxidative stress, inflammation, and cancer link. *Free Radic. Biol. Med.*, *49*, 1603–1616.

Ross, J.A. and Kasum, C.M. 2002. Dietary flavonoids: Bioavailability and safety. *Annu. Rev. Nutr.*, *22*, 19–34.

Russo, S.J., Murrough, J.W., Han, M.H., Charney, D.S. and Nestler, E.J. 2012. Neurobiology of resilience. *Nat. Neurosci.*, *15*, 1475–1484.

Samtiya, M., Aluko, R.E., Dhewa, T. and Moreno-Rojas. J.M. 2021. Plant food-derived bioactive components. *Foods*, *10*, 839.

Shaikh, S.A., Sahoo, S.K. and Periasamy, M. 2016. Regulators of SERCA calcium ATPase. *J. Mol. Cell. Cardiol.*, *91*, 81–91.

Shu, X.O., Zheng, Y., Cai, H., et al., 2009. Soy food intake and breast cancer survival. *JAMA*, *302*, 2437–2443.

Singh, G., Bala, S. and Singh, S. 2024a. Nutrigenomics review. *Afr. J. Biol. Sci.*, *6*, 599–699.

Slavin, J. 2003. Whole grains and protective mechanisms. *Proc. Nutr. Soc.*, *62*, 129–134.

Story, E.N., Kopec, R.E., Schwartz, S.J. and Harris, G.K. 2010. Health effects of lycopene. *Annu. Rev. Food Sci. Technol.*, *1*, 189–210.

Sung, H., Ferlay, J. and Siegel, R.L. 2021. Global cancer statistics 2020. *CA Cancer J. Clin.*, *71*, 209–249.

Suzuki, Y., Sugai, T., Fukui, N., Watanabe, J., Ono, S., Tsuneyama, N. et al., 2014. Undernutrition in schizophrenia patients. *Psychiatry Clin. Neurosci.*, *68*, 78–82.

Tanaka, M., Fernández-del Castillo, C., Adsay, V., Chari, S., Falconi, M., Jang, J.Y. et al., 2012. Consensus guidelines for IPMN and MCN. *Pancreatology*, *12*, 183–197.

Taylor, P.R. and P. Greenwald. 2005. Nutritional interventions in cancer prevention. *J. Clin. Oncol.*, *23*, 333–345.

Tezerji, S., Robati, F.N., Abdolazimi, H., Fallah, A. and Talaei, B. 2022. Quercetin effects in colon cancer rat model. *Clin. Nutr. ESPEN*, 48, 441–445.

Theinel, M.H., Nucci, M.P., Alves, A.H., Dias, O.F.M., Mamani, J.B., M.M. Garrigós, et al., 2023. Effects of omega-3 on breast cancer. *Nutrients*, *15*, 1310.

Thyagarajan, A., Forino, A.S., Konger, R.L. and R.P. Sahu. 2020. Polyphenols in pancreatic cancer chemoprevention. *Antioxidants*, *9*, 651.

Verhoeven, D.T., Goldbohm, R.A., van Poppel, G., Verhagen, H. and P.A. van den Brandt. 1996. Brassica vegetables and cancer risk. *Cancer Epidemiol. Biomarkers Prev.*, *5*, 733–748.

Visht, S. and G. Singh. 2023. Nutraceuticals in osteoporosis management. In: *Nutraceuticals in Osteoporosis.* CRC Press.

Voutilainen, S., Nurmi, T., Mursu, J. and Rissanen. T.H 2006. Carotenoids and cardiovascular health. *Am. J. Clin. Nutr.*, *83*, 1265–1271.

Wang, F., Kong, W., Wong, G., Fu, L., Peng, R., Z. Li, et al., 2016. MYB12 regulates flavonoid accumulation. *Mol. Genet. Genomics*, *291*, 1545–1559.

West, L., Yin, Y., Pierce, S.R., Fang, Z., Fan, Y., Sun, W. et al., 2020. DHA inhibits ovarian cancer metastasis. *Am. J. Cancer Res.*, *10*, 4450–4463.

Wilms, L.C., Hollman, P.C., Boots, A.W. and Kleinjans, J.C. 2005. Quercetin protection against DNA damage. *Mutat. Res.*, *582*, 155–162.

Wu, A.H. 2001. Soy and hormone-related cancers. *Adv. Exp. Med. Biol.*, *492*, 19–28.

Zhang, Y.J. 2010. Chemical carcinogens and genetic variation in liver cancer. *World J. Hepatol.*, *2*, 94.

CHAPTER 8

Fortification and Enrichment of Foods

Methods and Challenges

Aysha Sherin K.[1], Khushi Sharma[2] and Roji Waghmare[3*]

1. Introduction

Food fortification and enrichment are essential public health strategies aimed at combating micronutrient deficiencies; a global health issue commonly known as "hidden hunger". Fortification means adding essential vitamins and minerals, such as vitamin A, vitamin D, iron, zinc and iodine, into staple food items like milk, oil, rice, salt and wheat, while enrichment refers to bringing back those nutrients lost during food processing to their original level (Kruger et al., 2020). These interventions work to upgrade the nutritional profile of diets and tackle micronutrient malnutrition, affecting more than 2 billion people worldwide, especially in low-and middle-income countries (Olson et al., 2021).

Hindered cognitive growth, increased susceptibility to infectious illnesses, and elevated mortality rates are all consequences of micronutrient malnutrition, particularly for vulnerable groups like children and expectant mothers (Keats et al., 2021). For example, 1.2 billion people worldwide suffer from iron deficiency anemia, underscoring the pressing need for efficient interventions (Global Nutrition Report, 2021). After the implementation of both voluntary and later mandatory enrichment of bread and flour with yeast fortified with vitamins, a significant reduction in the occurrence of pellagra, linked to a deficiency in niacin, and beriberi, caused by a lack of thiamine, was recorded in the Southern United States and in Canada, respectively. (Fletcher et al., 2004; Osendarp et al., 2018). Furthermore, in the US, federal authorization for large-scale mandatory fortification of enriched cereal grain products with folic acid was granted in 1996, with nationwide implementation completed by 1998 (Osendarp et al., 2018).

In order to fight malnutrition, the Food Safety and Standards Authority of India (FSSAI) promotes fortifying five staple foods. These are wheat flour and rice (fortified with iron, vitamin B_{12} and folic acid), milk and edible oil (fortified with vitamins D and vitamin A) and double fortified salt (fortified with iodine and iron) which are represented in Figure 1. Without altering the flavor or texture, fortification raises the nutritional value of frequently consumed foods. It is an economical and safe way to lessen the micronutrient deficiencies of the population.

[1-3] Department of Food Engineering and Technology, Institute of Chemical Technology, Mumbai, India-400019.

* Corresponding author: rosewaghmare@gmail.com

DOI: 10.1201/9781003659808-8

Fig. 1 Fortified staples (Food Safety and Standards Fortification of Foods Regulations, 2018).

Restoring nutrients that may have been lost during the processing of food products or refinement is known as enrichment. Enrichment is utilized to replenish some of the nutrients that are often lost during the refining process (Mattar et al., 2022). To maintain nutrition during processing, for instance, iron and B-vitamin enrichment is added to refined wheat flour to replace iron lost during the conversion of whole wheat to white flour (Molino et al., 2021). When food is cooked at elevated temperatures, for extended periods, and with significant quantities of water or oil (which is later thrown away), some vitamins and minerals are lost in greater quantities (Mannar and Khan, 2015). Air and light exposure, as well as pH, are other elements that impact nutrient retention.

Accordingly, the goal of enrichment is to increase or restore the nutritional value of foods (Allen et al., 2006). During the grinding and refining process, rice, bread, pasta, and flour lose a significant amount of their natural nutrients, making them common foods in the enrichment category (Olson et al., 2021). They have the ability to improve the health and nutritional status of at-risk groups, such as pregnant women, children, and the elderly (Mattar et al., 2022). This preventative approach can help people and communities live better lives and reduce healthcare costs (Mkambula et al., 2020). Through popular products and staple foods, food fortification and enrichment are made affordable, sustainable, and scalable interventions that can reach a sizable population (Dwyer et al., 2015). Selecting supplements that are both effective and of high quality is essential, particularly as emerging research underscores the role of food fortification and enrichment in advancing global nutrition goals.

Nutrition and health have clearly improved in low- and middle-income countries (LMICs) as a result of industrial food fortification. It helped reduce cases of anemia, goiter and neural tube defects (NTD). In fact, population-wide interventions were linked to a 34% drop in anemia, a 74% lower risk of goiter and a 41% reduction in NTD benefits that were most pronounced among people at highest risk of deficiencies. Figure 2 highlights the advantages of food fortification at various phases of life (Olson et al., 2021).

2. Importance of Fortification and Enrichment

Food fortification is an inexpensive method of enhancing population nutrition and has significant economic benefits (Olson et al., 2021). Large-scale food fortification initiatives have greatly enhanced the population's micronutrient status, particularly when paired with additional nutrition interventions (Keats et al., 2021). Osendarp et al. (2018) also emphasized the importance of fortified foods in tackling the double burden of malnutrition, which occurs when under-nutrition coexists

	Pregnancy	Lactating mother	6–23 mo	2–5 years	5–18 years	WRA (15–49 years)	Adult men	Elderly
Micro-nutrient need	very high	very high	very high	high	moderate to high	moderate to high	low to moderate	moderate to high
Amount of food eaten	moderate	moderate	low	low, increasing with age	increases with age	moderate	high	moderate
Potential to benefit	high	high	low	low, increasing with age	increases with age	high	high	high
Potential to fully meet need	low	low	no	low, increasing with age	increases with age	high	high	high

Fig. 2 The advantages of food fortification at various phases of life (Olson et al., 2021).

with overweight and obesity, especially in areas that are rapidly urbanizing (Osendarp et al., 2018). Poniedziałek et al. (2020) demonstrated that fortification and enrichment can enhance public health outcomes and reduce health disparities (Poniedziałek et al., 2020).

Adding more essential vitamins and minerals to everyday foods is a practical and fair way to help everyone get the nutrients they need, especially in areas where resources are limited. Iron deficiency, which causes about 0.8 million deaths per year globally (1.5% of all deaths), and vitamin A deficiency, which causes a comparable number of deaths (WHO, 2013). Numerous studies suggest that widespread food fortification may have an impact on public health in both high-income and LMIC nations. Recent reviews of 50 studies carried out in LMICs show that fortification with iodine, folic acid, and vitamins significantly reduced serious illness (Mkambula et al., 2020). The establishment of population-wide fortification programs resulted in improved iron reserves, 74% decrease in the odds of goiter, 41% decrease in the odds of neural tube diseases, and 34% decrease in anemia, with the greatest benefits being experienced by those most at risk of deficiency (Kraemera and Zutphen, 2019).

The Sustainable Development Goals of the 2030 Agenda for Sustainable Development seek to end hunger and all types of malnutrition. Data and international cooperation will be used to achieve these objectives. Additionally, food fortification improves public-private partnerships and benefits society across multiple sectors (Kruger et al., 2020). When combined with social safety net programs such as school feeding programs, food fortification has been demonstrated to be an efficient means of disseminating diet-related information and supplying fortified food to vulnerable people when paired with social safety net programs like school feeding programs, distributions to the poor or vulnerable groups, food for work programs, and food aid during emergencies (Van Stuijvenberg, 2005).

3. Legal Requirements of Food Fortification

Standards for food fortification are established and enforced by a regulatory body in each nation. For instance, regulatory oversight is provided by the "Food and Drug Administration" (FDA) in the United States, the "European Food Safety Authority" in the European Union, and the "Food Safety and Standards Authority of India" in India. Food fortification laws and regulations are frequently a component of larger public health or food safety laws. The law mandates that certain foods be fortified with particular nutrients, such as to prevent iodine deficiency disorders, iodine is required in many countries to be added to salt. Folic acid must be added to wheat flour in certain countries in order to prevent neural tube defects in newborns (Thakur et al., 2023). Manufacturers have the option to add extra nutrients to foods, but they have to abide by legal requirements regarding acceptable nutrient levels. In order to guarantee food safety and effectiveness, legal standards frequently specify the minimum and maximum amounts of nutrients that can be added (Madhari et al., 2020).

Regulations governing labeling, including accurate representation of nutrient content, must be followed by fortified foods. Any health claims pertaining to fortified foods need to be supported by science and authorized by the appropriate authority (Orriss, 1998). Fortification must be properly labeled if it introduces possible allergens (such as vitamin E derived from soy). In order to standardize standards across nations, the FDA and WHO formed the Codex Alimentarius Commission, which offers international guidelines on food fortification (Marks et al., 2018). As part of international initiatives to address micronutrient deficiencies, the WHO offers recommendations on fortification. The regulatory agencies may carry out inspections to verify adherence to fortification standards and noncompliance may result in fines, product recalls or legal action. Food manufacturers must abide by these rules and governments are crucial in keeping an eye on and upholding these requirements to safeguard the health of consumers.

In Tamil Nadu, the double fortified salt program (DFS) started in 2004 with the primary goal of providing midday meals to school children. Currently, the Public Distribution System is used by states like Gujarat and Madhya Pradesh to deliver DFS. More than 12 million people in 20 districts can now purchase double-fortified salt at a heavily discounted cost. Through the Integrated Child Development Services program, between 2010 and 2014, Telangana children received funds from the Global Alliance for Improved Nutrition (GAIN) to produce and distribute Bal Amrutham, an additional take-home ration. (Thakur et al., 2023).

4. Difference between Fortification and Enrichment

The intentional addition of vital micronutrients, such as vitamins and trace elements, to food products is known as "food fortification". This process can serve dual purposes, as it may be driven by commercial motivations to improve food's nutritional profile or implemented as a public health strategy to reduce dietary deficiencies within a population (Pant and Chinwan, 2014). Because the conventional diet naturally lacks sufficient quantities of these micronutrients or because regional soil deficiencies impact the nutrient content of crops, monotonous or restricted diets frequently fail to provide adequate levels of specific nutrients (Sandhu et al., 2023). By adding the required nutrients to staple foods, fortification fills these gaps and enhances overall dietary quality and health outcomes.

An enriched food is one that has had certain nutrients added back in after they were lost during processing (Kraemer and Zutphen, 2019). These nutrients are usually found in the food in its natural state, but industrial processing methods remove them (Kolanowski and Laufenberg, 2025). White bread is a common example of this because it has vitamins added to it that are taken out during the bleaching of flour. Many people use the words fortification and enrichment interchangeably, but they don't mean the same thing. The WHO and FAO of the United Nations say that fortification is the intentional addition of essential micronutrients like vitamins, minerals, and trace elements to a food product, even if these nutrients were already there before processing (Mkambula et al., 2020). In contrast, enrichment refers specifically to the addition of micronutrients that were lost during processing. This brings the nutrient content of the food back to its original or close-to-original levels (Liyanage and Hettiarachchi, 2011). The Food Safety and Standards (Fortification of Foods) Regulations, 2018 standards are shown in Table 1. It provides a summary of the suggested fortification levels for different food items, making sure they satisfy dietary needs while staying within safe levels. By enhancing the nutritional value of frequently consumed foods, these recommendations aim to safely and consistently address the micronutrient deficiencies of the population.

Table 1 Standards for fortification of food (Food Safety and Standards): (Fortification of Foods Regulations, 2018).

Food item	*Nutrient*	*Level of Nutrients*	*Source of Nutrients*
Salt	Iodine (Manufacture level) Iodine (Distribution channel)	20–30 ppm	Potassium Iodate
	Iron content	850–1100 ppm	Ferrous sulphate or Ferrous Fumarate
Edible Oil	Vitamin A	6 μg RE–9.9 μg RE per gm of oil	Retinyl acetate or Retinyl palmitate
	Vitamin D	0.11 μg–0.16 μg per gm of oil	Cholecalciferol or Ergocalciferol
Milk	Vitamin A	270–450 μg RE	Retinyl acetate or Retinyl palmitate
	Vitamin D	5–7.5 μg	Cholecalciferol or Ergocalciferol
Wheat flour and maida	Iron	28 mg–42.5 mg	Electrolytic iron, ferrous citrate, ferrous lactate, ferrous sulfate, ferrous pyrophosphate, ferrous fumarate, or ferrous bisglycinate
		14 mg–21.25 mg	Sodium Iron (III) Ethylene diamine tetra Acetate Trihydrate (Sodium feredetate-Na Fe EDTA)
	Folic acid	75 μg–125 μg	Folic acid
	Vitamin B12	0.75 μg–1.25 μg	Cyanocobalamine or Hydroxycobalamine
	Zinc	10 mg–15 mg	Zinc Sulphate
	Vitamin B1	1 mg–1.5 mg	Thiamine hydrochloride or Thiamine mononitrate
	Vitamin B2	1.25 mg–1.75 mg	Riboflavin or Riboflavin 5''phosphate sodium
	Vitamin B3	12.5 mg–20 mg	Nicotinamide or Nicotinic acid
	Vitamin B6	1.5 mg–2.5 mg	Pyridoxine hydrochloride
Rice	Iron	28 mg–42.5 mg	Ferric pyrophosphate
		14 mg–21.25 mg	Sodium feredetate -Na Fe EDTA
	Folic acid	75 μg–125 μg	Folic acid
	Vitamin B12	0.75 μg–1.25 μg	Cyanocobalamine or Hydroxycobalamine
	Zinc	10 mg–15 mg	Zinc Oxide
	Vitamin A	500 μg RE–750 μg RE	Retinyl Palmitate
	Vitamin B1	1 mg–1.5 mg	Thiamine hydrochloride or Thiamine mononitrate
	Vitamin B2	1.25 mg–1.75 mg	Riboflavin or Riboflavin 5''phosphate sodium
	Vitamin B3	12.5 mg–20 mg	Nicotinamide or Nicotinic acid
	Vitamin B6	1.5 mg–2.5 mg	Pyridoxine hydrochloride

5. Types of Nutrients Used in Fortification and Enrichment

5.1 Vitamins

The vitamin content in foods is highly vulnerable to degradation or destruction during technological and cooking processes which is influenced by various factors (Keats et al., 2019). A significant portion of these losses is attributed to the water solubility of certain vitamins which is heavily dependent on the cooking method used (De Lourdes Samaniego-Vaesken et al., 2012). Additionally, specific vitamins exhibit distinct sensitivities, with B vitamins being particularly susceptible to heat and light exposure, while fat-soluble vitamins are prone to oxidative damage (Adinepour et al., 2022). The extent of nutrient loss varies depending on the food type and processing conditions. For instance, thiamine, natural folates, and vitamin C can experience losses of up to 100% under certain combined cooking and processing conditions (Fatemi et al., 2023).

5.1.1 Vitamin A

Vitamin A is necessary for fetal growth and development, immune response, and visual health. Vitamin A deficiency is a public health issue in many parts of the world, particularly in Africa and Southeast Asia (Brauw et al., 2019). It can also make a child more likely to get sick and die from diseases like measles and diarrhea and it can also cause vision problems like night blindness (Grasso et al., 2023). Adding vitamin A to staple foods could be a cheap way to make sure people get enough of it, especially when it's not possible to make the food more diverse by improving its quality. Several researches done by WHO has shown that eating foods fortified with vitamin A, like cereal grains, milk, refined sugar, condiments, and edible oils can help improve vitamin A status (WHO, 2023).

5.1.2 Vitamin E

Fat spreads, fat oils, margarine, and morning cereals are commonly fortified with vitamin E. It is usually added as tocopherol acetate (Adinepour et al., 2022). It has been demonstrated that consuming it enhances the absorption and bioconversion of dietary carotenoids into vitamin A and it is strongly correlated with the consumption of total dietary fat (Keats et al., in 2019). According to recent studies consuming more than 30 IU of vitamin E per day may offer protection against degenerative illnesses (Shen and Ji, 2012). In light of this scientific evidence the ideal levels of vitamin E fortification should be reassessed in order to optimize its health benefits (Redeuil et al., 2021).

5.1.3 Vitamin D

Absorption of calcium and parathyroid hormone are linked to vitamin D metabolism (Maurya et al., 2023). Marginal vitamin D levels are a concern for older adults, children under three, and any group whose exposure to sunlight is limited because of cultural practices or weather patterns (Maurya and Aggarwal, 2020). Currently, a lot of research is being done to assess how sunscreen affects vitamin D levels. However, it is crucial to remember that rapid, easy, accurate, and trustworthy techniques of testing vitamin D levels must be established (Souza et al., 2022). Without this, it is challenging to modify fortification procedures as needed in response to input from thorough analyses of how fortification affects vitamin D status (Vieira and Souza, 2022).

5.1.4 Vitamin C

Numerous technologies can be used to fortify dairy products, fruit juices, certain breakfast cereals, fruit juice drinks with vitamin C (Maurya et al., 2023). Vitamin C-rich foods are known to improve the absorption of iron. High moisture content (more than 7%) in the presence of oxygen is known to adversely affect the stability of vitamin C in cereals (Sobczak et al., 2021). More stable vitamin C molecules must be created, and their potential for fortifying various vehicles must be assessed. The potential of vitamin C to prevent chronic illnesses like cancer, cataracts, and cardiovascular disease is still being investigated (Fatemi et al., 2023).

5.1.5 B-Complex Vitamins

The technological aspects of adding B-complex vitamins to cereals and grains are well-understood and pose no significant challenges (Rathee et al., 2022). For instance adding thiamine after heat treatment during processing readily reduces its thermal instability which can result in off flavors. Although more investigation is required to elucidate this relationship riboflavin is thought to contribute to the prevention of anemia (Whitfield et al., 2021). Populations whose consumption of animal-based foods is restricted e.g., 3. B-vitamin status is often marginal in people who consume highly refined cereal products (milk, meat, liver) as well as in communities moving from rural to urban lifestyles (Bajaj and Singhal, 2021). B-complex vitamin fortification may be especially advantageous for these populations. Additionally, identifying target populations for fortification and tracking the efficacy of fortification would depend on the development of better methods for determining B-vitamin status (Zhu et al., 2021).

5.2 Minerals

In communities with restricted access to foods high in nutrients or dietary diversity, mineral fortification is an essential tactic to address micronutrient deficiencies (Manzoor et al., 2024). To enhance public health outcomes, essential minerals like iron, zinc, calcium, and iodine are frequently added to processed foods and staple foods (Jaiswal et al., 2022).

5.2.1 Iron

Iron deficiency is among the most prevalent micronutrient deficits worldwide which leads to anemia and stunts cognitive development. Staple foods like rice, wheat flour, and salt are frequently fortified with iron (Muthayya et al., 2012). Iron fumarate, iron sulfate, and elemental iron are a few examples of suggested iron fortifiers. The choice of iron fortifier is influenced by the food vehicle (Saffarionpour and Diosady, 2021). It is commonly recognized that non-heme iron forms are more bioavailable when vitamin C is present. Bovine hemoglobin concentrate, which has a very high bioavailability, is frequently used as a heme-iron fortifier in local feeding programs (Manzoor et al., 2024). Phytates, polyphenols, and calcium have been shown to adversely affect the bioavailability of non-heme iron fortifiers. Growing data suggests that sodium iron-EDTA could be a superior fortifier option in these situations in the future because iron from the EDTA complex is still available despite the presence of iron absorption inhibitors (Mattar et al. 2022).

5.2.2 Iodine

Iodates and iodides are the two chemical types of iodine now utilized for iodization (Jacobsen et al., 2022). The sodium and potassium salts of iodides and iodates can be used to iodize salt in accordance with the Codex Alimentarius standard for food-grade salt (Trumbo, 2016). Fortification has been used at concentrations between 30–200 parts per million. However, harmonizing these two forms of iodine in terms of equivalency is necessary to avoid trade barriers (Vatandoust and Diosady, 2022). The purity of the salt is a crucial factor to take into account while selecting the two compounds. Iodides break down more quickly in the presence of contaminants, although iodates stay stable in inferior salt. Bread, milk, wheat, sugar, and sauces are additional possible iodization carriers (Rigutto-Farebrother and Zimmermann, 2024).

Fortification of animal diets can also raise the amount of iodine in animal products. Salt can be iodized using suitable techniques in both large and small manufacturing facilities. Four primary technologies have been used to supplement salt with iodine. These include submersion, spray mixing, drip feed addition, and dry mixing (Ekott and Etukudo, 2025).

5.2.3 Zinc

Zinc is essential for numerous essential bodily processes, including growth, metabolism, immune defense and brain activity through its structural role and its function as a cofactor in various processes (Hall and King, 2022). Its deficiency affects many populations globally and may be the root cause of a significant amount of the global malnutrition issue (Singh et al., 2023). The focus of current zinc fortification initiatives, such as biofortification and zinc salt fortification, is on staple foods like wheat, rice, and their derivatives (Mazhar et al., 2022). However, zinc fortification presents a unique set of challenges. The zinc bioavailability of the complete diet should be taken into account while developing zinc fortification solutions because protein and phytate affect zinc absorption. The concentrations of zinc in plasma barely change when zinc is absorbed by diet (Tsang et al., 2021).

5.2.4 Calcium

Many communities have low dietary calcium, which increases the risk of various health problems and causes nutritional rickets and osteomalacia in both adults and children (Palacios et al., 2020). Calcium is a nutrient of concern thus it could be challenging to fulfill dietary requirements without dairy products. Offering culturally appropriate foods fortified with calcium may increase calcium intake when it is a cost-effective and practical solution (Manzoor et al., 2024). The amount added should be equal to the recommended level and bridge the gap between inadequate and adequate intake of calcium in order to maintain safe intake (Palacios et al., 2020). Additionally, it's critical to monitor the quality of fortified foods and use diet checks to determine how much calcium individuals are consuming. Calcium fortification of wheat flour is mandated only in the United Kingdom (Straeten et al., 2020). According to global data, this fortified staple component accounts for 13–14% of the calcium consumption by the British population. Water, rice, and maize flour are other products fortified with calcium (Adinepour et al., 2022).

5.3 Proteins and Amino Acids

The process of increasing the amount of protein in different food items to better satisfy the nutritional requirements of senior citizens is known as protein fortification. Sarcopenia, a disorder that causes a substantial decline in muscle mass, strength, and functional ability as people age, is particularly likely to affect this group. By promoting improved muscle maintenance and overall physical well-being, increasing the amount of protein in the diet can help this susceptible population age more healthily (Karin Wendin et al., 2017). The process of improving a diet by including essential amino acids, which are necessary for many body processes but cannot be produced by the body, is known as amino acid fortification. This strategy makes it possible to consume less protein overall while maintaining the body's amino All acid requirements are satisfied. Without depending on higher protein levels in their diet, people can maintain nutritional balance and promote optimal health by carefully incorporating these essential amino acids.

5.4 PUFA

The consumption of omega-6 fatty acids has increased over the past three decades, while the intake of saturated fats has decreased. This has resulted in a high ω-6/ω-3 ratio and a decreased intake of omega-3 fatty acids. Numerous lifestyle diseases have increased in tandem with this dietary shift. For human health, the ideal ω-6:ω-3 fatty acid ratio is believed to be between 1:1 and 5:1. The overly high ratios found in modern diets are caused by a decrease in ω-3 PUFA consumption and an increase in ω-6 PUFA consumption.

Therefore, the suggested strategy emphasizes increasing ω-3 intake instead of increasing ω-6 intake through fortification. The objective is to improve the overall fatty acid profile by increasing omega-3 consumption. A balanced omega-6/omega-3 ratio must be achieved through dietary choices, especially by increasing omega-3 intake through food fortification, in order to promote public health. Since the human body is unable to synthesize these fatty acids, they are vital as precursors to arachidonic acid (AA), omega-6 fatty acids, like linoleic acid (LA), are critical for inflammation management, blood cholesterol reduction, and brain development. Alpha-linolenic acid (ALA), eicosapentaenoic acid (EPA), and docosahexaenoic acid (DHA) are examples of omega-3 fatty acids, which are critical for neurological function, cardiovascular health, and lipid metabolism. They exhibit antidepressant, anti-inflammatory, and neuroprotective properties. Through its effects on body fat metabolism and general inflammation, a proper omega-6/omega-3 ratio ideally between 1:1 and 2:1 is essential for healthy development and the prevention of chronic diseases. However, ratios of 15–20:1 or even higher are commonly found in typical Western diets (Patel et al., 2022).

5.4.1 Methods for incorporating PUFAs

The omega-3 content of meat can be increased by providing animal feed that is supplemented with linseed oil, fish oil, or marine algae. It is also common to add encapsulated oils or walnuts to processed meats. The omega-3 content of eggs can be increased by feeding hens a diet containing flaxseed or algae oils. For dairy, linseed or marine sources can be added to ruminant diets to increase omega-3 content in milk. Also, some marketers have developed processes to directly enrich milk or yogurt with nutrients from fish or algae oils. Fortified yogurt quality can also be improved through microencapsulation. The long-chain PUFAs are also an important ingredient in infant formulas due to the importance of DHA (docosahexaenoic acid) and AA in infant development.

Other foods, such as breads, yogurts, yogurt-like drinks, margarines, spreads, juices, and even pasta, have been enhanced with omega-3 fatty acids using a variety of encapsulation processes. Omega-3 fatty acids in food items can be stabilized and shielded against oxidation by microencapsulation, which can also cover up off-flavors that could affect consumer acceptance.

Common encapsulation methods include emulsion-based approaches, freeze drying, and spray drying. (Alok Patel et al., 2022).

5.5 Phytochemicals and Functional Compounds

Phytochemical fortification refers to the process of taking naturally occurring plant compounds and adding them to foods to provide health benefits that go beyond basic nutrition. The main goal is to simply increase the antioxidants in food, to support overall health, and to reduce oxidative stress in the body. In addition to increasing antioxidants in food, which is a part of functional nutrition, phytochemical fortification may have no benefits to blood pressure normalizing, cholesterol levels, blood sugar regulation, and anti-inflammatory. Regarding fortification, there are potentially many sources of phytochemicals, such as whole plant powders, different herbs, spices, and fruit and vegetables extracts. These potential sources will also provide high concentrations of the beneficial compounds that are identified in these foods, two of the more recognized being flavonoids, carotenoids and polyphenols, that have each been related to provide various and different health benefits to humans. However, there are issues with successfully fortifying food items this way: the compounds may break down during food processing (like baking and cooking), and/or during storage (facing challenges of stability), and there may be questions about bioavailability. One approach to overcoming these challenges is through a technique described as encapsulation. Encapsulation protects highly sensitive compounds from degradation and increases their absorption in the digestive system. This innovative delivery tool can help food constituents that support human health to be used more efficiently.

Encapsulated phytochemicals and other beneficial compounds and fibers (e.g., dietary fiber, inulin, prebiotic oligosaccharides) could elevate the nutrition profile of foods. These additional components may positively affect biological processes in support of gut health through promoting good bacteria and aiding digestion. This serves to enhance the functional nutrition profile of these fortified foods. Essentially, the goal of food fortification with functional compounds will yield food items colloquially referred to as "designer foods". These foods are designed to provide an extra level of health benefits beyond the standard nutritional profile of foods to ensure that the consumer is benefiting from improved health without making major changes to their diet. The strategy of food fortification with functional compounds is also advantageous because it can be used to reach a great number of people, making a broad population healthy and providing nutrients and functional compounds that support health in a cost-effective manner. As research continues in this arena, it is likely that we will see more creative food fortification solutions that contribute to the public health environment (Hayta and Işcimen, 2019).

6. Methods and Technologies for Food Fortification

6.1 Biofortification

Biofortification is the enhancement of food crop nutritional value that is accomplished through post-harvest processes. Using plant breeding or agronomic techniques (mineral fertilizer), biofortification initiatives are primarily intended to boost iron, zinc, and provitamin A carotenoid in a variety of food crops; moreover, biofortified products have been created with protein and/or amino acids (Olson et al., 2021). Rice, beans, corn, and sweet potatoes can be biofortified with iron; wheat, rice, beans, sweet potatoes, and corn can be biofortified with zinc; and sweet potatoes, corn, and cassava can be biofortified with vitamin A (Brauw et al., 2019). Biofortification can improve the nutritional value of crops using agronomic biofortification procedures and traditional breeding methods. Nutrient-poor diets based solely on staple crops (cereals and tubers) are typically linked to a number of micronutrient deficits (Olson et al., 2021). Additionally, biofortification allows for the multi-use insertion of genetic markers that encourage current high-level micronutrient accumulation and higher levels of numerous micronutrients in a single food crop, so significantly addressing dietary inadequacies through genetic engineering (van der Straeten et al., 2020).

6.2 Home Fortification

Home fortification with multiple micronutrient powders (MNPs) refers to the addition of a powder containing vitamins and minerals to semi-solid food just prior to consumption. This process can occur at home or wherever food is being consumed. MNPs include zinc, iron, vitamin A, and other important minerals and vitamins. They increase the amounts of essential vitamins and minerals in the diets of infants and young children without changing their usual diet patterns. The goal of this intervention is to increase the consumption of micronutrients by children under two years of age (Suchdev et al., 2020).

6.3 Large Scale Food Fortification/Industrial Fortification

Large-scale food fortification (LSFF) means adding vitamins and minerals to commonly eaten staple foods during processing. Its goal is to improve the nutrition of people, particularly in LMIC countries, without asking them to change what they eat (Thakur et al., 2023). LSFF usually takes place through mandatory national programs that focus on foods like wheat flour, rice, maize flour, edible oils, and salt. It is seen as a cost-effective way to tackle micronutrient deficiencies, including iron deficiency anemia, iodine deficiency, and vitamin A deficiency (Olson et al., 2021). Fortification plays a powerful role in the fight against malnutrition. By adding essential nutrients to everyday foods, it helps strengthen our immune systems and protect us from illness and even death. It can prevent serious health issues caused by vitamin and mineral deficiencies—like rickets, birth defects, goitre, blindness, brain damage, and weakened immunity (Brauw et al., 2019). LSFF works alongside other nutrition strategies and stands out because it doesn't require people to change what they eat. Instead, it boosts the nutritional value of the staples and condiments that people already consume every day (Degerud et al., 2015).

7. Stability and Bioavailability of Fortified Nutrients

The ratio or percentage of the nutrient that is eventually accessible to be taken up or stored by the body after it has gone through the process of digestion, absorption, and distribution is termed bioavailability. In lay man's terms, it tells you the portion of a nutrient that is actually bioavailable to you.

Several related terms are arranged hierarchically under bioavailability, such as bioaccessibility, which refers to the fraction of a nutrient released from the food matrix (including interactions with other foods) and made available for absorption in the intestine. Digestion and binding of nutrients to other constituents will all impact this phase. Absorption refers to when the bioavailable nutrient enters into the blood through the intestinal wall. Both active transport and passive diffusion cause absorption.

Bioavailability comes under absorption and bioaccessibility. The quantity of accessible nutrients that the body transforms into their active form is referred to as bioconversion. Bioefficacy refers to the amount of nutrient that is bioavailable for biological effect and is therefore bioavailability plus bioconversion. Functional Bioefficacy represents the percentage of a consumed nutrient that has a measured metabolic function.

There are a number of factors that can have a significant effect on the bioavailability of nutrients including the food matrix that is nutrients may be bound with other food products and contained in cells from which they will need to be digested out. Nutrients released from a food matrix will depend on digestion of the food, which occurs in the mouth, stomach, and intestines. Some food products can enhance or inhibit the body's ability to absorb nutrients. For example, in the diet vitamin C, i.e., ascorbic acid, facilitates iron absorption, while both phytate and polyphenols inhibit the absorption of zinc and iron by the body.

The nutrients' own chemical and physical attributes like the nutrients form, i.e., soluble versus insoluble iron compounds, and/or particle size will have an influence on solubility and subsequent absorption. Nutrient interactions like the presence of nutrient A may inhibit/ or lead to absorption and/or utilization of nutrient B. The dual-fortified salt discussion covered the relationship of iron and iodine. The way that external nutrients are absorbed and used will depend on the host's physiological and nutritional status, including age, sex, inadequate nutrition, and illness. For instance, the body can control how much iron is absorbed by using its iron reserves.

A key issue for food fortification strategies is bioavailability. Effective strategies will involve ensuring the fortified nutrient is sufficiently bioavailable for the target population and selecting the right food vehicle (e.g., seasonings, condiments) and the correct form of micronutrient. Research and development, as discussed before, of fortifiers aims to optimize bioavailability while sustaining fortification food acceptability and stability. The evaluation of bioavailability can occur through a variety of methods, including studies of human interventions with measures of biological markers and functional outcomes, to *in vitro* digestion and absorption models (Degerud et al., 2015). Table 2 provides an overview of fortified food products available in the market. This information is collected from the official websites of various manufacturers. It shows the growing emphasis on food fortification to address micronutrient deficiencies and encourage better nutrition. It includes a variety of products, such as staples, dairy items, beverages, breakfast cereals, and snacks that are enriched with nutrients like iron, vitamin D, calcium, zinc, and folic acid. Overall, it illustrates how food companies are responding to consumer demand for healthier options and contributing to better dietary quality with nutrient-enriched products.

Table 2 List of commercially available fortified foods (from the manufacturer's official website).

Product	*Fortificant*	*Manufacturer*
Aashirvaad fortified chakki Atta	Iron, Vitamin B_{12}, and Folic acid	https://www.itcportal.com/brands-microsite
Annapurna Super Atta	Vitamin B_{12}, Iron, and folic acid	https://umaglobalfoods.com/
Chakki fresh atta	Vitamin B_{12}, Iron, and folic acid	https://www.gillcoagro.com/
Nestle a+ nourish	Vitamin A and D	https://www.nestle.in/
Muralya Dairy milk	Vitamin A and D	https://muralyadairy.com/
Amul Lite	Vitamin A and D	https://www.amul.com/
Silver 22K Basmathi rice	Iron, Vitamin A, Folic acid, Vitamin B_{12}, Vitamin B_1, B_2, B_3, Zinc	https://www.dcp-int.com/in/en
Corn flakes	Vitamin B_{12}, Iron, Zinc, Folic Acid	https://www.kelloggs.com
Biscuits	Iron, Vitamin A and D, Folic acid, Calcium	https://www.britannia.co.in/
Fortune edible oil	Vitamin A, D and E	https://www.adani.com
Patanjali sunflower oil	Vitamin A and D	https://www.patanjaliayurved.net/
Jagannath refined double fortified salt	Iron and Iodine	https://jubilantconsumer.com
TATA Salt Plus	Iron and Iodine	https://www.tataconsumer.com
Almond milk	Calcium, Vitamin B_{12}, Vitamin D	https://www.danone.com
Margarine	Vitamin D	https://www.upfield.com
Yogurt	Vitamin D, Calcium, Probiotics	https://www.danone.com
Energy bar	Iron, Vitamin B_{12}, Zinc, Folic acid	https://www.clifbar.com

8. Health Benefits and Impact on Public Health

Many studies demonstrate that food fortification can enhance micronutrient status and lower the prevalence of deficiency diseases. For example, fortification with multiple micronutrients has resulted in significant improvements in iron deficiency, vitamin A deficiency, and anemia. These decreases are often statistically and clinically significant. When compared to no intervention, home fortification with micronutrient powder decreased iron deficiency by 51% and anemia by 31% in newborns and young children (Olson et al., 2021).

There are also significant changes for both anemia and iron deficiency in young children after home fortification with micro nutrient powders. The reduction rate has been found to be similar to that of daily supplementation with iron. However, not every situation will benefit from food fortification equally. A number of variables, including the particular micronutrient, the food carrier, the degree of fortification, and the dietary habits or compliance of the population or group, will determine how successful it is.

There are some differences in quality of the evidence; some studies are poorly designed or have small sample sizes. In addition, the bioavailability of the fortified micronutrient is an important factor; absorption can be significantly influenced by components such as the food matrix and presence of other dietary factors. Evaluation of potential national programs, like in Costa Rica, demonstrate a consistent decrease in anemia after fortification strategies. Likewise, a study done in Indonesia shows that fortification with iron, vitamin A, and iodine improves several biochemical and anthropometric indicators. Each of these programmes showed success because people evaluated them and planned and executed the food fortification adequately (Olson et al., 2021).

9. Challenges and Controversies in Fortification

Food fortification has been beneficial, but there are a number of barriers that prevent its full effectiveness and widespread adoption. These barriers can be roughly classified into four categories—social/cultural, regulatory/policy, economic, and technical.

Technical barriers arise because the efficacy of food fortification depends on maintaining consistent micronutrient levels throughout processing and storage. Program effectiveness can also be compromised by variations in its nutrient levels introduced by environmental conditions or incorrectly managed processing. It is equally important to ensure that the added micronutrients are stable and bioavailable within the chosen food vehicle. In some cases, degradation or interaction with other food ingredients may lead to a reduction in the bioavailability of certain micronutrients. It is also critical that a food vehicle is selected that is appropriate for fortification, widely consumed, and well accepted by food vehicles.

The food vehicle needs to be palatable, be affordable, and be consumed in amounts by the target population that will provide them with adequate amounts of micronutrients.

Economic challenges are one of the major challenges relating to the high upfront costs of some fortification technologies and the infrastructure associated with food fortification, especially in LMICs since these types of investments can be expensive (as can the purchase, installation, and maintenance of capital equipment). It might also be important to make sure that consumers who need access to the fortification benefits, particularly very poor individuals, can afford products if their formulation is incrementally more expensive than the unfortified alternative. Furthermore, while the costs associated with monitoring and evaluation are often disregarded, these may be the most important components of a food fortification program "success".

Further, proper regulations and policies will have to be imposed or made clear prior to food fortification in order to establish safety and effectiveness for the use of fortified foods. Regulations will specify limits on maximum allowable levels to regulatory or government or "the consumer" (noting the definitions vary considerably), labeling requirements, and approved fortifiers. Adequate oversight and enforcement of regulations are equally important; since food manufacturers, if without enough monetary resources, will manufacture fortified foods which are unlikely to comply with fortification standards, resulting in low levels of fortification or the use of inappropriate fortifiers. For food fortification programs, ultimately adequate funds and political commitment are required.

The success of food fortification is dependent on consumers accepting fortified foods. Consumers could reject the use of a fortified food because they reject the change in taste, texture, or appearance of the fortified food. It is also especially important to address consumer concerns regarding the risk of potential adverse effects. To achieve social acceptability and compliance with fortification, the strategies must be tailored to local dietary habits and practices. Cultural customs and beliefs are critical in understanding, both the food that people eat and the acceptance of new foods. (Rebecca Olson et al., 2021)

10. Future Trends in Food Fortification

Future studies should focus on some key areas to maximize the effectiveness and sustainability of food fortification programs. It is important to develop new fortifiers and food vehicles to enable fortification strategies to maximize bioavailability, stability, and palatability of fortified foods. The consideration of new fortifiers such as apple pomace creates opportunities for enhancing the nutrient profile. Emerging nanotechnology offers additional opportunities to enhance delivery and bioavailability of micronutrients. Longer-term studies are required to better understand the health outcomes associated with fortification, including possible adverse effects. Further research is needed to investigate potential relationships between high folate levels and cancer risk. The evaluation of cost-effectiveness will be an important component in informing policy decisions and resource allocation.

A more robust monitoring and evaluation tool needs to be developed in order to support implementation of food fortification programs. This should include the analysis of the level of nutrients in the fortified food products, as well as information on food consumption trends and health outcomes for populations after consuming fortified foods. Stronger regulatory frameworks and enforcement would also help to ensure proper food fortification by producers and consistent levels of fortification. Multi-stakeholder partnerships, including collaboration with communities, businesses, government and researchers is essential for achieving the sustainability of food fortification. To create acceptance and truly promote widespread use of fortified foods, communication and consumer education are also important. Importantly, affordability and access to fortified foods for lower-income groups are critical to address equity and access; and in some cases, targeted approaches may be required in order to reach certain demographic groups that are known to have high prevalence of micronutrient deficiencies. Approaches to improve the micronutrient deficiencies in pregnant women and young children need to take into consideration the nutrition and eating habits of pregnant women and young children. Furthermore, studies are required to investigate the impact of fathers' diet and the possible advantages of food fortification on their children's health.

Exploring emerging technologies such as biofortification may introduce a long-term, sustainable method to increase the nutritional value of staple crops. Biofortification processes have their own unique challenges, but they hold promise in adding nutrient density at the source and reducing reliance on industrial fortification. To maximize impact, consideration of large-scale fortification, biofortification, food fortification to food should be evaluated (Venkatesh Manna et al., 2018).

11. Conclusion

With a focus on LMICs in particular, food fortification is a crucial public health intervention for addressing issues associated with micronutrient deficiencies and enhancing population health. While there has been significant progress, challenges remain in terms of improving fortification systems, enhancing implementation efforts, and ensuring equitable access. Partnerships, increased policy, and specific research are the avenues toward tackling these challenges, and there is work to be done. The highest priority research efforts must be devoted to developing inexpensive, culturally sensitive, and sustainable fortification approaches that respond to the diverse needs of populations and environments. If we are to systematically understand the strengths, weaknesses, and maximized effects of specific food fortification technologies, such as food-to-food fortification, biofortification, point-of-use fortification, and fortification at scale, we should conduct joint research on their unique adaptations.

A multi sectoral action will be essential for successful and sustainable food fortification initiatives in local and worldwide contexts, involving governments, industry, researchers, and communities together.

By overcoming existing challenges and implementing innovative solutions, food fortification can significantly improve nutrition and health outcomes globally.

References

Adinepour, F., S. Pouramin, A. Rashidinejad and S.M. Jafari. 2022. Fortification/enrichment of milk and dairy products by encapsulated bioactive ingredients. *Food Res. Int. 157*, 111212.

Allen, L. 2006. *Guidelines on Food Fortification with Micronutrients*. World Health Organization and Food and Agriculture Organization of the United Nations.

Bajaj, S.R. and R.S. Singhal. 2021. Fortification of wheat flour and oil with vitamins B12 and D3: Effect of processing and storage. *J. Food Compos. Anal.*, *96*, 103703.

de Brauw, A., M. Moursi and A.B. Munhaua. 2019. Vitamin A intakes remain higher among intervention participants 3 years after a biofortification intervention in Mozambique. *Br. J. Nutr.*, *122*(10), 1175–1181.

de Lourdes Samaniego-Vaesken, M., E. Alonso-Aperte and G. Varela-Moreiras. 2012. Vitamin food fortification today. *Food Nutr. Res.*, *56*(1), 5459.

Degerud, E.M., M.S. Manger, T.A. Strand and J. Dierkes. 2015. Bioavailability of iron, vitamin A, zinc and folic acid when added to condiments and seasonings. *Ann. N.Y. Acad. Sci.*, *1357*, 29–42.

Dwyer, J.T., K.L. Wiemer, O. Dary, C.L. Keen, J.C. King, K.B. Miller, M.A. Philbert, V. Tarasuk, C.L. Taylor, P.C. Gaine, A.B. Jarvis and R.L. Bailey. 2015. Fortification and health: Challenges and opportunities. *Adv. Nutr.*, *6*, 124–131.

Ekott, E. and U. Etukudo. n.d. Recent advances in stability of iodine in iodized salt. *Int. J. Sci. Res. Eng. Dev.*, *8*.

Fatemi, S.A., C.J. Williams, J. Deines and E.D. Peebles. 2023. Vitamin Compatibility with the Marek's Disease Vaccine. *Poultry*, *2*(4), 442–448.

Fletcher, R.J., I.P. Bell and J.P. Lambert. 2004. Public health aspects of food fortification: A question of balance. *Proc. Nutr. Soc.*, *63*, 605–614.

Food Safety and Standards Authority of India. 2018. *Food Safety and Standards* (*Fortification of Foods*) *Regulations*. India.

Global Nutrition Report. 2021. A world free from malnutrition: An assessment of progress towards the global nutrition targets (Chapter 1). Bristol, UK: Development Initiatives.

Grasso, A.C., J.J.F. Besselink, M. Tyszler and M.J. Bruins. 2023. The potential of food fortification as an enabler of more environmentally sustainable, nutritionally adequate diets. *Nutrients*, *15*(11), 2473.

Hall, A.G. and J.C. King. 2022. Zinc fortification: Current trends and strategies. *Nutrients*, *14*(19), 3895.

Hayta, M. and E.M. İşçimen. 2019. Effects of phytochemical fortification of flour and bread on human health. In: *Flour and Breads and Their Fortification in Health and Disease Prevention* (pp. 273–289). Elsevier, Amsterdam.

Jaiswal, D.K., R. Krishna, G.K. Chouhan, A.P. de Araujo Pereira, A.B. Ade, S. Prakash, S.K. Verma, R. Prasad, J. Yadav and J.P. Verma. 2022. Bio-fortification of minerals in crops: current scenario and future prospects for sustainable agriculture and human health. *Plant Growth Regul.*, *98*(1), 5–22.

Jakobsen, L.S., J.O. Nielsen, S.E. Paulsen, M. Outzen, A. Linneberg, L.T. Møllehave, T. Christensen and G. Ravn-Haren. 2022. Risk–benefit assessment of an increase in the iodine fortification level of foods in Denmark—A pilot study. *Foods*, *11*(9), 1281.

Keats, E.C., L.M. Neufeld, G.S. Garrett, M.N.N. Mbuya and Z.A. Bhutta. 2019. Improved micronutrient status and health outcomes in low- and middle-income countries following large-scale fortification: Evidence from a systematic review and meta-analysis. *Am. J. Clin. Nutr.*, *109*, 1696–1708.

Keats, E.C., K.D. Charbonneau, J.K. Das and Z.A. Bhutta. 2021. Large-scale food fortification has great potential to improve child health and nutrition. *Curr. Opin. Clin. Nutr. Metab. Care*, *24*, 271–275.

Kolanowski, W. and G. Laufenberg. 2006. Enrichment of food products with polyunsaturated fatty acids by fish oil addition. *Eur. Food Res. Technol.*, *222*(3–4), 472–477.

Kraemer, K. and K.G. van Zutphen. 2019. Translational and implementation research to bridge evidence and implementation. *Ann. Nutr. Metab.*, *75*, 144–148.

Kruger, J., J.R.N. Taylor, M.G. Ferruzzi and H. Debelo. 2020. What is food-to-food fortification? A working definition and framework for evaluation of efficiency and implementation of best practices. *Compr. Rev. Food Sci. Food Saf.*, *19*, 3618–3658.

Liyanage, C. and M. Hettiarachchi. 2011. Food fortification. *Ceylon Med. J.*, *56*(3), 124.

Madhari, R., Swetha, B., Palika, R., Jyrwa, Y., Boiroju, N., Pullakhandam, R., Mamidi, R., Arlappa, N., Kulkarni, B. and Longvah, T. (2020). High dietary micronutrient inadequacy in peri-urban school children from a district in South India: Potential for staple food fortification and nutrient supplementation. Maternal & Child Nutrition, 16. https://doi.org/10.1111/mcn.13065

Manzoor, M.F., A. Ali, H.B.U. Ain, S. Kausar, A.A. Khalil, R.M. Aadil and X.-A. Zeng. 2024. Bioaccessibility mechanisms, fortification strategies, processing impact on bioavailability, and therapeutic potentials of minerals in cereals. *Future Foods*, *10*, 100425.

Mannar, M.G.V. and N.A. Khan. 2015. Food fortification: Rationale and methods. In: *Encyclopedia of Food and Health* (pp. 27–34). Elsevier.

Mannar, M.G.V., G.S. Garrett and R.F. Hurrell. 2018. Future trends and strategies in food fortification. In: *Food Fortification in a Globalized World* (pp. 375–381). Elsevier, London.

Marks, K.J., Luthringer, C.L., Ruth, L.J., Rowe, L.A., Khan, N.A., De-Regil, L.M., López, X. and Pachón, H. (2018). Review of Grain Fortification Legislation, Standards, and Monitoring Documents. Retrieved www.ghspjournal.org

Mattar, G., A. Haddarah, J. Haddad, M. Pujola and F. Sepulcre. 2022. New approaches, bioavailability, and the use of chelates as a promising method for food fortification. *Food Chem.*, *373*, 131394.

Maurya, V.K., K. Bashir and M. Aggarwal. 2020. Vitamin D microencapsulation and fortification: trends and technologies. *J. Steroid Biochem. Mol. Biol.*, *196*, 105489.

Maurya, V.K., A. Shakya, D.J. McClements, R. Srinivasan, K. Bashir, T. Ramesh, J. Lee and E. Sathiyamoorthi. 2023a. Vitamin C fortification: Need and recent trends in encapsulation technologies. *Front. Nutr.*, *10*.

Maurya, V.K., A. Shakya, D.J. McClements, R. Srinivasan, K. Bashir, T. Ramesh, J. Lee and E. Sathiyamoorthi. 2023b. Vitamin C fortification: Need and recent trends in encapsulation technologies. *Front. Nutr.*, *10*.

Mazhar, Z., J. Akhtar, A. Alhodaib, T. Naz, M.I. Zafar, M.M. Iqbal, H. Fatima and I. Naz. 2023. Efficacy of ZnO nanoparticles in Zn fortification and partitioning of wheat and rice grains under salt stress. *Sci. Rep.*, *13*(1), 2022.

Mkambula, P., M.N.N. Mbuya, L.A. Rowe, M. Sablah, V.M. Friesen, M. Chadha, A.K. Osei, C. Ringholz, F.C. Vasta and J. Gorstein. 2020. The unfinished agenda for food fortification in low- and middle-income countries: Quantifying progress, gaps, and potential opportunities. *Nutrients*, *12*, 354.

Molino, S., Lerma-Aguilera, A., Jiménez-Hernández, N., Gosalbes, M.J., Rufián-Henares, J.Á. and Francino, M.P. (2021). Enrichment of Food With Tannin Extracts Promotes Healthy Changes in the Human Gut Microbiota. Frontiers in Microbiology, 12. https://doi.org/10.3389/fmicb.2021.625782

Muthayya, S., P. Thankachan, S. Hirve, V. Amalrajan, T. Thomas, H. Lubree, D. Agarwal, K. Srinivasan, R.F. Hurrell, C.S. Yajnik and A.V. Kurpad. 2012. Iron fortification of whole wheat flour reduces iron deficiency and anemia and increases body iron stores in Indian school-aged children. *J. Nutr.*, *142*(11), 1997–2003.

Olson, R., B. Gavin-Smith, C. Ferraboschi and K. Kraemer. 2021. Food fortification: the advantages, disadvantages and lessons from sight and life programs. *Nutrients*, *13*, 4118.

Orriss, G.D. (1998). Food fortification: Safety and legislation. Food and Nutrition Bulletin, 19(2), 109–116.

Osendarp, S.J.M., H. Martinez, G.S. Garrett, L.M. Neufeld, L.M. De-Regil, M. Vossenaar and I. Darnton-Hill. 2018. Large-scale food fortification and biofortification in low- and middle-income countries: A review of programs, trends, challenges and evidence gaps. *Food Nutr. Bull.*, *39*, 315–331.

Palacios, C., G. Cormick, G.J. Hofmeyr, M.N. Garcia-Casal, J.P. Peña-Rosas and A.P. Betrán. 2021. Calcium-fortified foods in public health programs: Considerations for implementation. *Ann. N.Y. Acad. Sci.*, *1485*(1), 3–21.

Pant, S. and D. Chinwan. n.d. Food fortification and enrichment. *Int. J. Basic Appl. Biol.*, *2*(3).

Patel, A., S.S. Desai, V.K. Mane, J. Enman, U. Rova, P. Christakopoulos, et al. 2022. Futuristic food fortification with a balanced ratio of dietary ω-3/ω-6 omega fatty acids for the prevention of lifestyle diseases. *Trends Food Sci. Technol.*, *120*, 140–153.

Poniedziałek, B., K. Perkowska and P. Rzymski. 2020. Food fortification: What's in it for the malnourished world? In: N. Benkeblia (Ed.), *Vitamins and Minerals Biofortification of Edible Plants* (pp. 27–44).

Rathee, S., Nayak, V., Singh, K.R. and Ojha, A. (2022). Nanofortification of vitamin B-complex in food matrix: Need, regulations, and prospects. In Food Chemistry: Molecular Sciences (Vol. 4). Elsevier B.V. https://doi.org/10.1016/j.fochms.2022.100100

Redeuil, K., A. Lévêques, J.-M. Oberson, S. Bénet, E. Tissot, K. Longet, A. de Castro, C. Romagny, L. Beauport, C.J. Fischer Fumeaux, J.-F. Tolsa, M. Affolter, E.C. Giménez, C.L. Garcia-Rodenas and S.K. Thakkar. 2021. Vitamins and carotenoids in human milk delivering preterm and term infants: Implications for preterm nutrient requirements and human milk fortification strategies. *Clin. Nutr.*, *40*(1), 222–228.

Rigutto-Farebrother, J. and M.B. Zimmermann. 2024. Salt reduction and iodine fortification policies are compatible: Perspectives for public health advocacy. *Nutrients*, *16*(15), 2517.

Saffarionpour, S. and L.L. Diosady. 2021. Multiple emulsions for enhanced delivery of vitamins and iron micronutrients and their application for food fortification. *Food Bioproc. Technol.*, *14*(4), 587–625.

Sandhu, R., N. Chaudhary, Bindia, R. Shams, K. Singh and V.K. Pandey. 2023. A critical review on integrating biofortification in crops for sustainable agricultural development and nutritional security. *J. Agric. Food Res.*, *14*, 100830.

Shen, L. and H.-F. Ji. 2012. Vitamin E: Supplement versus diet in neurodegenerative diseases. *Trends Mol. Med.*, *18*(8), 443–445.

Singh, A., Mishra, S., Choudhary, M., Chandra, P., Rai, A.K., Yadav, R.K. and Sharma, P. C. (2023). Rhizobacteria improve rice zinc nutrition in deficient soils. Rhizosphere, 25, Article 100646. https://doi.org/10.1016/j.rhisph.2022.100646

Singh, T., M. Kothari, S. Mishra, A. Veer Singh, A.K. Verma, D. Shankhdhar and S.C. Shankhdhar. 2025. Synergistic effect of zinc solubilizing bacteria and consortia on zinc marker enzymes and gaseous exchange parameters in rice (*Oryza sativa* L.) for zinc biofortification. *Plant Physiol. Biochem.*, *223*, 109807.

Sobczak, A., M. Danowska-Oziewicz, K. Ząbek, J. Miciński and A. Narwojsz. 2022. Effect of vitamin C fortification on the quality of cow's and goat's yoghurt. *Food Sci. Nutr.*, *10*(11), 3621–3626.

Souza, S.V.S., N. Borges and E.F. Vieira. 2022. Vitamin D-fortified bread: Systematic review of fortification approaches and clinical studies. *Food Chem.*, *372*, 131325.

Suchdev, P.S., M.E.D. Jefferds, E. Ota, K. da Silva Lopes and L.M. De-Regil. 2020. Home fortification of foods with multiple micronutrient powders for health and nutrition in children under two years of age. *Cochrane Database Syst. Rev.*, *2020*(2), CD008959.

Thakur, S., Singh, A., Insa, B. and Sharma, S. (2023). Food fortification in India as malnutrition concern: a global approach. In Sustainable Food Technology (Vol. 1, Issue 5, pp. 681–695). Royal Society of Chemistry. https://doi.org/10.1039/d3fb00079f

Trumbo, P.R. 2016. FDA regulations regarding iodine addition to foods and labeling of foods containing added iodine. *Am. J. Clin. Nutr.*, *104*, 864S–867S.

Tsang, B.L., E. Holsted, C.M. McDonald, K.H. Brown, R. Black, M.N.N. Mbuya, F. Grant, L.A. Rowe and M.S. Manger. 2021. Effects of foods fortified with zinc, alone or co-fortified with multiple micronutrients, on health and functional outcomes: A systematic review and meta-analysis. *Adv. Nutr.*, *12*(5), 1821–1837.

Van Stuijvenberg, M.E. 2005. Using the school feeding system as a vehicle for micronutrient fortification: Experience from South Africa. *Food Nutr. Bull.*, *26*(Suppl. 2), S213–S220.

van der Straeten, D., N.K. Bhullar, H. de Steur, W. Gruissem, D. MacKenzie, W. Pfeiffer, M. Qaim, I. Slamet-Loedin, S. Strobbe, J. Tohme, K.R. Trijatmiko, H. Vanderschuren, M. van Montagu, C. Zhang and H. Bouis. 2020. Multiplying the efficiency and impact of biofortification through metabolic engineering. *Nat. Commun.*, *11*(1), 5203.

Vatandoust, A. and L. Diosady. 2022. Iron compounds and their organoleptic properties in salt fortification with iron and iodine: An overview. *Curr. Opin. Food Sci.*, *43*, 232–236.

Vieira, E.F. and S. Souza. 2022. Formulation strategies for improving the stability and bioavailability of vitamin D-fortified beverages: A review. *Foods*, *11*(6), 847.

Wendin, K., Höglund, E., Andersson, M. and Rothenberg, E. (2017). Protein enriched foods and healthy ageing: Effects of protein fortification on muffin characteristics. Agro Food Industry Hi Tech, 5, 16–18.

Whitfield, K.C., T.J. Smith, F. Rohner, F.T. Wieringa and T.J. Green. 2021. Thiamine fortification strategies in low- and middle-income settings: A review. *Ann. N.Y. Acad. Sci.*, *1498*(1), 29–45.

WHO. 2023. *Vitamin A Fortification of Staple Foods*. World Health Organization.

WHO Library Cataloguing-in-Publication Data. 2013. World Health Organization copyright and licensing form.

Zhu, H., S. Mettu, F. Cavalieri and M. Ashokkumar. 2021. Ultrasonic microencapsulation of oil-soluble vitamins by hen egg white and green tea for fortification of food. *Food Chem.*, *353*, 129432.

CHAPTER 9

Functional Foods: Quality Control and Regulatory Considerations

Latika Yadav[1*]

1. Introduction

In recent years, the global food industry has witnessed a growing interest in functional foods—products that offer health benefits beyond basic nutrition. This trend is driven by increasing consumer awareness of the relationship between diet and health, coupled with a rising prevalence of lifestyle-related chronic diseases such as obesity, diabetes, and cardiovascular disorders. Functional foods, which include products enriched with bioactive compounds like vitamins, minerals, probiotics, and phytochemicals, have emerged as a promising solution to address these health challenges (Keservani et al., 2015; Rane et al., 2022).

Despite their potential, the development and commercialization of functional foods pose unique challenges, particularly in the areas of quality control and regulatory compliance. Ensuring the safety, efficacy, and consistency of functional foods is crucial not only for protecting consumer health but also for fostering trust in these products. Additionally, the regulatory frameworks governing functional foods vary significantly across regions, creating complexities for manufacturers seeking to market their products globally (Arora, et al., 2019; Ngan et al., 2023; Verma et al., 2023; Rane et al., 2023a; Rane et al., 2023b).

This chapter provides a comprehensive overview of the critical considerations for quality control and regulatory compliance in the functional food industry. It begins by defining functional foods and exploring their role in promoting health and preventing disease. The discussion then transitions to the scientific and technical aspects of quality control, including ingredient sourcing, analytical methods for detecting and quantifying bioactive compounds, and stability testing to ensure product integrity over time (Bendale et al., 2024). Figure 1 consists of functional foods quality control and regulatory considerations.

The regulatory landscape forms a significant part of this exploration. Functional foods occupy a gray area between conventional foods and pharmaceuticals, necessitating clear definitions, rigorous safety assessments, and substantiated health claims. This chapter highlights the differences and commonalities in these regulations, offering practical insights for addressing compliance challenges (Garud et al., 2023).

Moreover, the chapter examines the importance of transparency and traceability in the functional food supply chain. As consumers demand greater accountability, manufacturers must adopt robust

[1] Vijay Singh Pathik Government PG College, Kairana, Shamli, Uttar Pradesh-247774, India.
* Corresponding author: drlatika27@gmail.com

DOI: 10.1201/9781003659808-9

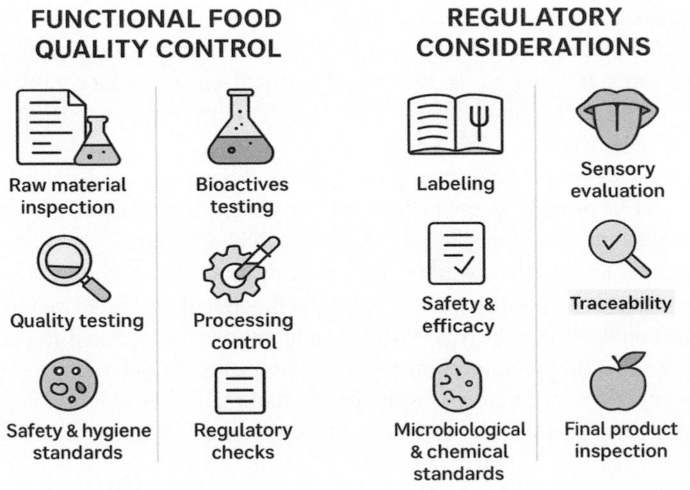

Fig. 1 Overview of Functional foods quality control and regulatory considerations.

quality assurance practices to meet expectations and adhere to regulatory standards. Emerging trends such as digital tools for supply chain monitoring and advances in analytical technology are also discussed as enablers of quality and regulatory compliance (Keservani et al., 2010; Keservani et al., 2017).

Finally, the chapter concludes with a forward-looking perspective on the future of functional foods. It emphasizes the need for an interdisciplinary approach that integrates scientific research, technological innovation, and regulatory foresight to drive the development of functional foods that are both effective and accessible. By addressing the challenges and opportunities in quality control and regulatory considerations, this chapter aims to serve as a valuable resource for researchers, industry professionals, and policymakers in the functional food sector.

2. Overview of Functional Foods

Functional foods represent a rapidly growing segment of the global food industry, offering a blend of nutrition and health benefits that appeal to increasingly health-conscious consumers. These foods are formulated to provide physiological advantages beyond their basic nutritional value, often targeting specific health conditions such as cardiovascular health, immune support, gut health, and cognitive function. Derived from both natural and fortified sources, functional foods incorporate bioactive compounds, probiotics, prebiotics, vitamins, minerals, and phytochemicals, contributing to their therapeutic potential (Walia et al., 2022; Keservani et al., 2010; Keservani et al., 2020).

The conceptual foundation of functional foods lies in their ability to bridge the gap between conventional foods and pharmaceuticals. Unlike dietary supplements, which are consumed in pill or capsule form, functional foods are typically integrated into regular diets, making them a more accessible and appealing option for promoting health and preventing disease (Ahire et al., 2024d).

Despite their popularity, the functional food industry faces significant challenges in ensuring product safety, efficacy, and consumer trust. Quality control measures play a crucial role in addressing these challenges, encompassing rigorous testing for bioactive compound levels, stability, and potential contaminants. These measures are critical not only for maintaining product integrity but also for complying with diverse regulatory requirements that vary across countries and regions (Tutelyan et al., 2019).

The regulatory environment for functional foods is inherently complex, reflecting differences in how nations define, classify, and evaluate these products. For example, the United States Food and Drug Administration (FDA) primarily monitors health claims on food labels, scientific evidence supporting those claims. In many cases, the lack of harmonized standards complicates international trade and creates hurdles for manufacturers attempting to expand globally (Singh et al., 2017a, b; Kesharwani et al., 2016; Suryawanshi et al., 2023).

This work provides an in-depth exploration of the functional food domain, with particular attention to the scientific principles, quality control practices, and regulatory frameworks that define this field. It highlights the challenges of validating health claims, ensuring safety and consistency, and meeting regulatory expectations while fostering innovation. By examining these critical aspects, this chapter aims to provide a holistic understanding of functional foods and their role in shaping the future of nutrition and health (Murthy, 1998). The integration of functional foods into mainstream diets offers immense potential for improving public health. However, achieving this potential requires collaboration among scientists, industry professionals, and regulatory authorities to establish robust systems that ensure product reliability and consumer confidence (Ahire et al., 2022).

3. Quality Control in Functional Foods

Quality food industry, ensuring the safety, efficacy, and consistency of products marketed for their health-promoting properties. Functional foods often contain bioactive compounds in concentrations that can influence physiological functions, making rigorous quality control essential to guarantee their benefits while minimizing risks to consumers. This section explores the critical elements of quality control in functional foods, covering raw material selection, ingredient authentication, manufacturing practices, analytical testing, and packaging considerations (Kim et al., 2008). Table 1 consists of the advances in quality control technologies for functional foods.

Table 1 Advances in Quality Control Technologies for Functional Foods.

Technology	*Description*	*Benefits for Functional Foods*
Artificial Intelligence (AI)	Use of machine learning algorithms, data analysis, and AI-powered sensors to monitor and optimize food quality.	• Real-time monitoring of key quality parameters • Predictive maintenance of production equipment • Personalized formulations based on consumer data and scientific research
Blockchain Technology	A decentralized and tamper-proof ledger that records the traceability of ingredients and products in the supply chain.	• Enhanced transparency and traceability • Verification of ingredient sourcing and bioactive content • Prevents fraud and mislabeling
Advanced Techniques	HPLC, mass spectrometry, and NMR for precise quantification of bioactive ingredients.	• Accurate quantification and validation of active compounds • Ensures product consistency and potency • Reduces reliance on manual testing methods
Automation and Robotics	Use of robotics and automated systems for sorting, packaging, and quality inspection during production.	• Increased efficiency and precision • Reduced human error • Streamlined production processes and batch consistency
Hyperspectral Imaging and Portable Sensors	Non-destructive imaging and portable sensors for on-site testing of functional foods for quality control.	• Rapid, on-site testing for ingredient verification • Minimizes downtime and lab costs • Facilitates real-time product quality adjustments

3.1 Raw Material Selection and Ingredient Authentication

The foundation of quality control begins with the careful selection and authentication of raw materials. Functional foods rely heavily on bioactive ingredients such as probiotics, prebiotics, vitamins, minerals, phytochemicals, and omega-3 fatty acids. Ensuring the purity, potency, and stability of these components is paramount. Implementing traceability systems to track the origin of raw materials helps mitigate risks associated with contamination and adulteration. Certification programs, such as organic or non-GMO (genetically-modified gene) labels, provide additional assurance to consumers (Ahire et al., 2023).

3.2 Good Manufacturing Practices

Adherence to good manufacturing practices ensures that functional foods are produced consistently and meet quality standards. Detailed protocols for handling, processing, and packaging to maintain product integrity. Strict measures to prevent microbial contamination during production.

Equipment Calibration: Regular maintenance and calibration of equipment to ensure accurate measurements and consistent manufacturing outcomes. Functional foods must deliver their promised health benefit, which requires precise quantification of bioactive compounds. Analytical testing is crucial for verifying the content, stability, and efficacy of these ingredients. Techniques such as HPLC (high performance liquid chromatography), mass spectrometry, and UV-visible spectroscopy are used to quantify vitamins, minerals, and phytochemicals. Probiotic products are tested for strain identity, viability, and colony-forming unit counts to ensure efficacy. Assessing the stability of bioactive compounds under various storage conditions, including temperature, humidity, and light exposure, helps determine shelf-life (Koirala et al., 2022).

3.3 Safety and Contaminant Testing

Ensuring the safety of functional foods involves comprehensive testing for potential contaminants that could compromise consumer health. Products are tested for pathogens such as *Salmonella, E. coli, and Listeria.* Analytical methods are used to detect pesticides, heavy metals (e.g., lead, mercury, cadmium), and mycotoxins. Products are screened for allergens, especially when cross-contamination during processing is possible (Keservani et al., 2014; Keservani et al., 2016a).

3.4 Packaging and Labeling Considerations

Packaging plays a critical role in preserving the quality and functionality of bioactive compounds. Proper labelling ensures transparency and compliance with regulatory standards. Packaging materials should prevent exposure to oxygen, light, and moisture, which can degrade sensitive bioactive compounds. Incorporating technologies such as oxygen scavengers or moisture absorbers can enhance product stability. Labels must accurately reflect the composition, health claims, and serving sizes, aligning with regulatory guidelines to avoid misleading consumers (Koirala et al., 2022).

3.5 Quality Assurance Programs

A robust quality assurance framework integrates quality control measures into a systematic process that encompasses all stages of functional food production. International standards like ISO 22000 ensure food safety management throughout the supply chain. Independent verification of product quality and compliance by accredited laboratories enhances consumer trust. Natural variability in raw materials can affect the concentration and efficacy of bioactive compounds. Developing cost-effective and rapid testing methods for complex bioactive matrices remains a challenge. Varying quality standards and testing protocols across regions can complicate global market access.

Advancements in technology and innovation are transforming quality control practices in functional foods. Emerging trends include enhancing supply chain transparency and traceability for both raw materials and finished products. By implementing robust quality control measures, the functional food industry can ensure the safety and efficacy of its products, maintain consumer trust, and comply with evolving regulatory requirements. A systematic approach to quality control not only safeguards public health but also drives innovation and sustainability in this growing sector (Hasler, 2002).

Quality control in functional foods is a multifaceted and evolving field, requiring the integration of scientific rigor, advanced technologies, and regulatory compliance. By adopting comprehensive quality control frameworks, manufacturers can ensure that functional foods are safe, effective, and aligned with consumer expectations. The ongoing advancements in quality control methodologies, coupled with global collaboration on regulatory harmonization, will play a crucial role in the future success of the functional food industry (Pol et al., 2024).

3.6 Key Quality Parameters

Functional foods are designed to offer health benefits beyond basic nutrition. To achieve this, they must meet specific quality parameters that ensure their safety, efficacy, and consumer acceptance. These parameters include nutritional content, bioavailability, stability, and sensory attributes. Each plays a vital role in the development and quality control of functional foods, ensuring they deliver on their intended purpose while maintaining appeal to consumers.

The nutritional content of functional foods is the foundation of their health-promoting claims. It includes the levels of macro- and micronutrients, as well as bioactive compounds that provide specific physiological benefits (Ahire, et al., 2020).

Nutritional analysis employs techniques such as proximate analysis, HPLC, and atomic absorption spectroscopy to quantify carbohydrates, proteins, fats, vitamins, and minerals. Bioactive ingredients are added to functional foods, such as omega-3 fatty acids, polyphenols, probiotics, and fiber, which need to be quantified to ensure compliance with label claims and regulatory standards. Fortification and enrichment involve ensuring the addition of nutrients like folic acid, vitamin D, or calcium in appropriate amounts without exceeding tolerable upper limits. Nutrient interactions involve evaluating the potential interactions between added nutrients and food matrices to prevent degradation or reduced efficacy. Products must adhere to regional and international standards for nutrient content claims, such as “high in vitamin C” or “source of fiber”. These claims require strict validation through analytical testing (Marone et al., 2016).

Bioavailability refers to the extent and rate at which the active ingredients in functional foods are absorbed and utilized by the body. It is a critical factor that determines the effectiveness of bioactive compounds. Food matrix refers to the interaction of bioactive compounds with other food components, which can enhance or inhibit their absorption. For instance, fats can improve the absorption of fat-soluble vitamins (A, D, E, and K). Techniques such as encapsulation, emulsification, and fermentation can improve bioavailability by protecting sensitive compounds and enhancing their delivery. Individual differences in digestion, metabolism, and gut microbiota also influence bioavailability (Kokande et al., 2024).

Microencapsulation and nanoencapsulation protect bioactive compounds from degradation and improve their absorption. Lipid-based carriers, such as micelles and liposomes, improve the solubility and uptake of hydrophobic compounds. Pre-treatment of certain compounds with enzymes (e.g., hydrolysis of protein into peptides) can increase their bioavailability. Simulated digestion models assess the release and solubility of bioactive compounds in gastrointestinal conditions. Human or animal studies measure the absorption and bioactivity of compounds (Arvanitoyannis et al., 2005).

Stability ensures that functional foods maintain their nutritional and functional properties over their shelf-life. It is a critical parameter for product safety, efficacy, and marketability. Ensuring that vitamins, minerals, and bioactive compounds do not degrade during storage or processing.

Environmental Factors: Exposure to light, oxygen, heat, and humidity can degrade bioactive compounds such as antioxidants and probiotics (Keservani et al., 2015). Interactions between ingredients can lead to instability or reduced efficacy. Poor packaging can compromise stability by allowing moisture or air ingress (Bharti et al., 2024; Gujarathi et al., 2024; Rai et al., 2024).

Simulating long-term storage conditions to predict product shelf-life. Sensory attributes determine the consumer acceptance of functional foods, making them a key parameter in quality control. These attributes include taste, texture, aroma, and appearance. Functional foods must balance health benefits with appealing sensory properties to encourage regular consumption. Products with poor sensory attributes risk rejection despite their functional benefits.

Bioactive Ingredients: Compounds like omega-3 fatty acids and certain probiotics can impart off-flavors or undesirable textures. Thermal and non-thermal processing can alter the sensory profile of functional foods. Trained panels evaluate specific sensory attributes to develop a detailed profile. Surveys and focus groups assess overall acceptability and preferences (Plaami et al., 2001). Devices like texture analyzers and electronic noses quantify sensory characteristics objectively. Key quality parameters—nutritional content, bioavailability, stability, and sensory attributes—are interdependent and collectively determine the overall quality and success of functional foods. Rigorous testing, innovative formulation strategies, and adherence to regulatory standards are essential to meet these quality benchmarks. By addressing these parameters holistically, the functional food industry can deliver products that not only meet consumer demands but also contribute to improved public health outcomes (Khairnar et al., 2024; Keservani, et al., 2016).

3.7 Analytical Techniques

The choice of analytical technique for assessing active ingredients in functional foods depends on the specific compound, food matrix, and desired outcomes. Chromatographic methods excel in separation and quantification, spectroscopic techniques offer rapid and non-destructive analysis, and microbiological assays are indispensable for evaluating probiotic functionality and safety. Emerging technologies like NMR (nuclear magnetic resonance) and biosensors promise to revolutionize the field by offering higher precision and real-time capabilities. A combination of these techniques ensures robust quality control, enabling the functional food industry to meet consumer expectations and regulatory standards effectively (Zawistowski, 2008).

Analytical techniques play a pivotal role in ensuring the safety, efficacy, and quality of functional foods. These methods are essential for identifying, quantifying, and validating the active ingredients that contribute to the health-promoting properties of functional foods. The use of advanced and precise analytical tools enables manufacturers and regulatory authorities to verify label claims, maintain consistency across batches, and comply with regulatory standards. This section provides a detailed overview of the key analytical techniques used to assess active ingredients in functional foods, including chromatographic methods, spectroscopic techniques, and microbiological assays (Pagar et al., 2024).

3.8 Chromatographic Techniques

Chromatography is a widely used technique for separating, identifying, and quantifying bioactive compounds in complex food matrices.

3.8.1 High-Performance Liquid Chromatography

HPLC is the process to separate components based on how they interact with a stationary phase and a mobile phase. Evaluation of polyphenols and vitamins (including vitamins C and E, for example). Analysis of a number of phytochemicals, including flavonoids and carotenoids, among others. Identification of contaminants, artificial sweeteners, and preservatives in the food. Precision and sensitivity of the highest. Applicability for a wide range of bioactive compounds available. Equipment that is expensive and staff that are knowledgeable are required. Putting together a sample can be a time-consuming process (Keservani et al., 2024).

3.8.2 Gas Chromatography

It can use the partitioning of compounds between a stationary phase and a gas-phase mobile phase to separate volatile and semi-volatile substances. Identification of flavorings, fatty acids, and essential oils. Omega-3 and omega-6 fatty acid analysis. The high repeatability and resolution for volatile substances are the advantages. It is restricted to substances that have been derivatized for analysis or are thermally stable (Kwak and Jukes, 2001).

3.8.3 Chromatography using Thin Layer

A solvent system is used as the mobile phase in thin-layer chromatography (TLC), which is used to separate compounds on a thin stationary phase (like silica gel). This is done by first screening for polyphenols, antioxidants, and alkaloids on the substance. The visual identification of compounds is possible through the use of specific staining agents. It does not cost much and is simple to operate. In addition, it is perfect for making short, preliminary assessments. Additionally, in comparison to GC (gas chromatography) and HPLC, it has a lower resolution and sensitivity than the former instrument (Pathan et al., 2024).

3.9 Methods of Spectroscopy

When it comes to identifying and quantifying active compounds, spectroscopy is a very important tool because it is based on how these substances interact with electromagnetic radiation. It measures the amount of visible and ultraviolet light that a substance is able to absorb. Measurement of phenolic compounds, antioxidants, and pigments (including carotenoids and chlorophyll, for example) is done. Quantification of the levels of nucleic acids and proteins in the sample is calculated. Effortless, speedy, and cost-effective is determined. An analysis of any element that does not cause any harm, and has restricted specificity; and is frequently needed for further separation procedures (Ahire et al., 2023).

3.9.1 Infrared Fourier-Transform Spectroscopy

The molecular fingerprint of the sample is determined by examining the amount of infrared light that is absorbed by the sample. The classification of functional groups in bioactive compounds is carried out. An analysis of the quantity of oxidation of lipids and carbohydrates present is done. Little to no preparation of the sample and non-destructive test is carried out. A restricted capacity to quantify without the need for additional calibration is determined (Randell and Race, 1996).

3.9.2 Mass Spectrometry

Through the process of mass spectrometry, the mass-to-charge ratio of ions is determined, which allows for the identification and measurement of compounds. An explanation of the structures of bioactive chemicals for clarity is provided. The concentration of trace bioactives, such as phytosterols and alkaloids, their precision and sensitivity of the highest level, and the ability to analyze complex mixtures for effectiveness is determined. The equipment is expensive, and it is necessary to have operators that are knowledgeable (Khairnar et al., 2024).

3.10 Microbiological Tests

Methods from the field of microbiology are required in order to conduct an analysis of the functional efficacy, pathogenic contamination, and viability of probiotics. A measurement of colony-forming units is taken after the cultivation of probiotics using selective media, using functional meals to determine whether or not probiotics are viable and taking measures to ensure that the probiotic content claims on labels are adhered to for measuring living microorganisms, it is trustworthy and widely used. This process is both time-consuming and prone to errors due to the cultural differences involved (Roberts and Rogerson, 2008).

4. Challenges in Quality Control

Quality control in functional foods is a complex process due to the dynamic nature of the ingredients, processing conditions, and regulatory requirements. Ensuring the consistent safety, efficacy, and quality of functional foods requires addressing several inherent challenges. Key issues include variability in raw materials, shelf-life considerations, and contamination risks, each of which poses unique obstacles to manufacturers and regulatory authorities. Raw materials, particularly those derived from natural sources, exhibit significant variability that can impact the quality and efficacy of functional foods. Bioactive compound concentrations can vary depending on the growing season, climate, and soil conditions. Different varieties of the same plant species may have varying levels of functional ingredients (Aher et al., 2024). Improper harvesting, storage, or preliminary processing can degrade the quality of raw materials. Variations in nutrient or bioactive content can lead to discrepancies between batches. Products may fail to deliver the promised health benefits due to variations in ingredient potency. Partnering with certified suppliers who follow good agricultural and collection practices (GACP). Performing rigorous quality checks using analytical techniques like HPLC or spectrophotometry. Mixing raw materials from different sources to minimize variability and standardize ingredient concentrations (Bahorun et al., 2019).

Shelf-life is critical in ensuring that functional foods retain their quality, safety, and bioactivity over time. Prolonged storage or adverse environmental conditions can lead to degradation. Sensitive ingredients like probiotics, omega-3 fatty acids, and vitamins can lose potency over time. Exposure to light, oxygen, humidity, and temperature accelerates spoilage and degradation. Poor barrier properties or compromised seals can allow moisture and air to enter, reducing stability. Degraded bioactives may render the product ineffective in delivering health benefits. Visible changes in texture, taste, or color can lead to product rejection. Products failing to meet label claims throughout their shelf-life can face regulatory action. Employing methods like encapsulation, freeze-drying, or natural antioxidants to enhance stability (Bidlack et al., 2000).

Pesticide residues, heavy metals, and mycotoxins can accumulate in raw materials or finished products. Foreign objects like glass, metal, or plastic fragments can inadvertently enter the food during processing. Contaminated products can lead to foodborne illnesses or long-term health consequences. Product recalls and reputational damage can result in significant financial losses for manufacturers. Failure to meet contamination limits can lead to regulatory fines or product bans. Identifying and monitoring critical points in the production process to prevent contamination is essential (Crawford, 2019).

Addressing the challenges of variability in raw materials, shelf-life considerations, and contamination risks is essential for maintaining the quality and safety of functional foods. Overcoming these challenges requires a proactive approach that combines advanced analytical techniques, robust quality control frameworks, and innovative technologies. By adopting these strategies, manufacturers can ensure that functional foods consistently meet consumer expectations and regulatory requirements, thus contributing to public health and market success (Kroes and Walker, 2004).

5. Regulatory Considerations

It is essential to have regulatory monitoring frameworks in place in order to guarantee that functional foods are not only safe but also effective and are presented to consumers in an appropriate manner.

These frameworks manage the processes of production, labelling, marketing, and quality control. They also guide the processes of quality control. Complex regulatory frameworks, which vary widely from region to region, must be navigated by manufacturing enterprises that manufacture functional foods. These frameworks are required to be satisfied. One of the key areas of attention is on the international regulatory framework, which includes the requirements for claims and labeling, as well as the obligations to conform to standards and recommendations (Gómez et al., 2023). Table 2 comprising the regulatory considerations for functional foods.

Table 2 Regulatory Considerations for Functional Foods.

Regulatory Consideration	*Description*	*Challenges and Opportunities*
Categorization of Functional Foods	The need for clear definitions of functional foods to differentiate them from supplements or pharmaceuticals.	• Clearer definitions will aid in regulation and labeling. • Uncertainty in classification could lead to inconsistent oversight and confusion.
Harmonization of International Regulations	Efforts to standardize regulations across countries to facilitate global trade and consumer safety.	• Variations in regulations can hinder trade. • Harmonization can simplify the global market but requires international cooperation and adaptation.
Health Claims and Labeling Guidelines	Development of guidelines for substantiating health claims and ensuring they are based on scientific evidence.	• Regulating health claims without stifling innovation. • Ensuring claims are substantiated can enhance consumer trust, but balancing proof and flexibility can be complex.
Novel Ingredient Safety Standards	Establishing protocols for the safety evaluation of new ingredients, such as nanotechnology, GMOs, and probiotics.	• Addressing new risks posed by innovative ingredients. • Updating safety standards regularly to ensure consumer health while encouraging innovation in the food sector.
Transparency and Traceability Requirements	Regulations to ensure the transparency of ingredient sourcing, production practices, and product traceability.	• Traceability can help build consumer trust and prevent fraud. • Implementing traceability systems can be costly for producers, but blockchain and AI can offer cost-effective solutions.

5.1 Global Regulatory Landscape

In order to shape the regulatory landscape for functional foods, it is recommended that the rules and regulations of certain countries or regions be utilized here. It is important to note that although there are some parallels, there are significant disparities in the manner in which healthy foods are defined, approved, and supervised in various regions. The Federal Food, Drug, and Cosmetic Act, which was amended by the Dietary Supplement Health and Education Act of 1994, provides the FDA (2016) of the United States with instructions on how to exercise control over items that are marketed as dietary supplements when they are manufactured. In accordance with the Food, Drug, and Cosmetic Act, a dietary supplement is defined as any product that is intended to improve the diet, with the exception of tobacco. Dietary supplements can consist of any of the following: vitamins, minerals, plants, amino acids, dietary components, or a combination of these. It is not possible to propose the use of dietary vitamins in any other manner simply because they can only be consumed in a single manner. Additionally, it is against the law to utilize medications that have been subjected to research or have been granted authorisation to be used in pharmaceuticals. Dietary supplements, on the other hand, are mostly monitored by post-market monitoring and do not require approval. This is in contrast to pharmaceuticals, which are regulated by the FDA (2019a) and require evidence of their safety and effectiveness before being approved. With a few notable exceptions, the New Dietary Ingredient Notification method's objective is to determine whether or not any dietary substances that were added after 1994 continue to be safe to consume. Dietary supplements can be sold in the United States even without first receiving pre-marketing approval. On the other hand, it is the responsibility of the producer to ensure that the product is safe for customers and to publish instructions on the label regarding how to use the product. In the United States of America, it is against the law to assert that dietary supplements may treat, cure, alleviate, or altogether prevent sickness (FDA, 2019b).

5.1.1 United States: Food and Drug Administration

Definition and Oversight

Functional foods are regulated as conventional foods, dietary supplements, or medical foods, depending on their intended use and claims.

The FDA does not specifically define "functional foods" but oversees their safety and labeling through the Food, Drug, and Cosmetic Act (FDCA).

All ingredients must be Generally Recognized as Safe (GRAS) or approved food additives. Manufacturers are responsible for substantiating health claims with scientific evidence. Dietary supplements require notification for new dietary ingredients but are not pre-approved by the FDA (FDA, 2019a; FDA, 2019b).

5.1.2 European Union: European Food Safety Authority (EFSA)

Definition and Oversight

Functional foods in the EU are regulated as food products with health claims rather than a separate category. EFSA evaluates the scientific validity of claims before approval.

Compliance with the EU Regulation No. 1924/2006 on nutrition and health claims is mandatory. Only pre-approved claims listed in the EU Register of Nutrition and Health Claims may be used on products (De Boer, 2021).

5.1.3 India: Food Safety and Standards Authority of India

Definition and Oversight

Functional foods are categorized under "Foods for Special Dietary Uses (FSDUs)" or "Nutraceuticals" in India.

FSSAI regulates their safety, quality, and claims under the Food Safety and Standards (Health Supplements, Nutraceuticals, Food for Special Dietary Uses, Food for Special Medical Purposes, Functional Foods, and Novel Food) Regulations, 2016. Pre-market approval is required for novel ingredients. Manufacturers must adhere to the labeling requirements, including a mandatory disclaimer stating that the product is not intended to diagnose, treat, cure, or prevent any disease (FDA Regulations, 2016).

5.2 Claims and Labeling Requirements

Claims that are accurate and labeling that is transparent are essential for gaining the trust of consumers and complying with regulations. Claim classification is determined by regulatory organizations according to the nature of the claim and the requirements for scientific substantiation. Regulatory action, product recalls, and a loss of consumer confidence are all potential outcomes that might result from failure to comply with these criteria. Therefore, it is necessary for manufacturers to remain current with the ever-changing requirements and to make certain that their claims and labelling methods are both true and fully transparent (Otles and Cagindi, 2012).

5.2.1 Types of Claims

5.2.1.1 Health claims

Describe the relationship between a food or its component and reduced risk of a disease or health-related condition. Example: "Adequate calcium intake may reduce the risk of osteoporosis." Require rigorous scientific evidence and pre-approval from regulatory authorities like EFSA or the FDA. The relationship between a food or its component and the reduced risk of a disease or health-related condition is typically expressed through health claims. These claims highlight the potential benefits of a food or its ingredient in decreasing the likelihood of developing a specific condition. To ensure credibility and consumer safety, such claims are subject to rigorous scientific scrutiny and regulatory approval. The claim must be supported by substantial scientific evidence, often derived from well-designed clinical trials, epidemiological studies, and meta-analyses. The evidence must demonstrate a clear cause-and-effect relationship between the food component and the claimed health outcome. Regulatory authorities such as the EFSA or the U.S. FDA evaluate the evidence to ensure it meets high scientific and methodological standards. Claims cannot be misleading or imply disease prevention without adequate proof. Manufacturers must submit a detailed dossier to regulatory bodies, including data on the active component, dosage, mechanism of action, and evidence from peer-reviewed studies. Once reviewed and approved, the claim is authorized for use under specific conditions and wording. Claims must be worded carefully to reflect the strength of the evidence and avoid misleading consumers (e.g., "may reduce the risk" instead of "prevents"). Misrepresentation can lead to regulatory actions, fines, or loss of consumer trust (Sezgin, 2020).

5.2.1.2 Structure-function claims

Describe how a nutrient or ingredient affects the normal structure or function of the body without referencing disease prevention. "Fiber supports healthy digestion". Generally require less stringent scientific substantiation than health claims. Structure-function claims describe the role of a nutrient or ingredient in maintaining or supporting the body's normal structure or functions without making explicit references to disease prevention or treatment. These claims are more focused on promoting general health and are often easier to substantiate compared to health claims. These claims emphasize the contribution of a nutrient or ingredient to a healthy body system or function. "Fiber supports healthy digestion" highlights the role of fiber in normal gastrointestinal processes without implying disease prevention or treatment. Structure-function claims must avoid suggesting that a product can diagnose, treat, cure, or prevent any disease. "Calcium helps build strong bones" is acceptable, while "Calcium prevents osteoporosis" is a health claim that requires more stringent substantiation. Scientific evidence supporting these claims typically does not need to meet the same rigorous standards as health claims. Substantiation may include published studies, historical use data, or evidence from recognized authoritative bodies. In some jurisdictions, such as the U.S., manufacturers must notify regulatory agencies (e.g., the FDA) of structure-function claims within a specified time frame after marketing a product. A disclaimer may also be required, stating that the claim has not been evaluated by the regulatory authority (Pascall et al., 2022).

5.2.1.3 Nutrient content claims

Highlight the presence or absence of specific nutrients in a product. Example: "Low in sodium", or "Rich in vitamin C". Must comply with specific thresholds defined by regulatory authorities. Nutrient content claims provide information about the presence, absence, or level of a specific nutrient in a food product. These claims help consumers make informed dietary choices by comparing products or identifying those that meet specific nutritional needs. Claims must use specific, approved phrases to prevent consumer confusion. For example, "light" can only refer to reduced fat or calories and must specify the context (e.g., "light in fat"). For consumers, nutrient content claims simplify product selection by clearly identifying foods that align with their dietary goals. For manufacturers, these claims enhance marketing opportunities, provided they adhere to strict regulatory standards to maintain trust and avoid penalties (Pascall et al., 2022; Sezgin, 2020).

5.2.2 Regulatory Considerations for Labeling

Functional foods must list all ingredients, including active compounds, along with their quantities. Claims must be supported by peer-reviewed studies and validated by relevant regulatory bodies. Misleading or exaggerated claims are prohibited. Labels must clearly distinguish between scientific facts and marketing language. Labeling functional foods requires strict adherence to regulatory guidelines to ensure consumer protection and maintain market integrity. Labels must include all ingredients, including active compounds or bioactive ingredients responsible for the functional benefits (e.g., omega-3 fatty acids, probiotics). The exact amounts of active compounds must be listed to provide transparency and assist consumers in understanding the product's efficacy and relevance to their dietary needs. Any health or functional claims made on the label must be backed by robust, peer-reviewed scientific studies demonstrating the claimed effects. Claims must be reviewed and approved by relevant regulatory bodies (e.g., EFSA, FDA) before use. EFSA requires a scientific opinion on the relationship between the food component and the claimed benefit. FDA mandates claims be either authorized health claims or qualified health claims, with the latter including disclaimers about the level of evidence. Labels must not overstate the benefits of a product or suggest it can prevent, treat, or cure diseases unless it meets drug-level regulatory standards. Failure to comply with these labeling standards can result in product recalls, fines, or legal action. Regulators routinely monitor for compliance and issue warnings or bans for non-conforming products (Koirala and Anal, 2021).

5.3 Standards and Guidelines

Adherence to international standards and guidelines is essential for ensuring the quality, safety, and consistency of functional foods. Adhering to international standards and guidelines is critical for maintaining the quality, safety, and consistency of functional foods. These frameworks ensure that functional foods meet consumer expectations and regulatory requirements while facilitating global trade. Functional foods must be produced in accordance with recognized quality standards to ensure product integrity and effectiveness. Enforced by bodies like the FDA and World Health Organization (WHO), GMP (good manufacturing practice) ensures consistent production, proper ingredient handling, and contamination prevention. For example, ISO 22000 provides a comprehensive framework for food safety management systems, applicable to functional foods (FDA, 2016; 2019a).

5.3.1 Codex Alimentarius

The Food and Agriculture Organisation (FAO) and the World Health Organisation (WHO) collaborated to produce the Codex Alimentarius, which is a set of standards, guidelines, and codes of practice that are recognized on a global scale. The purpose of this global reference is to ensure that food items, including functional foods, are safe, of high quality, and that they are traded in a fair manner. A collection of standards, recommendations, and codes of practice that have been produced by the FAO and the WHO and are recognized on a global scale. This document offers a structure for the evaluation of the safety of novel substances, as well as standards for labeling and claims for nutrients. In order to facilitate international trade and safeguard public health, a number of countries establish their own food legislation based on the requirements established by Codex (Siró, et al., 2008).

5.3.2 Good Manufacturing Practices

GMPs are a set of internationally recognized guidelines that ensure the consistent production and control of products, including functional foods, according to established quality standards. GMP is essential for safeguarding consumer health, maintaining product integrity, and complying with regulatory requirements. GMP ensures that products are consistently produced and controlled according to quality standards.

Products must be consistently manufactured to meet quality standards appropriate for their intended use. Processes must be validated to ensure uniformity and reliability across batches. Strict protocols for cleanliness in facilities, equipment, and personnel reduce the risk of contamination. Regular cleaning schedules, pest control, and waste management are mandatory (Karmacharya, 2014). Manufacturing areas must be designed to minimize errors and contamination risks, such as using proper ventilation, temperature control, and segregated spaces for different processes. Comprehensive records of every step in the production and quality control processes must be maintained. These records ensure traceability, facilitate audits, and help identify the root causes of issues. Employees must be adequately trained in GMP principles, hygiene, and the specific requirements of their roles. Ongoing education ensures that personnel remain up to date with best practices and regulatory updates. Equipment must be regularly inspected, maintained, and calibrated to ensure consistent performance. Faulty or outdated machinery must be repaired or replaced to prevent quality issues. Only approved and tested raw materials are used in production. Ingredients must be stored under appropriate conditions to maintain their quality and potency. Finished products undergo rigorous testing to ensure they meet predefined specifications, such as nutrient content and safety criteria. Deviations or non-conformities must be investigated and corrected promptly (Shukla, 2019).

5.3.3 Key Requirements

The documentation of procedures and controls in an appropriate manner. Maintaining a consistent schedule of equipment validation and calibration. Food safety standards are being taught to the staff. A decrease in the likelihood of contamination, adulteration, and discrepancies occurring. Improves both the confidence of consumers and compliance with regulations. A comprehensive management system for food safety that encompasses the full food supply chain for food. Pays attention to quality management methods in order to guarantee that the product quality is consistent. A dedication to quality and safety can be demonstrated through compliance with ISO standards, which improves the likelihood of receiving regulatory approval and gaining access to the market (Meléndez-Martínez et al., 2021).

6. Challenges and Opportunities

6.1 Current Challenges

6.1.1 Harmonization of regulations across regions

The global market for functional foods has seen significant growth, driven by increased consumer awareness of health and wellness. However, this expansion is hindered by the lack of harmonized regulations across different regions. Regulatory frameworks for functional foods vary widely in terms of definitions, permissible claims, testing requirements, and approval processes. For example, in the United States, the FDA regulates functional foods primarily as dietary supplements or conventional foods, depending on their intended use (Patel et al., 2008). In contrast, the European Union employs a more structured regulatory framework under the European Food Safety Authority, emphasizing stringent scientific substantiation of health claims. This disparity creates challenges for manufacturers and stakeholders seeking to market products internationally. Companies often face increased costs and complexities due to the need to meet diverse regulatory standards. Additionally, regulatory ambiguity in certain regions can lead to inconsistent enforcement, leaving room for non-compliance and potential public health risks. Addressing these discrepancies requires collaborative efforts between international regulatory bodies to establish unified standards and protocols (Binns and Howlett, 2009).

6.1.2 Balancing innovation with compliance

The functional food industry is at the forefront of scientific innovation, integrating advancements in nutrigenomics, bioactive compounds, and novel food technologies. While these developments hold immense potential to address global health challenges, they also present significant regulatory hurdles. The introduction of novel ingredients or processing methods often necessitates rigorous safety evaluations and compliance with established legal frameworks. For instance, the use of nanoparticles for enhanced bioavailability of nutrients or bioactive compounds is a promising area of research. However, the regulatory landscape for nanotechnology in foods remains fragmented and underdeveloped, leading to uncertainties for manufacturers (Ruckman, 2008). Similarly, the increasing consumer demand for clean-label and plant-based functional foods has prompted companies to explore alternative ingredients and sustainable production methods. These innovations, while appealing, must align with food safety and labeling requirements, which can stifle creativity and slow product development. Moreover, the dynamic nature of scientific research poses a challenge in aligning innovation timelines with regulatory approval processes. Scientific breakthroughs often outpace the capacity of regulatory bodies to evaluate and update existing guidelines. This gap can delay the introduction of potentially beneficial products to the market and discourage investment in research and development (Niva and Mäkelä, 2007).

6.1.3 Addressing Current Challenges

To overcome these challenges, stakeholders in the functional food industry must adopt a proactive and collaborative approach. Harmonization of regulations can be achieved through international dialogues facilitated by organizations such as the Codex Alimentarius Commission, which aims to develop global food standards. Additionally, fostering partnerships between academia, industry, and regulatory agencies can streamline the translation of scientific innovations into practical and compliant applications. Investing in robust research to substantiate health claims and ensuring transparency in product labeling can also enhance consumer trust and regulatory acceptance. Ultimately, balancing innovation with compliance will require adaptive regulatory frameworks that can accommodate emerging technologies while safeguarding public health and ensuring market fairness (Mellentin and Heasman, 2014).

6.2 Future Opportunities

The functional food industry is experiencing rapid growth, driven by consumer demand for products that not only nourish but also offer health benefits. As the scope of functional foods expands to include novel ingredients, such as probiotics, prebiotics, phytochemicals, and nutraceuticals, ensuring quality and safety has become a central focus for manufacturers, regulators, and consumers alike. Quality control (QC) technologies are evolving alongside innovations in food science, and regulatory frameworks are also adapting to accommodate these advancements. In this chapter, we explore the future opportunities in QC technologies and the potential for regulatory updates to support the growing functional food sector (Wells et al., 2017).

6.2.1 Advances in quality control technologies

6.2.1.1 Artificial Intelligence (AI) in quality control

AI is revolutionizing the food industry by offering advanced solutions for monitoring and improving product quality. In functional foods, AI can be utilized to optimize production processes, detect contaminants, and ensure consistency across batches. Machine learning algorithms can analyze large datasets from quality assurance tests and identify patterns or anomalies that human inspectors might overlook. These technologies are also capable of predictive maintenance for equipment, reducing downtime and enhancing operational efficiency (Anninou and Foxall, 2017).

AI-powered sensors and imaging systems can perform real-time monitoring of key quality parameters, such as texture, color, and chemical composition. These technologies offer a level of precision that traditional quality control methods cannot match. For example, in functional foods containing bioactive compounds, AI can assist in measuring the potency of active ingredients, ensuring they meet label claims without the need for extensive manual testing. Furthermore, AI can be used to improve product formulations by analyzing vast amounts of consumer preference data and scientific literature to suggest new ingredient combinations or processing techniques. This ability to personalize formulations based on real-time data will be critical as the demand for tailored functional foods continues to rise (Wells et al., 2017).

6.2.1.2 Blockchain technology for traceability

One of the greatest challenges in functional food production is ensuring transparency and traceability, especially as products often contain multiple ingredients with diverse sources. Blockchain technology, which provides a decentralized and tamper-proof ledger, is emerging as a powerful tool for traceability in the food supply chain. By securely recording every step in the production, processing, and distribution of functional foods, blockchain enables manufacturers, regulators, and consumers to track the origin and journey of ingredients (Anninou and Foxall, 2017). Blockchain's role in quality control is particularly vital when dealing with sensitive ingredients like probiotics or plant-based compounds, which are prone to degradation. The ability to trace these ingredients from farm to table ensures that products maintain their integrity and safety throughout their lifecycle. Additionally, blockchain could be pivotal in preventing fraud or mislabeling, which is an ongoing concern in the functional food sector. For example, products that claim to contain specific bioactive compounds can be verified through blockchain, ensuring that consumers get what they pay for. In the future, blockchain could also serve as a tool for certification and regulatory compliance. By integrating blockchain with regulatory databases, manufacturers can provide real-time access to verification documents, such as testing reports, certifications, and compliance records, making the quality assurance process more transparent and efficient (Blandon et al., 2008).

6.2.1.3 Advanced analytical technologies for functional ingredient verification

As functional foods become more complex, ensuring the authenticity and concentration of bioactive ingredients is a growing challenge. Advanced analytical technologies, such as high-performance liquid chromatography, mass spectrometry, and NMR spectroscopy, will play an increasing role in the QC of functional foods. These methods allow for the precise quantification of active compounds, confirming their presence and concentration in finished products (Betoret et al., 2011).

Additionally, emerging technologies such as hyperspectral imaging and portable sensor devices are making it possible to conduct rapid on-site testing of functional foods for quality control. These tools can be particularly valuable for verifying the potency and stability of bioactive ingredients in real-time, reducing the need for expensive and time-consuming laboratory tests (Vlaicu et al., 2023).

6.2.1.4 Automation and robotics in quality control

The integration of automation and robotics in food production is already underway, but as demand for functional foods increases, these technologies will become even more critical. Robotics can streamline packaging, sorting, and inspection processes, ensuring that only products meeting the highest standards reach the consumer. Furthermore, automated systems can be integrated with AI to provide continuous feedback during production, enabling the immediate adjustment of parameters to maintain product consistency. Automation will also enhance the reproducibility of testing procedures, reducing human error, and ensuring that each batch of functional food meets stringent quality standards (Sato et al., 2024).

6.2.2 Potential regulatory updates for functional foods

6.2.2.1 Adapting regulatory frameworks for novel functional foods

The regulatory landscape for functional foods is evolving, but existing frameworks in many countries may not fully accommodate the complexity and innovation in this field. Governments and regulatory bodies, such as the U.S. FDA, EFSA, and others, will need to revise and expand current regulations to address the novel nature of functional foods and the integration of new technologies. A key challenge is the categorization of functional foods. Many functional foods overlap with dietary supplements or pharmaceuticals, and the lack of clear definitions can lead to confusion regarding regulation. To support the growing industry, regulators will need to create distinct categories for novel functional foods that clearly define their purpose, claims, and safety standards (Astrini et al., 2020).

6.2.2.2 Harmonization of international regulations

With the global nature of the food industry, there is an increasing need for harmonized regulatory standards across borders. The global trade of functional foods is hindered by varying regulations between countries, particularly when it comes to claims about health benefits, labeling, and ingredient sourcing. Efforts toward international harmonization of regulations would enable functional food manufacturers to streamline their production and marketing processes while ensuring that consumers worldwide can trust the safety and efficacy of these products (Blandon et al., 2008).

Regulatory bodies may also need to collaborate more closely with stakeholders in the functional food industry to develop updated safety standards that reflect the latest scientific research. For example, as more evidence emerges about the health benefits of probiotics, prebiotics, or certain plant-based ingredients, regulatory agencies will need to assess new safety and efficacy thresholds based on the evolving body of evidence (Sato et al., 2024).

6.2.2.3 Clearer guidelines for health claims and marketing

Health claims associated with functional foods are an area of increasing scrutiny by regulatory authorities. In many countries, the use of terms like “immune-boosting”, “anti-inflammatory”, or “cholesterol-lowering” is strictly regulated to prevent misleading advertising. However, as the understanding of functional ingredients deepens and scientific research supports more health benefits, the regulatory framework must evolve to allow for new health claims that are substantiated by credible evidence. Future regulations may need to include specific guidelines for substantiating claims, including the use of clinical trials, meta-analyses, and other forms of scientific evidence. This would allow consumers to make more informed decisions while ensuring that functional food companies are held accountable for the claims they make (Blandon et al., 2008).

6.2.2.4 Safety standards for novel ingredients

As functional foods incorporate novel ingredients, there is a growing need for updated safety assessments. The use of GMOs, nanotechnology, and other innovative approaches in functional food formulations presents unique challenges for regulators. For example, nanomaterials used in food products may require special safety evaluations due to their potential to behave differently from their bulk counterparts. Regulatory agencies will need to establish clear protocols for evaluating the safety of new functional ingredients, considering factors like bioavailability, potential interactions with other ingredients, and long-term health effects. Regular updates to these safety standards will be essential to ensure that novel functional foods do not pose risks to consumers (Sato et al., 2024).

The future of functional foods lies in the continued advancement of QC technologies and the evolution of regulatory frameworks to accommodate novel ingredients and formulations. AI, blockchain, advanced analytical techniques, and automation hold significant potential for improving product quality, safety, and traceability. At the same time, regulatory agencies must adapt to keep pace with innovation, providing clear guidelines and safety standards for functional food products. By fostering collaboration between the food industry, regulatory bodies, and technology developers, we can ensure that functional foods fulfill their promise to promote health while safeguarding consumer well-being (Astrini et al., 2020).

7. Conclusion

The functional food industry is rapidly evolving due to increasing consumer demand for health-promoting products. The integration of technologies like AI and blockchain improves product consistency, safety, and transparency. However, regulatory frameworks must be flexible and comprehensive to ensure safety and efficacy. Accurate labeling, substantiated claims, and well-defined safety protocols are crucial for consumer trust and industry growth. Collaboration between the food industry, regulatory agencies, and technology innovators is essential for navigating these challenges. The functional food sector is poised to address modern health concerns, such as chronic disease prevention and gut health. By embracing technological advancements and adapting regulatory standards, the industry can support the safe development of functional foods.

References

Aher, P.R., Surana, K.R., Ahire, E.D., Patil, D.M. and Kshirsagar, S.J. 2024. Functional Foods for Autism. In: *Applications of Functional Foods in Disease Prevention* (pp. 33–50). Apple Academic Press.

Ahire, E.D., Keservani, R.K., Balekundri, A., Rajora, A.K. and Surana, K.R. 2024. Effect of nutrients on brain function. In: *Nutraceutical Fruits and Foods for Neurodegenerative Disorders* (pp. 237–248). Academic Press.

Ahire, E.D., Keservani, R.K., Surana, K.R., Singh, S. and Kesharwani, R.K. (Eds.). 2023. *Vitamins as Nutraceuticals: Recent Advances and Applications*. John Wiley and Sons.

Ahire, E.D., Sharma, N., Gupta, P.C., Khairnar, S., Surana, K., Ahire, B., Sonawane, V., Laddha, U., Sonkamble, S., Sabale, R. and Kshirsagar, S. 2022. Developing formulations of prebiotics and probiotics. In: *Prebiotics and Probiotics in Disease Regulation and Management* (pp. 271–290). Wiley Scrivener.

Ahire, E.D., Surana, K.R., Patil, C.D., Shah, H.S., Sonwane, G.B. and Talele, S.G. 2020. Role of omega-3 fatty acids in different neurodegenerative disorders. In: *Applied Pharmaceutical Science and Microbiology* (pp. 173–194). Apple Academic Press.

Anninou, I. and Foxall, G.R. 2017. Consumer decision-making for functional foods: Insights from a qualitative study. *Journal of Consumer Marketing*, *34*(7), 552–565.

Arora, M., Arora, M., Bansal, P. and Baldi, A. 2019. Need and recommendations for universal guidelines on regulatory status and quality control/safety of probiotic products. *Applied Clinical Research, Clinical Trials and Regulatory Affairs*, *6*(3), 231–249.

Arvanitoyannis, I.S. and Van Houwelingen-Koukaliaroglou, M. 2005. Functional foods: A survey of health claims, pros and cons, and current legislation. *Critical Reviews in Food Science and Nutrition*, *45*(5), 385–404.

Astrini, N.J., Rakhmawati, T., Sumaedi, S. and Bakti, I.G.M.Y. 2020. Identifying objective quality attributes of functional foods. *Quality Assurance and Safety of Crops & Foods*, *12*(2), 24–39.

Bahorun, T., Aruoma, O.I. and Neergheen-Bhujun, V.S. 2019. Phytomedicines, nutraceuticals, and functional foods regulatory framework: The African context. In: *Nutraceutical and Functional Food Regulations in the United States and Around the World* (pp. 509–521). Academic Press.

Bendale, A.R., Ahire, E.D. and Khairnar, S.J. 2024. Nutrigenomics Research-based New Approaches to Food Functions. In: *Nutrigenomics and Nutraceuticals* (pp. 89–106). Apple Academic Press.

Betoret, E., Betoret, N., Vidal, D. and Fito, P. 2011. Functional foods development: Trends and technologies. *Trends in Food Science & Technology*, *22*(9), 498–508.

Bharti, Sanjay K., Shukla, D., Mahapatra, D.K., Asati, V. and Keservani, Raj K. 2024. Therapeutic potential of nutraceuticals for the management of neurodegenerative disorders, In: Keservani, Raj K., Kesharwani, Rajesh K., Emerald, M., Sharma and Anil K. (Eds.), *Nutraceutical Fruits and Foods for Neurodegenerative Disorders* (Chap 26, pp. 525–535). Academic Press, Elsevier,. ISBN: 9780443189517.

Bidlack, W.R. and Wang, W. 2000. Designing functional foods to enhance health. In: *Phytochemicals as Bioactive Agents* (pp. 241–270). CRC Press.

Binns, N. and Howlett, J. 2009. *Functional Foods in Europe: International Developments in Science and Health Claims.* Summary report of an international symposium held 9–11 May 2007, Portomaso, Malta.

Blandon, J., Henson, S. and Cranfield, J. 2008. Functional foods and natural health products: A review of consumer acceptance and regulatory issues. *CABI Reviews*, *2008* , 1–11.

Crawford, W.M. 2019. The Food Safety Modernization Act (FSMA)'s role in the safety of functional foods and their ingredients. In: *Nutraceutical and Functional Food Regulations in the United States and Around the World* (pp. 61–74). Academic Press.

De Boer, A. 2021. Fifteen years of regulating nutrition and health claims in Europe: The past, the present, and the future. *Nutrients*, *13*(5), 1725.

FDA, 2019a. New Dietary Ingredients (NDI) Notification Process. 2017. https://www. fda.gov/food/dietary-supplements/new-dietary-ingredients-ndi-notification-process.

FDA, 2019b. Dietary Supplements. https://www.fda.gov/food/dietary-supplements.

FDA.https://www.fda.gov/food/dietary-supplements/new-dietary-ingredients-ndinotification-process.

FDA https://www.fda.gov/food/dietary-supplements.

Garud, S.R., Lamdande, A.G. and Gholap, S.R. 2023. Regulations on functional foods and nutraceuticals. In: *Industrial Application of Functional Foods, Ingredients, and Nutraceuticals* (pp. 785–823). Academic Press.

Gómez Gómez, C.V., Castillo Cortéz, I.G., Martínez Montenegro, I. and Ibañez San Martín, L. 2023. The regulatory status of functional foods in the economic integration organizations of Latin America and the Caribbean. *Archivos Latinoamericanos de Nutrición*, *73*(4), 297–312.

Gujarathi, Nayan A., Aher, Abhijeet A., Sukhia, A., Patil, Tulshidas S., Goyal, Yogeeta S., Keservani and Raj K. 2024. Nutraceutical interventions in Alzheimer's disease, In: Keservani, Raj K., Kesharwani, Rajesh K., Emerald, M., Sharma and Anil K. (Eds), *Nutraceutical Fruits and Foods for Neurodegenerative Disorders* (Chap 20, pp. 379–404). Academic Press, Elsevier.

Hasler, C.M. 2002. Functional foods: Benefits, concerns, and challenges: A position paper from the American Council on Science and Health. *The Journal of Nutrition*, *132*(12), 3772–3781.

Karmacharya, J.B. 2014. Good manufacturing practices (GMP) for medicinal products. Promising Pharmaceuticals, In: Purusotam Basnet (Ed.), *Promising Pharmaceuticals* (Chap 7, pp.101–148). Intechopen.

Keservani, R.K., Kesharwani, R.K., Vyas, N., Jain, S., Raghuvanshi, R. and Sharma, A.K. 2010. Nutraceutical and functional food as future food: A review. *Der Pharmacia Lettre*, *2*(1), 106–116.

Keservani, R.K., Sharma, A.K., Ahmad, F. and Baig, M.E. 2014. Nutraceutical and functional food regulations in India. In: *Nutraceutical and Functional Food Regulations in the United States and Around the World* (pp. 327–342). Academic Press. ISBN: 978-0-12-405870-5.

Keservani, Raj K., Kesharwani, Rajesh K., Sharma, Anil K. and Jarouliya, U. 2015. Dietary Supplements, Nutraceutical, and Functional Foods in Immune Response (Immunomodulators). In: Debasis Bagchi, Harry G. Preuss, and Anand Swaroop (Eds.), *Nutraceutical and Functional Foods in Human Life and Disease Prevention* (Chap 20, pp. 343–358). Boca Raton, Fl. CRC Press, Taylor and Francis, ISBN: 9781482237214.

Keservani, Raj K., Kesharwani, Rajesh K., Sharma, Anil K., Gautam, Surya Prakash., Verma and Santosh Kumar. 2016. Nutraceutical Formulations and Challenges. In: Debasis Bagchi and S. Nair (Eds.), *New Product Development in Nutraceuticals and Functional Foods: From Concept to Market* (Chap 9, pp. 161–177). Academic Press, Elsevier.

Keservani, Raj K., Kesharwani, Rajesh K., Sharma and Anil K. 2016a. Naturally Occurring Toxicants as Etiologic Agents of Food-borne Disease. In: Debasis Bagchi and Swaroop, Anand (Eds.), *Food Toxicology* (Chap 12, pp. 245–262). CRC Press, Taylor and Francis. ISBN: 9781498708746.

Keservani, R.K., Sharma, A.K. and Kesharwani, R.K., 2017. An overview and therapeutic applications of nutraceutical and functional foods. In: *Recent Advances in Drug Delivery Technology* (pp. 160–201). Bentham Science Publishers. ISBN:9781522507543.

Keservani, R.K., Sharma, A.K. and Kesharwani, R.K. (Eds.). 2020. *Nutraceuticals and Dietary Supplements: Applications in Health Improvement and Disease Management* (pp. 1–344). Apple Academic Press, CRC Press. ISBN: 9781771888738.

Keservani, R.K., Tung, B.T., Kesharwani, R.K. and Ahire, E.D. (Eds.). 2024. *Plant Metabolites and Vegetables as Nutraceuticals*. CRC Press.

Khairnar, S.J., Rudrapal, M., Ahire, E.D., Jagtap, M.R. and Kshirsagar, S.J. 2024. Overview of Functional Foods. In: *Applications of Functional Foods in Disease Prevention* (pp. 1–31). Apple Academic Press.

Kim, J.Y., Kim, D.B. and Lee, H.J. 2008. Regulations on health/functional foods in Korea. In: *Nutraceutical and Functional Food Regulations in the United States and Around the World* (pp. 281–290). Academic Press.

Koirala, S. and Anal, A.K. 2021. Probiotics-based foods and beverages as future foods and their overall safety and regulatory claims. *Future Foods*, *3*, 100013.

Koirala, S., Thapa, S.P. and Anal, A.K. 2022. Overall safety considerations and regulatory oversight for probiotics-based foods and beverages. In: *Probiotics, Prebiotics, and Synbiotics: Technological Advancements towards Safety and Industrial Applications* (pp. 441–461). Newark: John Wiley and Sons.

Kokande, A.M., Surana, K.R., Mahajan, S.K., Ahire, E.D. and Patil, S.J. 2024. Functional Foods for Microbial Infections. In: *Applications of Functional Foods in Disease Prevention* (pp. 165–189). Apple Academic Press.

Kroes, R. and Walker, R. 2004. Safety issues of botanicals and botanical preparations in functional foods. *Toxicology*, *198*(1–3), 213–220.

Kwak, N.S. and Jukes, D.J. 2001. Functional foods. Part 2: The impact on current regulatory terminology. *Food Control*, *12*(2), 109–117.

Marone, P.A., Birkenbach, V.L. and Hayes, A.W. 2016. Newer approaches to identify potential untoward effects in functional foods. *International Journal of Toxicology*, *35*(2), 186–207.

Meléndez-Martínez, A.J., Böhm, V., Borge, G.I.A., Cano, M.P., Fikselová, M., Gruskiene, R., Lavelli, V., Loizzo, M.R., Mandić, A.I., Brahm, P.M. and Mišan, A.Č. 2021. Carotenoids: Considerations for their use in functional foods, nutraceuticals, nutricosmetics, supplements, botanicals, and novel foods in the context of sustainability, circular economy, and climate change. *Annual Review of Food Science and Technology*, *12*(1), 433–460.

Mellentin, J. and Heasman, M. 2014. *The Functional Foods Revolution: Healthy People, Healthy Profits.* Routledge.

Murthy, M. 1998. Nutraceuticals, functional foods, and medical foods: Commentary and caveats. *Journal of Nutraceuticals, Functional & Medical Foods*, *1*(3), 73–99.

Ngan, D.K., Tung, B.T., Keservani, R.K., Rane and Bhushan R. 2023. Breast Cancer Prevention by Nutraceuticals, In: Raj K. Keservani, Bui Thanh Tung, Sippy Singh and Rajesh K. Kesharwani (Eds.), *Nutraceuticals in Cancer Prevention, Management, and Treatment* (Chap 6, pp. 113–132). Apple Academic Press, CRC Press, Taylor & Francis. ISBN: 9781774913277.

Niva, M. and Mäkelä, J. 2007. Finns and functional foods: Socio-demographics, health efforts, notions of technology, and the acceptability of health-promoting foods. *International Journal of Consumer Studies*, *31*(1), 34–45.

Otles, S. and Cagindi, O. 2012. Safety considerations of nutraceuticals and functional foods. In: *Novel Technologies in Food Science: Their Impact on Products, Consumer Trends, and the Environment* (pp. 121–136). Springer Nature Link.

Pagar, K., Gunjal, S., Surana, K.R., Lokhande, T., Mahajan, S. and Ahire, E.D. 2024. Functional Foods for Viral Infections. In: *Applications of Functional Foods in Disease Prevention* (pp. 241–260). Apple Academic Press.

Pascall, M.A., DeAngelo, K., Richards, J. and Arensberg, M.B. 2022. Role and importance of functional food packaging in specialized products for vulnerable populations: Implications for innovation and policy development for sustainability. *Foods*, *11*(19), 3043.

Patel, D., Dufour, Y. and Domigan, N. 2008. Functional food and nutraceutical registration processes in Japan and China: A diffusion of innovation perspective. *Journal of Pharmacy & Pharmaceutical Sciences*, *11*(4), 1–11.

Pathan, A.S., Bendale, K.A., Pandagale, P.M., Patil, C.D. and Ahire, E.D. 2024. Role of Functional Foods in Psychotic Disorders. In: *Applications of Functional Foods in Disease Prevention* (pp. 223–240). Apple Academic Press.

Plaami, S.P., Dekker, M., van Dokkum, W. and Ockhuizen, T. 2001. *Functional Foods: Position and Future Perspectives* (No. 2000/15). Nationale Raad voor Landbouwkundig Onderzoek.

Pol, M.P., Surana, K.R., Patil, D.M., Ahire, E.D. and Kshirsagar, S.J. 2024. Approaches for Treating Ulcer with Functional Foods. In: *Applications of Functional Foods in Disease Prevention* (pp. 191–210). Apple Academic Press.

Rai, S., Keservani, Raj K., Kumar, P., Nikam, V.K., Kachave, R.N., Kumar, Y., Kesharwani and Rajesh K. 2024. Importance of functional foods in the management of autism, In: Keservani, Raj K., Kesharwani, Rajesh K., Emerald, M., Sharma and Anil K. (Eds.), *Nutraceutical Fruits and Foods for Neurodegenerative Disorders* (Chap 8, pp. 151–171). Academic Press, Elsevier. ISBN: 9780443189517.

Randell, A. and Race, J. 1996. Regulatory and legal aspects of functional foods: An international perspective. *Nutrition Reviews*, *54*(11), S152.

Rane, B.R., Bharat, M. and Keservani, R.K. 2023a. COVID-19: Infection and Recent Investigations and Future Prospects. In: Rajesh K. Kesharwani and Raj K. Keservani (Eds.), *COVID-19 and Immunomodulation With Special Emphasis on Nutraceutical and Herbal Formulation* (Chap 9, pp. 223–261). Apple Academic Press, CRC Press, Taylor & Francis. ISBN: 9781774913178.

Rane, B.R., Patil, R.R., Jain, A.S., Keservani, R.K. and Kesharwani, R.K. 2022. Nutraceuticals as Disease Preventive Food and Immunity Boosters. In: Kesharwani, R.K., Keservani, R.K. and Sharma, A.K. (Eds.), *Nutraceuticals and Functional Foods in Immunomodulators*. Springer, Singapore.

Rane, B.R., Telange, T.S., Waghchaure, P.S., Keservani, Raj K., Jain and Ashish S. 2023b. COVID-19 Pandemic: Role of Ayurveda for the Prevention and Treatment, In: Rajesh K. Kesharwani and Raj K. Keservani (Eds.), *COVID-19 and Immunomodulation With Special Emphasis on Nutraceutical and Herbal Formulation* (Chap 4, pp. 107–138). Apple Academic Press, CRC Press, Taylor & Francis. ISBN: 9781774913178.

Roberts, A. and Rogerson, R. 2008. Chinese approach on regulating food additives, novel foods, functional foods, and dietary supplements. In: *Nutraceutical and Functional Food Regulations in the United States and Around the World* (pp. 291–303). Academic Press.

Ruckman, S.A. 2008. Regulations for nutraceuticals and functional foods in Europe and the United Kingdom. In: *Nutraceutical and Functional Food Regulations in the United States and Around the World* (pp. 221–238). Academic Press.

Sato, K., Kodama, K. and Sengoku, S. 2024. Driving Innovation through Regulatory Design and Corporate Behaviour: A Case Study of Functional Food Industry in Japan. *Foods*, *13*(20), 3302.

Shukla, J. 2019. Good manufacturing practice (GMP): An overview. Research Gate.

Sezgin, D., 2020. Developments and regulations about functional foods in Turkey: A literature review. *Akademik Gıda*, *18*(1), 79–86.

Singh, P., Kesharwani, R.K. and Keservani, R.K. 2017a. Antioxidants and Vitamins: Roles in Cellular Function and Metabolism. In: *Sustained Energy for Enhanced Human Functions and Activity* (pp. 385–407). Academic Press. ISBN:9780128054130.

Singh, P., Kesharwani, R.K. and Keservani, R.K. 2017b. Protein, carbohydrates, and fats: Energy metabolism. In: *Sustained Energy for Enhanced Human Functions and Activity* (pp. 103–115). Academic Press. ISBN:9780128054130.

Siró, I., Kápolna, E., Kápolna, B. and Lugasi, A. 2008. Functional food. Product development, marketing, and consumer acceptance: A review. *Appetite*, *51*(3), 456–467.

Suryawanshi, Hemant P., Rane, Bhushan R. and Keservani, Raj K. 2023. Beneficial Effect of Dietary Fibers in Metabolic Induced Obesity, In: Keservani, Raj K., Yadav, Durgavati, Kesharwani, Rajesh K., Singh, Sippy, and Sandeep, Kumar (Eds.), *The Metabolic Syndrome: Dietary Supplements and Food Ingredients* (Chap 12, pp. 251–279). Apple Academic Press, CRC Press, Taylor & Francis Group. ISBN: 978-1-77491-111-2.

Tutelyan, V.A., Sukhanov, B.P., Kochetkova, A.A., Sheveleva, S.A. and Smirnova, E.A. 2019. Russian regulations on nutraceuticals, functional foods, and foods for special dietary uses. In: *Nutraceutical and Functional Food Regulations in the United States and Around the World* (pp. 399–416). Academic Press.

Verma, M., Ahire, ED. and Keservani, R.K. 2023. Colon Cancer Prevention by Medicinal Plants, In: Raj K. Keservani, Bui Thanh Tung, Sippy Singh, Rajesh K. Kesharwani (Eds.), *Nutraceuticals in Cancer Prevention, Management, and Treatment* (Chap 9, pp. 163–189). Apple Academic Press, CRC Press, Taylor & Francis. ISBN: 9781774913277.

Vlaicu, P.A., Untea, A.E., Varzaru, I., Saracila, M. and Oancea, A.G. 2023. Designing nutrition for health: Incorporating dietary by-products into poultry feeds to create functional foods with insights into health benefits, risks, bioactive compounds, food component functionality, and safety regulations. *Foods*, *12*(21), 4001.

Walia, A., Mehra, R., Kumar, N., Singh, T.P. and Kumar, H. 2022. Good manufacturing practices and safety issues in functional foods and nutraceuticals. In: *Bioactive Components: A Sustainable System for Good Health and Well-Being* (pp. 167–182). Singapore: Springer Nature Singapore.

Wells, M.L., Potin, P., Craigie, J.S., Raven, J.A., Merchant, S.S., Helliwell, K.E., Smith, A.G., Camire, M.E. and Brawley, S.H. 2017. Algae as nutritional and functional food sources: Revisiting our understanding. *Journal of Applied Phycology*, *29*, 949–982.

Zawistowski, J. 2008. Regulation of functional foods in selected Asian countries in the Pacific rim. In: *Nutraceutical and Functional Food Regulations in the United States and Around the World* (pp. 365–401). Academic Press.

Chapter 10

Consumer Awareness and Acceptance of Functional Foods

Yuna Seo[1*]

1. Introduction

Functional foods are increasingly discussed in the context of public health, not merely as nutrient-enhanced products but as dietary tools positioned between everyday foods and medical interventions. These foods are specifically designed to promote wellness and prevent diseases by incorporating bioactive compounds, vitamins, probiotics, and other beneficial substances (Vignesh et al., 2024). The rising prevalence of chronic diseases such as diabetes, cardiovascular conditions, and obesity, coupled with the aging global population, underscores the growing need for innovative solutions in public health. In this context, functional foods hold significant promise, offering both preventive and complementary measures to traditional healthcare. These foods can improve overall well-being by enhancing immune function, supporting digestive health, and even promoting mental health, thus playing a multifaceted role in health promotion (López-Varela et al., 2002). To illustrate these relationships, Figure 1 presents a conceptual model linking functional foods, health pathways, and public health outcomes.

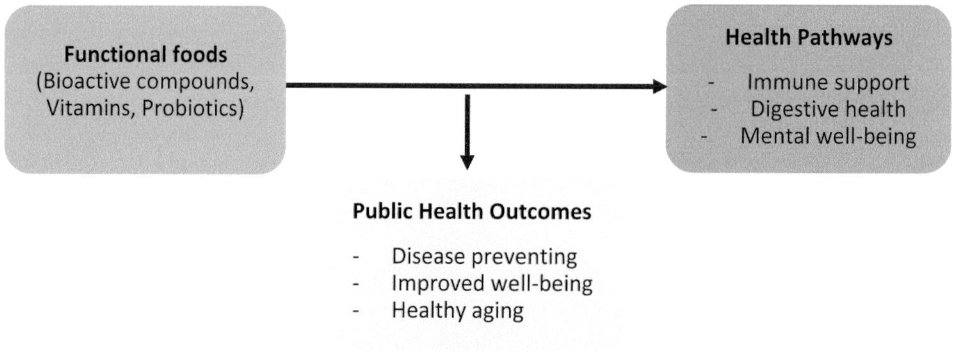

Fig. 1 Conceptual model of functional foods and health outcomes.

[1] Department of Industrial and Systems Engineering, Tokyo University of Science, 2651, Yamazaki, Noda, Chiba, JPN (278-8510).

* Corresponding author: yuna.s@rs.tus.ac.jp

DOI: 10.1201/9781003659808-10

The surge in health-consciousness worldwide has fueled increased interest in functional foods, as consumers seek products that align with their dietary preferences and health goals (Baker et al., 2022). The COVID-19 pandemic further amplified this trend, as individuals turned to health-enhancing products to boost immunity and support healthier lifestyles (Das et al., 2022). Despite the promising potential of functional foods, their development and promotion face several challenges. These include ensuring scientific validation of health claims, addressing consumer skepticism, navigating complex regulatory frameworks, and meeting consumer demand for natural and sustainable products (Madududu et al., 2024; Ketkaew et al., 2024). A summary of key examples of functional foods, their active components, and corresponding health benefits is provided in Table 1.

Table 1 Examples of functional foods and their health benefits.

Food Type	*Active Component*	*Health Benefit*	*Reference*
Iron-fortified cereals	Iron	Helps prevent anemia	WHO, 2020
Vitamin D-enriched milk	Vitamin D	Supports bone health, reduces risk of osteoporosis	Holick, 2007
Iodized salt	Iodine	Prevents iodine deficiency, supports thyroid health	Zimmermann, 2009
Probiotic yogurt	Probiotics (Lactobacillus, Bifidobacterium)	Improves gut microbiota, supports digestive health	Marco et al., 2017
Omega-3 enriched eggs	Omega-3 fatty acids	Supports cardiovascular health, reduces inflammation	Swanson et al., 2012
Oatmeal / Oat-bran products	β-glucan (≥3 g/day)	Helps lower LDL cholesterol and total cholesterol, reducing cardiovascular disease risk	Whitehead et al., 2014
Barley-based foods	β-glucan	Contributes to lowering LDL-C and non-HDL-C, improving lipid profile for cardiovascular health	Ho et al., 2016
Plant-sterol/stanol-fortified spreads or drinks	Phytosterols (plant sterols/stanols)	Reduces LDL cholesterol in a dose-dependent manner, supporting cardiovascular health	Demonty et al., 2009
Folic-acid-fortified flour/ rice	Folic acid (Vitamin B9)	Prevents neural tube defects when consumed periconceptionally	De-Regil et al., 2015
Calcium-fortified soy milk	Calcium (carbonate, citrate, etc.)	Provides calcium absorption comparable to cow's milk, supporting bone health	Heaney et al., 2000

Consumer awareness and acceptance play a central role in determining how functional foods are perceived and adopted within the market. A clear understanding of consumer attitudes, preferences, and knowledge structures is therefore important for supporting market development and facilitating broader uptake of functional food products (Ponte et al., 2025; Sgroi et al., 2024). At the same time, levels of awareness and acceptance differ markedly across regions, reflecting variations in cultural values, socioeconomic conditions, and educational contexts. While relatively high awareness has been reported in developed regions such as North America and Europe, consumers in developing countries often encounter barriers related to information availability and price sensitivity (Baker et al., 2022; Ponte et al., 2025). These regional contrasts suggest that context-specific approaches are required to reduce disparities and to ensure that functional foods are perceived as both accessible and relevant across diverse populations.

Beyond demographic and cultural influences, trust in health claims and perceived product effectiveness constitutes a key determinant of consumer acceptance. Such trust is closely linked to transparent communication practices and the presence of robust scientific evidence. In the absence of credible information, consumers may remain cautious toward functional foods, even when potential health benefits are acknowledged (Hosseini et al., 2025). In addition, sensory attributes—including taste, texture, and visual appeal—strongly shape consumer satisfaction and influence repeat purchase behavior (Forde and Graaf, 2022; Xiao et al., 2015). Accordingly, successful functional food products are expected to balance health-related benefits with sensory quality and everyday usability.

Against this background, the present manuscript aims to review and integrate existing studies on consumer awareness and acceptance of functional foods. By examining key themes such as conceptual definitions, classification frameworks, health effects, market developments, and consumer perceptions, this review seeks to clarify current knowledge and identify areas requiring further investigation. In doing so, it aims to offer insights relevant to researchers, policymakers, and industry stakeholders, and to support the design of strategies that enhance the effective promotion of functional foods and their contribution to public health.

2. Understanding Functional Foods

2.1 Definitions and Classification

Functional foods are not easily captured by a single definition, as they encompass a wide range of products with differing nutritional roles and health-related functions. Rather than constituting a uniform category, they are more appropriately viewed as a continuum, extending from foods with restored or enhanced nutrient content to products designed to influence specific physiological processes through bioactive components, such as probiotics. Across this spectrum, functional foods are commonly developed with the intention of supporting health maintenance, addressing particular health concerns, and contributing to disease prevention.

One prominent group within this continuum consists of fortified foods, in which conventional food products are supplemented with additional nutrients to reduce the risk of deficiency in targeted populations (Taofiq et al., 2017). Typical examples include iron-fortified cereals, which are intended to lower the prevalence of anaemia, and vitamin D–enriched milk, which is associated with improved bone health and a reduced risk of osteoporosis (Whittaker et al., 2001). Such fortification strategies are frequently embedded in public health policies, especially in settings where micronutrient deficiencies remain widespread (Olson et al., 2021). The long-standing use of iodized salt to prevent iodine deficiency and related thyroid disorders provides a well-established illustration of this approach (Wu et al., 2002).

Closely related, but conceptually distinct, are enriched foods. In this case, nutrients lost during industrial processing are intentionally reintroduced in order to restore the food's original nutritional profile (FAO, 2018). Enriched cereal products, such as flour supplemented with iron and B vitamins including niacin and riboflavin, exemplify this practice. These foods play a particularly important role in dietary contexts where processed products constitute a substantial proportion of daily intake, helping to sustain adequate micronutrient consumption at the population level (Dwyer et al., 2015).

A further category comprises probiotic and prebiotic foods, which are primarily associated with gastrointestinal health. Probiotic products, including yogurt and kefir, contain live microorganisms such as *Lactobacillus* and *Bifidobacterium*, whereas prebiotic foods supply substrates that selectively stimulate the growth of beneficial gut bacteria. Ingredients such as garlic, onions, bananas, and whole grains are frequently cited in this context. Evidence indicates that the combined action of probiotics and prebiotics can contribute to improved gut microbiota balance, enhanced immune responses, and the management of gastrointestinal conditions, including irritable bowel syndrome (Latif et al., 2023; Kumar et al., 2022).

In contrast to these technologically modified products, natural functional foods derive their health-promoting properties from bioactive compounds that are inherently present in the food matrix, without the need for fortification or enrichment (Vignesh et al., 2024). Commonly cited examples include nuts, berries, fatty fish such as salmon and mackerel, and green tea, all of which have been linked to antioxidant activity, anti-inflammatory effects, and cardiovascular or cognitive benefits. The longstanding incorporation of these foods into traditional dietary patterns underscores their perceived contribution to health and longevity.

Taken together, these categories demonstrate the diversity of functional foods and their capacity to address health objectives ranging from micronutrient sufficiency to digestive and cardiovascular support. Clarifying these classifications is essential for facilitating research, informing policy discussions, and supporting the development and dissemination of functional foods that align with population-specific nutritional needs.

2.2 Health Benefits

Functional foods have been discussed in the literature as dietary components capable of influencing multiple dimensions of health, encompassing both physical and psychological outcomes. Their effects are generally attributed to the presence of bioactive substances, including vitamins, minerals, fatty acids, and other functional constituents, which interact with specific biological pathways relevant to disease prevention and health maintenance.

Cardiovascular health represents one of the most frequently examined domains in relation to functional food consumption (Coates et al., 2008). In particular, foods containing omega-3 fatty acids—such as fatty fish (e.g., salmon, mackerel, and sardines) and certain nuts including walnuts and almonds—have been associated with a lower incidence of cardiovascular disease (Gupta et al., 2015). Evidence suggests that omega-3 fatty acids contribute to favorable lipid metabolism by reducing triglyceride concentrations and increasing high-density lipoprotein levels (Cartolano et al., 2022), while also attenuating systemic inflammatory processes that are implicated in atherosclerosis and related cardiovascular disorders (Chang and Deckelbaum, 2014). In addition, fiber-rich foods such as oats and barley have been linked to improvements in blood pressure regulation and cholesterol profiles, further supporting cardiovascular function.

Digestive health constitutes another major area in which functional foods have attracted attention, particularly through the consumption of probiotic- and prebiotic-containing products. Probiotic foods, including yogurt, kimchi, and sauerkraut, introduce live microorganisms that contribute to the stabilization of gut microbiota composition and support nutrient digestion and absorption (Leeuwendaal et al., 2022). Prebiotic components, commonly found in bananas, garlic, onions, and whole grains, serve as substrates that promote the activity and proliferation of beneficial intestinal bacteria. The combined intake of probiotics and prebiotics has been associated with improved bowel function, modulation of inflammatory responses, and potential effects on psychological well-being via the gut–brain axis (Ansari et al., 2023).

Support of immune function is also frequently highlighted in discussions of functional foods. Fortified products supplying nutrients such as vitamins C and D, zinc, and selenium have been linked to enhanced immune responsiveness and a reduced vulnerability to infectious diseases (Thirumdas et al., 2021). Vitamin D–fortified dairy products and cereals, for example, have been examined for their role in immune regulation and respiratory health. Similarly, vitamin C–rich fruits and vegetables, including citrus fruits, strawberries, and bell peppers, are known to contribute to leukocyte production and immune defense mechanisms (Mitra et al., 2022).

Beyond these functions, functional foods have been examined for their relevance in the prevention and management of chronic conditions. Antioxidant-rich foods such as berries, green tea, and dark chocolate have been shown to counteract oxidative stress by scavenging free radicals, thereby lowering the risk of chronic diseases including cancer and neurodegenerative disorders (Lobo et al., 2010). Calcium- and vitamin D-enriched products, such as fortified milk and orange juice, are commonly discussed in relation to bone health and osteoporosis prevention (Sunyecz, 2008). In addition, dietary fiber from plant-based sources—legumes, whole grains, and vegetables—plays a role in glycaemic control and may reduce the likelihood of developing Type 2 diabetes (McMacken and Shah, 2017). Certain foods, including garlic and flaxseeds, have also been associated with blood pressure regulation, contributing to hypertension prevention (Toulabi et al., 2022).

Mental health outcomes have increasingly been incorporated into the discussion of functional foods. Nutrients such as omega-3 fatty acids and antioxidant compounds are thought to support cognitive function and emotional regulation, with potential implications for delaying cognitive decline and reducing the risk of disorders such as Alzheimer's disease (Thomas et al., 2015). In parallel, fermented foods that influence gut microbiota composition, including yogurt and kimchi, have been associated with reductions in symptoms of anxiety and depression, further underscoring the relevance of diet–microbiome interactions for mental well-being (Selhub et al., 2014).

Overall, the health-related effects attributed to functional foods extend across multiple physiological systems, ranging from cardiovascular and metabolic regulation to immune defense and mental health. The accumulation of scientific evidence in this area underscores the growing relevance of functional foods within contemporary dietary patterns and highlights their potential contribution to improved quality of life when integrated into everyday nutrition.

2.3 Market Growth and Trends

The functional food market has expanded substantially in recent years, reflecting a broader shift in consumer priorities toward health maintenance and disease prevention. As awareness of the long-term relationship between dietary patterns and health outcomes increases, foods that provide specific physiological benefits have attracted growing interest. This expansion has been associated with multiple structural factors, including population aging, rising healthcare expenditures, and the increasing burden of lifestyle-related conditions such as obesity, diabetes, and cardiovascular disease (Ruze et al., 2023). In parallel, changes in consumer preferences, progress in food science and technology, and regulatory frameworks governing health claims have collectively contributed to shaping current market dynamics (Priya and Alur, 2023).

Among recent developments, plant-based functional foods have gained particular prominence. Motivated by ethical considerations, environmental concerns, and perceived health advantages, consumers have shown increasing interest in plant-based dietary options, stimulating the introduction of a wide range of related functional products (Alcorta et al., 2021). These include protein-enhanced plant-based beverages, fortified alternatives to dairy milk such as almond, soy, and oat drinks, as well as plant-derived bioactive compounds, notably polyphenols and flavonoids, which are widely discussed for their antioxidant and anti-inflammatory properties. This trend appears especially pronounced among younger demographic groups, including Millennials and Generation Z, who often associate plant-based choices with sustainability and personal health goals (Raptou et al., 2024).

Growth in the functional food sector has also been supported by sustained interest in products targeting gut health, particularly those containing probiotics and prebiotics. Increasing scientific attention to the gut microbiome and its links to digestion, immune function, and mental health has heightened consumer awareness of these products (Cunningham et al., 2021; Obayomi et al., 2024). As a result, items such as probiotic yogurts, kefir, fermented vegetables including kimchi and sauerkraut, and fiber-based prebiotic formulations have gained market visibility. In response to this demand, manufacturers have begun incorporating probiotic components into a broader range of formats, including snacks, beverages, and confectionery products, to better align with contemporary consumption patterns (Okpiaifo et al., 2023).

Another notable development is the increasing emphasis on personalized nutrition. Technological advances in areas such as genetic analysis, artificial intelligence, and wearable health monitoring devices have enabled more individualized dietary recommendations and product formulations (Donovan et al., 2025). Within this context, functional food companies are adopting data-driven approaches to design products and services tailored to specific health needs, including metabolic health, digestive function, and micronutrient status. Examples include customized supplement bundles, DNA-informed dietary guidance, and nutrition plans based on gut microbiome profiles (Bassagnanya-Riera et al., 2021; Wastyk et al., 2022). This shift is closely aligned with broader developments in precision medicine, where nutrition is increasingly recognized as a key component of preventive healthcare.

Market expansion has further been facilitated by the incorporation of functional ingredients into conventional food categories. Everyday products such as breakfast cereals, snack bars, and beverages are increasingly formulated with added protein, dietary fiber, omega-3 fatty acids, or vitamins, enabling consumers to obtain functional benefits without substantially altering established eating habits. This approach has been particularly visible within the beverage sector, where products such as vitamin-fortified waters, adaptogenic teas, and probiotic soft drinks have experienced rapid diversification.

Regulatory environments and health claim governance also play an important role in shaping market trajectories. In many regions, public authorities have strengthened requirements for functional food labeling and the substantiation of health-related claims in order to enhance consumer protection and scientific credibility. Regulatory bodies, including the U.S. Food and Drug Administration, the European Food Safety Authority, and Japan's Consumer Affairs Agency, have contributed to the harmonization and enforcement of industry standards (Yamamoto et al., 2024). In response, companies have increasingly invested in clinical research and evidence generation to support product claims and reinforce consumer trust.

In addition, digital distribution channels and marketing strategies have become increasingly influential. The growth of e-commerce platforms, direct-to-consumer business models, and online wellness communities has enabled functional food brands to reach wider and more diverse audiences (Chen and Martirosyan, 2021). Social media engagement, influencer-led promotion, subscription-based services, and algorithm-driven recommendations have reshaped how consumers discover and evaluate functional food products (Farivar and Wang, 2022). The convenience associated with online purchasing has further supported demand, particularly among time-constrained and health-oriented consumers (Maalouf et al., 2025).

Sustainability considerations and clean-label expectations have also emerged as central elements of market development. Consumers are increasingly attentive to issues such as ingredient transparency, production practices, and environmental impact, prompting manufacturers to prioritize natural, organic, and minimally processed formulations (Dove et al., 2020). Alongside ingredient choices, efforts related to sustainable packaging, ethical sourcing, and carbon footprint reduction have gained strategic importance, reflecting broader consumer concern for environmental responsibility.

Overall, current trends indicate that the functional food market is being shaped by a combination of health-oriented consumer behavior, technological innovation, regulatory oversight, and sustainability considerations. Developments such as the expansion of plant-based products, growing attention to gut health, advances in personalized nutrition, and the mainstreaming of functional ingredients are likely to continue influencing the sector. As interest in health-enhancing foods persists, the functional food market is expected to remain a dynamic space characterized by ongoing innovation and diversification.

3. Consumer Awareness and Acceptance of Functional Foods

3.1 General Awareness Levels

The extent to which consumers recognize and understand functional foods is unevenly distributed across regions and population segments. These differences are shaped by broader social structures, including cultural norms, economic capacity, and educational exposure. Research conducted in developed regions—such as North America, Europe, and selected Asian countries—suggests relatively high levels of awareness, a pattern that coincides with long-standing policy engagement, market penetration, and institutional support for functional foods. In these contexts, coordinated actions by regulatory authorities, food producers, and health-related institutions have facilitated information diffusion through certification schemes, public communication, and labeling systems. The Japanese Foods for Specified Health Uses (FOSHU) program illustrates how regulatory endorsement can contribute to both visibility and consumer confidence in functional food products (MHLW, n.d.).

By contrast, in many developing regions, consumer familiarity with functional foods remains limited. Constraints related to income, information access, and infrastructural development often direct attention toward basic nutritional sufficiency rather than value-added health functions (Durao et al., 2020). In addition, the absence of sustained educational initiatives and targeted marketing reduces opportunities for consumers to engage with the concept of functional foods. Nonetheless, ongoing transformations associated with globalization, urban expansion, and the widespread use of digital communication technologies are gradually reshaping information environments. As a result, both multinational firms and public institutions have begun to frame functional foods as potential instruments for addressing nutrition-related public health challenges, including micronutrient deficiencies and non-communicable diseases, leading to incremental gains in consumer awareness (Uusitalo et al., 2002).

Despite these contextual differences, ambiguity surrounding the definition of functional foods represents a shared issue across markets. Divergent interpretations arise when consumers are exposed to inconsistent terminology, promotional claims, or non-uniform labeling practices. While some consumers demonstrate a clear conceptual distinction between functional and conventional foods, others experience uncertainty that hinders informed choice. These observations reinforce the importance of continued efforts aimed at improving nutrition literacy and strengthening transparency in product communication.

3.2 Sources of Information

Consumer perceptions of functional foods are shaped through interaction with multiple information sources, each characterized by distinct levels of authority, reach, and persuasive capacity. The relative influence of these sources affects not only awareness but also the degree of trust consumers place in functional food products.

3.2.1 Media and Advertising

Mass communication channels, encompassing broadcast media, social networking services, and web-based platforms, play a prominent role in circulating information about functional foods (Avelino et al., 2024). Marketing messages commonly frame functional foods in relation to specific health objectives and are frequently customized for defined audience groups, including physically active individuals, older adults, or consumers concerned with immune or digestive health. The expansion of influencer-led promotion and sponsored digital content has further intensified exposure, particularly among younger cohorts who actively engage with lifestyle-oriented health narratives on platforms such as Instagram, TikTok, and YouTube (Fan et al., 2023).

At the same time, the informational quality of media-based communication remains uneven. Promotional content varies widely in its reliance on scientific evidence, and exaggerated or selectively framed claims may undermine consumer confidence (Walter et al., 2024). Although regulatory frameworks aimed at ensuring truthful advertising have been introduced in many jurisdictions, inconsistencies in enforcement persist. Consequently, consumers who depend heavily on media sources are often advised to corroborate marketing claims through independent and credible information channels (Aade et al., 2022).

3.2.2 Health Professionals

Advice provided by health professionals—including dietitians, nutritionists, and physicians—continues to hold a high degree of credibility in the context of dietary decision-making. Their recommendations are particularly influential for individuals managing health conditions or seeking evidence-based guidance. Functional foods are frequently incorporated into dietary counseling as complementary strategies for addressing cardiovascular risk, glycemic control, or gastrointestinal health (Alissa and Ferns, 2012; Vignesh et al., 2024). Educational campaigns organized by medical associations further contribute to legitimizing functional foods within public health discourse.

However, access to professional nutritional guidance is not evenly distributed. Urban populations with well-resourced healthcare systems typically benefit from greater availability of expert consultation, whereas individuals in rural or underserved areas face structural limitations (Elliott and Gibson, 2023). The growing adoption of telemedicine and online health services has the potential to mitigate these disparities by extending expert advice beyond conventional clinical settings (Anawade et al., 2024).

3.2.3 Product Labeling

Information presented directly on product packaging constitutes a critical interface between functional food producers and consumers. Nutritional declarations and health-related claims serve as immediate reference points that can support product evaluation and comparison (Penzavecchia et al., 2022). To safeguard consumer interests, regulatory authorities—including Japan's FOSHU system, the European Food Safety Authority (EFSA), and the U.S. Food and Drug Administration (FDA)—have established criteria governing the use and substantiation of health claims. Labels highlighting attributes such as probiotic content, omega-3 levels, or immune-related benefits are intended to facilitate informed purchasing decisions.

Nevertheless, the effectiveness of labeling as an informational tool is moderated by consumer literacy. Technical terminology and complex ingredient lists may impede comprehension, particularly among individuals with limited background knowledge in nutrition (Miller and Cassady, 2016). To address these challenges, strategies such as simplified language, standardized visual indicators, and digital extensions—most notably QR codes linking to explanatory resources—have been proposed as means of enhancing clarity and trust.

3.3 Factors Influencing Awareness

Differences in consumer awareness of functional foods are closely linked to demographic and socioeconomic conditions, which shape how information is accessed, interpreted, and translated into dietary behavior. Examining these factors provides a basis for designing interventions that improve both awareness and access across diverse population groups.

3.3.1 Education

Educational attainment has repeatedly been associated with variation in awareness and comprehension of functional foods (Baker et al., 2022). Individuals with higher levels of education tend to actively seek health-related information, assess product claims more critically, and incorporate such information into dietary decision-making. Formal and informal educational settings—including universities, research institutions, and public health initiatives—contribute to this process through nutrition-related curricula, workshops, and community-based outreach activities (Silva et al., 2023).

At the same time, unequal access to education generates persistent knowledge gaps. In contexts characterized by lower literacy levels or limited health education infrastructure, consumers may depend disproportionately on anecdotal sources or promotional messages, increasing the likelihood of misunderstanding. Integrating functional food concepts into school education and lifelong learning programs has therefore been suggested as a means of reducing these disparities.

3.3.2 Income

Economic resources also play a substantial role in shaping awareness and exposure to functional foods. Consumers with higher income levels generally have greater access to information and are more likely to prioritize health-related attributes when making purchasing decisions. This group is often engaged with premium health-oriented products, wellness events, and professional dietary consultations. By contrast, individuals with lower incomes may encounter limited exposure to functional foods due to budget constraints and competing priorities related to basic food needs.

Price differentials further influence awareness and adoption. Although functional foods are commonly perceived as beneficial, their relatively higher cost compared with conventional alternatives can discourage uptake among price-sensitive consumers (Baker et al., 2022; Tandon et al., 2021). Policy measures aimed at improving affordability—such as subsidies, support for local production, or collective purchasing schemes—have been proposed as mechanisms to broaden access and reduce socioeconomic disparities.

3.3.3 Age

Age-related differences also shape awareness and engagement with functional foods. Younger cohorts, particularly Millennials and Generation Z, tend to demonstrate heightened interest, driven by health-oriented lifestyles and extensive use of digital media (Chen et al., 2024). Social networking platforms, wellness-focused content, and fitness-oriented communities play a prominent role in influencing their perceptions. These consumers are often more willing to experiment with novel products, adopt plant-based diets, and incorporate supplements into everyday routines.

Older adults, in contrast, often approach functional foods through the lens of health maintenance and disease prevention. Their level of awareness may depend strongly on previous exposure to nutrition education and healthcare advice (Vella et al., 2014; Vignesh et al., 2024; Silva et al., 2023). This group tends to rely more heavily on medical professionals and established dietary habits when evaluating functional food options (Szakos et al., 2022). Adapting communication strategies to different age groups—combining digital tools with traditional information channels—can therefore enhance awareness across the life course.

Taken together, these demographic and socioeconomic influences highlight the need for integrated approaches that combine educational initiatives, transparent communication, and economic support to promote informed engagement with functional foods.

3.4 Attitudes toward Functional Foods

Consumer attitudes toward functional foods emerge from the interaction of perceived health value, confidence in manufacturer claims, cultural norms, and individual beliefs about health and nutrition. Individuals who prioritize preventive health and lifestyle management often express favorable attitudes, viewing functional foods as complementary components of balanced diets that support long-term well-being (Nguyen and Phan, 2022; Silva et al., 2023).

Nevertheless, positive attitudes are not universal. Skepticism persists among certain consumer segments, particularly when health claims are perceived as overstated or insufficiently supported by scientific evidence. In some cases, functional foods are interpreted as marketing-driven responses to transient health trends rather than as credible dietary options (Hailu et al., 2009). Cultural preferences also play a role, as traditional food practices and reliance on familiar or locally sourced foods may reduce receptivity to technologically positioned products (Siegrist et al., 2015; Baker et al., 2022). These dynamics underscore the multifaceted and context-dependent nature of consumer attitudes.

3.5 Willingness to Pay for Functional Foods

Consumers' willingness to pay for functional foods is shaped by perceptions of benefit, product quality, and price considerations. Empirical research suggests that many consumers accept price premiums when health claims are regarded as credible and scientifically substantiated (Alsubhi et al., 2022; Baker et al., 2022). Products associated with tangible outcomes—such as immune support or weight management—are often perceived as offering value that justifies higher costs, particularly among individuals seeking preventive or therapeutic benefits (Kruk et al., 2018).

Willingness to pay, however, varies substantially across income groups. Consumers with higher disposable incomes generally demonstrate greater readiness to invest in functional foods and are more inclined to prioritize health-related expenditures (Siegrist et al., 2008). In contrast, lower-income consumers frequently emphasize affordability and convenience, which can limit adoption despite perceived benefits (Andika and Setyanta, 2024). Achieving broader market penetration therefore requires strategies that align product pricing with the financial realities of price-sensitive populations.

3.6 *Perceived Efficacy and Trust in Functional Foods*

Perceptions of efficacy and trust constitute central determinants of functional food acceptance. Consumers are more likely to engage with functional foods when they believe that claimed health effects are authentic and reliably supported. This trust is shaped by several interrelated elements, including the credibility of health claims, brand reputation, and the presence of regulatory oversight.

Health claims grounded in scientific evidence, clinical research, or professional endorsement tend to enhance consumer confidence (Tønnesen et al., 2022). Transparent communication regarding the scope and limitations of claimed benefits further reinforces trust (Arnot et al., 2016). In parallel, brand reputation influences perceptions of reliability, as established brands are often assumed to have stronger incentives to maintain quality and safety standards (Wu et al., 2021; Handoyo, 2024).

Regulatory mechanisms also contribute to trust formation. Certification schemes, governmental regulations, and independent testing procedures provide signals of compliance with safety and quality requirements (Gupta et al., 2025). When functional foods are endorsed or monitored by recognized authorities, consumer confidence in both safety and efficacy is strengthened. Figure 2 illustrates how these dimensions interact to shape awareness, trust, and purchasing behavior.

Figure 2 highlights how these factors—medical efficacy, health claims, certifications, and health benefits—interact to shape consumer awareness and acceptance of functional foods. As seen in Figure 2, these elements play a critical role in fostering trust, which ultimately drives purchasing decisions.

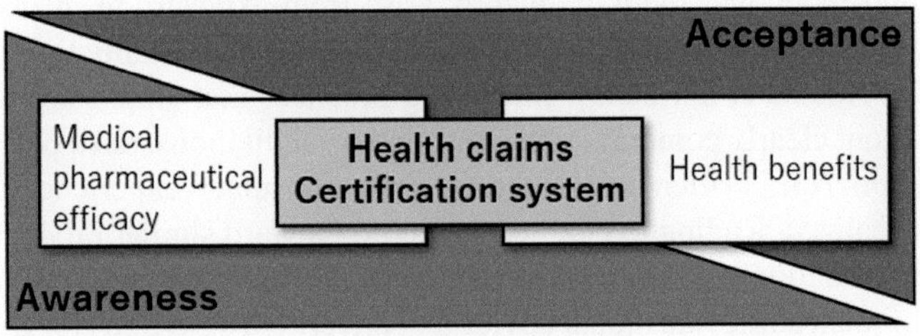

Fig. 2 Importance of medial pharmaceutical efficacy, health claims, certification, and health benefits on awareness and acceptance of functional foods.

3.7 *Sensory Preferences and Taste*

Sensory characteristics remain a decisive factor in consumer acceptance of functional foods. Regardless of perceived health value, products that fail to meet expectations regarding taste, texture, or appearance are unlikely to achieve sustained market success. Many consumers demonstrate limited willingness to compromise sensory enjoyment in exchange for health benefits, rendering sensory performance a critical determinant of acceptance (Imtiyaz et al., 2021).

The inclusion of functional ingredients—such as probiotics, dietary fibers, or plant-derived extracts—can introduce challenges related to flavor and mouthfeel. In response, manufacturers increasingly employ technological solutions, including flavor-masking techniques, texture modification, and novel formulation approaches, to improve palatability (Jeong and Lee, 2021).

Sensory preferences are also culturally embedded. Variations in preferred flavor intensity, texture, and familiarity across regions influence how functional foods are perceived and adopted. Recognizing and accommodating these differences is essential for successful market entry. Ultimately, the alignment of sensory quality with consumer expectations plays a pivotal role in determining whether functional food products achieve widespread acceptance.

4. Barriers and Challenges to Acceptance

Consumer acceptance of functional foods is constrained by a set of interrelated psychological, economic, and cultural factors. To conceptualize these constraints, Figure 3 provides an integrated overview of key challenges that influence consumer decision-making. Figure 3 illustrates how skepticism and misinformation, financial considerations, and cultural or regional contexts interact to shape perceptions of functional foods. The following subsections examine each of these dimensions in greater detail, with particular attention to their underlying mechanisms and implications for acceptance.

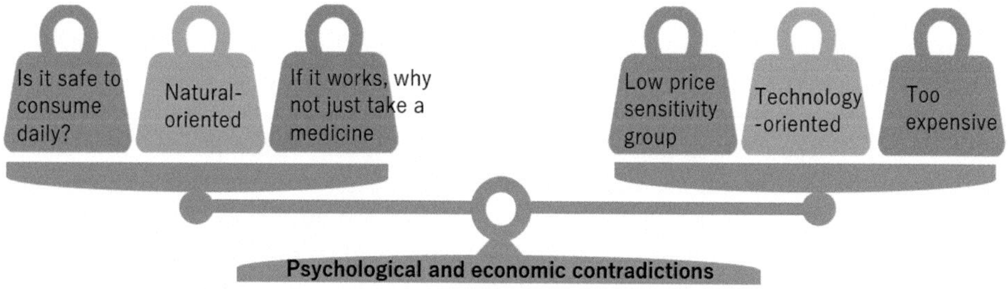

Fig. 3 The psychological and economic contradictions of consumer perceptions of functional foods.

4.1 Skepticism and Misinformation

Doubts surrounding functional foods frequently arise from exposure to misinformation and from the circulation of health claims that lack adequate scientific substantiation (Ponte et al., 2024; Baker et al., 2022; Adams et al., 2023). Marketing practices that emphasize dramatic or absolute health effects, without clearly communicating evidentiary limitations, have contributed to persistent mistrust. This skepticism is often reinforced by prior consumer experiences in which promised benefits were not realized, leading to heightened caution toward similar products.

The digital information environment further amplifies this challenge. Online platforms enable the rapid spread of misleading nutritional advice and unverified assertions regarding functional food efficacy, making it difficult for consumers to distinguish credible information from promotional content (Someko et al., 2024). As a result, uncertainty persists regarding whether health claims reflect scientific consensus or marketing-driven narratives aligned with prevailing wellness trends.

Ambiguities in product labeling intensify these concerns (Maeda-Yamamoto, 2021). When ingredient lists or claimed benefits are difficult to interpret, consumers may hesitate to purchase functional foods. This issue is particularly evident when broadly defined terms such as "natural" or "superfood" are used without standardized definitions, allowing for multiple interpretations. Moreover, differences in regulatory standards across countries mean that identical products may be subject to varying claim requirements, further complicating consumer understanding (Farid et al., 2019).

Addressing skepticism therefore requires coordinated action at multiple levels. Strengthening regulatory oversight and enforcing clearer standards for health claims have been identified as essential steps (Katan, 2004). Equally important are educational efforts that explain the scientific basis of functional foods in accessible terms (Baker et al., 2022). Through transparent communication and evidence-based public health messaging, it may be possible to counter misinformation and gradually restore consumer confidence.

4.2 Cost Concerns

Financial considerations represent another major obstacle to the broader adoption of functional foods. Price premiums associated with these products often limit access, particularly among low- and middle-income consumers (Michel et al., 2024). Although functional foods are frequently perceived as offering health advantages, their cost can position them as discretionary or luxury items, reinforcing the notion that they are intended primarily for affluent or highly health-oriented consumers (Bech-Larsen and Grunert, 2003). This perception narrows the potential consumer base and restricts market expansion.

For price-sensitive households, higher product costs may discourage experimentation with functional foods, even when interest in potential health benefits exists. In such contexts, purchasing decisions tend to prioritize staple foods that satisfy immediate nutritional needs over higher-priced products with added health functions (Nassar et al., 2021). As a consequence, cost-related barriers affect not only individual adoption but also the overall diffusion of functional foods within the market.

Reducing these barriers requires approaches that address affordability without compromising quality. Strategies such as improving production efficiency, optimizing supply chains, and offering smaller or entry-level product formats have been proposed as ways to lower consumer price thresholds. In addition, the development of competitively priced functional foods may help broaden appeal among consumers with limited purchasing power.

Public-sector involvement may further support affordability. Subsidies or collaborations between governments and health organizations could ease financial constraints, particularly for functional foods with demonstrated benefits in managing chronic health conditions (An, 2012). By alleviating cost pressures, such measures may contribute to wider access and more equitable health outcomes.

4.3 Cultural and Regional Factors

Acceptance of functional foods is also shaped by cultural norms, dietary traditions, and regional food practices (Baker et al., 2022; Ponte et al., 2024). Consumers often evaluate new food products through the lens of established eating habits, which can either facilitate or hinder adoption. Where functional foods align with traditional dietary patterns, acceptance is generally higher. Fermented foods such as kimchi and miso, for instance, are long-standing components of East Asian cuisines and are widely perceived as health-promoting, contributing to their acceptance as functional foods (Hongu et al., 2017).

Conversely, resistance may arise when functional foods diverge from familiar food cultures. In Western contexts where meat and dairy products have historically dominated diets, plant-based functional foods may face skepticism, despite recognized health or environmental benefits (Clemente-Suárez et al., 2023). Similar hesitation can be observed in regions where exposure to ingredients such as algae or fermented soy products is limited, reducing consumer willingness to try unfamiliar formulations.

Cultural interpretations of health further influence acceptance. In some societies, greater value is placed on traditional remedies or holistic health practices than on commercially produced foods supported by scientific claims (Mingay et al., 2021). In such settings, functional foods may be perceived as overly industrialized or incompatible with local conceptions of "natural" or "authentic" eating.

Overcoming these barriers requires sensitivity to cultural context. Product adaptation that reflects regional tastes, ingredients, and culinary practices can enhance relevance and acceptance. Educational initiatives that frame functional foods in relation to local health beliefs, combined with engagement from trusted community figures or health professionals, may further support adoption. By aligning product development and communication strategies with cultural expectations, acceptance of functional foods can be strengthened across diverse regional settings.

5. Strategies to Improve Consumer Awareness and Acceptance

5.1 Effective Marketing and Education

Efforts to strengthen consumer awareness and acceptance of functional foods depend heavily on how information is communicated and contextualized. Marketing and educational activities are most effective when they reflect the heterogeneity of consumer groups and address differences in values, expectations, and constraints (Chen and Martirosyan, 2021). Factors such as age, income, educational background, health orientation, and cultural context shape how messages are received and interpreted. Accordingly, communication strategies should not only highlight potential health benefits but also openly address consumer concerns related to product efficacy, safety, and cost.

Collaboration among multiple actors can substantially enhance the reach and credibility of these initiatives. Health professionals—including dietitians, nutritionists, and physicians—are particularly influential, as their evidence-based guidance can legitimize information about functional foods and reduce uncertainty (Karelakis et al., 2020). In parallel, wellness-oriented influencers play an important role in engaging younger and digitally connected audiences by presenting functional foods in accessible and relatable formats (Şenyapar, 2024). Community organizations focused on nutrition and public health further contribute by embedding information within local contexts and facilitating grassroots-level education.

Educational interventions are most effective when scientific information is translated into formats that are accessible, engaging, and culturally appropriate (Stuart and Achterberg, 1997; Silva et al., 2023). Public health campaigns that simplify complex evidence through clear language, visual aids, and practical examples can enhance comprehension. Interactive approaches—such as workshops, online courses, and webinars—also enable more personalized engagement. Importantly, education should extend beyond health benefits to include issues of safety, sensory quality, and affordability, thereby fostering informed dialogue and strengthening consumer confidence (Alsubhi et al., 2022).

5.2 Role of Health Claims and Labeling

Health claims and labeling practices constitute a central interface between functional food producers and consumers. Claims that are transparent, specific, and grounded in scientific evidence support informed decision-making and contribute to trust formation. When health benefits are substantiated through rigorous research and recognized by regulatory authorities, consumers are more likely to perceive products as both effective and safe (Maleki Varnosfaderani and Forouzanfar, 2024). In contrast, vague or exaggerated statements risk generating confusion and reinforcing skepticism.

Beyond claims themselves, labeling functions as a key informational tool by conveying details about ingredients, sourcing, and production processes. Clear explanations of how specific components contribute to health outcomes can demystify functional foods, particularly for consumers concerned about additives or artificial substances (Novais et al., 2022). Certifications such as organic or non-GMO (genetically modified organism) labeling further assist consumers in aligning purchases with personal values. Presenting nutritional information in a concise and intuitive format also enables consumers to evaluate products in relation to individual dietary goals.

Regulatory oversight remains essential in ensuring the integrity of labeling practices. By establishing and enforcing standards for permissible claims, governments and independent agencies can reduce misleading practices and enhance transparency within the functional food market. Such frameworks not only protect consumers but also create conditions under which credible innovation can flourish.

5.3 Industry and Policy Recommendations

Enhancing awareness and acceptance of functional foods requires coordinated action from both industry stakeholders and policymakers. Key areas of intervention include consumer education, affordability, and cultural responsiveness.

Consumer education. Long-term educational efforts are necessary to build sustained understanding of functional foods and their role in health promotion. Industry-led initiatives that explain functional food concepts and mechanisms can complement academic and public-sector activities (McConnon et al., 2002). Partnerships with universities, research institutes, and health organizations enable the dissemination of evidence-based information and strengthen scientific credibility (Fekete et al., 2025). Collaboration with healthcare providers further ensures that consumers receive accurate guidance from trusted sources. Educational materials that clearly describe product types, ingredients, and potential benefits empower consumers to make informed dietary choices (Michel et al., 2024).

Affordability. Cost remains a major constraint on adoption, particularly among lower-income groups (Baker et al., 2022). Addressing this challenge requires efforts to reduce production and distribution costs through efficiency gains, technological innovation, and optimized supply chains (Hariyani et al., 2024). Public–private partnerships and targeted subsidies may further enhance access for economically disadvantaged populations, especially where functional foods offer demonstrable public health benefits (An, 2013). Providing products across a range of price points can also help accommodate diverse consumer budgets.

Cultural adaptation. Cultural relevance is critical for acceptance across different regions (Neupane et al., 2019). Dietary traditions, taste preferences, and local health beliefs shape how functional foods are perceived. Engaging local experts and consumer groups during product development can help ensure alignment with regional expectations (Hossain et al., 2025). Incorporating locally sourced ingredients or familiar recipes may enhance resonance, while culturally sensitive marketing messages can strengthen trust and legitimacy (Chen and Martirosyan, 2021). In parallel, policymakers play an enabling role by supporting research and development, funding public health communication, and maintaining regulatory environments that balance consumer protection with innovation (Sato et al., 2025). Alignment between industry initiatives and policy frameworks is therefore essential for improving awareness, expanding access, and sustaining growth within the functional food sector.

In addition to these strategies, policymakers should play a supportive role by creating an enabling environment for the functional food industry. This includes providing incentives for research and development, supporting public health campaigns, and ensuring that regulatory frameworks are designed to protect consumers while promoting innovation in the functional food sector (Sato et al., 2025). By aligning industry efforts with policy support, stakeholders can work together to enhance consumer awareness, improve accessibility, and ultimately drive the growth of the functional food market.

6. Research Gaps and Future Directions

6.1 Limitations in Current Studies

Despite the growing body of literature on functional foods, several structural limitations continue to constrain a comprehensive understanding of consumer behavior and long-term impacts. A substantial

6.3 Hybrid Distribution: Online, Pharmacies, and Grocery Retail

Distribution strategies for functional foods must reconcile credibility, convenience, and customer trust. Traditionally, these products were found in pharmacies, health stores, or premium grocery aisles, where their health claims could be taken more seriously. However, the COVID-19 pandemic dramatically accelerated the shift to direct-to-consumer (DTC) channels, allowing brands to control narratives, collect consumer data, and offer personalized experiences. For example, Seed Health and Ritual have built successful DTC subscription models that emphasize convenience and customization (Pandey and Joseph, 2025). Meanwhile, hybrid distribution strategies, where functional foods are available online and in trusted physical retail outlets, are becoming the norm. NielsenIQ (2023) reports that 72% of global functional food consumers prefer a hybrid discovery-to-purchase path, beginning with online education and ending with in-store or online checkouts (NielsenIQ, 2023). Importantly, distribution strategy also influences brand credibility. Placement in pharmacies or wellness clinics signals authority, while placement in discount stores can undermine perceived value. Retail partnerships with gyms, dieticians, or wellness platforms are emerging as effective alternatives for niche targeting.

6.3.1 Distribution and equity in access

Access to functional foods is not merely a matter of distribution logistics but a critical equity concern. The World Health Organization (2022) has emphasized that the rising costs associated with functional and preventive nutrition can exacerbate existing health inequalities, particularly by limiting access for lower-income and marginalized populations (WHO, 2022). Without inclusive strategies, functional nutrition may inadvertently become a domain of the privileged, deepening disparities in health outcomes. To address this, a range of policy and market-based interventions have been proposed. Governments and international organizations are increasingly advocating for mechanisms such as subsidies, targeted food fortification, and integration of functional foods into public health schemes, including school feeding programs and maternal-child health initiatives (FAO, 2021; WHO, 2022). In the private sector, companies such as Nestlé and Unilever have piloted inclusive distribution models aimed at improving affordability and accessibility (Nestlé, 2020; Unilever, 2021). These include offering functional food products in low-cost sachets, leveraging rural kiosks, and employing mobile vendors in underserved regions across India and Latin America (Nestlé, 2020; Unilever, 2021).

As functional nutrition becomes embedded in global health and sustainability agendas, equitable distribution must be viewed as both a moral imperative and a strategic priority. Business models must balance profitability with social responsibility, ensuring that nutritional innovations reach those who need them most.

7. Commercialization of University-Originated Functional Food Technologies

While universities are often seen as the originators of functional food innovations, their role doesn't stop at discovery. In today's fast-moving wellness economy, academic institutions, and especially their technology transfer offices (TTOs), increasingly serve as strategic marketing enablers. By shaping not just the science but also the story, positioning, and credibility of functional food innovations, these institutions play a pivotal role in translating research into consumer-ready solutions. From crafting evidence-backed value propositions to guiding health claims and brand development, commercialization efforts rooted in universities are most successful when marketing insights are embedded from the start.

7.1 Marketing Begins in the Lab and Moves from Discovery to Desirability

Contrary to the common view of marketing as a "post-launch" activity, the most successful functional food technologies are developed with market orientation built into the research process. For example, when a university lab discovers a new prebiotic compound or bioactive ingredient, the TTO often assesses not just patentability, but also:

- What consumer problem does it solve?
- Who is the ideal user?
- How differentiated is it from current offerings?
- Can the health benefits be explained simply?

7.2 Technology Transfer Offices as Brand Architects and Communication Strategists

A modern TTO is no longer just a legal and licensing team, it is a branding incubator. It helps researchers understand how their innovation will be perceived, how the product's functional claim aligns with consumer language, and how to present scientific validity without jargon. This marketing-savvy support includes:

- Helping create marketable names and narratives (moving from "Polyphenol-RX129" to something like "HeartBloom").
- Guiding the use of visual identity, health symbols, and storytelling elements in compliance with regulations.
- Aligning the academic value proposition with competitive positioning, for instance, showing how an Indian university's turmeric extract might outperform mass-market curcumin supplements in terms of absorption.

One successful model is 'StartLife' at Wageningen, which uses design thinking workshops with spin-offs to craft unique brand personalities and marketing messages around functional claims (Bonants, 2020). Similarly, Pusa Krishi, IARI, and BIRAC-supported incubators in India now coach startups on how to convert science into story, especially for niche but growing categories like women's health, gut microbiome, and mental wellness (Ahmed, 2024). This evolving ecosystem emphasizes that marketing is not a downstream add-on, but an integral part of making science accessible, relatable, and desirable, especially in the functional food space, where trust and clarity drive adoption. From a marketing lens, university-originated functional foods represent an underleveraged source of white-space innovation. These products often come with strong scientific validation, yet they need the touch of marketers to convert clinical data into consumer relevance. Brands that partner early with academic spin-offs, or scout university pipelines, gain access to low-competition, high-trust innovations that can be launched with authentic origin stories, regulatory backing, and strong first-mover positioning. Marketers should therefore view TTOs as not just tech gateways but marketing collaborators, capable of co-developing products that are differentiated, defensible, and deeply resonant with consumer needs.

8. Regulatory and Ethical Marketing Challenges

Marketing functional foods requires navigating a tightrope between scientific accuracy, regulatory compliance, and persuasive communication. As these products often make health-related claims, they fall under stricter scrutiny than conventional foods, both legally and ethically. Marketers must not only avoid misleading consumers but also ensure that the communicated health benefits are substantiated, culturally appropriate, and accessible. Inadequate compliance or ethical missteps can lead to legal penalties, consumer backlash, and long-term brand erosion.

8.1 Global Regulatory Frameworks: European Food Safety Authority, Food and Drug Administration, and Foods for Specialized Health Uses

Globally, functional foods are regulated under different frameworks based on how health claims are defined. In the European Union, the EFSA maintains a central register of approved health claims. To use a claim like "supports immune health ", a product must be backed by human clinical data and pass EFSA's strict scientific evaluation (EFSA, 2022). In the United States, the FDA distinguishes between structure-function claims (e.g., "supports digestion") and disease claims, the latter requiring drug-level evidence (Borchers et al., 2016). Claims must carry a qualified disclaimer if not FDA-evaluated. Japan's FOSHU system allows selected products to be labelled as "foods for specified health uses ", provided that government review certifies their efficacy and safety for conditions like cholesterol or blood sugar control (Farid et al., 2019). These regulatory bodies aim to prevent exaggerated or false advertising. For marketers, this means that creativity must coexist with compliance. Storytelling, design, and testimonials must be framed carefully, translating science for consumers without distortion.

8.2 India's Regulatory Landscape: Food Safety and Standards Authority of India Guidelines

In India, the Food Safety and Standards Authority of India (FSSAI) has stepped up enforcement in functional foods (Gulati et al., 2021). The FSSAI (Packaging and Labelling) Regulation, 2020 and Nutraceutical Regulations, 2022 require that, functional ingredients (e.g., probiotics, omega-3s, vitamins) must conform to specified purity and dosage limits, health claims must be substantiated by peer-reviewed studies, preferably in Indian populations, and Products must avoid therapeutic, curative, or misleading terms such as "prevents diabetes" or "cures cancer" (Gulati et al., 2021).

FSSAI, 2022 also introduced the Eat Right Scientific Panel, which works on standardizing claims and verifying misleading advertising (Gulati et al., 2021). Indian brands like Kapiva, Setu, and Wellbeing Nutrition now include scientific citations or disclaimers on their packaging to build transparency and align with FSSAI norms (Gulati et al., 2021). However, awareness and enforcement remain uneven across rural and D2C sectors, creating a gray zone for startups.

8.3 Ethical Marketing: Beyond Compliance

Even when legally compliant, a functional food ad can be ethically questionable if it plays on fear, exaggerates outcomes, or exploits vulnerable populations. For instance, advertising iron-fortified products with images of malnourished children without disclosing efficacy data can be misleading and emotionally manipulative. Ethical marketing requires clarity over confusion: using simple, scientifically accurate language, empowerment over fear: promoting wellness, not fear of illness, and inclusivity over elitism: avoiding aspirational cues that marginalize lower-income groups. Hieke et al. (2016) report that approximately 64% of consumers distrust functional food brands they perceive as "overpromising" (Hieke et al., 2016). This highlights the importance of ethical storytelling and transparent labeling to build trust, not just compliance.

8.4 Recommendations for Ethical and Compliant Marketing

To align with both regulatory and ethical standards, functional food marketers must begin by conducting a thorough compliance audit of all claims, packaging, and marketing content in consultation with legal experts. They should incorporate third-party certifications, such as FSSAI logos, NSF approvals, or EFSA health claim identifiers, to enhance brand credibility and consumer trust. It is equally important to train marketing teams in the fundamentals of nutrition science and labelling laws to ensure responsible communication. When referring to clinical studies or scientific data, marketers should clearly disclose data sources, highlight study limitations, and indicate the population relevance to avoid misleading interpretations. Furthermore, they must refrain from using emotive or exaggerated claims such as "Feel younger overnight!", that cannot be scientifically validated (FEEDAP, 2012). An emerging best practice is the use of QR-code-based traceability, allowing consumers to access detailed information about clinical evidence, ingredient sourcing, and product approvals, thereby transforming regulatory compliance into a strategic trust-building tool.

9. Consumer Behavior and Trust Building

In the functional food domain, consumer behaviour is shaped by a complex interplay of perceived benefits, trust in science, cultural values, and communication clarity. Unlike traditional food products, functional foods often require the consumer to process novel information, about bioactives, gut health, immunity, or brain function, and to rely on scientific claims they cannot verify themselves. As such, the success of marketing efforts depends not only on visibility and pricing but also on deep psychological and emotional factors, especially trust.

9.1 Factors Driving Consumer Purchase Decisions

Consumer purchase decisions for functional foods are shaped by a combination of perceived health benefits, product safety, and the credibility of the producer, consistently identified as top global drivers (Baker et al., 2022). Consumers typically evaluate both intrinsic cues, such as ingredients, nutritional content, and product format (e.g., fermented, plant-based, organic), and extrinsic cues, including branding, packaging, health symbols, and endorsements. In the Indian context, ICAR-NIANP (2023) the urban consumers reported strong preferences for Ayurvedic-inspired formulations, gut health supplements, and immunity-enhancing products—preferences that have intensified in the post-COVID era (ICAR-NIANP, 2023). However, consumer skepticism remains notable, especially when scientific language is ambiguous or when health claims appear exaggerated. This aligns with global findings that consumer trust and clarity of health information significantly impact acceptance (Baker et al., 2022).

Price sensitivity also exhibits generational variation. Indian millennials and Gen Z consumers are more willing to pay premium prices for functional foods linked to fitness, beauty, or mental well-being. In contrast, older consumers prioritize cost-effectiveness and associate functional food purchases with managing chronic health conditions. These behavioural and psychological determinants reflect broader patterns in functional food acceptance globally, where lifestyle alignment and perceived relevance play a crucial role (Baker et al., 2022).

9.2 Role of Trust in Functional Food Adoption

Trust is not merely a contributing factor, it is the foundation of successful functional food marketing. Unlike pharmaceuticals, where efficacy is often immediate and clinically verifiable, functional foods occupy a complex space where perceived benefits are gradual, preventative, and often subjective (Baker et al., 2022). As a result, consumers must rely heavily on trust, i.e., trust that the product contains what it claims, that the brand operates transparently and ethically, and that the scientific rationale behind health claims is both credible and relevant. Research shows that consumer trust increases significantly when brands communicate scientific backing through citations, clinical trials, or the involvement of medical professionals (Baker et al., 2022). Global brands like Olipop, Seed Health, and Yakult have successfully leveraged this strategy by foregrounding evidence-based claims and scientific transparency in their marketing communication (Baker et al., 2022).

In the Indian context, trust is often mediated through culturally specific signals. Regulatory approvals such as FSSAI certification, AYUSH endorsements, and visible associations with traditional healers or health influencers contribute to perceived authenticity and credibility. Additionally, packaging in local languages and narratives that blend traditional wisdom with modern science such as referencing "grandmother's remedies enhanced by clinical research" have been shown to increase brand trust, particularly among rural and semi-urban consumers (ICAR-NIANP, 2023).

9.3 Strategies for Building Consumer Trust in Functional Food Marketing

To cultivate lasting trust in a highly competitive space, brands must focus on:

- **Science Communication:** Translate complex ingredients (e.g., "*Bacillus coagulans*") into relatable benefits ("boosts gut strength") using lay language and visuals.
- **Transparency:** Show dosage, sourcing, regulatory compliance, and testing on-pack or via QR codes.
- **Consistency:** Maintain formulation integrity, visual identity, and claims across platforms, mismatched messages erode trust.
- **Human Voices:** Use testimonials, doctor interviews, and real-world case studies to reinforce believability.
- **Responsive Education:** Offer interactive content (e.g., videos, blogs, podcasts) that explain how functional ingredients work.

One emerging best practice is "interactive packaging", where brands include QR codes that link to clinical trial summaries, nutrition education, or usage guidance, creating an educational moment at the point of sale.

9.4 The Trust-Behavior Feedback Loop

Trust is both a cause and effect of consumer behavior. When a product delivers results, whether increased energy, reduced bloating, or better sleep, trust is reinforced, loyalty grows, and word-of-mouth marketing expands. However, if expectations are poorly set or the benefit is delayed, trust quickly erodes. Functional food marketers must therefore manage expectation Vs. experience alignment, particularly in new markets like India where misinformation and wellness fads are also widespread. In this context, ethical branding, thoughtful education, and regulatory visibility are not optional, they are strategic imperatives.

10 Digital and Emerging Marketing Trends

The functional food industry is evolving beyond health-focused product development into a dynamic arena of digital marketing, personalized nutrition, and sustainability-driven innovation. As consumer expectations grow more sophisticated, fueled by online information, wellness influencers, and tech-enabled lifestyles, brands must adopt strategies that are not only effective but also adaptive to shifting behaviour, especially in younger and urban demographics.

10.1 Digital-First Engagement and Direct-to-Consumer Evolution

The rise of direct-to-consumer (D2C) platforms has dramatically changed how functional foods are marketed. Brands now use digital channels not just to sell but to educate, engage, and personalize consumer experiences. Globally, companies like Seed Health and Ritual use their websites and social platforms to provide content-rich, science-backed stories that foster trust. These strategies also allow brands to bypass traditional retail bottlenecks and connect with niche audiences through tailored subscription models, loyalty programs, and AI-driven recommendations.

In India, startups such as Setu and Wellbeing Nutrition have emerged as pioneers of digital-first approach. Setu markets its functional supplements, targeting vision, immunity, and sleep health, through a combination of Instagram influencers, wellness quizzes, and gamified web tools that personalize product suggestions. Similarly, Wellbeing Nutrition has adopted QR-code-enabled packaging, linking consumers directly to content that explains the science behind their probiotics, adaptogens, and vitamins. These methods reflect a broader trend: digital storytelling is now central to trust-building and brand recall in functional foods.

10.2 Personalization and Predictive Nutrition Technologies

One of the most exciting developments in functional food marketing is the integration of personalized nutrition, where products are tailored based on individual health data, such as gut microbiome profiles, genetic markers, or lifestyle inputs. Internationally, ZOE (UK/US) exemplifies this trend, offering AI-based food scores and tailored dietary plans based on at-home tests. In India, the market is beginning to respond. Emerging ventures like BillionCheers and Genefitletics are exploring AI-powered wellness diagnostics and personalized supplement subscriptions, combining Indian nutraceutical wisdom with modern health-tech. While these are early-stage efforts, they reflect growing demand among urban Indian consumers for precision nutrition, particularly in areas like digestive health, metabolic wellness, and cognitive performance. Personalized marketing based on digital diagnostics allows brands to shift from mass messaging to micro-targeted communication, offering greater perceived value and higher engagement.

10.3 Sustainability as a Functional Proposition

Today's health-conscious consumer is also deeply aware of environmental sustainability. This shift has made sustainable sourcing, eco-packaging, and plant-based innovation not just ethical choices but also compelling marketing narratives. Brands like Huel and VitaCup incorporate sustainability into their brand identity, appealing to younger demographics that seek purpose-driven consumption.

In India, startups like Oziva and Kapiva have embedded sustainability into both product and promotion. Oziva emphasizes its plant-based, clean-label credentials, with third-party verification of its safety and environmental practices. Kapiva integrates traditional Indian ingredients sourced from sustainable farms into modern wellness formats, communicating both authenticity and responsibility. These brands demonstrate how sustainability is becoming a strategic differentiator in functional food marketing.

10.4 Future-Forward Trends: Artificial Intelligence, Voice Commerce, and Functional Snacking

Looking ahead, functional food marketing will be increasingly shaped by emerging technologies and consumption formats. The adoption of Artificial Intelligence (AI) for dynamic product suggestions, voice-assisted ordering (via smart devices like Alexa), and gamified health tracking will enable brands to meet consumers where they are, across digital ecosystems. Another high-growth area is functional snacking, the integration of health benefits into convenient, flavourful snacks like nootropic nut bars, collagen-enriched gummies, or prebiotic chocolate bites. Euromonitor International (2023) projects that this category will grow at a global CAGR of over 16%, fueled by rising demand for convenient, health-focused products (Euromonitor International, 2023). Indian brands can capitalize on this shift by developing culturally resonant yet modernized formats, such as fortified millet cookies, herbal energy laddus, or mini chyawanprash shots, packaged and marketed for the urban professional or fitness enthusiast. When supported by smart labeling, influencer-led storytelling, and compliance-backed claims, these innovations hold global potential.

References

Ahmed, N. 2024. Empowering women in India through innovative incubators and accelerators for energy entrepreneurship. In: *University Incubators and Their Role in the Entrepreneurial Ecosystem* , (pp. 293–322). University Press.

Baker, M.T., Lu, P., Parrella, J.A. and Leggette, H.R. 2022. Consumer acceptance toward functional foods: A scoping review. *Int. J. Environ. Res. Public Health, 19*, 1217.

Berto, G. 2024. *Engaging Generation Z in an Institutional Project with a Forward-Looking Perspective: The Case of VeniSia.* Ph.D. Thesis, Venice.

Bhaiswar, R., D. Chawla and H. Joshi. 2025. Segmenting electronic word-of-mouth consumers: A lifestyle and behaviour-based study. *Global Business Review*, 09721509251374405.

Bhasale, A.L., A. Sarpatwari, M.L. De Bruin, J. Lexchin, R. Lopert, P. Bahri and B.J. Mintzes. 2021. Postmarket safety communication for protection of public health: A comparison of regulatory policy in Australia, Canada, the European Union, and the United States. *Clin. Pharmacol. Ther., 109*, 1424–1442.

Bigliardi, B. and F. Galati. 2013. Innovation in functional foods: Consumer demand and market dynamics. *Trends Food Sci. Technol., 31*, 118–129.

Bonants, R.B.J. 2020. *The Role of Startups in Collective Innovation Agendas: Case-study of Dutch Top Sectors Agri&Food and Horticulture.* Master's Thesis, Department Innovation, Environment and Energy Sciences, Faculty of Geosciences, Utrecht University, Utrecht.

Borchers, A.T., C.L. Keen and M.E. Gershwin. 2016. The basis of structure/function claims of nutraceuticals. *Clin. Rev. Allergy Immunol., 51*, 370–382.

Burges Watson, D., T. Moreira and M. Murtagh. 2009. Little bottles and the promise of probiotics. *Health, 13*, 219–234.

Cioeta, R., P. Muti, M. Rigoni, A. Cossu and E. Giovagnoni. 2024. Actitan: A natural complex for managing diarrhea: Insights from cross-sectional survey research involving patients, pharmacists, and physicians. *Gastrointest. Disord., 6*, 753–764.

Chhabra, E. 2022. San Diego startup turns to custom probiotics to solve health issues. *Forbes*, Online Article, New York, USA.

Danuser, D.J. 2024. *Evoking & Embodying Science: Representing Science and Social Differences in Early 21st Century Television Commercials.* Ph.D. Thesis, University of Pittsburgh, Pittsburgh, USA.

EFSA. (2022). Health Claims and Scientific Substantiation: Updated List of Authorized Claims. Retrieved from, https://www.efsa.europa.eu/en/topics/topic/health-claims.

EFSA Panel on Additives and Products or Substances used in Animal Feed (FEEDAP). 2012. Scientific opinion on the safety and efficacy of L-carnitine as a feed additive for all animal species based on a dossier submitted by EUROPE-ASIA Import Export GmbH. *EFSA J., 10*, 2677.

Euromonitor International. 2023. Global Market Shares in Functional Dairy: Activia Case Study. Retrieved from, https://www.euromonitor.com.

FAO. (2021). *Food Systems and Nutrition: Building Pathways to Sustainable Development.* Rome: Food and Agriculture Organization of the United Nations, Rome, Italy.

Farid, M., Kodama, K., Arato, T., Okazaki, T., Oda, T., Ikeda, H. and Sengoku, S. 2019. Comparative study of functional food regulations in Japan and globally. *Glob. J. Health Sci., 11*, 132.

FMCG Gurus. (2022). Post-COVID-19: The evolution of health and wellness trends. *FMCG Gurus*. Retrieved from https://fmcggurus.com.

FSSAI. 2022. *Food Safety and Standards (Health Supplements, Nutraceuticals, Food for Special Dietary Use, Food for Special Medical Purpose, Functional Food, and Novel Food) Regulations.* Retrieved from, https://www.fssai.gov.in.

Grand View Research. 2024. *Functional Foods Market Size, Share, & Trends Analysis Report 2024–2030.* Retrieved from https://www.grandviewresearch.com/industry-analysis/functional-food-market.

Grelet, G., Lang, S., Merfield, C., Calhoun, N., Robson-Williams, M., Horrocks, A. and Kerner, W. 2021. Regenerative agriculture in Aotearoa New Zealand: Research pathways to build science-based evidence and national narratives. New Zealand. FAO.

Grunert, K.G. 2005. Food quality and safety: Consumer perception and demand. *Eur. Rev. Agric. Econ., 32*, 369–391.

Gulati, K., Thokchom, S.K., Joshi, J.C. and Ray, A. 2021. Regulatory guidelines for nutraceuticals in India: An overview. *Nutraceuticals*, 1273–1280.

Haider, R. and G. Das. 2021. Marketing and regulatory issues for functional foods and nutraceuticals. *Int. J. Econ. Bus. Manag. Acc., 3*, 109–122.

Hieke, S., Cascanette, T., Pravst, I., Kaur, A., Van Trijp, H., Verbeke, W. and Grunert, K.G. 2016. The role of health-related claims and symbols in consumer behaviour: The CLYMBOL project. *Agro Food Industry Hi-Tech., 27*(3), 26–29.

Huel (2023). *Annual Brand Strategy and Growth Report.* Retrieved from, https://huel.com.

ICAR-NIANP. 2023. *Top Three Food Trends Shaping Demand in Indian Cities.*

Retrieved from, https://www.euromonitor.com/article/top-three-food-trends-shaping-demand-in-indian-cities

Jesuola, G. D. 2024. *Impact of the Nigerian Government's Fuel Subsidy Removal: Data Analysis and Suggestions* (Master's thesis, Texas A&M University-Kingsville). Retrieved from https://www.proquest.com/openview/cf33c0e6f013cb3b6415edb1ce5c8c8f/1?pq-origsite=gscholar&cbl=18750&diss=y.

Katan, M.B. 2004. Health claims for functional foods: Regulations vary between countries and often permit vague claims. *BMJ*, *328*, 180–181.

Keller, K.L. 2001. *Building Customer-Based Brand Equity: A Blueprint for Creating Strong Brands*. Marketing Science Institute Working Paper. Cambridge, MA, USA.

Khatun, R. and M. Bhattacharya. 2024. Probiotic drinks: Balancing the benefits with possible side-effects. *Int. J. Biotechnol. Trends Technol.*, *14*, 6–14.

Kotler, P. and K.L. Keller. 2016. Marketing Management (15th Edn.). Pearson Education, London.

Kotler, P., Keller, K.L., Ancarani, F. and M. Costabile. 2016. *Marketing Management* (European Edition). Pearson Education, London.

Magnier, L., J. Schoormans and R. Mugge. 2016. Judging a product by its cover: Packaging sustainability and perceptions of quality in food products. *Food Quality and Preference*, *53*, 132–142.

McKinsey. (2022). *The Future of Wellness: Functional Foods and Digital Health Convergence*. Retrieved from, https://www.mckinsey.com.

Mellentin, J. 2007. *Danone Actimel: Innovation Builds a Probiotic Mega-brand*. New Nutrition Business, London.

Mintel. 2023. *Positioning in Functional Food Brands: Bio&Me and the Rise of Science-Led Marketing*. Retrieved from, https://www.mintel.com.

Nestlé. 2020. *Creating Shared Value Progress Report*. Retrieved from https://www.nestle.com/csv.

NielsenIQ. 2023. *Who Buys Functional Foods and Why: Global Functional Food Consumer Report*. NielsenIQ, Global Insights Division.

Nielsen, K.E. 2016. Health beneficial consumer products: Status and trends. In: Developing Food Products for Consumers with Specific Dietary Needs (pp. 15–42). Woodhead Publishing.

Niva, M. 2007. 'All foods affect health': Understanding of functional foods and healthy eating among health-oriented Finns. *Appetite*, *48*, 384–393.

O'Hagan, L.A. 2022. Modernity, beauty and the Swedish "way of life": Lifestyle marketing in Stomatol toothpaste advertisements, 1910–1940. *J. Hist. Res. Mark.*, *14*, 424–452.

Oliveira, D., L. Machín, R. Deliza, A. Rosenthal, E.H. Walter, A. Giménez and G. Ares. 2016. Consumers' attention to functional food labels: Insights from eye-tracking and change detection in a case study with probiotic milk. *LWT – Food Science and Technology*, *68*, 160–167.

Pandey, A. and Joseph, J. 2025. Analyzing the evolving landscape of digital food platforms and technology: The digital plate. In: *Food in the Metaverse and Web 3.0 Era: Intersecting Food, Technology, and Culture* (pp. 1–18). IGI Global Scientific Publishing.

Plasek, B., Z. Lakner and Á. Temesi. 2019. I believe it is healthy: Impact of extrinsic product attributes in demonstrating healthiness of functional food products. *Nutrients*, *13*(10), 1–16.

Qi, D. and B.E. Roe. 2018. Creating value for customers, clients, partners, and society: An introduction to the special issue on food waste. *J. Food Prod. Mark.*, *24*, 495–499.

Rubin, J., and Axe, J. 2023. *The Probiotic Diet: Improve Digestion, Boost Your Brain Health, and Supercharge Your Immune System*. Destiny Image Publishers, United Kingdom.

Satine, L.A. 2008. Is my yogurt lying? Developing and applying a framework for determining whether wellness claims on probiotic yogurts mislead. *Food & Drug L.J.*, *63*, 537._

Tang, Z., X. Xu, Y. Song, and H. Yang. 2022. Data analytics applications in the soda industry. In: *Int. Conf. Bus. Policy Stud.* (pp. 677–688). Springer Nature Singapore.

Unilever. (2021). *Making Sustainable Living Commonplace: Annual Report and Accounts 2021*. Retrieved from, https://www.unilever.com/investors/annual-report-and-accounts.

Verbeke, W. 2005. Consumer acceptance of functional foods: Socio-demographic, cognitive, and attitudinal determinants. *Food Qual. Prefer.*, *16*, 45–57.

Verma, S. 2025. The rise of green consumerism and consumer protection trust issues in India: An analysis of Patanjeli Ayurved Ltd's eco-claims. *East Afr. J. Law Policy Glob.*, *1*, 1–27.

Vijaykumar, S., J. McCready, P.L. Graham, and D. Morris. 2022. That gut feeling: Public perceptions of media coverage and science surrounding probiotic products. *Br. Food J.*, *124*, 3428–3446.

Wedel, M. and W.A. Kamakura. 2012. Market Segmentation: Conceptual and Methodological Foundations (2nd Edn.). Springer Science & Business Media, New York.

Wills, J. (2022). Foundations for Health Promotion-E-Book. Elsevier health sciences.

World Health Organization. (2022). *Equity in Health and Nutrition: A Framework for Action*. Geneva: WHO, Geneva.

Wu, W., A. Zhang, R.D. van Klinken, P. Schrobback, and J.M. Muller. 2021. Consumer trust in food and the food system: A critical review. *Foods*, *10*, 2490.

Yousaf, A., I. Amin and A. Gupta. 2017. Conceptualising tourist-based brand-equity pyramid: An application of Keller brand pyramid model to destinations. *Tourism and Hospitality Management*, *23*, 119–137.

CHAPTER 12

The Role of Biotechnology in the Development of Functional Foods

Surabhi Pandey[1], Kritika Bajpai[2] and Anurag Singh[3*]

1. Introduction

The fast-developing multidisciplinary discipline of biotechnology uses biological systems, living things, or their derivatives to create novel goods and technologies that improve both the environment and human well-being. It integrates concepts from biology, chemistry, physics, engineering, and computer science to create solutions across various sectors, including healthcare, agriculture, industry, and environmental management. Through the development of vaccines, gene therapy, stem cell therapy, and biopharmaceuticals like insulin and monoclonal antibodies, biotechnology has transformed illness diagnosis, treatment, and prevention in the medical profession (Pham et al., 2018). Agricultural biotechnology seeks to improve crop yield and tolerance to pests and environmental issues by creating genetically modified organisms (GMOs), biofertilizers, and disease-resistant plant varieties (Das et al., 2023). By using biological processes to produce biofuels, biodegradable polymers, and environmentally benign chemicals, industrial biotechnology, also referred to as white biotechnology minimizes pollution and lessens dependency on fossil fuels (Tang and Zhao, 2009). Environmental biotechnology employs bioremediation and waste management techniques using microorganisms to clean up contaminated sites and treat wastewater, contributing to sustainable environmental practices (Gavrilescu, 2010). Food biotechnology has improved food production and preservation through fermentation, the creation of functional foods, and enhanced food safety measures. Critical tools driving these advancements include genetic engineering, CRISPR-Cas9 gene editing, tissue culture, fermentation technology, and bioinformatics for analyzing complex biological data (Wikandari et al., 2023).

1.1 Historical Background

From antiquated customs to state-of-the-art scientific discoveries, the history of biotechnology is a colorful tale of human ingenuity. Early humans unknowingly practised biotechnology through the domestication of plants and animals, selecting and breeding species with desirable traits for improved yield, taste, and resilience. As far back as 6000 BCE, civilizations in Mesopotamia and Egypt harnessed the power of fermentation to cultivate essential crops like bread, beer, and wine, relying on naturally occurring microorganisms (Skinner, 2022).

[1-3] Department of Food Technology, Harcourt Butler Technical University, Kanpur-208002.
* Corresponding author: a.singh@hbtu.ac.in

DOI: 10.1201/9781003659808-12

The scientific era of biotechnology began in the 19th century when foundational discoveries unveiled the biological mechanisms behind these ancient practices. Louis Pasteur's work in the mid-1800s established the role of microbes in fermentation and disease, revolutionizing food safety and medical hygiene (Smith, 2012). Gregor Mendel laid the foundation for genetics at the same time by discovering the fundamental principles of heredity via his groundbreaking research with pea plants (Mendel, 1866). The word "biotechnology" was first used in 1919 by Hungarian agricultural engineer Károly Ereky, who envisioned it as an industrial process that could biologically turn raw materials into useful products (Selvakumar et al., 2023).

The 20th century marked a turning point in biotechnology thanks to developments in molecular biology and genetics. James Watson and Francis Crick discovered the double helix structure of DNA in 1953, revealing the molecular blueprint of life and opening the door to genetic modification (Selvakumar et al., 2023). Stanley Cohen and Herbert Boyer's invention of recombinant DNA technology in 1973, which enabled scientists to cut and splice DNA from different species, making genetic engineering conceivable (Cohen et al., 1972). The treatment of diabetes was completely changed by this discovery, which resulted in the creation of the first genetically modified medication, synthetic human insulin (Humulin), which was authorized in 1982 (Johnson, 1983). Global agriculture was revolutionized in the 1990s when genetically modified (GM) crops, such as pest-resistant corn and herbicide-tolerant soybeans, were commercially available (Walia et al., 2024).

With the conclusion of the Human Genome Project in 2003, which mapped all human genes and greatly increased our understanding of genetic illnesses, the 21st century saw tremendous improvements (Keller, 2022). The CRISPR-Cas9 gene-editing technique, which was developed in 2012, revolutionized the industry and created opportunities for the advancement of medical therapies, agricultural improvement, and the treatment of genetic illnesses by making it possible to precisely, effectively, and economically modify DNA (Wikandari et al., 2021). The accelerated development of mRNA COVID-19 vaccines by Pfizer-BioNTech and Moderna in 2020 demonstrates how genetic engineering may be used to urgent public health problems and is an example of how biotechnology has recently played a key role in global health (Uddin and Roni, 2021).

1.2 Modern Biotechnology

Modern biotechnology is a dynamic and transformative field that leverages advanced scientific techniques to manipulate biological systems for the benefit of healthcare, agriculture, industry, and environmental sustainability. Unlike traditional biotechnology, which relied on natural processes like fermentation and selective breeding, modern biotechnology incorporates cutting-edge tools such as genetic engineering, molecular biology, bioinformatics, and synthetic biology to develop innovative products and solve complex global challenges.

One of the most groundbreaking advancements in modern biotechnology is CRISPR-Cas9 gene-editing technology, introduced in 2012 (Wikandari et al., 2021). This tool allows scientists to make accurate and specific alterations to DNA, enabling the correction of genetic disorders, the treatment of inherited diseases, and the creation of enhanced crop varieties. In the realm of medical biotechnology, revolutionary treatments and diagnostic tools have been developed to combat diseases more effectively. Pfizer-BioNTech and Moderna's quick development of mRNA vaccines against COVID-19 showed how effective biotechnology can be in addressing international health crises. These vaccines used synthetic messenger RNA to trigger an immune response without using live pathogens, setting the stage for future vaccines and treatments (Patel et al., 2022). Similarly, by altering a patient's immune cells to recognize and destroy cancer cells, CAR-T cell therapy has transformed the treatment of cancer (Marco et al., 2023). The quality of life for millions of people worldwide is being improved by potential therapies for genetic abnormalities, organ failure, and tissue damage that are made possible by developments in gene therapy and regenerative medicine.

Biotechnology in agriculture has been central in contributing to the overall food security in the world by producing GM crops which yield more food, improved nutrition, and can withstand environmental stresses. Genetic engineering has produced crops that are GM to overcome malnutrition (Golden Rice), which is a vitamin-enriched dust-resistant maize that overcomes food shortages (Beyer et al., 2002). Pest-resistant crops and disease-resistant livestock are engineered using gene editing tools, and pesticides and antibiotics are minimized, thus promoting sustainable agriculture methods (Ricroch, 2019).

Industrial biotechnology, also known as white biotechnology, employs microorganisms and enzymes to sustainably produce bio-based products. This encompasses the production of biofuels such as ethanol and biodiesel, which lessen dependence on fossil fuels, as well as biodegradable plastics that aid in reducing plastic pollution (Tang and Zhao, 2009). Synthetic biology is further enabling the design of microbes that can produce high-value chemicals, textiles, and pharmaceuticals, making industrial processes eco-friendlier and more efficient (Horner et al., 2006).

In addressing environmental challenges, environmental biotechnology provides sustainable solutions for pollution control and resource management. Techniques like bioremediation use engineered microorganisms to clean up oil spills, detoxify hazardous waste, and degrade plastic pollutants (Tang and Zhao, 2009). Synthetic biology is advancing modern biotechnology by integrating biology with engineering principles to design and build entirely new biological systems. Researchers are developing synthetic organisms that can produce biofuels, pharmaceuticals (Cameron et al., 2014) and even lab-grown meat, providing sustainable alternatives to traditional industries (Post, 2012). Synthetic biology also has potential applications in space exploration, where engineered microbes could produce oxygen, food, and essential materials for long-term space missions (Verseux et al., 2016).

1.3 Functional Foods

Functional foods are foods that have been fortified or altered in some way to deliver health benefits other than the mere nutritional value (Temple, 2022). Biotechnology is essential in the preparation of these foods, as complex methods are employed to add or boost the concentration of these bioactive compounds, including genetic engineering, fermentation, and microbial transformations. The compounds have been revealed to provide therapeutic benefits, including enhancing the immune system, gut health, and lowering the risk of chronic disease occurrence, including heart disease, diabetes, and cancer. Functional foods can be designed according to the specific health requirements through biotechnology and provide more specific nutritional solutions to the consumer. In addition, biotechnology can also be used to fortify the foods that are generally consumed with useful elements so that they can help in the health enhancement of the population on a greater scale. The role of food biotechnology in the development of functional foods is very important, as these foods are foods that are beneficial in terms of their health advantage over and above normal nutrition. Biotechnology has made it easy to produce food that is not only essential in terms of nutrition but also overall well-being through genetic modification, fermentation processes and through the use of advanced bioengineering processes. Biotechnology is leading the challenge to solve global health problems, as well as enhancing the health of the population by improving the nutritional profile of food products. The review examines the current application of biotechnological innovations to the production of bioactive-enriched foods like vitamins, minerals, and probiotics that can be used to prevent or manage the occurrence of chronic diseases. Table 1 demonstrates the role of various functional foods in preventing diseases and enhancing well-being by providing various types of bioactive compounds to overall health.

2. Scope of Biotechnology in Food Industry

Biotechnology is swiftly becoming a significant participant in the food industry thanks to its innovative solutions in a number of sectors, such as the production and processing of food, sustainability enhancement, and health challenges. The scope of biotechnology in food systems

Table 1 Major Sources of Functional Foods and Associated Health Benefits.

Source of Functional Food	*Health Benefits*
Fatty Fish (Salmon, Mackerel, Sardines)	Rich in heart-healthy omega-3s, which lower inflammation and boost cognitive function.
Yogurt (with live cultures)	Probiotics aid digestion, immunity, and infection prevention.
Whole Grains (Oats, Brown Rice, Quinoa)	Rich in dietary fiber, which promotes digestive health, reduces cholesterol, and helps control blood sugar
Leafy Greens (Spinach, Kale, Swiss Chard)	Rich in antioxidants (vitamins A, C, and K), they reduce oxidative stress and may help protect against cancer and heart disease.
Nuts and Seeds (Almonds, Walnuts, Flaxseeds)	Rich in fiber, protein, and healthy fats, they support heart health, lower cholesterol, and enhance cognition.
Berries (Blueberries, Strawberries, Raspberries)	Rich in fiber, vitamins, and antioxidants, berries support immunity, reduce inflammation, and boost cognition.
Tomatoes	Rich in lycopene, an antioxidant that may lower the risk of heart disease, promote prostate health, and fight cancer.
Green Tea	Has catechins, potent antioxidants that may enhance metabolism, aid in weight loss, and lower the risk of heart disease.
Garlic	Contains sulfur compounds that may strengthen the immune system, decrease blood pressure, enhance heart health, and minimize the risk of several types of cancer.
Fermented Foods (Kimchi, Sauerkraut, Kefir)	Give probiotics that boost immunity, enhance digestion, and support a healthy gut microbiota.
Turmeric (Curcumin)	Promote brain health, lessen arthritic symptoms, and function as an antioxidant.
Citrus Fruits (Oranges, Lemons, Grapefruits)	Rich in vitamin C and flavonoids, they support immunity, skin health, and heart health.
Beans and Legumes	Rich in antioxidants, fiber, and protein, they help lower cholesterol, regulate blood sugar, and support digestion.

Source: Shaikh, 2022; Tafese Awulachew, 2024; Dixit et al., 2023.

is vast, ranging from traditional techniques like fermentation to cutting-edge genetic engineering, with applications extending to food safety, functional foods, waste valorisation, and alternative protein sources, as shown in Figure 1. These advancements hold promise for enhancing food quality, improving efficiency in production, and ensuring food security as the global population continues to grow (Tonukari and Omotor, 2010).

2.1 Improvement of Food Production and Processing Efficiency

Biotechnology plays a critical role in enhancing food production by introducing new methods to boost yields, enhance quality, and reduce the environmental effect of food systems. One of the primary applications of biotechnology in this area is crop genetic modification (GM). GMOs can increase productivity and reduce the need for chemical pesticides by making crops more resilient to diseases, pests, and environmental stresses (Beyer et al., 2002). Furthermore, biotechnology is transforming food processing through the development of specialized enzymes that facilitate various processes, like fermentation, texturization, and preservation (Singh et al., 2016). In the dairy industry, for instance, enzymes like rennet, which are traditionally derived from animal sources, can now be produced through recombinant DNA technology. These biocatalysts are used to optimize cheese production, improve yield, and reduce production costs (Amiri et al., 2021). Additionally, enzymes used in the brewing industry facilitate the breakdown of starches into fermentable sugars, enhancing the fermentation process for beer production (Sharma et al., 2017). Enzymatic solutions are also utilized to strengthen the texture of bakery products, such as cakes and bread, and to enhance the shelf-life of packaged foods (Souza, 2010).

2.2 Fermentation and Microbial Biotechnology

One of the most traditional and well-established biotechnological methods utilized in the food sector is fermentation. Traditional fermentation techniques are being improved, and their uses are being expanded by recent developments in microbial biotechnology. Probiotics like *Lactobacillus* and *Bifidobacterium* are added to fermented foods like cheese, kefir, and yogurt in the dairy industry to support gut health. Biotechnology enables the targeted cultivation of these beneficial microorganisms, improving the quality and functionality of probiotic foods (Dahiya and Nigam, 2022). Beyond probiotics, biotechnology has advanced the use of microorganisms in producing bioactive compounds like vitamins, antioxidants, and flavours. For example, genetically engineered strains of *Saccharomyces cerevisiae* (brewer's yeast) can produce B vitamins such as folate, which are then added to functional foods (Rai et al., 2019). Similarly, specialized strains of fungi and bacteria are being used to enhance the flavours and textures of fermented products, creating new varieties of cheese or fermented meats with unique tastes and qualities (Kumar et al., 2017). Another key application of microbial biotechnology is in the development of alternative proteins. The fermentation process is increasingly being used to cultivate plant-based proteins or even lab-grown proteins from microorganisms, such as fungal protein (mycoprotein), which is used in products like Quorn (Majumder et al., 2024). The efficient large-scale manufacturing of these proteins made possible by biotechnology contributes to the rising demand for plant-based diets and meat alternatives, providing a more sustainable alternative to conventional animal agriculture.

2.3 Food Safety, Quality Control, and Traceability

Food safety is the most crucial factor in the food sector, and biotechnology is essential to improving food safety protocols. Biosensors that can quickly identify foodborne pathogens, including *Salmonella, E. coli,* and *Listeria,* have been made possible via genetic engineering. These biological sensors have biological components, e.g., antibodies or enzymes that are used to react with the pathogen to create a measurable signal. The capacity to identify pathogens quickly and precisely reduces the risk of contamination and improves food security throughout the production chain (Horner et al., 2006). Another impact of biotechnology to quality control is that it offers ways of determining food adulteration, determining food freshness and tracking the condition of food storage (Horner et al., 2006).

2.4 Functional Foods and Nutraceuticals

The growing demand for functional foods, which provide health benefits beyond basic sustenance, is driving biotechnology's expansion in this field. Biotechnology can improve food nutritional profiles by genetically modifying microbes and crops. To address vitamin A shortage in underdeveloped nations, for instance, golden rice, a genetically engineered rice enhanced with provitamin A (beta-carotene), was created (Beyer et al., 2002). To address the pervasive nutritional deficits, biofortified crops that are high in micronutrients like iron or zinc are also being produced. Nutraceuticals, or foods with therapeutic qualities, are another product of biotechnology. Biotechnology has revolutionised the nutraceutical sector by improving the nutritional quality, yield, and functionality of crops and microbes through techniques like genetic engineering. Examples include *Golden Rice* (vitamin A-enriched) and the *Flavr-Savr tomato* (enhanced shelf-life). It also enables large-scale production of bioactive compounds such as flavonoids, alkaloids, saponins, and essential oils, as well as microbial-based therapeutic agents (Rajak et al., 2023). For example, biotechnology may be used to include functional elements like antioxidants, phytochemicals, and omega-3 fatty acids into food items (Reynolds and Martirosyan, 2016). It is feasible to boost the concentration of advantageous substances by fermenting microbes or genetically modifying plants, producing meals that strengthen the immune system, enhance heart health, and fend off chronic illnesses (Beyer et al., 2002; Yadav et al., 2012).

2.5 Sustainable and Alternative Food Sources

The creation of sustainable food supplies is one of the most exciting uses of biotechnology in the food sector. Biotechnology is offering ways to fulfill the growing need for food produced in an environmentally appropriate way, despite the mounting strain on land, water, and other natural resources (Husaini and Sohail, 2023). One potential way to lessen the environmental effect of cattle farming is through cellular agriculture, which includes growing animal cells in bioreactors to make lab-grown meat (Djisalov et al., 2021). This technology helps address problems with resource utilization, greenhouse gas emissions, and animal welfare by doing away with the need to raise and slaughter animals. In addition to cultured meat, biotechnology is being used to create vegan meat alternatives that resemble the taste and texture of animal products, such as those made from soy, peas, or mycoprotein (Sharma et al., 2022). These plant-based foods not only give consumers a sustainable option, but they also contribute to reducing the carbon footprint of food production.

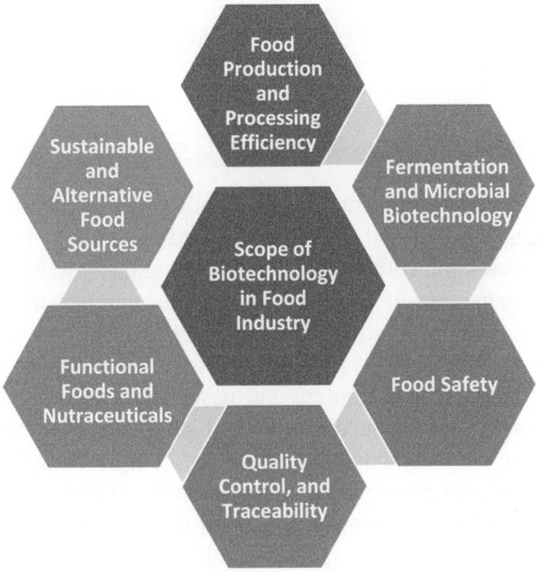

Fig. 1 Scope of Biotechnology in the Food Industry.

3. Biotechnology Tools in the Enhancement of Functional Foods

3.1 Genetic Engineering

Genetic engineering is the state-of-the-art science that deals directly with modifying the genetic composition of an organism with the intent to modify its characteristics. This makes it possible to introduce new traits or improve on old ones, or remove nuisance genetic characters through manipulating the DNA codes. Genetic engineering has transformed many sectors, such as industrial biotechnology, environmental science, medicine, agricultural industry and has given solutions to several of the biggest issues the world is grappling with nowadays (Figure 2). Genetic engineering

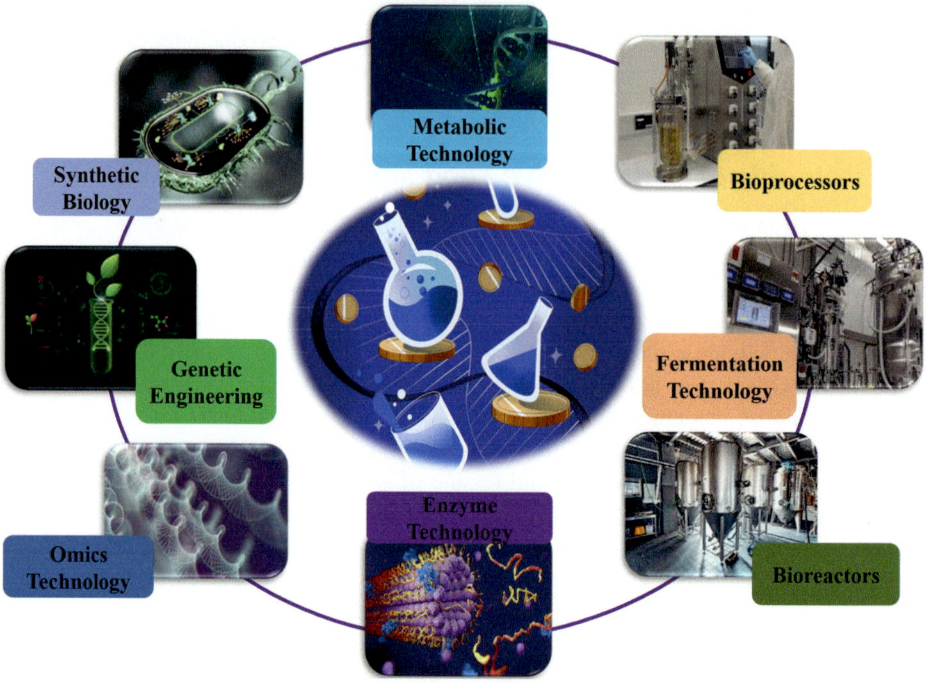

Fig. 2 Various biotechnology tools applied for the enhancement of functional foods.

revolves around the use of advanced techniques for modifying genetic information. Recombinant DNA technology is among the first and most widely used techniques. It involves the integration of the DNA of various species and embedding it in a host organism to bring about novel qualities. Indicatively, bacteria can be genetically engineered to secrete human insulin (Johnson, 1983), which is an effective and cheaper way of treating diabetes. Another new technology is the gene-editing technology CRISPR-Cas9, which allows making very specific and precise changes to the DNA (Redman et al., 2016). Since CRISPR can cut and paste genes with great precision, contrary to other technologies, the researchers can study the role of genes, create disease-resistant agricultural products, and repair genetic defects (Wikandari et al., 2021). Agriculture has been largely affected by genetic engineering in the issue of production and security of food across the globe. GM crops are those whose genetic composition has been altered using genetic engineering techniques in order to add or improve features that the species would not normally have. These modifications enable the crops to withstand pests, diseases, herbicides, and other unfavorable environmental elements including drought and high temperatures (Kumar et al., 2020). For instance, Bt corn has been genetically modified to produce a natural insecticide, doing away with the need for chemical pesticides, and golden rice has been genetically modified to increase vitamin A content to combat the common cases of deficiency in most parts of the developing world. In order to prevent related health problems including blindness and weakened immune systems, rice will provide a handy source of this vital vitamin (Beyer et al., 2002). In the food sector, genetic engineering has fundamentally altered how food is produced, processed, and consumed. Scientists and agricultural businesses have been able to improve food quality, boost output, and strengthen resilience to pests, illnesses, and environmental difficulties by directly manipulating the genetic composition of crops and livestock. Global issues like hunger, food security, and the environmental effects of agriculture may all be solved with this technology. In order to increase food quality, nutritional content, and production efficiency, genetic engineering has been required in the food processing business. As an illustration, genetic engineering microbes such as bacteria and yeast are used to produce enzymes that enhance the fermentation and cheesemaking processes (Bharadwaj and Singh, 2022). The production of rennet to make cheese with the help of genetically modified microorganisms is an example because now sources of animal origin are not needed (Mendley et al., 2021). GM plants, such as soybean and canola, are used in oil production to produce oils that have a healthier composition, such as a high-oleic acid oil, good as a source of reductions in trans-fat intake (Bansal and Kour, 2022). Moreover, GM yeast strains find extensive application in baking and brewing procedures to enhance the efficiency of the fermentation process, facilitate the production process, and improve the flavour (Zhou et al., 2021). It is also possible to create vitamins and nutrients (vitamin B12 or folic acid) using genetic engineering and add them to processed foods to increase their nutritional value (Balabanova et al., 2021). Furthermore, GM microorganisms are used to produce synthetic sweeteners such as stevia and sucralose, which is a more efficient and scalable way of producing sugar substitutes in food and beverages (Hashmi et al., 2024). These advancements help to meet consumer demands for healthier, more sustainable food options while optimizing processing techniques.

3.2 Fermentation Technology

A key component of contemporary biotechnology is fermentation technology, which includes a broad range of procedures that use enzymes and microbes to create several useful compounds. The English word fermentation comes from the Latin verb "fevere", which means "boil" (Ezemba et al., 2022). Fermentation was once used to produce and preserve food, but thanks to developments in microbiology, biochemistry, and engineering, it is now a sophisticated industrial technology with uses in the food industry, pharmaceuticals, agriculture, biofuels, and environmental management. Fundamentally, fermentation uses microbes' metabolic activity to biochemically transform organic substrates like sugars and starches into desirable products. Depending on the procedure and the kind of product being made, this transformation may take place in anaerobic (without oxygen) or aerobic (with oxygen) environments. Key parameters influencing the efficiency of fermentation comprise pH, oxygen supply, nutrient availability, temperature, and agitation (Stanbury et al., 2013). Maintaining optimal conditions ensures maximum productivity and quality. Fermentation is powered by microorganisms, and the type of microorganism used depends on the final output. In addition to adding flavor, microorganisms that ferment food items also create bioactive chemicals that increase the foods' functionality, making them functional foods (Sun et al., 2022). The necessity to develop novel microbial strains for fermented food production is highlighted by the rise in chronic illnesses such obesity, diabetes, cardiovascular disorders, and cancer, which are frequently associated with poor dietary practices. These specialized microbes can enhance the health-promoting properties of foods, offering protection against these serious health issues (Yadav et al., 2012).

Biotechnology has revolutionized microbial strain development, a cornerstone of fermentation technology. Microorganisms may be precisely GM to increase product production, improve substrate consumption, and strengthen resistance to fermentation-related stressors thanks to advanced genetic engineering techniques like CRISPR-Cas9 and recombinant DNA technology (Wikandari et al., 2021). Mutagenesis and adaptive evolution further contribute to developing robust strains for industrial-scale fermentation. These approaches have significantly improved the efficiency and consistency of producing fermented foods like yogurt, cheese, beer, and soy-based products.

In the food sector, functional foods enriched with nutrients that promote health have gained attention. By fermenting dairy and plant-based substrates, biotechnology makes it easier to produce probiotics and prebiotics (Gupta and Abu-Ghannam, 2012). The production of vitamins, vital fatty acids, and bioactive substances is improved by engineered microbial strains. Fortified foods and beverages, such as kombucha, probiotic yogurts, and fermented plant-based drinks, have been developed as a result of this innovation, improving the health and wellness of consumers (Singh et al., 2016). Customized fermentation methods have also been developed to improve the bioavailability of minerals and the digestion of plant-based vegan proteins in response to the growing demand for plant-forward diets (Kussman et al., 2023).

Development of dairy products has been revolutionized by fermentation technology, which is powered by biotechnology. Lactic acid bacteria and other microbial starter cultures are cultivated to produce cheese and yogurt with better flavour, texture, and nutritional qualities. Furthermore, by fermenting nuts, grains, and legumes, biotechnology enables easier production of plant-based dairy alternatives, providing a variety of choices for customers who are lactose intolerant or vegan. Fermentation technology is used in the developing field of cultured meat production to cultivate animal cells under regulated conditions, providing ethical and sustainable meat substitutes (Post et al., 2020). Mycoprotein, derived from fungal fermentation (e.g., *Fusarium venenatum*), serves as a high-protein, low-fat meat substitute (Majumder et al., 2024). Hybrid products combining plant-based and cultured components are advancing, offering improved sensory experiences and addressing the growing demand for sustainable protein sources.

3.3 Enzyme Technology

Enzyme technology has become an integral component of food technology, revolutionising food processing, preservation, and enhancement. As biological catalysts, enzymes speed up biochemical reactions in mild environments, which makes them perfect for a range of food industry uses (Thapa et al., 2019). Refinements in biotechnology have enabled the production and optimisation of enzymes through microbial fermentation, genetic engineering, and protein engineering, resulting in more efficient and sustainable food production and processing practices (Patel et al., 2023). Enzyme technology aligns with the growing demand for cleaner and greener food production technologies by offering specificity, eco-friendliness, and the ability to replace harsh chemical processes. Food enzymes are majorly produced via microbial fermentation using bacteria, fungi, and yeast (Raveendran et al., 2018). GM of microorganisms has further enhanced enzyme yield, specificity, and stability. Although both plants and animals may provide enzymes, microbial production provides a more scalable and sustainable option. Singh et al., (2016). Generally used enzymes in the food industry include amylases, proteases, lipases, lactase (β-galactosidase), pectinases, glucose oxidase and cellulases. These enzymes contribute to various processes such as breaking down starches, hydrolyzing proteins, enhancing flavors, and improving product texture (Chaudhary et al., 2015). Enzymes play diverse and critical roles in improving food processing efficiency, product quality, and sustainability. In the baking industry, enzymes like amylases and proteases improve dough handling and bread volume, while lipases extend shelf life (Souza, 2010). In the dairy industry, rennet enzymes and lactase facilitate cheese production and lactose-free dairy. Lipases have long been added to food to change its flavor by synthesizing alcohols and short-chain fatty acid esters, which are known to have flavor and aromatic properties (Hasan et al., 2006). Brewing and distilling benefit from enzymes that enhance sugar conversion and clarity. Juice and beverage production utilizes pectinases and cellulases for better extraction and clarity (Sharma et al., 2017), and proteases tenderize meat (Bekhit et al., 2014). Oil extraction processes use enzymes to increase yield and reduce chemical usage (Hasan et al., 2006). Innovations in biotechnology have facilitated the innovation of highly efficient enzymes tailored for industrial food processing. Protein engineering improves enzyme stability and activity, while immobilized enzymes allow reuse and continuous processing (Rocha et al., 2022). Directed evolution creates enzymes with enhanced properties, making them more robust and adaptable to industrial conditions. Enzyme technology also facilitates the extraction and modification of food components, enhancing product functionality and nutritional value. Lipases and proteases develop complex flavours in cheese and soy products, phytases improve mineral bioavailability, and cellulases modify dietary fibres for better texture and health benefits (Kumar et al., 2010). Enzymatically modified starches and proteins are used in low-fat food formulations, offering healthier alternatives without compromising quality (Bangar et al.,

2022). Enzymes facilitate sustainable food production by reducing waste and optimizing resource utilization. Enzymes are employed to create biodegradable packaging materials from agricultural waste, further promoting sustainability (Pandey et al., 2024).

3.4 Metabolic Engineering

Mallikarjuna and Yellamma (2019) define metabolic engineering as the controlled modification of an organism's metabolic pathways to improve knowledge and utilization of cellular processes related to energy conversion, chemical transformations, and the synthesis of supramolecular structures. Genetic engineering, synthetic biology, and systems biology are used in metabolic engineering to modify an organism's metabolism in order to produce useful food-related molecules (Helmy et al., 2020). It enables the exact modification of biological systems to meet the growing need for healthy, sustainable, and high-quality food products. One of the most well-known applications of metabolic engineering in the food industry is fermentation-based food manufacturing. Microorganisms like bacteria, yeast, and fungus are often modified to enhance the synthesis of natural taste compounds, amino acids, and vitamins—all necessary components in a range of food items (Bution et al., 2015). Metabolic engineering has made it possible to manufacture more essential amino acids, such as tryptophan and lysine, which are used in animal feed and as food additives (Mallikarjuna and Yellamma, 2019). Probiotics' capacity to endure in the human digestive system and increase their health benefits can be improved through metabolic engineering (Yadav et al., 2018). Another important use of metabolic engineering is the synthesis of substitute proteins, which is motivated by the world's need for plant-based and sustainable foods. Through genetic modifications, microorganisms such as yeast or bacteria can be engineered to produce single-cell proteins that serve as nutritious and eco-friendly alternatives to animal-based proteins (Li et al., 2024). These proteins can be integrated into plant-based meat substitutes or other food items, helping to lessen the environmental impact of traditional animal husbandry and address food security problems.

3.5 Omics Technology

In the food sector, omics technology, also referred to as "foodomics" technology, is a bioinformatics tool that effectively organizes and interprets data obtained from food screening concerning taste, flavour, food quality, and safety (Zhang et al., 2024). By exposing their interactions with targeted genes and their effects on metabolites and proteins, this method improves our comprehension of the molecular functions of certain food additives, such as nutritional agents and preservatives. Food products with preservatives and functional features might be more successfully created with this knowledge. Transcriptomics, proteomics, metabolomics, and genomes are examples of omics approaches that provide the essential data required to systematically address food safety concerns (Davis, 2010). Owing to the development of cutting-edge genomics technology like next-generation sequencing, sometimes referred to as whole-genome sequencing, studying microbial genomes now requires much less time and money. Whole-genome sequencing can be used to detect, monitor, quickly identify, and analyze foodborne diseases in order to help avoid them and forecast future outbreaks (Prakash et al., 2020). Recent developments in sequencing technologies, such as high-throughput sequencing, restriction-site associated DNA sequencing, and genotyping-by-sequencing of medicinal plants, have illuminated the biosynthetic mechanisms underlying the synthesis of secondary metabolites in plants that can be used as functional and preservative agents (Boutanaev et al., 2015; Hao and Zhao, 2015; Mehta and Hasija, 2018). Global interest in bioinformatics applications in food science and nutrition, especially in foodomics, has increased due to recent studies that have demonstrated the beneficial relationship between functional food ingredients and health/disease (Prakash et al., 2020).

3.6 Synthetic Biology

Synthetic biology is the study of creating new biological components, instruments, and systems that are not found in nature as well as modifying biological systems that currently exist to perform certain functions. In 1980, Hobom used the term "synthetic biology" to describe microorganisms that have undergone genetic modification through the application of recombinant DNA technology (Tyagi et al., 2014). Synthetic biology has the potential to alter the global food system. It has a significant impact on agriculture and food systems in a number of ways, including innovations in agricultural practices, food safety protocols, disease treatment delivery methods, molecular and cellular biology tools, biosensors for pathogen detection, pesticides, packaging materials, environmental protection strategies, and public and workforce education (Horner et al., 2016). The synthetic biology technique has led to the discovery of new antibacterial chemicals that are effective against a variety of food spoilage and hazardous pathogens (Tyagi et al., 2014). Additionally, the coupling of nanotechnology and synthetic biology has facilitated the development of sensors and detection systems for the timely identification of product spoilage, quality control, and abuse detection both at the source and along the production chain (Horner et al., 2016). Synthetic biology is rapidly growing in importance globally in the food and dairy sectors because of its many fascinating applications. These include making products like chips, ice cream, chocolate, and soft drinks that are marketed as healthier substitutes by reducing calories, fat, or carbohydrates and increasing protein, fiber, or vitamins (McClements, 2019). Furthermore, synthetic biology, which speeds up manufacturing and lowers the cost of ingredients and processing, is enabling greater flavors, colorings, nutritional additives, and processing aids. Another crucial area of focus is developing meals that may alter their color, flavor, or nutritional characteristics in response to a person's dietary needs, allergies, or taste preferences. For big food companies, this is a key research objective. Furthermore, by detecting bacteria, degradation, or nutrient loss and responding by releasing flavors, antimicrobials, nutritional supplements, or colors, synthetic biology is creating smart packaging that may extend food's shelf-life (Tyagi et al., 2014).

3.7 Bioprocessing and Bioreactors

The utilization of biological systems or creatures, such as enzymes, bacteria, or even plant and animal cells, to process raw materials and create desired food products is known as bioprocessing. Utilizing these biological agents' inherent capacities to turn foodstuffs into food items with better nutritional value, longer shelf lives, and distinctive flavors or textures is the aim. Fermentation, the synthesis of enzymes, and bio-preservation are just a few of the food industries where bioprocessing is essential. A bioreactor is a system or tank that provides a controlled environment for the growth of microorganisms, animal cells, or plant cells to support various bioprocessing procedures (Allman, 2018). Bioreactors are necessary for increasing the production of food and drink products that rely on fermentation or other biochemical processes. These systems are used to optimize factors such as temperature, pH, oxygen levels, and nutrient availability in order to ensure maximum efficiency and production of the desired product (Martin and Vermette, 2005). Stirred tank and airlift bioreactors are the most popular types of bioreactors used in food production. In order to guarantee complete mixing of the culture medium, provide consistent nutrient distribution, and stop cells or microbes from settling, a stirred tank bioreactor is fitted with a mechanical stirrer. Because continuous agitation and aeration are necessary to maintain ideal conditions for microbial growth, these bioreactors are crucial for the manufacture of fermented goods such as wine, beer, and yogurt. Owing to their great scalability, stirred tank bioreactors are perfect for both large-scale industrial food production and laboratory-scale food production. The airlift bioreactor creates a circulating flow of liquid through the movement of air bubbles rising through the medium. This gentle mixing minimizes shear forces, making it particularly useful for processes that require careful handling of sensitive microorganisms. Airlift bioreactors are often employed in the production of delicate bioproducts like probiotics or in the fermentation of food ingredients such as mushrooms, where minimal damage to cells is crucial. Both stirred tank and airlift bioreactors are essential in food production, each serving different needs depending on the nature of the fermentation and the fragility of the microorganisms involved (Spier et al., 2011).

The use of bioprocessing and bioreactors in the food industry offers several significant benefits. One of the most important advantages is efficiency and scalability. Bioreactors allow for the controlled, large-scale production of food products, ensuring consistent quality and yield. The ability to replicate small-scale lab results in an industrial setting is crucial for mass production, especially when dealing with biological processes that are sensitive to environmental conditions (Mitra and Murthy, 2022). More recent innovations in bioreactor technology include photobioreactors that use light to cultivate microalgae rich in bioactive compounds. These bioreactors enhance the nutritional value and functional characteristics of food ingredients such as omega-3 fatty acids and antioxidants. These microalgae are increasingly incorporated into functional beverages and dietary supplements (Shekh et al., 2022). Novel encapsulation techniques developed in bioreactor environments improve the stability and bioavailability of these compounds. Another major benefit bioprocessing offers is sustainability. Bioprocessing often involves renewable resources such as plant-based materials, and the processes themselves tend to be more environmentally friendly compared to traditional chemical or synthetic methods (Seth and Meena, 2024). Fermentation and enzyme-based processes, for instance, can produce food products with minimal waste, and bioreactors help ensure that resources are used efficiently. Additionally, bioprocessing makes it possible to create novel food items. A breakthrough facilitated by bioprocessing is precision fermentation. Precision fermentation technology uses genetically engineered microorganisms to produce animal-free proteins such as dairy whey, collagen, and egg white proteins. By encoding protein synthesis instructions into microbial DNA, precision fermentation allows scalable, sustainable production of complex proteins with identical functionality to their animal-derived counterparts, thus effectively reducing environmental footprints compared to traditional agriculture (Knychala et al., 2024). Food producers can produce innovative goods with new flavors, textures, or health advantages by manipulating microbes and enzymes. For instance, developments in bioprocessing technologies that enable the creation of meat-like textures and flavors using plant proteins have contributed to the rising demand for plant-based meat substitutes (Djisalov et al., 2021). Similarly, a major step toward lessening the environmental effect of meat production may be the capacity to generate cultured meat in bioreactors (Munteanu et al., 2021).

4. Applications of Biotechnology in the Development of Functional Foods

4.1 Nutrient Enhancement

"Nutrient enhancement through biotechnology" refers to the use of genetic modification or other biotechnological methods to increase the nutritional value of food. The addition of vital vitamins, minerals, and amino acids to staple crops is a common example. Crops with improved nutritional profiles, such as golden rice, which is designed to generate more vitamin A, have been created by genetic manipulation (Tang et al., 2009). Furthermore, biotechnological methods are employed to improve the bioavailability of micronutrients, which enhances the body's ability to absorb nutrients (Das et al., 2020). Biotechnology also contributes to fortifying food products with essential nutrients. For instance, biotechnologically produced yeast and microorganisms can be used to synthesize vitamins such as B12, folic acid, and riboflavin, which are often lacking in certain diets (Revuelta et al., 2018). These techniques are particularly beneficial for addressing micronutrient deficiencies in developing countries.

Table 2 Tools of biotechnology with their descriptions and applications.

Tools	*Description*	*Applications*	*References*
Genetic Engineering	The alteration of an organism's genome using biotechnology techniques as recombinant DNA technology, gene cloning, and CRISPR.	Genetically modified plants like Bt cotton, golden rice, drought-resistant corn, etc.	Wikandari et al., 2021; Beyer et al., 2002; Ricroch, 2019
Fermentation Technology	The use of microorganisms such as bacteria, yeast, and fungi for the conversion of raw materials into beneficial products like biofuels, antibiotics, and functional ingredients.	Probiotic foods such as yogurt, kefir, fortified juices, etc.	Stanbury et al., 2013; Sun et al., 2022; Gupta, and Abu-Ghannam, 2012; Raveendran et al., 2018
Enzyme Technology	The use of natural or engineered enzymes to catalyse biochemical reactions for various industrial applications.	Meat tenderization, fortification of juices, starch modification in bakery, etc.	Thapa et al., 2019; Patel et al., 2023; Chaudhary et al., 2015
Metabolic Engineering	Modification of an organism's metabolic pathways to raise the output of desired products or promote the creation of particular metabolites.	Vitamin production, amino acid synthesis, antioxidant extraction, etc.	Mallikarjuna and Yellamma, 2019; Helmy et al., 2020
Omics Technology	A collection of techniques to study biological data at large scales, including genomics, proteomics, transcriptomics, and metabolomics.	Fermented soy products, cheese ripening, dough fermentation, juice processing, etc.	Davis, 2010; Zhang et al., 2024
Synthetic Biology	The creation and assembly of new biological components, devices, and systems, along with the re-engineering of biological functions for targeted purposes.	Genetically engineered yeast, vitamins, omega-3 fatty acid, vegetables, etc.	Tyagi et al., 2014; Horner et al., 2016
Bioprocessing and Bioreactors	The use of bioreactors to cultivate microorganisms or cells under controlled conditions for large-scale production of biological products.	Production of amylase, lysine, lactic acid bacteria, flavoring compounds, etc.	Sharma et al., 2022; Mitra and Murthy, 2022

4.2 Production of Probiotics and Prebiotics

Probiotics are living microorganisms that, when consumed in adequate amounts, improve the health of the host, whereas prebiotics are indigestible food components that encourage the development and activity of beneficial bacteria in the gut. Biotechnology has a major impact on the development and production of probiotics and prebiotics, which provide potential treatments for improving gut health and preventing gastrointestinal illnesses (Davani-Davari et al., 2019). Dairy products like kefir and yogurt are fermented using probiotic bacteria like *Lactobacillus* and *Bifidobacterium*. The creation of strains with greater health advantages, such as increased tolerance to environmental stressors and the capacity to provide larger quantities of bioactive chemicals, has been made possible by genetic engineering of these microorganisms (Shah et al., 2024). Furthermore, strains that have undergone GM might be engineered to generate certain advantageous substances, including antimicrobial peptides, which could support gut health and general well-being (Mugwanda et al., 2023). Apart from probiotics, prebiotics are also made via biotechnology from plant ingredients like fibers and oligosaccharides. To encourage the growth of good gut bacteria and enhance digestive health, prebiotics are being added to a wider range of food products, such as drinks, cereals, and snack foods (Dwivedi et al., 2014).

4.3 Allergen Reduction

Food allergies are a growing concern worldwide, with millions of people suffering from adverse reactions to certain proteins in food. Biotechnology offers several approaches for reducing allergens in food products, primarily through GM and enzyme treatments. The development of hypoallergenic food varieties, such as genetically modified peanuts and wheat, has shown promise in mitigating allergic reactions (Dunn et al., 2017). Genetic engineering can facilitate the silencing or modification of specific allergenic proteins in food crops, reducing their potential to trigger allergic responses in sensitive individuals. Additionally, enzyme-based methods, such as enzymatic hydrolysis, can be employed to break down allergenic proteins into smaller, less-reactive fragments, thereby reducing the allergenic potential of food (Calcinai et al., 2022).

4.4 Development of Functional Foods

Functional foods are goods that provide health benefits beyond basic sustenance. Biotechnology has made it feasible to create functional meals that can promote certain health outcomes, including improved heart health, enhanced immunity, and reduced inflammation. One such instance is the production of omega-3 fatty acids by biotechnology. Omega-3 fatty acids may be created in greater amounts and added to a variety of food items, including margarine, dairy products, and functional drinks, by using genetically engineered microbes such as *Yarrowia lipolytica* (Madzak, 2021). Additionally, biotechnology has made it easier to create dietary components with extra health advantages, such as antioxidants, plant sterols, and bioactive peptides, which may be added to food to support good health (Zaky et al., 2021). These functional components give customers easy and accessible ways to enhance their health and can be obtained from fermented foods, GM crops, or engineered microbes.

4.5 Food Safety and Shelf Life

In the food sector, food safety is a major problem. Biotechnology has been used to improve food safety through improved contamination control, preservation technologies, and detection methods. By making it possible to quickly identify contaminants and pathogens in food products, rapid molecular diagnostic technologies like polymerase chain reaction (PCR) and enzyme-linked immunosorbent assays (ELISA) have enhanced food safety (Lao et al., 2014). Additionally, biotechnology is used to extend the shelf-life of food products by producing natural preservatives such as bacteriocins and antimicrobial peptides that are produced by lactic acid bacteria and other microbes. These bioprotective substances can help preserve the food's safety and quality by stopping the growth of diseases and spoilage microbes (Verma et al., 2022). Biotechnology also plays a major role in improving food packaging technologies. For example, biotechnologically produced biodegradable packaging materials have been developed to enhance food product safety while minimizing the environmental impact of food packaging (Santos et al., 2024).

5. Role of Biotechnology in the Development of Functional Foods

Biotechnology has played a major role in the development of functional foods, which offer health benefits in addition to nutritional value. Particularly in underdeveloped countries, this technology is being used more and more to improve food quality, nutritional content, and amounts of physiologically active substances. In particular, biotechnology has great promise for improving food crops, addressing micronutrient deficiencies, and creating food products that aid in the prevention of conditions like cancer, diabetes, and heart disease.

5.1 Improving Phytochemicals and Bioactive Components

Biotechnology, particularly genetic modification, enables the optimization of food plants to produce increased levels of bioactive components known for their health-promoting effects (Kamal et al., 2024). Phytochemicals such as antioxidants, carotenoids, and flavonoids, which are often found in vegetables, fruits, and legumes, play crucial roles in protecting the body against oxidative stress, inflammation, and various chronic diseases (Sharma et al., 2018). For instance, the development of high-vitamin A rice, a genetically modified product, has been a significant breakthrough. By increasing the beta-carotene content in rice, which is converted to vitamin A, this modification addresses deficiencies in vitamin A, especially in children and pregnant women (Tang et al., 2009). Furthermore, bioactive compounds such as lycopene, found in tomatoes and carrots, have been linked to cancer prevention and promoting eye health (Collins et al., 2022). These advancements underscore biotechnology's potential in creating foods that enhance health outcomes, particularly for vulnerable populations.

5.2 Improving Starch and Glycemic Index of Staple Crops

Foods with a high glycemic index are produced by the high amylopectin content of several staple crops, including yams and cassava. High-glycemic meals are linked to diet-related conditions, including Type 2 diabetes and raise blood sugar levels. Biotechnology offers the possibility to modify the starch composition of these crops, increasing the amylose to amylopectin ratio, thereby improving their glycemic index and making them healthier for populations relying on starchy foods. By improving the starch synthesis pathway, crops like potatoes, cassava, and yams can be optimized for better glycemic control, leading to healthier dietary options (Chen et al., 2021).

5.3 Enhancing Protein Quality and Availability

Protein is an essential nutrient for human health, yet in many developing countries, access to high-quality protein is limited, especially animal protein. Biotechnology has the potential to address this challenge by improving the protein quality in plants, particularly legumes and grains (Le et al., 2016). For instance, through genetic modification, the amino acid composition of crops such as soybean and canola has been enhanced to increase the levels of essential amino acids, such as lysine (Kishore and Shewmaker, 1999). This not only improves the protein profile of these crops but

Table 3 Overview of key biotechnological applications in food, highlighting their techniques, examples, and associated benefits.

Application Area	*Examples*	*Biotech. Techniques used*	*Benefits*
Nutrient Enhancement	Golden Rice (Vitamin A enhancement) - Biofortified crops (e.g., rice, maize, wheat)	- GM (e.g., CRISPR, transgenics) - Genetic engineering of plants - Fermentation	Increases essential nutrient content (vitamins, minerals) in staple crops
	Biotechnologically enhanced yeast for vitamin production (e.g., Vitamin B12)	Yeast and microbial fermentation for nutrient production	Addresses micronutrient deficiencies, especially in developing regions
Production of Probiotics and Prebiotics	Probiotic dairy products (e.g., yogurt, kefir) - Prebiotic foods (e.g., inulin, oligosaccharides)	Microbial fermentation - GM of microorganisms (e.g., Lactobacillus, Bifidobacterium)	Promotes gut health by balancing microbiota - Improves digestion and immune function
	Engineered probiotics with enhanced health properties (e.g., Lactobacillus producing antimicrobial peptides)	Metabolic engineering - Genome editing (e.g., CRISPR, RNA interference)	Can be tailored to produce specific health-promoting compounds
Allergen Reduction	Genetically modified hypoallergenic crops (e.g., peanuts, wheat) - Reduced allergenic milk	GM (e.g., RNA silencing, transgenics) - Enzymatic hydrolysis of allergens	Reduces allergenic potential of foods - Makes foods safer for people with food allergies
	Enzyme-based treatments to degrade allergens (e.g., soy or milk proteins)	Protease treatments - CRISPR-Cas9 gene editing - Recombinant DNA technology (to reduce allergenic proteins)	Improved consumer safety and product accessibility for allergic individuals
Development of Functional Foods	Omega-3 enriched foods (e.g., fish oil supplements, fortified margarine) - Functional snacks (e.g., fortified cereals)	Microbial fermentation - Genetic engineering (e.g., algae or yeast for omega-3 production)	Provides health benefits beyond basic nutrition - Targets specific health issues (e.g., heart health, immunity)
	Bioactive peptides (e.g., from fermented dairy products) - Functional beverages (e.g., antioxidant-rich drinks)	peptide synthesis (enzymatic or microbial) - GM of microorganisms and plants	Enhances disease prevention and overall well-being
Food Safety and Shelf Life	Antimicrobial-treated food (e.g., bacteriocins in cheese)	Biopreservation (using bacteriocins, antimicrobial peptides)	Extends shelf-life by preventing spoilage and contamination - Reduces reliance on chemical preservatives
	Rapid pathogen detection in food (e.g., PCR-based kits for Salmonella, E. coli)	PCR and qPCR (polymerase chain reaction) - ELISA (enzyme-linked immunosorbent assay)	Enhances food safety by enabling fast identification of contaminants
	Biodegradable films and coatings (e.g., from lactic acid bacteria)	Microbial fermentation - Biopolymer production (e.g., alginate, chitosan)	Reduces the environmental impact of food packaging - Preserves food quality

Source: Adapted from Revuelta et al., 2018; Davani-Davari et al., 2019; Calcinai et al., 2022; Zaky et al., 2021; Verma et al., 2022.

also provides a more accessible source of high-quality protein, crucial for the nutrition of children, pregnant women, and populations in areas with limited access to animal protein.

5.4 Biofortification and Micronutrient Enhancement

In regions with limited access to diverse diets, micronutrient deficiencies are prevalent, leading to various health issues such as anemia, stunted growth, and weakened immune systems. Biotechnology offers a solution through biofortification, the process of increasing the micronutrient content of staple foods. For example, genetic modifications have been used to increase the iron content in rice by introducing the soybean ferritin gene, which enhances iron absorption in the body (Slamet-Loedin et al., 2015). Similarly, genetic modifications have targeted reducing anti-nutritional factors like phytates in crops such as maize. Phytates interfere with mineral absorption, particularly iron and zinc, making these minerals less available to the body. By reducing phytate levels, the bioavailability of essential micronutrients in grains and legumes can be significantly improved, thus addressing deficiencies and promoting better health (Duraiswamy et al., 2023).

5.5 Fermentation and Bio-processing for Functional Foods

Fermentation is a key process in many traditional food products, especially in developing countries, where fermented foods play an essential role in nutrition and preservation. Fermentation not only enhances the flavor and digestibility of food but also improves its nutritional value. For instance, fermentation can reduce anti-nutritional factors like cyanogenic glycosides in cassava and yams, making them safer for consumption (Bolarinwa et al., 2016). Additionally, fermentation can generate beneficial metabolites, such as short-chain fatty acids, which have been associated with improved gut health and protective effects against colon cancer (Liu et al., 2016). Probiotic microorganisms, such as *Lactobacillus* and *Bifidobacteria*, also play a role in improving gastrointestinal health by enhancing the immune system and combating harmful bacteria (Maziotta et al., 2023). The application of biotechnology in optimizing fermentation processes can improve the quality, digestibility, and health benefits of fermented foods, making them more effective as functional foods.

6. Benefits of Biotechnology in Functional Foods

A major factor in changing the way food is produced and consumed is biotechnology. Functional foods are the foods that offer health advantages beyond basic nutrition; they have become more efficient, affordable, and sustainable thanks to a variety of biotechnological applications. The main advantages of biotechnology in functional foods, including better health outcomes, more food accessibility, and environmental sustainability, are discussed in this section.

6.1 Enhanced Health Outcomes

One of the most significant contributions of biotechnology to functional foods is the enhancement of their health-promoting properties. Foods enhanced with bioactive substances that provide certain health advantages can be developed thanks to biotechnology. These meals can boost immunity, support general health, and aid in the management of chronic illnesses. Food items may be made more nutritious by fortifying them with vital vitamins, minerals, and other nutrients thanks to biotechnology. For example, it is possible to create GM crops to have increased quantities of certain vitamins, such as Omega-3 fatty acids in plant-based diets (as in GM soybeans) or Vitamin A (Golden Rice) (Tang et al., 2009). Biotechnology also helps to incorporate bioactive compounds into foods that have therapeutic effects. Examples include the creation of meals with phytochemicals like antioxidants that help lower inflammation and oxidative stress in the body, or probiotic foods like yogurt that support gut health (Sharifi-Rad et al., 2020).

6.2 Increased Food Accessibility

Biotechnology plays a crucial role in addressing food security by improving the availability and accessibility of functional foods, particularly in regions facing food scarcity or nutrient deficiencies. Biotechnology can increase the yield of nutrient-dense crops, ensuring a steady supply of functional foods. GM crops, such as drought-resistant rice and enhanced soybean varieties, can thrive in diverse climatic conditions, helping to alleviate food shortages in developing regions (Kobayashi et al., 2023). By increasing the bioavailability of nutrients in meals, genetic alterations can guarantee that the body absorbs vital nutrients more effectively. For instance, numerous nations have successfully adopted biofortification, which uses biotechnology to boost the nutritional value of staple crops, to assist battle against vitamin and mineral shortages in underprivileged people (Ofori et al., 2022). The cost of functional foods can be decreased as biotechnology increases food production efficiency, making them more accessible and affordable for a larger spectrum of customers. This is especially critical for marginalized communities in low-income regions who might otherwise lack access to expensive supplements or specialized health foods (Kishore and Shewmaker, 1999).

6.3 Environmental Sustainability

Biotechnology contributes significantly to environmental sustainability in addition to improving human health and food security. Biotechnology helps guarantee that functional meals may be produced in a way that is both resource-efficient and environmentally benign by lowering the environmental impact of food production. Genetically engineered crops that are resistant to pests and diseases reduce the need for chemical pesticides, which can harm ecosystems. This leads to a decrease in the environmental pollution associated with conventional agriculture (Romeis et al., 2019). For instance, it has been demonstrated that Bt crops, which generate a toxin that is toxic to pests, greatly lessen the need for pesticides. The development of crops that require less land, water, and fertilizer is made possible by biotechnology, which is essential for protecting natural resources. In order to assist in preserving water and lessen the strain on water resources, drought-tolerant crop types have been created, such as rice and corn that can withstand drought (Dar et al., 2020). The carbon footprint of food production can be reduced by using GM crops and other biotechnological advancements. According to Smyth et al., (2024), these crops may be cultivated more effectively and usually need fewer inputs (water, energy, and chemicals), which helps to lower greenhouse gas emissions.

7. Ethical and Regulatory Implications

As biotechnology continues to advance, particularly in the development of functional foods, it introduces both significant opportunities and complex challenges. The integration of biotechnology into the food system requires careful navigation of regulatory frameworks and ethical boundaries to ensure safety, transparency, and consumer trust. Functional foods developed through biotechnological methods, including genetic modification and metabolic engineering, hold great potential for improving nutrition, health, and sustainability (Granato et al., 2023). However, their acceptance and commercialization depend heavily on compliance with international regulations and ethical standards that govern their production, distribution, and consumption.

The laws governing biotechnologically improved foods and GMOs exhibit considerable complexity and differ greatly between nations. Strict safety review procedures have been created by regulatory organizations including the Food and Agriculture Organization (FAO), the European Food Safety Authority (EFSA), and the U.S. Food and Drug Administration (FDA). These include assessments for allergenicity, toxicity, and nutritional equivalence to ensure that biotech-derived foods are safe for human consumption (Naegeli et al., 2017). Environmental risk assessments are also conducted to determine the potential ecological effects, such as biodiversity loss or gene

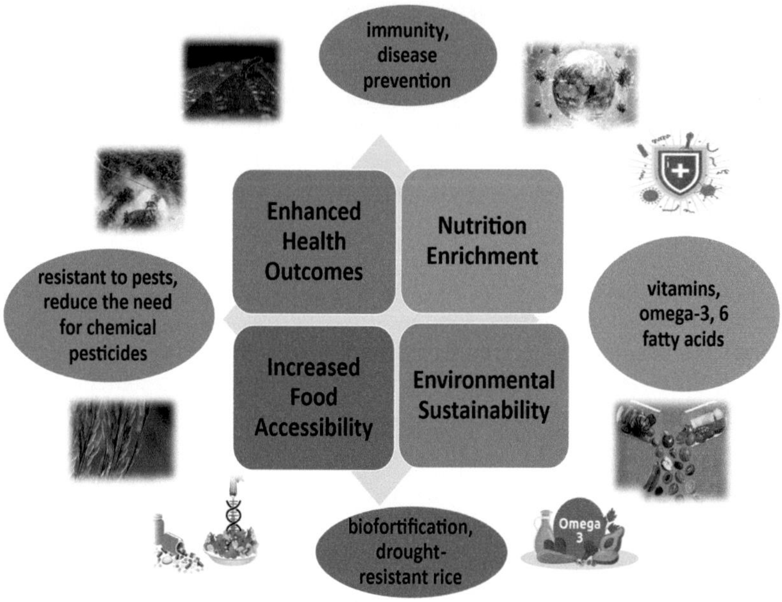

Fig. 3 Various benefits of biotechnology on food and health.

transfer to wild species. Despite these efforts, regulatory inconsistencies between countries create trade barriers and slow down innovation. For instance, while the European Union enforces strict precautionary labeling and approval processes, countries such as the United States, Brazil, and Argentina maintain more permissive frameworks that promote rapid adoption of biotech crops (Dey, 2025). This lack of global harmonization complicates international commercialization and often leads to market fragmentation.

Beyond regulatory complexities, economic and legal barriers also affect the development of functional foods through biotechnology. The high cost of research and development, coupled with lengthy approval procedures, often discourages smaller enterprises and startups from entering the market. Intellectual property rights present another challenge, as balancing innovation protection with public accessibility remains a critical issue. Affordable access to biotechnological advancements, especially in developing countries, is necessary to prevent widening the socio-economic gap between nations and populations (Dey, 2025).

The ethical dimensions of biotechnology are equally important in shaping its responsible application. Bioethics plays a primary role in guiding research and policymaking by addressing concerns related to human intervention in natural processes. The manipulation of genetic material, whether in crops, animals, or microorganisms, raises questions about what is morally acceptable in altering life forms for human benefit (Chhikara and Maurya, 2025). Ethical considerations in biotechnology encompass socio economic, environmental, cultural, religious, and legal aspects that influence public perception and policy decisions. From an equity standpoint, the uneven distribution of benefits from biotechnology can lead to socio economic imbalances. Wealthier nations and large corporations often dominate the field, while small-scale farmers and marginalized communities struggle to access biotech tools and innovations. Ensuring fair access and equitable benefit-sharing is therefore essential for ethical progress.

Environmental ethics also play a significant role in biotechnology. The modification of crops and organisms may have unintended consequences on ecosystems and biodiversity. Transgenic plants, for example, can potentially transfer modified genes to wild relatives, altering natural populations or reducing genetic diversity. Ethical frameworks must balance technological benefits with environmental sustainability to avoid long-term ecological harm. Similarly, religious and cultural beliefs shape public attitudes toward biotechnology. In some societies, manipulating living organisms is viewed as an infringement on natural or divine order. Respecting these beliefs through transparent communication and culturally sensitive policymaking is crucial to achieving societal acceptance of biotechnological innovations (Dey, 2025).

Public perception remains one of the most influential factors in determining the success of biotechnology in functional food development (Dey, 2025). Consumers' trust in regulatory institutions, understanding of biotechnology, and perceptions of safety largely determine their acceptance of genetically modified or bioengineered foods (Wunderlich and Gatto, 2015). In many cases, misinformation and lack of awareness contribute to scepticism and fear. Therefore, transparent labeling, public education, and open dialogue between scientists, regulators, and consumers are essential to building confidence in biotechnology (Jimenez et al., 2022). Ethical implications surrounding food safety, environmental impact, and genetic modification can be addressed effectively through inclusive communication and responsible governance.

The development of functional foods using biotechnology seems to have a very bright future despite the difficulties. Precise genetic alterations that can improve nutritional value, boost the bioavailability of vital nutrients, and decrease toxic chemicals are made possible by emerging technologies like CRISPR-Cas9 and other gene-editing techniques (Ansori et al., 2023). Personalized functional meals based on each person's genetic and health profile are made possible by these developments. Furthermore, research in medication development, genetic analysis, and process optimization is being revolutionized by the confluence of biotechnology and artificial intelligence, resulting in more effective and long-lasting advancements. The field's ability to solve global environmental and health issues is further demonstrated by sustainable biotechnological solutions, including the manufacturing of biofuels, biodegradable materials, and resource-efficient food systems.

8. Challenges and Future Prospects

There are a number of obstacles to overcome when using biotechnology to formulate functional products. One of the key hurdles is regulatory approval. The approval process for GMOs and biotechnologically enhanced foods is often lengthy and complex, involving rigorous safety assessments for both human health and environmental impact. These regulations vary across countries, creating barriers to market entry and international trade. Additionally, public skepticism and concerns about the safety of GM foods can hinder consumer acceptance, influenced by misinformation, lack of understanding, and ethical concerns surrounding genetic modification.

Despite its immense potential, biotechnology faces additional challenges that could hinder its broader economic impact. Regulatory hurdles often slow down the approval of biotech products and delay market entry, while the high costs of research and development can be prohibitive for startups and smaller companies. Ethical and social concerns, including those surrounding genetic modification and environmental safety, require careful management to ensure public acceptance. Intellectual property rights present further complexities, as balancing innovation protection with affordability remains a key issue.

The future outlook for biotechnology in economic growth and food innovation is highly promising. New opportunities in agriculture, bioengineering, and disease therapy are being made possible by developments in gene editing technologies like CRISPR. Personalized medicine, based on genetic profiling, promises more effective healthcare and improved economic efficiency. Biotechnological solutions focused on sustainability, including biofuels and biodegradable materials, are poised to meet the growing demand for eco-friendly alternatives. Furthermore, integration with artificial intelligence accelerates research in drug discovery, genetic analysis, and process optimization, enhancing the overall efficiency and impact of biotech innovations.

Advances in biotechnology, particularly in genome editing techniques like CRISPR, could revolutionize the creation of functional foods by allowing for precise modifications that meet specific nutritional or health-related goals. This could lead to the development of personalized functional foods tailored to individual genetic profiles or health needs, enhancing their effectiveness. Moreover, as consumer education and awareness about biotechnology improve, public trust in GM foods is expected to grow, fostering greater acceptance of biotechnological innovations. The integration of biotechnology in functional foods could also benefit from evolving regulatory frameworks, promoting global collaboration and standardization, which would facilitate the widespread adoption of these products and their potential to improve health outcomes worldwide.

9. Conclusion

The development of functional meals is greatly aided by biotechnology, which provides creative answers to global health issues like food security, chronic illnesses, and malnutrition. Biotechnology has the potential to transform the food sector and enhance public health outcomes by improving food safety, nutritional value, and bioavailability. However, there are a number of obstacles to the broad use of biotechnologically improved functional meals, such as complicated regulatory approval procedures, public mistrust, and high research and development expenses. Furthermore, cultural and ethical concerns surrounding GM continue to shape consumer perceptions and acceptance. By offering solutions that improve food products' safety, nutritional value, and health advantages, biotechnology has greatly enhanced the food sector. Among the major areas where biotechnology has had an influence are nutrient improvement, the manufacture of probiotics and prebiotics, the reduction of allergens, the development of functional foods, and food safety. Global food security, health issues, and sustainability difficulties may be resolved by the ongoing incorporation of biotechnological advancements into food production systems. To fully realize the advantages of biotechnology while guaranteeing consumer safety and acceptability, however, more study and legislative developments are required.

References

Ahmad Kamal, A.H., Mohd Hamidi, N.F., Zakaria, M.F., Ahmad, A., Harun, M.R., Chandra Segaran, T. and Jusoh, M. 2024. Genetically engineered microalgae for enhanced bioactive compounds. *Discov. Appl. Sci. 6*, 482.

Allman, T. 2018. Bioreactors: Design, operation, and applications. In: *Fermentation Microbiology and Biotechnology* (4th Edn., pp. 283–308). CRC Press.

Amiri, S., Aghamirzaei, M., Mostashari, P., Sarbazi, M., Tizchang, S. and Madahi, H. 2021. The impact of biotechnology on the dairy industry. In: *Microbial Biotechnology in Food and Health* (pp. 53–79). Academic Press.

Ansori, A.N., Antonius, Y., Susilo, R.J., Hayaza, S., Kharisma, V.D., Parikesit, A.A., Zainul, R., Jakhmola, V., Saklani, T., Rebezov, M., Ullah, M.E., Maksimiuk, N., Derkho, M. and Burkov, P. 2023. Application of CRISPR-Cas9 genome editing technology in various fields: A review. *Narra J.* **3**(2), e184.

Balabanova, L., Averianova, L., Marchenok, M., Son, O. and Tekutyeva, L. 2021. Microbial and genetic resources for cobalamin (vitamin B12) biosynthesis: From ecosystems to industrial biotechnology. *Int. J. Mol. Sci.*, *22*(9), 4522.

Bangar, S.P., Ashogbon, A.O., Singh, A., Chaudhary, V. and Whiteside, W.S. 2022. Enzymatic modification of starch: A green approach for starch applications. *Carbohydr. Polym.*, *287*, 119265.

Bansal, R. and Kour, J. 2022. Genetically modified canola. In: *Genetically Modified Crops and Food Security: Commercial, Ethical, and Health Considerations* (Chapter 11). Routledge.

Bekhit, A.A., Hopkins, D.L., Geesink, G., Bekhit, A.A. and Franks, P. 2014. Exogenous proteases for meat tenderization. *Crit. Rev. Food Sci. Nutr.*, *54*(8), 1012–1031.

Beyer, P., Al-Babili, S., Ye, X., Lucca, P., Schaub, P., Welsch, R. and Potrykus, I. 2002. Golden rice: Introducing the β-carotene biosynthesis pathway into rice endosperm by genetic engineering to defeat vitamin A deficiency. *J. Nutr.*, *132*(3), 506S–510S.

Bhardwaj, P.K. and Singh, K. 2022. Genetically modified microorganisms for enhancing the nutritional properties of food. In: *Current Developments in Biotechnology and Bioengineering* (pp. 323–336). Elsevier.

Bolarinwa, I.F., Oke, M.O., Olaniyan, S.A. and Ajala, A.S. 2016. A review of cyanogenic glycosides in edible plants. InTech. doi: 10.5772/64886.

Boutanaev, A.M., Moses, T., Zi, J., Nelson, D.R., Mugford, S.T., Peters, R.J. and Osbourn, A. 2015. Investigation of terpene diversification across multiple sequenced plant genomes. *Proc. Natl. Acad. Sci. USA*, *112*(1), E81–E88.

Bution, M.L., Molina, G., Abrahão, M.R. and Pastore, G.M. 2015. Genetic and metabolic engineering of microorganisms for the development of new flavor compounds from terpenic substrates. *Crit. Rev. Biotechnol.*, *35*(3), 313–325.

Calcinai, L., Bonomini, M.G., Leni, G., Faccini, A., Puxeddu, I., Giannini, D., Petrelli, F., Prandi, B., Sforza, S. and Tedeschi, T. 2022. Effectiveness of enzymatic hydrolysis for reducing the allergenic potential of legume by-products. *Sci. Rep.*, *12*(1), 16902.

Cameron, D.E., Bashor, C.J. and Collins, J.J. 2014. A brief history of synthetic biology. *Nat. Rev. Microbiol.*, *12*(5), 381–390.

Chaudhary, S., Sagar, S., Kumar, M., Sengar, R.S. and Tomar, A. 2015. The use of enzymes in food processing: A review. *South Asian J. Food Technol. Environ.*, *1*(3&4), 190–210.

Chhikara, Manisha, and Maurya, Prabhakar. 2025. Biotechnology Ethics and Regulatory Frameworks. In: *The Spectrum of Discovery: Multidisciplinary Research in the Modern Era* (Vol. 1, pp. 203–213). Redshine Publication.

Cohen, S.N., Chang, A.C. and Hsu, L. 1972. Nonchromosomal antibiotic resistance in bacteria: genetic transformation of Escherichia coli by R-factor DNA. *Proceedings of the National Academy of Sciences*, *69*(8), 2110–2114.

Collins, E.J., Bowyer, C., Tsouza, A. and Chopra, M. 2022. Tomatoes: An Extensive Review of the Associated Health Impacts of Tomatoes and Factors that Can Affect Their Cultivation. Feb 4; *Biology* (*Basel*) , *11*(2), 239. doi: 10.3390/biology11020239.

Dahiya, D. and Nigam, P.S. 2022. Probiotics, prebiotics, synbiotics, and fermented foods as potential biotics in nutrition improving health via microbiome-gut-brain axis. *Fermentation*, *8*(7), 303.

Dar, M.H., Waza, S.A., Shukla, S., Zaidi, N.W., Nayak, S., Hossain, M., Kumar, A., Ismail, A.M. and Singh, U.S. 2020. Drought Tolerant Rice for Ensuring Food Security in Eastern India. *Sustainability*, *12*(6), 2214. https://doi.org/10.3390/su12062214.

Das, P., Adak, S. and Lahiri Majumder A. 2020. Genetic Manipulation for Improved Nutritional Quality in Rice. Jul 24; *Front Genet.*, *11*, 776. doi: 10.3389/fgene.2020.00776.

Das, S., Ray, M.K., Panday, D. and Mishra, P.K. 2023. Role of biotechnology in creating sustainable agriculture. *PLOS Sustainability and Transformation*, *2*(7), e0000069.

Davani-Davari, D., Negahdaripour, M., Karimzadeh, I., Seifan, M., Mohkam, M., Masoumi, S.J., Berenjian, A. and Ghasemi, Y. 2019. Prebiotics: Definition, Types, Sources, Mechanisms, and Clinical Applications. Mar 9; *Foods*, 8(3), 92. doi: 10.3390/foods8030092.

Davies, H. 2010. A role for "omics" technologies in food safety assessment. *Food Control*, *21*(12), 1601–1610.

Dey, J. (2025). Biotechnology: Advancements, challenges, and future prospects in global health, agriculture. ResearchGate.

Djisalov, M., Knežić, T., Podunavac, I., Živojević, K., Radonic, V., Knežević, N.Ž., ... and Gadjanski, I. 2021. Cultivating multidisciplinarity: Manufacturing and sensing challenges in cultured meat production. *Biology*, *10*(3), 204.

Dunn, S. and Vicini; John and Glenn; Kevin and Fleischer; David and Greenhawt; and Matthew. 2017. The allergenicity of genetically modified foods from genetically engineered crops. *Annals of Allergy, Asthma, & Immunology*, *119*, 214–222. e3. 10.1016/j.anai.2017.07.010.

Duraiswamy, A., Sneha, A., N.M., Jebakani, K.S., Selvaraj, S., Pramitha, J.L, Selvaraj, R., Petchiammal, K.I., Kather Sheriff, S., Thinakaran, J., Rathinamoorthy, S. and Kumar, P.R. 2023. Genetic manipulation of anti-nutritional factors in major crops for a sustainable diet in future. Feb 15; *Front. Plant Sci.*, *13*, 1070398. doi: 10.3389/fpls.2022.1070398.

Ezemba, C., Ekwegbalu, E. A. and Ezemba, A.S. 2022. *Fermentation, Types of Fermenters, Design & Uses of Fermenters and Optimization of Fermentation Process*. Department of Microbiology, Chukwuemeka Odumegwu Ojukwu UniversityPress.

Gavrilescu, M. 2010. Environmental biotechnology: Achievements, opportunities, and challenges. *Dynamic Biochemistry, Process Biotechnology, and Molecular Biology*, *4*(1), 1–36.

Granato, D., Zabetakis, I. and Koidis, A. 2023. Sustainability, nutrition, and scientific advances of functional foods under the new EU and global legislation initiatives. *Journal of Functional Foods, 109*, 105793. https://doi.org/10.1016/j.jff.2023.105793.

Gupta, S. and Abu-Ghannam, N. 2012. Probiotic fermentation of plant based products: Possibilities and opportunities. *Critical Reviews in Food Science and Nutrition*, *52*(2), 183–199.

Hao, D.C. and Xiao, P.G. 2015. Genomics and evolution in traditional medicinal plants: Road to a healthier life. *Evolutionary Bioinformatics*, *11*, EBO–S31326.

Hasan, F., Shah, A.A. and Hameed, A. 2006. Industrial applications of microbial lipases. *Enzyme and Microbial technology*, *39*(2), 235–251.

Hashmi, Z., Idriss, I.M., Khalid, D., Abbas, S.H., Ali, S.O., Bozdar, M.M., ... and Solangi, N.H. 2024. Production of Microorganism-Based Low-Calorie Sugars. In: *Application of Bio-Additives for the Food Industry* (pp. 89–129). Cham: Springer Nature Switzerland.

Helmy, M., Smith, D. and Selvarajoo, K. 2020. Systems biology approaches integrated with artificial intelligence for optimized metabolic engineering. *Metabolic Engineering Communications*, *11*, e00149.

Horner, S.R., Mace, C.R., Rothberg, L.J. and Miller, B.L. 2006. A proteomic biosensor for enteropathogenic E. coli. *Biosensors and Bioelectronics*, *21*(8), 1659–1663.

Humbert, O., Davis, L. and Maizels, N. 2012. Targeted gene therapies: Tools, applications, optimization. *Critical Reviews in Biochemistry and Molecular Biology*, *47*(3), 264–281.

Husaini, A.M. and Sohail, M. 2023. Robotics-assisted, organic agricultural-biotechnology based environment-friendly healthy food option: Beyond the binary of GM versus Organic crops. *Journal of Biotechnology*, *361*, 41–48.

Jiawen Chen, Erica Hawkins and David Seung. 2021. Towards targeted starch modification in plants. *Current Opinion in Plant Biology*, *60*, 102013. ISSN 1369-5266, https://doi.org/10.1016/j.pbi.2021.102013.

Jimenez, J., Gamble-George, J., Danies, G., Hamm, R.L. and Porras, A.M. 2022. Public Engagement with Biotechnology Inside and Outside the Classroom: Community-Focused Approaches. Aug; *GEN Biotechnol.*, *1*(4), 346–354. doi: 10.1089/genbio.2022.0024.

Johnson, I.S. 1983. Human insulin from recombinant DNA technology. *Science*, *219*(4585), 632–637.

Jörg Romeis, Steven E. Naranjo, Michael Meissle and Anthony M. Shelton. 2019. Genetically engineered crops help support conservation biological control. *Biological Control*, *130*, 136–154. ISSN 1049-9644, https://doi.org/10.1016/j.biocontrol.2018.10.001.

Keller, E.F. 2022. Nature, nurture and the human genome project. In: *The Ethics of Biotechnology* (pp. 335–354). Routledge.

Kishore G.M. 1999. Shewmaker C. Biotechnology: enhancing human nutrition in developing and developed worlds. May 25; *Proc. Natl. Acad. Sci. U S A.*, *96*(11), 5968–5972. doi: 10.1073/pnas.96.11.5968.

Knychala, M.M., Boing, L.A., Ienczak, J.L., Trichez, D. and Stambuk, B.U. 2024. Precision fermentation as an alternative to animal protein: A review. *Fermentation*, *10*(6), 315.

Kobayashi, K., Wang, X. and Wang, W. 2023. Genetically Modified Rice is Associated with Hunger, Health, and Climate Resilience. Jul 21; *Foods*, *12*(14), 2776. doi: 10.3390/foods12142776. PMID: 37509868; PMCID: PMC10379675.

Kumar, K., Gambhir, G., Dass, A., Tripathi, A.K., Singh, A., Jha, A.K., ... and Rakshit, S. 2020. Genetically modified crops: Current status and future prospects. *Planta*, *251*(4), 91.

Kumar, M., Yadav, A.K., Verma, V., Singh, B., Mal, G., Nagpal, R. and Hemalatha, R. 2016. Bioengineered probiotics as a new hope for health and diseases: An overview of potential and prospects. *Future Microbiology*, *11*(4), 585–600.

Kumar, P., Chatli, M.K., Verma, A.K., Mehta, N., Malav, O.P., Kumar, D. and Sharma, N. 2017. Quality, functionality, and shelf life of fermented meat and meat products: A review. *Critical Reviews in Food Science and Nutrition*, *57*(13), 2844–2856.

Kumar, V., Sinha, A.K., Makkar, H.P. and Becker, K. 2010. Dietary roles of phytate and phytase in human nutrition: A review. *Food Chemistry*, *120*(4), 945–959.

Kussmann, M., Abe Cunha, D.H. and Berciano, S. 2023. Bioactive compounds for human and planetary health. *Frontiers in Nutrition*, *10*, 1193848.

Le, D.T., Chu, H.D. and Le, N.Q. 2016. Improving Nutritional Quality of Plant Proteins Through Genetic Engineering. Jun; *Curr. Genomics*, *17*(3), 220–229. doi: 10.2174/1389202917666160202215934. PMID: 27252589; PMCID: PMC4869009.

Li, Y. P., Ahmadi, F., Kariman, K. and Lackner, M. 2024. Recent advances and challenges in single cell protein (SCP) technologies for food and feed production. *NPJ Science of Food*, *8*(1), 66. doi: 10.1038/s41538-024-00299-2. PMID: 39294139; PMCID: PMC11410949.

Liu, G., Tang, J., Zhou, J. and Dong M. 2024. Short-chain fatty acids play a positive role in colorectal cancer. Sep 10; *Discov Oncol.*, *15*(1), 425. doi: 10.1007/s12672-024-01313-5. PMID: 39256239; PMCID: PMC11387572.

Madzak C. 2021. *Yarrowia lipolytica* Strains and Their Biotechnological Applications: How Natural Biodiversity and Metabolic Engineering Could Contribute to Cell Factories Improvement. Jul 10; *J. Fungi* (*Basel*). *7*(7), 548. doi: 10.3390/jof7070548. PMID: 34356927; PMCID: PMC8307478.

Majumder, R., Miatur, S., Saha, A. and Hossain, S. 2024. Mycoprotein: Production and nutritional aspects: A review. *Sustainable Food Technology*, *2*(1), 81–91.

Mallikarjuna, N. and Yellamma, K. 2019. Genetic and metabolic engineering of microorganisms for the production of various food products. In: *Recent Developments in Applied Microbiology and Biochemistry* (pp. 167–182). Academic Press.

Martin, Y. and Vermette, P. 2005. Bioreactors for tissue mass culture: Design, characterization, and recent advances. *Biomaterials*, *26*(35), 7481–7503.

Mazziotta, C., Tognon, M., Martini, F., Torreggiani, E. and Rotondo, J.C. 2023. Probiotics Mechanism of Action on Immune Cells and Beneficial Effects on Human Health. Jan 2; *Cells*, *12*(1), 184. doi: 10.3390/cells12010184.

McClements, D.J. and McClements, D.J. 2019. The science of foods: Designing our edible future. *Future Foods: How Modern Science is Transforming the Way We Eat* (pp. 1–25). ResearchGate.

Mehta, A. and Hasija, Y. 2018. Bioinformatics approaches for genomics and post genomics applications of anticancer plants. In: *Anticancer Plants: Mechanisms and Molecular Interactions*. (Vol. 4, pp. 283–317).

Mendel, G. 1866. Experiments on Plant Hybrids. (Trans.), Müller-Wille, S. and Hall, K. http://www.bshs.org.uk/bshs-translations/mendel/2016?page=15.

Mendly-Zambo, Z., Powell, L.J. and Newman, L.L. 2021. Dairy 3.0: Cellular agriculture and the future of milk. *Food, Culture & Society*, *24*(5), 675–693.

Mistry, K., Sardar, S.D., Alim, H., Patel, N., Thakur, M., Jabbarova, D. and Ali, A. 2022. Plant based proteins: Sustainable alternatives. *Plant Science Today*, *9*(4), 820–828.

Mitra, S. and Murthy, G.S. 2022. Bioreactor control systems in the biopharmaceutical industry: A critical perspective. *Systems Microbiology and Biomanufacturing* 1–22.

Mugwanda, K., Hamese, S., Van Zyl, W.F., Prinsloo, E., Du Plessis, M., Dicks, L.M.T. and Thimiri Govinda Raj, D.B. 2023. Recent advances in genetic tools for engineering probiotic lactic acid bacteria. Jan 31; *Biosci Rep.*, *43*(1), BSR20211299. doi: 10.1042/BSR20211299.

Munteanu, C., Mireşan, V., Răducu, C., Ihuţ, A., Uiuiu, P., Pop, D., ... and Groza, I. 2021. Can cultured meat be an alternative to farm animal production for a sustainable and healthier lifestyle. *Frontiers in Nutrition*, *8*, 749298.

Naegeli, H., Birch, A., N, Casacuberta, J., De Schrijver, A., Gralak, M.A., Guerche, P., Jones, H., Manachini, B., Messéan, A., Nielsen, E.E., Nogué, F., Robaglia, C., Rostoks, N., Sweet, J., Tebbe, C., Visioli, F., Wal, J.M., Eigenmann, P., Epstein, M., Hoffmann-Sommergruber, K., Koning, F., Lovik, M., Mills, C., Moreno, F.J., van Loveren, H., Selb, R. and Fernandez Dumont A. 2017. Guidance on allergenicity assessment of genetically modified plants. Jun 22; *EFSA J.*, *15*(6), e04862. doi: 10.2903/j.efsa.2017.4862.

Niba, L.L. 2003. The relevance of biotechnology in the development of functional foods for improved nutritional and health quality in developing countries. *African Journal of Biotechnology*, *2*(12), 631–635.

Ofori, K.F., Antoniello, S., English, M.M. and Aryee ANA. 2022. Improving nutrition through biofortification-A systematic review. Dec 9; *Front Nutr.*, *9*, 1043655. doi: 10.3389/fnut.2022.1043655. PMID: 36570169; PMCID: PMC9784929.

Pandey, V.K., Shafi, Z., Tripathi, A., Singh, G., Singh, R. and Rustagi, S. 2024. Production of biodegradable food packaging from mango peel via enzymatic hydrolysis and polyhydroxyalkanoates synthesis: A review on microbial intervention. *Current Research in Microbial Sciences*, *7*, 100292.

Patel, A.K., Dong, C.D., Chen, C.W., Pandey, A. and Singhania, R.R. 2023. Production, purification, and application of microbial enzymes. In *Biotechnology of Microbial Enzymes* (pp. 25–57). Academic Press.

Patel, R., Kaki, M., Potluri, V.S., Kahar, P. and Khanna, D. 2022. A comprehensive review of SARS-CoV-2 vaccines: Pfizer, Moderna & Johnson & Johnson. *Human Vaccines & Immunotherapeutics*, *18*(1), 2002083.

Pham, P.V. 2018. Medical biotechnology: Techniques and applications. In: *Omics Technologies and Bio engineering* (pp. 449–469). Academic Press.

Post, M.J. 2012. Cultured meat from stem cells: Challenges and prospects. *Meat Science*, *92*(3), 297–301.

Post, M.J., Levenberg, S., Kaplan, D.L., Genovese, N., Fu, J., Bryant, C.J. and Moutsatsou, P. 2020. Scientific, sustainability, and regulatory challenges of cultured meat. *Nature Food*, *1*(7), 403–415.

Prakash, B., Singh, P.P., Kumar, A. and Gupta, V. 2020. Prospects of omics technologies and bioinformatics approaches in food science. In: *Functional and Preservative Properties of Phytochemicals* (pp. 317–340). Academic Press.

Rai, A.K., Pandey, A. and Sahoo, D. 2019. Biotechnological potential of yeasts in functional food industry. *Trends in Food Science & Technology*, *83*, 129–137.

Rajak, N., Tiwari, A., Kumar, P. and Garg, N. 2023. Overview on nutraceuticals and biotechnology. In: Inamuddin, T. Altalhi and J. N. Cruz (Eds.), *Nutraceuticals* (pp. 175–191). Academic Press. https://doi.org/10.1016/B978-0-443-19193-0.00002-2.

Raveendran, S., Parameswaran, B., Ummalyma, S.B., Abraham, A., Mathew, A.K., Madhavan, A., ... and Pandey, A. 2018. Applications of microbial enzymes in food industry. *Food Technology and Biotechnology*, *56*(1), 16.

Redman, M., King, A., Watson, C. and King D. 2016. What is CRISPR/Cas9? Aug; *Arch Dis Child Educ Pract Ed.*, *101*(4), 213–215. doi: 10.1136/archdischild-2016-310459. Epub 2016 Apr 8.

Revuelta, J.L., Serrano-Amatriain, C., Ledesma-Amaro, R. and Jiménez, A. 2018. Formation of folates by microorganisms: towards the biotechnological production of this vitamin. Oct; *Appl Microbiol Biotechnol.*, *102*(20), 8613–8620. doi: 10.1007/s00253-018-9266-0.

Reynolds, T.J. and Martirosyan, D.M. 2016. Nutrition by design: A review of biotechnology in functional food. *Functional Foods in Health and Disease*, *6*(2), 110–120.

Ricroch, A. 2019. Global developments of genome editing in agriculture. In: *Transgenic research* (Vol. 28, pp. 45–52). Springer International Publishing.

Rocha, R.A., Speight, R.E. and Scott, C. 2022. Engineering enzyme properties for improved biocatalytic processes in batch and continuous flow. *Organic Process Research & Development*, *26*(7), 1914–1924.

Sangam Dwivedi, Kanwar Sahrawat, Naveen Puppala and Rodomiro Ortiz. 2014. Plant prebiotics and human health: Biotechnology to breed prebiotic-rich nutritious food crops. *Electronic Journal of Biotechnology*, *17* (5), 238–245. ISSN 0717-3458, https://doi.org/10.1016/j.ejbt.2014.07.004.

CHAPTER 13

Functional Foods Nutritional Insights, Bioactive Molecules, and Future Directions

Pesaru Sai Laya[1] and Priya Prasad[2*]

1. Introduction

Functional foods are a dynamic and rapidly evolving domain formed by the intersection of the principles of nutrition and medicine. These foods offer additional health benefits over basic nutrition. Functional foods may be occurring naturally or designed by the fortification of essential specific nutrients. They contain various bioactive compounds that influence specific physiological functions, boost immune response, maintain gut health, regulate glucose and lipid metabolism, enhance cognitive performance, reduce the risk of likelihood of disease, and improve overall health (Kaur and Das, 2011).

1.1 Defining Functional Foods

Functional foods are described in a wide variety of ways due to the lack of a universal scientific consensus, as functional foods represent the propositions of nutritional science, food science, and medicine. The regulatory differences across countries and the evolving knowledge over time have made the principles of functional foods dynamic (Kaur and Das, 2011). Below are several definitions of functional foods as described by standard organizations.

i. *Japanese Ministry of Health and Welfare* (*1984*): The Japanese Ministry of Health and Welfare defined functional foods as: "The foods that have been scientifically proven to benefit one or more target functions in the body, beyond adequate nutritional effects, in a way that is relevant to either improved state of health and well-being and /or reduction of the risk of disease." (Arai, 1996.)
ii. *European Commission Concerned Action on Functional Food Science in Europe* (*FUFOSE, 1999*): The FUFOSE defined functional foods as: "A food which affects beneficially one or more target functions in the body, beyond adequate nutritional effects, in a way that is relevant to either improved health or reduction of disease risk." ("Scientific Concepts of Functional Foods in Europe Consensus Document," 1999.)

[1,2] Department of Biotechnology, National Institute of Technology Warangal, Warangal, Telangana, India-506004.
* Corresponding author: priyap@nitw.ac.in

DOI: 10.1201/9781003659808-13

iii. *Institute of Medicine* (*IOM*), *USA* (*2005*): The definition given by IOM is: "Functional foods are similar in appearance to conventional foods and are consumed as part of a usual diet but have demonstrated physiological benefits and/or reduce risk of chronic disease beyond basic nutritional functions." (Henry, 2010)

iv. *International Life Sciences Institute* (*ILSI, 2002*): ILSI defined functional foods as: "The foods that provide health benefits beyond basic nutrition, either by enhancing a function or by reducing the risk of disease." (Henry, 2010)

v. *Health Canada* (*2009*): Health Canada described functional foods as: "Foods similar in appearance to conventional foods and are consumed as a part of a usual diet. They demonstrate physiological benefits and/or reduce the risk of chronic disease beyond basic nutritional functions." (Henry, 2010)

vi. The U.S. Food and Drug Administration (FDA) does not have any official definition for functional foods, but it considers them as "Conventional foods that provide health benefits and reduce disease risk as a part of a healthful diet." (Williams et al., 2006)

vii. The World Health Organization, does not use any standard definition to describe functional foods, but they mention functional foods as: "The foods that have a potentially positive effect on health beyond basic nutrition. They promote optimal health and help reduce the risk of disease." (Williams et al., 2006)

1.2 Historical Background

The concept of functional foods has undergone wide development over recent years, emerging as a bridge between traditional dietary customs and modern nutritional science. The idea of functional foods can be dated back to the ancient civilizations where garlic, turmeric, onions, and many other naturally occurring, edible ingredients are used for their medicinal properties (DeFelice, 1995). However, the Japanese government in the 1980s had introduced the term "functional food". Then a regulatory framework FOSCHU foods for specified health use, was introduced in 1991 by the Japanese Ministry of Health and Welfare to designate foods containing specific functional ingredients that have demonstrated health benefits (Arai, 1996). This concept, first introduced in Japan to improve public health, quickly gained global attention and was soon adopted by several European nations and Western countries in North America. In 1999, the European commission had launched a program on functional foods notably called FUFOSE program and defined functional foods and formulated regulations on these foods within the European union (Walter, 2008). At present, advancements in biotechnology, nutrigenomics, and microbiome research have driven the growth of functional foods into a major global sector at the intersection of health, medicine, and industry. This chapter introduces the concept and historical background of functional foods, subsequently discussing their classification and key bioactive components. It highlights the nutritional and health-promoting effects of functional foods with illustrative examples. The role of functional foods in managing chronic diseases is discussed in detail. Finally, current research trends and future directions are explored to provide insights into emerging opportunities in this field.

2. Classification of Functional Foods

The categorization of functional foods is important for a systematic understanding of the products that confer health benefits beyond basic nutrition. Classification provides a structural framework that helps in dealing with the concept in various aspects. It assists in understanding the current progress in the domain and enables more systematic studies on novel bioactive substances and health impacts. It aids doctors in providing evidence-based recommendations to patients. It also aids regulatory bodies in structuring appropriate guidelines for safety assessment, approval of products, and verification of effectiveness.

2.1 Based on Source

2.1.1 Naturally Occurring Functional Foods

These are the whole foods that naturally contain biologically active compounds that support health promotion beyond basic nutrition. These foods contain vital bioactive compounds like antioxidants, omega-3 fatty acids, dietary fibers, flavonoids, etc., which often showcase therapeutic effects (DeFelice, 1995). Such components are naturally inherited without the need of any fortification or processing. Examples of a few naturally occurring functional foods include oats, salmon fish, berries, cruciferous vegetables (Kaur and Kapoor, 2001; Roberfroid, 2002).

Because of enriched nutrient contents, these naturally occurring functional foods have been a part of traditional dietary practices over a long time and contribute significantly towards preventive nutrition. Traditional diets across cultures are built upon a rich diversity of natural foods, such as whole grains, vegetables, legumes, dairy, and fermented products. These support nutritional balance and also maintain cultural identity through shared culinary traditions. Clinical studies demonstrate that regular consumption of these foods lowered the rates of chronic diseases and improved longevity (Essa et al., 2023).

2.1.2 Modified or Fortified Functional Foods

The process of improving nutritional quality of the food by adding one or more essential nutrients is called fortification (Allen et al., 2006). The traditional foods may not possess all the nutritional values or the now-existing modern lifestyle may demand for increased uptake of nourishment. Food fortification or nutritional modification strategies bridge this gap by enhancing nutritional profile which promotes consumer well-being by improving public health and disease prevention (Thakur et al., 2023). The nutrient enhancement of conventional foods involves the addition of essential or beneficial compounds or by modifying the existing components (Hurrell, 2002). In recent years, the field of food fortification has evolved significantly, by mass fortification of staple crops, targeted fortification for vulnerable groups, and biofortification techniques to enhance nutrient density of crops by agronomic or genetic means (Kaur et al., 2022). Common examples include folic acid enriched cereals, iron and zinc fortified wheat flour, vitamin A enriched 'Golden rice', and calcium and Vitamin D fortified milk (Beyer et al., 2002; Calvo et al., 2004; Oakley, 2010).

2.1.3 Formulated Functional Foods

Formulated functional foods are specially engineered food products, designed by incorporating specific bioactive components into the food matrices to achieve targeted health effects (Granato et al., 2010). These foods differ from conventional and fortified foods as these are specifically developed through targeted formulation strategies to improve health and well-being. These formulation strategies not only focus on nutrient enrichment but also enhance the functionality through optimizing bioavailability and bioactive synergy. Examples include probiotic fortified yogurts for gut health, high protein foods for muscle growth, and meal replacement bars for weight management (Hurrell, 2002; Phillips, 2016).

2.2 Based on Targeted Health Benefits

Functional foods are structured in such a way that they meet specific physiological requirements and improve health status. In contrast to conventional foods, which provide only basic nutrition, targeted functional foods provide functional ingredients intended to affect particular biological functions or improve specific health conditions (Serafini, M., and Peluso, I., 2016). Functional foods are classified based on the particular health conditions they aimed to enhance. The main categories include:

2.2.1 Cardiovascular Health Products

Cardiovascular diseases (CVDs) are one of the major health threats significantly impacting global health by causing a large number of deaths worldwide and diet plays a vital role in the prevention and management of CVDs (Mozaffarian et al., 2016). The main goal of the functional foods developed targeting CVDs is to maintain normal blood pressure, lower cholesterol levels, and minimize the risk of heart strokes. The foods designed aiming CVDs contain bioactive compounds that include soluble fibers, omega-3 fatty acids, antioxidants, polyphenols, plant sterols, and stanols (Roberfroid, 2002). These influence lipid metabolism, enhance arterial elasticity, reduce inflammation, and regulate blood clotting mechanisms (Calder, 2012; Mukuddem-Petersen et al., 2005).

2.2.2 Digestive Health Products

Functional foods that are modeled to enhance gut health are formulated to augment nutrient absorption, maintain gut microbiome, reduce constipation, control gastro-esophageal acid refluxes, and prevent digestive disorders. The key bioactive components that help in the sustenance of digestive health include probiotics, prebiotics, dietary fibers, and polyphenols (Hill et al., 2014a; Valdes et al., 2018).

2.2.3 Immune Function Products

The immune system is always conditioned by nutrient uptake and functional foods are designed by including bioactive components like antioxidants, carotenoids, β-glucans, flavonoids, vitamins, and minerals like zinc and selenium to boost the immune system (Calder, 2020; Hill et al., 2014a). Foods like prebiotics, probiotics, citrus fruits, and dairy products strengthen the immune system, reduce susceptibility of infections, helps to recover from illness, manage inflammation and several autoimmune conditions (Carr and Maggini, 2017; Gombart et al., 2020; Lomax and Calder, 2008).

2.2.4 Bone Health Products

Bone is a metabolically active tissue that undergoes constant reconstruction and requires adequate nutritional resources to maintain its strength and structural robustness. Functional foods good for bone health are formulated with calcium, magnesium, phosphorus, vitamin D, vitamin K, and certain peptides and flavonoids (Rizzoli, 2014). These bioactive compounds aid in maintaining integrity of bones by regulating calcium absorption and enhancing bone mineralization thereby decreasing the risk of osteoporosis, osteomalacia, and fractures (Ma et al., 2008; Weaver et al., 2016).

2.2.5 Cognitive Function Products

Cognitive foods are designed with an aim to enhance cognitive functions and improve overall brain health. Omega-3 fatty acids (mainly EPA and DHA), vitamins B, C, E, antioxidants, flavonoids, and specific polyphenols are the key compounds that boost and maintain neurocognitive wellness (Gómez-Pinilla, 2008). These bioactive compounds improve blood flow to the brain, reduce oxidative damage to neurons, support neurotransmission, promote neuroplasticity and lower inflammation, and help enhance the well-being of the brain (Bazinet and Layé, 2014; Vauzour et al., 2008).

3. Functional Foods and Bioactive Components

Bioactive compounds are the non-nutritive, naturally occurring chemical constituents present in food, in very meager quantities, and influence various biological activities in the body. Although not essential, these components impact physiological responses like immune modulation, inflammation control, antioxidant effects, etc. Bioactive compounds are naturally present in plants, animals and microbial organisms and can be taken automatically when consumed as food (Guaadaoui et al.,

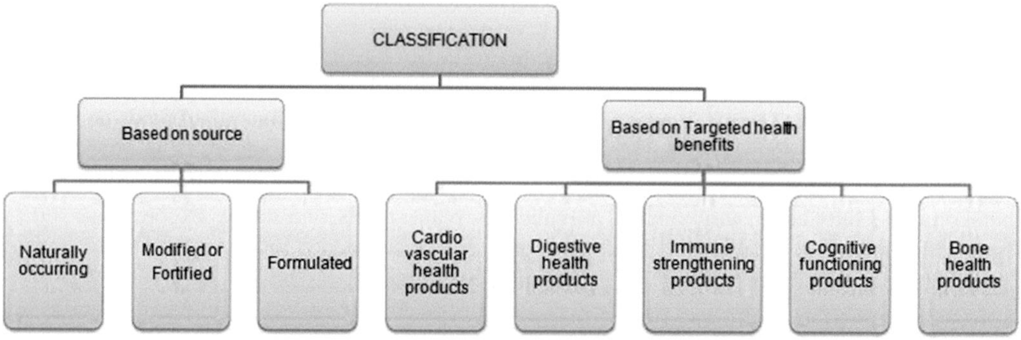

Fig. 1 Classification of functional foods based on source and targeted health benefits.

2014). Due to their preventative or therapeutic roles, foods are being fortified with bioactive entities like alkaloids, sterols, carotenoids, and polyphenols, to enhance physiological benefits, providing added functional properties and reduce the disease risk (Liu, 2003; Shahidi and Ambigaipalan, 2015).

There exists a wide variety of bioactive compounds with unique structural and functional features, each contributing to the health benefits and becoming key players in functional foods.

3.1 Polyphenols

Polyphenols are a large class of phytochemicals typified by multiple phenolic units. These exhibit anti-inflammatory, antioxidant, and strong cardioprotective properties. Polyphenols are very diversified and the subclasses include tannins, flavonoids (anthocyanins, catechins, quercetin), phenolic acids, and stilbenes. These are abundant in fruits, vegetables, cereals, tea, and coffee (Zheng et al., 2024). Polyphenols influence gut microbial community by suppressing harmful bacteria and by promoting beneficial species, which contributes to their anti-inflammatory and anti-cancer properties, offering protection against diseases like colorectal cancer and colitis (Zhao and Jiang, 2021). Anthocyanin rich *Vaccinium myrtillus* and *Vaccinium uliginosum* commonly called as bilberry and bog bilberry proclaims potential anti-cancer and anti-apoptosis activities (Kopystecka et al., 2023). Chlorogenic acid (CGA) is a food polyphenol present in various plant-based foods, including coffee beans. CGA influences the microbiota-gut-brain axis by regulating energy metabolism, neurotransmitters, and by controlling insulin resistance, obesity, synaptic dysfunction, and preventing cognitive-behavioral disturbances (Zhang et al., 2022).

3.2 Carotenoids

Carotenoids are the bioactive pigments with significant health promoting properties mostly present in bright fruits and vegetables (like carrots, apricots, tomatoes, sweet potatoes, pumpkins), eggs, dark leafy vegetables, and some vegetable oils. These can be categorized into carotenes – the hydrocarbon containing carotenoids (carotenes and lycopene), and xanthophylls – oxygenated carotenoids (Terao, 2023). Carotenoids, particularly β-carotene, lycopene, and astaxanthin exert strong antioxidant, anti-inflammatory, and immunomodulatory properties. Clinical evidence shows that dietary intake of carotenoids lowered the risks of obesity, Type 2 diabetes, aging-related issues, metabolic syndromes, and various cancers. Many carotenoids like β-carotene, α-carotene, and β-cryptoxanthin (known as provitamin A carotenoids) acts as the precursors of vitamin A which is prominent for cell growth regulation, immune function, and maintaining skin health and vision. The dietary uptake of non-provitamin A carotenoids like lutein and zeaxanthin mitigate the oxidative stress and inflammation thereby delaying age-related macular degradation (Bone and Landrum, 2010). Lycopene prevents

ROS-induced sperm dysfunction in males. Carotenoids contribute significantly to the maintenance of skin health (Yamamoto et al., 2017). They prevent aging by stimulating the production of collagen and elastin by fibroblasts, inhibiting the activity of MMPs and exhibiting UV-filtering and anti-inflammatory effects (Michalak et al., 2021). Carotenoids also exhibit hepatoprotective effect by regulating lipid metabolism, and neuroprotective effect by mitigating oxidative damage in neurodegenerative conditions.

3.3 Alkaloids

Alkaloids are the most diverse class of bioactive molecules comprising chemically and physiologically prominent compounds, with extensive disease prevention and health promoting properties. Based on their structure and biosynthetic precursors alkaloids can be classified into Indoles, tropane, purine, imidazole, piperidine, pyrrolidine, pyrrolizidine, quinolizidine and isoquinolines. These are multifaceted molecules with potent pharmacological properties and exhibiting a wide range of antimicrobial, antioxidant, anti-inflammatory, anti-cancer, antidiabetic, cardioprotective and hepatoprotective effects. Alkaloids are primarily produced as secondary metabolites from plants, besides several bacterial, fungal, and animal species also produce alkaloids as a part of defense, surviving mechanisms, or for interspecies communication. The ergot alkaloids, derivatives from L-tryptophan, from the fungus *Claviceps purpurea* are used for the production of lysergic acid diethyl amine, a hallucinogenic, used in the treatment of schizophrenia (Seneca, 2007). Caffeine and theophylline from coffee and tea functions as a stimulant of the central nervous system and metabolic regulator, increasing alertness and supporting metabolism thereby lowering the risk of Type 2 diabetes. The capsaicin present in chili peppers exhibit thermogenic and anti-inflammatory effects promoting weight loss and relieving the pain. The piperine present in the *Piper nigrum* increases nutrient absorption by enhancing bioavailability of nutrients present in the food and also boosts antioxidant enzyme activity (Letchuman et al., 2025). Berberine activates AMP-activated protein kinase, which improves glucose and lipid metabolism and also maintains cardiovascular health by improving endothelial function, lowering blood pressure, and preventing atherosclerosis. It also exhibits antimicrobial, anti-cancer, and neuroprotective effects (Neag et al., 2018).

3.4 Saponins

Saponins constitute a class of naturally occurring glycosides with a hydrophobic aglycone backbone (either steroidal or triterpenoid) linked to hydrophilic sugar moieties. This amphipathic structure favors their surface-active properties like emulsifying and foaming in plant biology and food sciences. Saponins are widely distributed in various plant species like legumes, quinova, ginseng, asparagus, and yam and some marine animals like sea stars, sea cucumbers and sponges. Due to their role in the food industry and physiological biology in maintaining health and preventing disease, saponins are considered as prominent bioactive molecules in functional foods and nutraceuticals (Chen et al., 2024). Saponins exhibit hypocholesterolemic effects by binding to bile acids in the gastrointestinal tract, promoting their excretion and also stimulating hepatic biosynthesis of bile acids from cholesterol thereby contributing to cardiovascular health (Chen et al., 2019). *Pulsatilla* saponins, from the herb *Pulsatilla chinensis*, widely used in Chinese traditional medicine to manage bacterial diseases, exhibits anti-cancer activity by inducing cancer cell apoptosis, inhibiting tumor angiogenesis along with antioxidant and anti-inflammatory effects (Zhong et al., 2022).

3.5 Phytosterols

Plant sterols and stanols, commonly called phytosterols, are naturally occurring steroids found in different parts of plants. They belong to the triterpene family and their major sources are vegetable oils, nuts, cereals, and bread (Li et al., 2022). Their wide range of physiological functions include antioxidant, anti-inflammatory, antipyretic, and hormone-like effects. Phytosterols remarkably reduce the cholesterol and low-density lipoprotein cholesterol (LDL-C) in the blood plasma and play a role in controlling and inhibiting cardiovascular and non-alcoholic fatty liver diseases. This is done by the competitive inhibition of phytosterol, which shares a similar structure with cholesterol. Similarly, they can improve lipid metabolism, insulin resistance, and can reduce the risk of Alzheimer's and cancer (Cabral and Klein, 2017). In addition to their health benefits, phytosterols are incorporated into foods as natural additives, where they help stabilize emulsions and enhance the texture of foods.

3.6 Bioactive Peptides

Bioactive peptides are short chains of amino acids obtained from food during food processing mechanisms like digestion, enzymatic hydrolysis, or fermentation. These peptides are inactive in their parental form and get activated during food processing (Akbarian et al., 2022). The natural sources of these peptides include milk, meat, eggs, dairy, fish, cereals, legumes, and a few marine organisms. Their remarkable structural diversity and sequence specific bioactivity endow them with multifunctional capabilities, enabling an extensive range of physiological and therapeutic activities like antioxidant, anti-inflammatory, antihypertensive, antimicrobial, immunomodulatory and anti-cancer properties (Zaky et al., 2022). Milk is a complete food richly packed with bioactive components like vitamins, minerals, immunoglobulins, growth factors, lactoferrins, cytokines proteins, lipids, and peptides. The peptides present in milk inhibit Angiotensin converting enzymes, thereby regulating blood pressure and reducing the risk of cardiovascular diseases (Singh and Gaur, 2024). Bioactive peptides also play a role in the management of obesity and diabetes by regulating metabolic health. The soybean-derived peptides, egg white proteins, and rice bran protein peptides have anti-obesity effects and influence lipid metabolism and reduce adipogenesis (Hamadou, 2025; W. Wang et al., 2023). The opioid peptides present in various dairy products and cereals have analgesic properties with a specific impact on the neurological system. The CTLEW peptides from defatted walnuts and BG-4 peptides from bitter gourd seeds increases autophagy and apoptosis exhibiting potent anti-cancer effects (Bidram and Ganjalikhany, 2024; Hamdi et al., 2025).

For a better understanding of these health-promoting constituents, the table provided below provides an organized outline of major bioactive components.

Table 1 Types of Bioactive compounds, their sub-types and sources.

Name of the Bioactive Compound	*Types*	*Sources*	*References*
Polyphenols	Flavonoids, Phenolic acids, Lignans	Fruits, vegetables, tea, flax seeds	(Shahidi and Ambigaipalan, 2015)
Carotenoids	β-carotene, Lycopene, Lutein, Zeaxanthin	carrots, tomatoes, spinach, sweet potatoes	(Krinsky and Johnson, 2005)
Glucosinolates	Indoles, Isothiocyanates	Cruciferous vegetables like broccoli, cabbage, sprouts	(Fahey et al., 2001)
Dietary fiber	Soluble fiber, Insoluble fiber	Whole grains, vegetables, fruits and legumes	(Anderson et al., 2009)
Saponins	Steroidal saponins, Triterpenoid saponins	Soybeans, chickpeas, legumes	(Sparg et al., 2004)
Prebiotics	Inulin, Fructooligosaccharides	Garlic, onions, leeks, bananas	(Roberfroid, 2007)
Probiotics	Lactobacillus spp, Bifidobacterium spp	Yogurt, kefir, fermented foods like kimchi	(Hill et al., 2014b)

Contd.

Table 1 *Contd.*

Name of the Bioactive Compound	*Types*	*Sources*	*References*
Bioactive peptides	Antihypertensive peptides, Antioxidant peptides, Antimicrobial peptides	Milk, fish, soy, eggs	(Korhonen and Pihlanto, 2006)
Omega-3 fatty acids	Eicosapentaenoic acid (EPA), Docosahexaenoic acid (DHA)	Fatty fish like salmon, mackerel, flaxseeds, walnuts	(Simopoulos, 2002)
Plant sterols and stanols	Sitosterol, Campesterol, Stigmasterol	Nuts, seeds, legumes, fortified margarine	(Amir Shaghaghi et al., 2013)
Vitamins and minerals	Vitamins C, E, D, B-complex; Minerals-selenium, zinc	Fruits, vegetables, dairy, cereals, meat, seafood	(Ames, 2004)
Alkaloids	Caffeine, Theobromine, Theophylline	Coffee, cocoa, tea	(Heckman et al., 2010)
Organosulfur compounds	Allicin, Diallyl disulfide	Garlic, onions, leeks	(Iciek et al., 2009)
Anthocyanins	Cyanin, Delphinidin, Malvidin	Berries, red grapes, purple corn	(Khoo et al., 2017)

4. Health Benefits of Functional Foods

Functional foods are considered as health-promoting substances because of their contribution beyond providing basic nutrition. They play a vital role in the improvement of health and the reduction of the exacerbation of disease. These foods may be rich in nutritional value or may be fortified to enhance their nutritional content during processing. Functional foods help in the regulation of metabolic activities, weight management, hormonal balance, immune modulations, healthy skin, bones, and support mental health (Roberfroid, 2002). Regular consumption of functional foods aids in improving gut health, reducing the susceptibility of chronic diseases like diabetes, cardiovascular disorders, and certain cancers (Williams et al., 2006). Below are a few well-documented benefits of functional foods.

4.1 Gut Health

Functional foods contribute substantially to gut health by maintaining the composition and activity of gut microbiota, which is crucial in keeping up the digestive balance and overall well-being of the organism. Foods containing prebiotics and probiotics promote digestive health. The prebiotic compounds like inulin, galactooligosaccharides, and fructooligosaccharides and probiotic foods like kefir, yogurt, and fermented vegetables and foods aids in the maintenance of gut microbiota enhancing the immunological functions in digestive system. It facilitates improved digestion, vitamin production, nutrient absorption, and defense against pathogens. Besides physical health, it also augments psychological well-being by positively impacting gut-brain axis, influencing mental health outcomes like anxiety and depression through neuroactive components like serotonin precursors, short-chain fatty acids, and gamma-amino butyric acid (GABA) (Kumar et al., 2023; Valdes et al., 2018). The dietary polyphenols from cocoa, green tea and berries also exhibit prebiotic-like effects by nurturing beneficial bacterial growth and reducing pathogenic species, favoring gut and systemic health (Wang et al., 2022).

4.2 Cardiovascular Health

Foods abundant in omega-3 fatty acids (fatty fish), soluble fibers (barley and oats), phytosterols (nuts and seeds) and polyphenols (berries) are called "cardioprotective foods". These foods help in maintaining the key cardiovascular factors like blood pressure, endothelial function, cholesterol level, and inflammation which aids in cardiovascular disease prevention. Many studies reveal that the consumption of fatty fish rich in omega-3 fatty acids, DHA and EPA has reduced the mortality risk by 36%. Similarly, intake of plant sterols on a daily basis can reduce the LDL cholesterol by 8–15% (Arnoldi and Williams, 2004).

4.3 Supporting Immune System

Immune functions are often boosted by the bioactive compounds present in functional foods. Nutrients like vitamin A, vitamin C, vitamin D, selenium, zinc, and polyphenols are essential in maintaining the immune system. These nutrients enhance immune functions, regulate inflammatory responses, and reduce oxidative stress (Calder et al., 2020). Prebiotic foods like garlic, onions, bananas, and probiotic foods like yogurt, kimchi improve the gut bacteria, which strengthens the gut immune defense (Roberfroid et al., 2010). Polyphenols from plant-derived foods like berries and green tea, and β-glucans from supplements, the immunomodulatory functions and antiviral properties by inhibiting pro-inflammatory cytokines and supporting activity of natural killer cells and macrophages (Cory et al., 2018; Shakoor et al., 2021). B-glucans and the polysaccharides present in oats, barley, and mushrooms also enhance immune functions by activating various components of immune system and improving vaccine responses (Zhong et al., 2023).

4.4 Improving Brain Health and Cognitive Function

The bioactive compounds found in foods like fatty fish, nuts, seeds and leafy vegetables are vital for maintaining the well-being of the brain and protection against neurodegenerative diseases. They impact cognitive function, memory, mood regulation and secretion of neurotransmitters (Gómez-Pinilla, 2008). Components like omega-3 fatty acids are essential for maintaining the fluidity, functionality and integrity of neural membranes. Deficiency in these essential fats may increase the risk of Alzheimer's and other neurodegenerative diseases (Bazinet and Layé, 2014). Polyphenol-rich foods offer neuroprotective effects by reducing oxidative stress, neuroinflammation, synaptic plasticity, enhancing cerebral blood flow modulating signaling pathways and neurogenesis (Godos et al., 2024). Vitamin B decreases homocysteine levels responsible for brain atrophy and dementia; probiotics influence gut-brain axis by modulating the production of neurotransmitters like serotonin and GABA impacting mood and cognitive function (Bazinet and Layé, 2014; Kumar et al., 2023).

4.5 Metabolic Health and Weight Management

Functional foods have a significant role in managing metabolic disorders like type-2 diabetes and obesity. Various bioactive compounds occur in foods that improve metabolic health and keep a check on weight management by regulating blood glucose levels, insulin sensitivity, lipid levels, and controlling appetite. The dietary fibers present in the food slowdowns gastric emptying and glucose absorption, leading to control in postprandial glucose and cholesterol levels (Mocanu and Madsen, 2024). Caffeine and catechins increase lipid oxidation and energy expenditure, aiding weight reduction. Polyphenols exhibit antioxidant and anti-inflammatory effects that improve insulin sensitivity and reduce adipose tissue inflammation (Arnoldi and Williams, 2004; Gasmi et al., 2022; Heckman et al., 2010; Shahidi and Ambigaipalan, 2015).

4.6 Bone Health

Functional foods support bone mineralization and bone formation by providing essential nutrients and also prevent conditions like osteoporosis and fractures by maintaining bone density. The bioactive compounds present in green leafy vegetables, nuts, seeds, and fortified dairy products are plentiful in calcium, vitamins, and minerals that are needed for maintaining bone health. Milk and milk products are valuable sources of calcium, and they promote bone health by facilitating calcium bioavailability and reducing inflammation. Vitamin D regulates bone remodeling by enhancing calcium absorption (Ma et al., 2008; Rizzoli, 2014). Magnesium works synergistically with calcium and regulates bone remodeling. Vitamin K promotes osteocalcin activation and bone matrix mineralization (Su et al., 2023). Polyphenols help in preserving bone mass through their antioxidant and anti-inflammatory properties and isoflavones present on foods like soy function as phytoestrogens, thereby offering the bone-protective benefits by imitating the actions of estrogen and are mostly advantageous to post-menopausal women (Ma et al., 2008; Sharma et al., 2023).

4.7 Cancer Prevention

Functional foods are gaining attention for their promising role in reducing the risk of cancer. The antioxidant, anti-inflammatory, and immunomodulatory effects of the bioactive compounds are making functional foods in marking themselves as effective agents of cancer prevention (Aghajanpour et al., 2017). Components such as polyphenols, carotenoids, glucosinolates, and dietary fibers remove the free radicals, enhancing detoxification pathways and preventing the proliferation of malignant cells (Scalbert et al., 2005; Tanaka et al., 2012). Sulforaphane, a bioactive component found in cruciferous vegetables, triggers apoptosis and has a role in removing the cancerous cells (Clarke et al., 2008). Similarly, isoflavones lower the risk of hormone-sensitive cancers; and prebiotics and probiotics also contribute to cancer prevention by maintaining gut microbiota composition and accelerating immune functions (Messina, 2016; O'Keefe, 2016). Bioactive phytochemicals also involve in modulating epigenetic mechanisms that suppress oncogene expression and promote tumor suppressor gene activity and a few of them are Indole-3-carbinol, apigenin, corosolic acid, curcumin, fucoxanthin, and delphidin (Bhattacharjee and Dashwood, 2020).

5. Examples of Functional Foods

5.1 Golden Rice

Golden rice is a biofortified variety of rice, designed by genetically engineering a gene responsible for the synthesis of β-carotene, which is the precursor of vitamin A. This variety is developed to combat the deficiency of vitamin A (Bouis and Saltzman, 2017). Golden rice serves as a functional food by providing health benefits beyond basic nutrition, compensating fortifying the micronutrient deficiency. This variety of rice addresses the temporary blindness and immune dysfunction caused by the deficiency of β-carotene. By incorporating β-carotene. Golden rice serves as an affordable and sustainable nutrient solution aimed at increasing the intake of vitamin A (Paine et al., 2005). It also highlights the effectiveness of biofortified crops as potent tools in enhancing food and nutrition security, and promoting preventive healthcare.

5.2 Iodized Salt

Iodized salt is a prime example of a functional food that showcases how nutrient enhancement can significantly improve public health on a global scale. Iodized salt is produced by the fortification of regular table salt with iodine. Iodine is an essential micronutrient necessary for the synthesis of thyroid hormones, which regulate the growth, metabolism, development of the brain, and cognitive development of individuals. The deficiency of thyroid hormones causes growth defects, cognitive impairment and also can lead to iodine deficiency disorders (IDD). Prior to the global adoption of the use of iodized salt, iodine deficiency disorders were prevalent in more than 100 countries all over the world, labeling it to be a public health concern. In response to this issue, the World Health Organization and UNICEF promoted the use of iodized salt. This strategy has successfully decreased the incidence of IDDs remarking iodization of salt as one of the best examples for food-based micronutrient intervention (Andersson et al., 2012; Zimmermann, 2009).

5.3 Cooking Oil

From being the basic kitchen ingredients, cooking oils have become essential functional foods due to the bioactive compounds present in them. Oils obtained from olives, rice bran, canola, sunflower, and flaxseeds are rich in monounsaturated and polyunsaturated fatty acids, antioxidants, omega-3 fatty acids, and phytosterols. These contribute to the management of cholesterol, metabolic balance, maintenance of cardiovascular health, inflammation control, etc. The omega-3 fatty acids present in the oil of flaxseeds regulate metabolic balance. The bioactive compounds including polyphenols like oleuropein present in extra virgin olive oil improves endothelial function, reduce oxidative stress, and the risk of heart disease (Kindernay et al., 2023). The fortification of cooking oils with fat-soluble vitamins like vitamin A and D helps deal with rickets, night blindness, and immune regulations (Covas et al., 2009; Das et al., 2013; Goyal et al., 2014; Patel et al., 2004; Perdomo et al., 2015).

5.4 Yogurt

Yogurt is a commonly consumed fermented milk product which offers numerous health benefits. Yogurt is produced by the fermentation of milk by *Lactobacillusbulgacus* or Streptococcus thermophilus, thus carrying a wide range of live probiotic cultures (Abdi-Moghadam et al., 2023; Meybodi et al., 2020; Mostafai et al., 2019). These probiotic organisms not only maintain gut microbiota, but also aid in nutrient absorption, immune modulation, and strengthening of defense mechanisms of the gut. Besides these, yogurt is a rich supplier of calcium, potassium, phosphorus, vitaminB12, and proteins and helps in tackling constipation, irritable bowel syndrome, lactose intolerance, and antibiotic associated diarrhea. Yogurt fortified with fiber, vitamin D, or plant sterols have a therapeutic potential to maintain bone health, reduce hypercholesterolemia, or hypertension (Meybodi et al., 2020; Mostafai et al., 2019).

5.5 Plant-based Milk

Oats, almonds, coconut, rice, and soy are a few sources for the plant-based milk. Besides serving as an alternative to dairy milk, plant-based milk is naturally fortified with numerous bioactive chemicals serving as functional foods. The milk extracted from plants is rich in calcium, omega-3 fatty acids, vitamin B12, and vitamin D and plays a crucial role in maintaining bone health, cardiovascular health, and modulation of immune actions. In addition to the above-mentioned nutrients, certain milks contain compounds that are specific to the plant from which the milk is isolated (Chalupa-Krebzdak et al., 2018; McClements et al., 2019). For example, isoflavones present in soy milk regulate estrogen activity and help to reduce the risk of heart diseases and hormone-related cancers (Messina, 2016). The β-glucans present in oat milk contribute to cholesterol reduction and glycemic control. The quinoa milk is nutritionally dense and a wholesome protein source, comprising all the essential nine amino acids. The medium chain triglycerides present in coconut milk play a prominent role in energy metabolism and weight management (Rehal et al., 2025).

6. Functional Foods in Management of Chronic Diseases

Functional foods provide a non-pharmacological approach in managing several chronic diseases, mostly the diseases caused by metabolic disruption, like phenylketonuria, Type-2 diabetes, hypercholesterolemia, obesity, etc. The bioactive components present in the functional foods help to regulate nutrient uptake, glucose and lipid metabolism, enhance hormonal sensitivity, and reduce oxidative stress (Hasler, 2002). A few conditions that can be controlled by dietary management are explained below.

6.1 Phenylketonuria (PKU)

Phenylketonuria is a rare metabolic disorder characterized by the deficiency of the enzyme phenylalanine hydroxylase. This enzyme catalyzes the conversion of the amino acid phenylalanine to tyrosine. Lack of this enzyme leads to the accumulation of phenylalanine in the blood and brain, causing intellectual disability, behavioral problems, and neurological complications (Blau et al., 2014). The keystone to managing PKU is to consume a restricted diet, and functional foods play a critical role in this context.

The uptake of phenylalanine free foods, specially formulated with low or no-phenylalanine protein but by providing all other essential amino acids and nutrients for proper functioning of organisms (Rocha and MacDonald, 2016). Consumption of Glycomacropeptide (GMP) based foods which are specially designed for patients with PKU, free from phenylalanine helps to manage PKU. GMP is a naturally occurring peptide derived from cheese whey and GMP-based foods available include pasta, cheese and a variety of beverages (Ney et al., 2016). The functional foods designed for PKU are often fortified with calcium, zinc, iron, vitaminB12, and vitamin D to overcome micronutrient deficiency (Meli and Bianca, 2002; Singh et al., 2014).

6.2 Type-2 Diabetes mellitus (T2DM)

Type-2 Diabetes mellitus is a chronic metabolic disorder marked by insulin resistance and impaired glucose metabolism. This is often associated with inflammation and oxidative stress (Kahn and Flier, 2000). Along with pharmacological treatment, dietary intervention is essential for the control of diabetes to check glycemic levels and improve insulin sensitivity (Kayode et al., 2023; Reaven, 1995). In this scenario, functional foods have emerged as promising agents to control T2DM by dietary management.

The functional foods designed for controlling T2DM are often rich in soluble dietary fibers like β-glucan, psyllium, and inulin. These slow down the digestion and glucose absorption, which parallelly reduces postprandial blood glucose levels. Oats, barley, chicory root, and legumes are good sources of soluble dietary fiber (Kayode et al., 2023; Slavin et al., 2009). Prebiotic-rich foods impact insulin sensitivity, and probiotics help to manage inflammation and glucose homeostasis (Prokisch et al., 2021). Polyphenol-rich foods like berries, green tea, cinnamon, and turmeric exert insulin-sensitizing effects by enhancing insulin receptor signaling (Williamson and Sheedy, 2020).

6.3 Hypercholesterolemia

Hypercholesterolemia is characterized by high levels of cholesterol in the body, particularly LDL-C (Rader et al., 2003). This increases the significant risk of cardiovascular diseases, atherosclerosis, obesity, poor weight management etc. functional foods effectively help in managing hypercholesterolemia by modulating cholesterol absorption, enhancing liver function, promoting excretion and protecting the cells from oxidative stress (Peña-Vázquez et al., 2024).

The functional foods tailored for managing hypercholesterolemia are rich in soluble dietary fibers, omega-3 fatty acids, soy protein, polyphenols and plant sterols and stanols (Lo et al., 1987). Soy protein-containing foods enhances the hepatic LDL receptor activity; oats and barley, rich in dietary fiber, promotes the use of circulating LDL-C in liver and reduces serum cholesterol levels (Caponio et al., 2020; Ho et al., 2016).

6.4 Alzheimer's Disease

Alzheimer's disease (AD) is a progressive neurodegenerative disorder marked by the gradual decline in memory, thinking abilities, and behavioral patterns, ultimately disrupting daily life activities. It is the most common cause of dementia, affecting elderly population worldwide. The underlying cause for Alzheimer's is a multifactorial disease involving complex interactions among genetic predispositions, environmental factors, and lifestyle. AD is characterized by the accumulation of amyloid-beta plaques, hyperphosphorylation of tau proteins, synaptic dysfunction, oxidative damage, and neuroinflammation (Breijyeh and Karaman, 2020).

In recent years, functional foods have been serving as great supportive therapies for AD by leveraging bioactive compounds that target key disease mechanisms. Several bioactive molecules like polyphenols, omega-3 fatty acids, phytosterols, and vitamins offer neuroprotective effects and maintain the integrity of the neural tissue. The polyphenols exert antioxidant and anti-inflammatory effects, reducing oxidative damage and inflammation in the brain (Bukhari, 2022). Omega-3 fatty acids like EPA and DHA maintain fluidity of the neuronal membrane and modulate neuroinflammation. Certain vitamins like Vitamins D, B (B6, B9, B12), and E offer neuroprotection by regulating homocysteine metabolism and neuronal functions (Mielech et al., 2020; Wei et al., 2023).

Table 2 List of functional foods effective against various diseases and their health benefits.

Disease/Condition	*Functional Food Component*	*Health Benefit*	*References*
Cardiovascular disease	Omega-3 fatty acids, Sterols, Oats	Lowers blood pressure, regulates lipid metabolism, and anti-inflammatory effects	(Kris-Etherton et al., 2002)
Irritable bowel syndrome(IBS)	Low-FODMAP foods, Probiotics, Peppermint oil	Reduces bloating, abdominal pain, ease bowel movements	(Marasco et al., 2025)
Osteoporosis	Calcium, Vitamin D, Vitamin K, Isoflavanoids	Enhance bone density and mineralization	(Faienza et al., 2024)
Inflammatory Bowel Disease (IBD)	Probiotics, Omega-3 fatty acids	Modulates inflammation, strengthens gut-barrier	(Mijan and Lim, 2018)
Obesity	Green tea, conjugated linoleic acid, capsaicin	Enhance lipid oxidation, metabolic regulation, suppress appetite	(Brown et al., 2015)
Alzheimer's disease	Omega-3 fatty acids, curcumin, flavonoids	Reduces oxidative stress, inflammation and support neural function	(Hasnieza M., et al., 2020)
Hypertension	Potassium and magnesium-rich foods	Promotes vasodilation, regulates blood pressure	(Chiu et al., 2020)
Constipation	Prebiotics, insoluble fiber	Improves bowel movements, maintains gut microbiome	(Tabrizi et al., 2020)
Non-alcoholic fatty liver disease	Green tea extract, Omega-3 fatty acids	Reduces fat accumulation around liver, inflammation	(L. Wang et al., 2023)
Cancer	Polyphenols, crucifer vegetables	Antioxidant, anti-inflammatory	(Aghajanpour et al., 2017)

7. Current Research Trends and Future Prospectus

The domain of functional foods has become the focal point for food science and medicine, as these are formulated in a way to provide more health benefits and to keep a check on chronic diseases whose root cause lies in the dietary zone. These foods generally promote the overall health and well-being by enhancing the nutrient uptake and regulating the physiological functions of an individual. The rising incidence in lifestyle related health conditions is accelerating the development and integration of functional foods in promoting public health and preventing chronic diseases (Rashidinejad, 2024).

7.1 Current Research Trends

7.1.1 Targeted Nutrition Fortification

The eminent trend in the field of functional foods is the fortification of staple foods with essential micronutrients to overcome the all-embracing nutritional deficiencies (Milner J. A. (2000)). A biofortified rice variety, Golden rice, which is crafted with the vitamin A precursor, β-carotene, has been efficient in preventing blindness and other deficiency-related disorders (Beyer et al., 2002). Similarly, iron-enriched flour, zinc-fortified wheat, and iodine supplemented salt are found to show impressive results world-wide in reducing the hidden hunger (Olson et al., 2021). Innovations in biofortification now focus on integrating multiple essential micronutrients into a single food matrix to achieve synergistic nutritional effects. For example, biofortified maize and pearl millet have been developed with increased levels of iron, zinc, and provitamin A carotenoids, addressing several nutrient deficiencies concurrently (Goredema-Matongera et al., 2021).

7.1.2 Personalized Nutrition

Advances in the areas of nutrigenomics and metabolomics are driving the innovation of designing dietary compounds formulated suitable to a person's unique biological and cultural prospects. Such customization enables targeted nutritional strategies thereby enhancing the overall effectiveness of nutrient interventions (Oak and Jha, 2019). This approach is particularly important in managing health conditions caused by the metabolic imbalances like Type-2 diabetes, cardiovascular disease, obesity, etc. whose cause lies in nutritional patterns and eating habits. The integration of genomic data along with metabolic and microbiome analysis facilitates the development of customized functional foods that address personal nutrient requirements and health risks. For example, lactose-free probiotic yogurt containing live *Lactobacillus* or *Bifidobacterium* strains is tailored for people with lactose intolerance (Oak and Jha, 2019). Vitamin D fortified products are developed based on individual serum 25-hydroxy vitamin D levels for individuals with vitamin D deficiency to optimize bone health and improve immune function (Alnafisah et al., 2024).

7.1.3 Discovery of Novel Bioactive Components

The discovery and characterization of novel bioactive components is a vital and rapidly growing area in the field of functional foods research. The development high analytical tools like mass-spectrometry, nuclear magnetic resonance (NMR) spectroscopy and high performance liquid chromatography (HPLC) are facilitating the study of a wide variety of phytochemicals and various metabolites from different organisms and discovery of compounds with significant health promoting properties (Marina and Francesco, 2022). The compounds derived from various plants, marine organisms, fungi, and fermented foods and exhibiting antioxidant, anti-diabetic, anti-inflammatory, antihypertensive, immunomodulatory and anticancer properties are selected and used in tailoring novel food varieties (Freile-Pelegrín and Robledo, 2013). Ongoing research into the discovery of bioactive components is continuously illuminating new categories of compounds, broadening the spectrum of health-promoting agents available for functional foods. A few recently discovered bioactive compounds are fucoxanthin from brown algae that display potent anti-obesity properties by enhancing thermogenesis and lipid metabolism (Lourenço-Lopes et al., 2021). Sulforaphane from broccoli acts as an efficient chemoprotective agent by activating phase-II detoxification enzymes and inducing apoptosis in cancer cells (Myzak and Dashwood, 2006).

7.1.4 Development of Delivery Systems for Functional Foods

As many functional ingredients like polyphenols, fatty acids, vitamins and small peptides are labile to environmental factors like pH, heat, oxidation, and enzymatic degradation which can affect their effectiveness (Kumar et al., 2024). So specific delivery systems are essential to ensure proper delivery and targeted release of these components. To address these challenges, a variety of encapsulation technologies like nano encapsulation, microencapsulation, liposomes, and emulsions are developed (Zabot et al., 2022). These delivery systems ensure the targeted delivery, controlled release alongside enhancing their stability and prolonging bioactivity. Nanoemulsion to deliver lipophilic compounds like curcumin is an example for an emulsion- based delivery system (Odriozola-Serrano et al., 2014). Hydrophilic and hydrophobic bioactive components such as vitamins D and E are encapsulated and efficiently delivered by liposomal delivery systems. The delivery of sensitive compounds like resveratrol and coenzyme 10 is favored by Solid lipid nanoparticles which facilitates controlled release and enhances its bioactivity (Pamunuwa and Karunaratne, 2022).

7.1.5 Functional Foods for Gut Microbiota Modulation

The human gastrointestinal tract harbors a wide variety of microbes which are involved in digestion, nutrient absorption, immune regulation and maintaining homeostasis. Dysbiosis in this microbial ecosystem results in a variety of health conditions like obesity, diabetes, inflammatory bowel disease and a few autoimmune diseases (Valdes et al., 2018). The microbial balance can be restored by the foods containing probiotics, prebiotics, postbiotics and synbiotics. Probiotic rich foods like cheese, yogurt, fermented vegetables introduce beneficial strains to the gut like *Lactobacillus, Bifidobacterium*, and *Saccharomyces* which enhance gut immunity. Prebiotic foods deliver and stimulate and enhance the growth of gut microbiota. These organisms colonize the gut and competitively inhibit the pathogen growth. Postbiotics are the non-viable microbial metabolites with numerous health promoting properties. They enhance the gut barrier integrity and stimulate mucosal immune response. The metabolites such as short chain fatty acids like butyrate, exert anti-inflammatory effects and contribute physiologically in the maintenance of gut integrity (Hill et al., 2014a; Kim et al., 2018). Microbiota modulation is mainly essential for gut health (Valdes et al., 2018).

7.2 Future Perspectives

The future perspectives of functional foods is to improve the nutrient profile and overall performance of crops by integrating the techniques of genetic engineering, traditional breeding, and agronomic management. Genetic engineering allows the precise addition of genes that boost nutrient levels, traditional breeding enhances traits such as yield and disease resistance, whereas agronomic practices manage nutrient uptake and accumulation. The integration of these strategies creates a comprehensive approach that addresses multiple aspects of crop development, resulting in sustainable and more effective biofortification (B. M., Rathnakumari, 2020). The other perspectives include expansion into immunonutrition, how specific nutrients and bioactive molecules modulate immune function to prevent and manage infectious and chronic inflammatory diseases (Ashaolu, 2020). The development of psychobiotics (Foster and Zhou, 2015), and integration of dietary consumption pattern with digital health platforms facilitate to track nutrient intake and rule out the dietary requirements of the body and personalize recommendations and tailoring sustainable and ethical functional foods with personalized nutrition and advancements in delivery systems (de Hoogh et al., 2021; Mohanty and Singhal, 2018).

References

Abdi-Moghadam, Z., Darroudi, M., Mahmoudzadeh, M., Mohtashami, M., Jamal, A.M., Shamloo, E. and Rezaei, Z. 2023. Functional yogurt, enriched and probiotic: A focus on human health. *Clin. Nutr. ESPEN*, *57*, 575–586. https://doi.org/10.1016/j.clnesp.2023.08.005.

Aghajanpour, M., Nazer, M.R., Obeidavi, Z., Akbari, M., Ezati, P. and Kor, N.M. 2017. Review Article Functional foods and their role in cancer prevention and health promotion: A comprehensive review. *Am. J. Cancer Res.*

Akbarian, M., Khani, A., Eghbalpour, S. and Uversky, V.N. 2022. Bioactive Peptides: Synthesis, Sources, Applications, and Proposed Mechanisms of Action. *Int. J. Mol. Sci.*, *23*, 1445. https://doi.org/10.3390/ijms23031445.

Allen, L., De Benoist, B., Dary, O. and Hurrell, R. 2006. Guidelines on food fortification with micronutrients A T. 24–37. WHO.

Alnafisah, R.Y., Alragea, A.S., Alzamil, M.K. and Alqahtani, A.S. 2024. The Impact and Efficacy of Vitamin D Fortification. *Nutrients*, *16*, 4322. https://doi.org/10.3390/nu16244322.

Ames, B.N. 2004. A role for supplements in optimizing health: The metabolic tune-up. Arch *Biochem. Biophys.*, *423*, 227–234. https://doi.org/10.1016/j.abb.2003.11.002.

Amir Shaghaghi, M., Abumweis, S.S. and Jones, P.J.H. 2013. Cholesterol-Lowering Efficacy of Plant Sterols/Stanols Provided in Capsule and Tablet Formats: Results of a Systematic Review and Meta-Analysis. *J. Acad. Nutr. Diet*, *113*, 1494–1503. https://doi.org/10.1016/j.jand.2013.07.006.

Anderson, J.W., Baird, P., Davis Jr, R.H., Ferreri, S., Knudtson, M., Koraym, A., Waters, V. and Williams, C.L. 2009. Health benefits of dietary fiber. *Nutr. Rev.*, *67*, 188–205. https://doi.org/10.1111/j.1753-4887.2009.00189.x.

Andersson, M., Karumbunathan, V. and Zimmermann, M.B. 2012. Global Iodine Status in 2011 and Trends over the Past Decade. *J. Nutr.*, *142*, 744–750. https://doi.org/10.3945/jn.111.149393.

Arai, S. 1996. Studies on Functional Foods in Japan—State of the Art. *Biosci. Biotechnol. Biochem.*, *60*, 9–15. https://doi.org/10.1271/bbb.60.9.

Arnoldi, A. and Williams, C.M. 2004. Functional Foods, Cardiovascular Disease, and Diabetes (pp. 215–239). Elsevier Science.

Ashaolu, T.J. 2020. Immune boosting functional foods and their mechanisms: A critical evaluation of probiotics and prebiotics. *Biomedicine & Pharmacotherapy*, *130*, 110625. https://doi.org/10.1016/j.biopha.2020.110625.

Bazinet, R.P. and Layé, S. 2014. Polyunsaturated fatty acids and their metabolites in brain function and disease. *Nat. Rev. Neurosci.*, *15*, 771–785. https://doi.org/10.1038/nrn3820.

Beyer, P., Al-Babili, S., Ye, X., Lucca, P., Schaub, P., Welsch, R. and Potrykus, I. 2002. Golden Rice: Introducing the β-Carotene Biosynthesis Pathway into Rice Endosperm by Genetic Engineering to Defeat Vitamin A Deficiency. *J. Nutr.*, *132*, 506S–510S. https://doi.org/10.1093/jn/132.3.506S.

Bhattacharjee, S. and Dashwood, R.H. 2020. Epigenetic Regulation of NRF2/KEAP1 by Phytochemicals. *Antioxidants*, *9*, 865. https://doi.org/10.3390/antiox9090865.

Bidram, M. and Ganjalikhany, M.R. 2024. Bioactive peptides from food science to pharmaceutical industries: Their mechanism of action, potential role in cancer treatment, and available resources. *Heliyon*, *10*, e40563. https://doi.org/10.1016/j.heliyon.2024.e40563.

Blau, N., Shen, N. and Carducci, C. 2014. Molecular genetics and diagnosis of phenylketonuria: State of the art. *Expert Rev. Mol. Diagn.*, *14*, 655–671. https://doi.org/10.1586/14737159.2014.923760.

Bone, R.A. and Landrum, J.T. 2010. Dose-dependent response of serum lutein and macular pigment optical density to supplementation with lutein esters. *Arch. Biochem. Biophys.*, *504*, 50–55. https://doi.org/10.1016/j.abb.2010.06.019.

Bouis, H.E. and Saltzman, A. 2017. Improving nutrition through biofortification: A review of evidence from HarvestPlus, 2003 through 2016. *Glob. Food Sec.*, *12*, 49–58. https://doi.org/10.1016/j.gfs.2017.01.009.

Breijyeh, Z., Karaman, R., 2020. Comprehensive Review on Alzheimer's Disease: Causes and Treatment. *Molecules*, *25*. https://doi.org/10.3390/molecules25245789.

Brown, L., Poudyal, H. and Panchal, S.K. 2015. Functional foods as potential therapeutic options for metabolic syndrome. *Obesity Reviews*, *16*, 914–941. https://doi.org/10.1111/obr.12313.

Bukhari, S.N.A. 2022. Dietary Polyphenols as Therapeutic Intervention for Alzheimer's Disease: A Mechanistic Insight. *Antioxidants* (*Basel*). *11*. https://doi.org/10.3390/antiox11030554.

Cabral, C.E. and Klein, M.R.S.T. 2017. Phytosterols in the Treatment of Hypercholesterolemia and Prevention of Cardiovascular Diseases. *Arq. Bras. Cardiol.*, *109*, 475–482. https://doi.org/10.5935/abc.20170158.

Calder, P.C. 2012. Mechanisms of Action of (n-3) Fatty Acids. *J. Nutr.*, *142*, 592S–599S. https://doi.org/10.3945/jn.111.155259.

Calder, P., Carr, A., Gombart, A. and Eggersdorfer, M. 2020a. Optimal Nutritional Status for a Well-Functioning Immune System is an Important Factor to Protect against Viral Infections. *Nutrients*, *12*, 1181. https://doi.org/10.3390/nu12041181.

Calder, P.C. 2020b. Nutrition, immunity, and COVID-19. *BMJ Nutr. Prev. Health*, *3*, 74–92. https://doi.org/10.1136/bmjnph-2020-000085.

Calvo, M.S., Whiting, S.J. and Barton, C.N. 2004. Vitamin D fortification in the United States and Canada: Current status and data needs. *Am. J. Clin. Nutr.*, *80*, 1710S–1716S. https://doi.org/10.1093/ajcn/80.6.1710S.

Caponio, G.R., Wang, D.Q.-H., Di Ciaula, A., De Angelis, M. and Portincasa, P. 2020. Regulation of Cholesterol Metabolism by Bioactive Components of Soy Proteins: Novel Translational Evidence. *Int. J. Mol. Sci.*, *22*, 227. https://doi.org/10.3390/ijms22010227.

Carr, A. and Maggini, S. 2017. Vitamin C and Immune Function. *Nutrients*, *9*, 1211. https://doi.org/10.3390/nu9111211.

Chalupa-Krebzdak, S., Long, C.J. and Bohrer, B.M. 2018. Nutrient density and nutritional value of milk and plant-based milk alternatives. *Int. Dairy J.*, *87*, 84–92. https://doi.org/10.1016/j.idairyj.2018.07.018.

Chen, M.L., Takeda, K. and Sundrud, M.S. 2019. Emerging roles of bile acids in mucosal immunity and inflammation. *Mucosal Immunol.*, *12*, 851–861. https://doi.org/10.1038/s41385-019-0162-4.

Chen, S., Zhou, Y., Li, H., You, L., Pedisić, S. and Shao, P. 2024. Saponins Based on Medicinal and Edible Homologous Plants: Biological Activity, Delivery Systems, and Its Application in Healthy Foods. *Food Bioengineering*, *3*, 464–481. https://doi.org/10.1002/fbe2.12111.

Chiu, H.-F., Venkatakrishnan, K. and Wang, C.-K. 2020. Nutraceuticals and functional foods in the prevention of hypertension induced by excessive intake of dietary salt. In: *Dietary Sugar, Salt, and Fat in Human Health* (pp. 423–450). Elsevier, https://doi.org/10.1016/B978-0-12-816918-6.00020-2.

Clarke, J.D., Dashwood, R.H. and Ho, E. 2008. Multi-targeted prevention of cancer by sulforaphane. *Cancer Lett.*, *269*, 291–304. https://doi.org/10.1016/j.canlet.2008.04.018.

Cory, H., Passarelli, S., Szeto, J., Tamez, M. and Mattei, J. 2018. The Role of Polyphenols in Human Health and Food Systems: A Mini-Review. *Front. Nutr.*, *5*. https://doi.org/10.3389/fnut.2018.00087.

Covas, M.-I., Konstantinidou, V. and Fitó, M. 2009. Olive Oil and Cardiovascular Health. *J. Cardiovasc. Pharmacol.*, *54*, 477–482. https://doi.org/10.1097/FJC.0b013e3181c5e7fd.

Das, J.K., Salam, R.A., Kumar, R. and Bhutta, Z.A. 2013. Micronutrient fortification of food and its impact on woman and child health: A systematic review. *Syst. Rev.*, *2*, 67. https://doi.org/10.1186/2046-4053-2-67.

de Hoogh, I.M., Winters, B.L., Nieman, K.M., Bijlsma, S., Krone, T., van den Broek, T.J., Anderson, B.D., Caspers, M.P.M., Anthony, J.C. and Wopereis, S. 2021. A Novel Personalized Systems Nutrition Program Improves Dietary Patterns, Lifestyle Behaviors, and Health-Related Outcomes: Results from the Habit Study. *Nutrients*, *13*, 1763. https://doi.org/10.3390/nu13061763.

DeFelice, S.L. 1995. The nutraceutical revolution: Its impact on food industry R&D. *Trends Food Sci. Technol.*, *6*, 59–61. https://doi.org/10.1016/S0924-2244(00)88944-X.

Diplock, A.T., Aggett, P., Ashwell, M., Bornet, F.R., Fern, E.B. and Roberfroid, M.B. 1999. Scientific Concepts of Functional Foods in Europe Consensus Document. *British Journal of Nutrition*, *81*, S1–S27. https://doi.org/10.1017/S0007114599000471.

Essa, M.M., Bishir, M., Bhat, A., Chidambaram, S.B., Al-Balushi, B. and Hamdan, H. 2023. Functional foods and their impact on health. *J. Food Sci. Technol.*, *60*, 820–834. https://doi.org/10.1007/s13197-021-05193-3.

Fahey, J.W., Zalcmann, A.T. and Talalay, P. 2001. The chemical diversity and distribution of glucosinolates and isothiocyanates among plants. *Phytochemistry*, *56*, 5–51. https://doi.org/10.1016/S0031-9422(00)00316-2.

Faienza, M.F., Giardinelli, S., Annicchiarico, A., Chiarito, M., Barile, B., Corbo, F. and Brunetti, G. 2024. Nutraceuticals and Functional Foods: A Comprehensive Review of Their Role in Bone Health. *Int. J. Mol. Sci.*, *25*, 5873. https://doi.org/10.3390/ijms25115873.

Foster, J. and Zhou, L. (Linda). 2015. Psychobiotics and the gut–brain axis: In the pursuit of happiness. *Neuropsychiatr. Dis. Treat.*, *715*. https://doi.org/10.2147/NDT.S61997.

Freile-Pelegrín, Y. and Robledo, D. 2013. Bioactive Phenolic Compounds from Algae. In: *Bioactive Compounds from Marine Foods* (pp. 113–129). Wiley, pp. 113–129. https://doi.org/10.1002/9781118412893.ch6.

Gasmi, A., Mujawdiya, P.K., Noor, S., Lysiuk, R., Darmohray, R. and Piscopo, S. 2022. Polyphenols in Metabolic Diseases. *Molecules*, *27*, 6280. https://doi.org/10.3390/molecules27196280.

Godos, J., Micek, A., Mena, P., Del Rio, D., Galvano, F. and Castellano, S. 2024. Dietary (Poly)phenols and Cognitive Decline: A Systematic Review and Meta-Analysis of Observational Studies. *Mol. Nutr. Food Res.*, *68*. https://doi.org/10.1002/mnfr.202300472.

Gombart, A.F., Pierre, A. and Maggini, S. 2020. A Review of Micronutrients and the Immune System: Working in Harmony to Reduce the Risk of Infection. *Nutrients*, *12*, 236. https://doi.org/10.3390/nu12010236.

Gómez-Pinilla, F. 2008. Brain foods: The effects of nutrients on brain function. *Nat. Rev. Neurosci.*, *9*, 568–578. https://doi.org/10.1038/nrn2421.

Goredema-Matongera, N., Ndhlela, T., Magorokosho, C., Kamutando, C.N., van Biljon, A. and Labuschagne, M. 2021. Multinutrient Biofortification of Maize (*Zea mays* L.) in Africa: Current Status, Opportunities and Limitations. *Nutrients*, *13*, 1039. https://doi.org/10.3390/nu13031039.

Granato, D., Branco, G.F., Nazzaro, F., Cruz, A.G., Faria, J.A.F. 2010. Functional Foods and Nondairy Probiotic Food Development: Trends, Concepts, and Products. *Compr. Rev. Food Sci. Food Saf.*, *9*, 292–302. https://doi.org/10.1111/j.1541-4337.2010.00110.x.

Guaadaoui, A., Benaicha, S., Elmajdoub, N., Bellaoui, M. and Hamal, A. 2014. What is a Bioactive Compound? A Combined Definition for a Preliminary Consensus. *International Journal of Nutrition and Food Sciences*, *3*, 174. https://doi.org/10.11648/j.ijnfs.20140303.16.

Hamadou, M. 2025. Bioactive peptides and metabolic health: A mechanistic review of the impact on insulin sensitivity, lipid profiles, and inflammation. *Applied Food Research*, *5*, 101056. https://doi.org/10.1016/j.afres.2025.101056.

Hamdi, M., Kilari, B.P., Mudgil, P., Nirmal, N.P., Ojha, S. and Ayoub, M.A. 2025. Bioactive peptides with potential anticancer properties from various food protein sources: Status of recent research, production technologies, and developments. *Crit. Rev. Biotechnol.*, *45*, 1076–1097. https://doi.org/10.1080/07388551.2024.2435965.

Hasler, C.M. 2002. Functional Foods: Benefits, Concerns, and Challenges: A Position Paper from the American Council on Science and Health. J. Nutr., *132*, 3772–3781. https://doi.org/10.1093/jn/132.12.3772.

Hasnieza M, N., Mastura Ya, H., Shahar, S., Wahida Ibr, F. and Fadilah Ra, N. 2020. Alzheimer's Disease and Functional Foods: An Insight on Neuroprotective Effect of its Combination. *Pakistan Journal of Biological Sciences*, *23*, 575–589. https://doi.org/10.3923/pjbs.2020.575.589.

Heckman, M.A., Weil, J. and De Mejia, E.G. 2010. Caffeine (1, 3, 7-trimethylxanthine) in Foods: A Comprehensive Review on Consumption, Functionality, Safety, and Regulatory Matters. *J. Food Sci.*, *75*. https://doi.org/10.1111/j.1750-3841.2010.01561.x.

Henry, C.J. 2010. Functional foods. *Eur. J. Clin. Nutr.* https://doi.org/10.1038/ejcn.2010.101.

Hill, C., Guarner, F., Reid, G., Gibson, G.R., Merenstein, D.J., Pot, B., Morelli, L., Canani, R.B., Flint, H.J., Salminen, S., Calder, P.C. and Sanders, M.E. 2014. The International Scientific Association for Probiotics and Prebiotics consensus statement on the scope and appropriate use of the term probiotic. *Nat. Rev. Gastroenterol. Hepatol.*, *11*, 506–514. https://doi.org/10.1038/nrgastro.2014.66.

Ho, H.V.T., Sievenpiper, J.L., Zurbau, A., Blanco Mejia, S., Jovanovski, E., Au-Yeung, F., Jenkins, A.L. and Vuksan, V. 2016. The effect of oat β -glucan on LDL-cholesterol, non-HDL-cholesterol and apoB for CVD risk reduction: A systematic review and meta-analysis of randomised-controlled trials. *British Journal of Nutrition*, *116*, 1369–1382. https://doi.org/10.1017/S000711451600341X.

Hurrell, R.F. 2002. Fortification: Overcoming Technical and Practical Barriers. *J. Nutr.*, *132*, 806S–812S. https://doi.org/10.1093/jn/132.4.806S.

Iciek, M., Kwiecień, I. and Włodek, L. 2009. Biological properties of garlic and garlic-derived organosulfur compounds. *Environ. Mol. Mutagen*, *50*, 247–265. https://doi.org/10.1002/em.20474.

Kahn, B.B. and Flier, J.S. 2000. Obesity and insulin resistance. *Journal of Clinical Investigation*, *106*, 473–481. https://doi.org/10.1172/JCI10842.

Kaur, C. and Kapoor, H.C. 2001. Antioxidants in fruits and vegetables: The millennium's health. *Int. J. Food Sci. Technol.*, *36*, 703–725. https://doi.org/10.1046/j.1365-2621.2001.00513.x.

Kaur, N., Agarwal, A. and Sabharwal, M. 2022. Food fortification strategies to deliver nutrients for the management of iron deficiency anaemia. *Curr. Res. Food Sci.*, *5*, 2094–2107. https://doi.org/10.1016/j.crfs.2022.10.020.

Kaur, S. and Das, M. 2011. Functional foods: An overview. *Food Sci. Biotechnol.* https://doi.org/10.1007/s10068-011-0121-7.

Kayode, A.A.A., Adebodun, G.O., Fabunmi, D.I. and Kayode, O.T. 2023. Role of functional foods in the management of diabetes mellitus: A concise review. *Bioactive Molecules and Pharmaceuticals*, *2*, 29. https://doi.org/10.31989/bmp.v2i7.1132.

Khoo, H.E., Azlan, A., Tang, S.T. and Lim, S.M. 2017. Anthocyanidins and anthocyanins: Colored pigments as food, pharmaceutical ingredients, and the potential health benefits. *Food Nutr. Res.*, *61*, 1361779. https://doi.org/10.1080/16546628.2017.1361779.

Kim, Y.A., Keogh, J.B. and Clifton, P.M., 2018. Probiotics, prebiotics, synbiotics, and insulin sensitivity. *Nutr. Res. Rev.*, *31*, 35–51. https://doi.org/10.1017/S095442241700018X.

Kindernay, L., Ferenczyová, K., Farkašová, V., Duľová, U., Strapec, J. and Barteková, M. 2023. Beneficial Effects of Polyphenol-Rich Food Oils in Cardiovascular Health and Disease. *Rev. Cardiovasc. Med.*, *24*. https://doi.org/10.31083/j.rcm2407190.

Kopystecka, A., Kozioł, I., Radomska, D., Bielawski, K., Bielawska, A. and Wujec, M. 2023. Vaccinium uliginosum and Vaccinium myrtillus-Two Species-One Used as a Functional Food. *Nutrients*, *15*. https://doi.org/10.3390/nu15194119.

Korhonen, H. and Pihlanto, A. 2006. Bioactive peptides: Production and functionality. *Int. Dairy J.*, *16*, 945–960. https://doi.org/10.1016/j.idairyj.2005.10.012.

Krinsky, N.I. and Johnson, E.J. 2005. Carotenoid actions and their relation to health and disease. *Mol. Aspects Med.*, *26*, 459–516. https://doi.org/10.1016/j.mam.2005.10.001.

Kris-Etherton, P.M., Harris, W.S. and Appel, L.J. 2002. Fish Consumption, Fish Oil, Omega-3 Fatty Acids, and Cardiovascular Disease. *Circulation*, *106*, 2747–2757. https://doi.org/10.1161/01.CIR.0000038493.65177.94.

Kumar, A., Pramanik, J., Goyal, N., Chauhan, D., Sivamaruthi, B.S. and Prajapati, B.G. 2023. Gut Microbiota in Anxiety and Depression: Unveiling the Relationships and Management Options. *Pharmaceuticals*, *16*, 565. https://doi.org/10.3390/ph16040565.

Kumar, P., Vaishnav, A., Chauhan, S., S., L., Lal, J., Singh, N.S.C. and Pati, B.K. 2024. Encapsulation techniques for enhancing the stability and bioavailability of food: A review. *International Journal of Advanced Biochemistry Research*, *8*, 16–23. https://doi.org/10.33545/26174693.2024.v8.i3Sa.681.

Letchuman, S., Madhuranga, H.D.T., Madhurangi, B.L.N.K., Premarathna, A.D. and Saravanan, M. 2025. Alkaloids unveiled: A comprehensive analysis of novel therapeutic properties, mechanisms, and plant-based innovations. *Intelligent Pharmacy*, *3*, 268–276. https://doi.org/10.1016/j.ipha.2024.09.007.

Li, X., Xin, Y., Mo, Y., Marozik, P., He, T. and Guo, H. 2022. The Bioavailability and Biological Activities of Phytosterols as Modulators of Cholesterol Metabolism. *Molecules*, 27. https://doi.org/10.3390/molecules27020523.

Liu, R.H. 2003. Health benefits of fruit and vegetables are from additive and synergistic combinations of phytochemicals. *Am. J. Clin. Nutr.*, *78*, 517S–520S. https://doi.org/10.1093/ajcn/78.3.517S.

Lo, G.S., Evans, R.H., Phillips, K.S., Dahlgren, R.R. and Steinke, F.H. 1987. Effect of soy fiber and soy protein on cholesterol metabolism and atherosclerosis in rabbits. *Atherosclerosis*, *64*, 47–54. https://doi.org/10.1016/0021-9150(87)90053-0.

Lomax, A.R. and Calder, P.C. 2008. Prebiotics, immune function, infection, and inflammation: A review of the evidence. *British Journal of Nutrition*, *101*, 633–658. https://doi.org/10.1017/S0007114508055608.

Lourenço-Lopes, C., Fraga-Corral, M., Jimenez-Lopez, C., Carpena, M., Pereira, A.G., Garcia-Oliveira, P., Prieto, M.A. and Simal-Gandara, J. 2021. Biological action mechanisms of fucoxanthin extracted from algae for application in food and cosmetic industries. *Trends Food Sci. Technol.*, *117*, 163–181. https://doi.org/10.1016/j.tifs.2021.03.012.

Ma, D.-F., Qin, L.-Q., Wang, P.-Y. and Katoh, R. 2008. Soy isoflavone intake increases bone mineral density in the spine of menopausal women: Meta-analysis of randomized controlled trials. *Clinical Nutrition*, *27*, 57–64. https://doi.org/10.1016/j.clnu.2007.10.012.

Marasco, G., Cremon, C., Salvi, D., Meacci, D., Dajti, E., Colecchia, L., Barbaro, M.R., Stanghellini, V. and Barbara, G. 2025. Functional Foods and Nutraceuticals in Irritable Bowel Syndrome. *J. Clin. Med.*, *14*, 1830. https://doi.org/10.3390/jcm14061830.

Marina Russo and Francesco Cacciola. 2022. *Characterization of Bioactive Compounds in Foods and Plants Using Advanced Analytical Techniques*. MDPI. https://doi.org/10.3390/books978-3-0365-2854-0.

McClements, D.J., Newman, E. and McClements, I.F. 2019. Plant-based Milks: A Review of the Science Underpinning Their Design, Fabrication, and Performance. *Compr. Rev. Food Sci. Food Saf.*, *18*, 2047–2067. https://doi.org/10.1111/1541-4337.12505.

Meli, C. and Bianca, S. 2002. Dietary control of phenylketonuria. *The Lancet*, *360*, 2075–2076. https://doi.org/10.1016/S0140-6736(02)11958-1.

Messina, M. 2016. Soy and Health Update: Evaluation of the Clinical and Epidemiologic Literature. *Nutrients*, *8*, 754. https://doi.org/10.3390/nu8120754.

Meybodi, N.M., Mortazavian, A.M., Arab, M. and Nematollahi, A. 2020. Probiotic viability in yoghurt: A review of influential factors. *Int. Dairy J.*, *109*, 104793. https://doi.org/10.1016/j.idairyj.2020.104793.

Michalak, M., Pierzak, M., Kręcisz, B. and Suliga, E. 2021. Bioactive Compounds for Skin Health: A Review. *Nutrients*, *13*. https://doi.org/10.3390/nu13010203.

Mielech, A., Puścion-Jakubik, A., Markiewicz-Żukowska, R. and Socha, K. 2020. Vitamins in Alzheimer's Disease-Review of the Latest Reports. *Nutrients*, *12*. https://doi.org/10.3390/nu12113458.

Mijan, M. Al and Lim, B.O. 2018. Diets, functional foods, and nutraceuticals as alternative therapies for inflammatory bowel disease: Present status and future trends. *World J. Gastroenterol.*, *24*, 2673–2685. https://doi.org/10.3748/wjg.v24.i25.2673.

Milner J. A. 2000. Functional foods: The US perspective. *The American Journal of Clinical Nutritio*n, *71*(6 Suppl), 1654S–5S. https://doi.org/10.1093/ajcn/71.6.1654S

Mocanu, V. and Madsen, K.L., 2024. Dietary fibre and metabolic health: A clinical primer. *Clin. Transl. Med.*, *14*. https://doi.org/10.1002/ctm2.70018.

Mohanty, S. and Singhal, K. 2018. Functional Foods as Personalised Nutrition: Definitions and Genomic Insights. In: *Functional Food and Human Health*. Springer Singapore, Singapore, pp. 513–535. https://doi.org/10.1007/978-981-13-1123-9_22.

Mostafai, R., Nachvakc, S.M., Mohammadi, R., Rocha, R.S., da Silva, M.C., Esmerino, E.A., Nascimento, K.O., Cruz, A.G. and Mortazavian, A.M. 2019. Effects of vitamin D-fortified yogurt in comparison to oral vitamin D supplement on hyperlipidemia in pre-diabetic patients: A randomized clinical trial. *J. Funct. Foods*, *52*, 116–120. https://doi.org/10.1016/j.jff.2018.10.040.

Mozaffarian, D., Benjamin, E.J., Go, A.S., Arnett, D.K., Blaha, M.J., Cushman, M., et al., 2016. Heart Disease and Stroke Statistics: 2016 Update. *Circulation*, *133*. https://doi.org/10.1161/CIR.0000000000000350.

Mukuddem-Petersen, J., Oosthuizen, W. and Jerling, J.C. 2005. A Systematic Review of the Effects of Nuts on Blood Lipid Profiles in Humans. *J. Nutr.*, *135*, 2082–2089. https://doi.org/10.1093/jn/135.9.2082.

Myzak, M.C. and Dashwood, R.H. 2006. Chemoprotection by sulforaphane: Keep one eye beyond Keap. *Cancer Lett.*, *233*, 208–218. https://doi.org/10.1016/j.canlet.2005.02.033.

Neag, M.A., Mocan, A., Echeverría, J., Pop, R.M., Bocsan, C.I., Crişan, G. and Buzoianu, A.D. 2018. Berberine: Botanical Occurrence, Traditional Uses, Extraction Methods, and Relevance in Cardiovascular, Metabolic, Hepatic, and Renal Disorders. *Front. Pharmacol.*, *9*. https://doi.org/10.3389/fphar.2018.00557.

Ney, D.M., Stroup, B.M., Clayton, M.K., Murali, S.G., Rice, G.M., Rohr, F. and Levy, H.L. 2016. Glycomacropeptide for nutritional management of phenylketonuria: A randomized, controlled, crossover trial. *Am. J. Clin. Nutr.*, *104*, 334–345. https://doi.org/10.3945/ajcn.116.135293.

Oak, S.J. and Jha, R. 2019. The effects of probiotics in lactose intolerance: A systematic review. *Crit. Rev. Food Sci. Nutr.*, 59, 1675–1683. https://doi.org/10.1080/10408398.2018.1425977.

Oakley, G.P. 2010. Folic Acid–Preventable Spina Bifida. *Am. J. Prev. Med.*, *38*, 569–570. https://doi.org/10.1016/j.amepre.2010.02.002.

Odriozola-Serrano, I., Oms-Oliu, G. and MartÃn-Belloso, O. 2014. Nanoemulsion-Based Delivery Systems to Improve Functionality of Lipophilic Components. *Front. Nutr.*, *1*. https://doi.org/10.3389/fnut.2014.00024.

O'Keefe, S.J.D. 2016. Diet, microorganisms and their metabolites, and colon cancer. *Nat. Rev. Gastroenterol. Hepatol.*, *13*, 691–706. https://doi.org/10.1038/nrgastro.2016.165.

Olson, R., Gavin-Smith, B., Ferraboschi, C. and Kraemer, K. 2021. Food Fortification: The Advantages, Disadvantages and Lessons from Sight and Life Programs. *Nutrients*, *13*, 1118. https://doi.org/10.3390/nu13041118.

Paine, J.A., Shipton, C.A., Chaggar, S., Howells, R.M., Kennedy, M.J., Vernon, G., Wright, S.Y., Hinchliffe, E., Adams, J.L., Silverstone, A.L. and Drake, R. 2005. Improving the nutritional value of Golden Rice through increased pro-vitamin A content. *Nat. Biotechnol.*, *23*, 482–487. https://doi.org/10.1038/nbt1082.

Pamunuwa, G.K. and Karunaratne, D.N. 2022. Liposomal Delivery of Plant Bioactives Enhances Potency in Food Systems: A Review. *J. Food Qual.*, *2022*, 1–11. https://doi.org/10.1155/2022/5272592.

Patel, Madhumita, Narayan Naik, S., Patel, M. and Naik, S.N. 2004. Gamma-Oryzanol from rice bran oil: A review. *Journal of Scientific & Industrial Research*,. *63*, 569–578. https://www.researchgate.net/publication/239785419.

Peña-Vázquez, G.I., Arredondo-Arenillas, A., Serrano-Sandoval, S.N. and Antunes-Ricardo, M. 2024. Functional foods lipids: Unraveling their role in the immune response in obesity. *Crit. Rev. Food Sci. Nutr.*, 1–22. https://doi.org/10.1080/10408398.2024.2382942.

Perdomo, L., Beneit, N., Otero, Y.F., Escribano, Ó., Díaz-Castroverde, S., Gómez-Hernández, A. and Benito, M. 2015. Protective role of oleic acid against cardiovascular insulin resistance and in the early and late cellular atherosclerotic process. *Cardiovasc. Diabetol.*, *14*, 75. https://doi.org/10.1186/s12933-015-0237-9.

Phillips, S.M. 2016. The impact of protein quality on the promotion of resistance exercise-induced changes in muscle mass. *Nutr. Metab.* (*Lond*), *13*, 64. https://doi.org/10.1186/s12986-016-0124-8.

Prokisch, J., Badgar, K. and El-Ramady, H. 2021. Fortification of Functional Foods for Human Health: A Case Study of Honey and Yogurt for Diabetes. *Environment, Biodiversity, and Soil Security*, *5*, 1–2. https://doi.org/10.21608/jenvbs.2021.110812.1154.

Rader, D.J., Cohen, J. and Hobbs, H.H. 2003. Monogenic hypercholesterolemia: New insights in pathogenesis and treatment. *Journal of Clinical Investigation*, *111*, 1795–1803. https://doi.org/10.1172/JCI18925.

Rashidinejad, A., 2024. The road ahead for functional foods: Promising opportunities amidst industry challenges. *Future Postharvest and Food* , *1*(2), 266–273. https://doi.org/10.1002/fpf2.12022.

Rathna Kumari, B.M. 2020. Biofortification Strategies for Enhancing Nutrient Content in Staple Crops: Progress, Challenges, and Future Directions. *Plant Science Archives*, *5*, 07–10. https://doi.org/10.51470/PSA.2020.5.1.07.

Reaven, G.M. 1995. Pathophysiology of insulin resistance in human disease. *Physiol. Rev.*, *75*, 473–486. https://doi.org/10.1152/physrev.1995.75.3.473.

Rehal, J., Thapa, A. and Kaushal, P. 2025. Benefits and challenges in plant-based milk alternatives. *Journal of Innovative Agriculture* (Review Article), *12*, 1. https://doi.org/10.37446/jinagri/ra/12.2.2025.1-11.

Rizzoli, R. 2014. Nutritional aspects of bone health. *Best Pract. Res. Clin. Endocrinol. Metab.*, *28*, 795–808. https://doi.org/10.1016/j.beem.2014.08.003.

Roberfroid, M. 2007. Prebiotics: The Concept Revisited. *J. Nutr.*, *137*, 830S–837S. https://doi.org/10.1093/jn/137.3.830S.

Roberfroid, M., Gibson, G.R., Hoyles, L., McCartney, A.L., Rastall, R., Rowland, I., et al.,2010. Prebiotic effects: metabolic and health benefits. *British Journal of Nutrition*, *104*, S1–S63. https://doi.org/10.1017/S0007114510003363.

Roberfroid, M.B. 2002. Global view on functional foods: European perspectives. *British Journal of Nutrition*, *88*, S133–S138. https://doi.org/10.1079/BJN2002677.

Rocha, J.C. and MacDonald, A. 2016. Dietary intervention in the management of phenylketonuria: Current perspectives. *Pediatric Health Med. Ther.*, *7*, 155–163. https://doi.org/10.2147/PHMT.S49329.

Scalbert, A., Johnson, I.T., Saltmarsh, M. 2005. Polyphenols: Antioxidants and beyond. *Am. J. Clin. Nutr.*, *81*, 215S–217S. https://doi.org/10.1093/ajcn/81.1.215S.

Seneca, 2007. Definition, Typology and Occurrence of Alkaloids. In: *Alkaloids: Secrets of Life* (pp. 1–59). Elsevier. https://doi.org/10.1016/B978-044452736-3/50003-9.

Serafini, M. and Peluso, I. 2016. Functional Foods for Health: The Interrelated Antioxidant and Anti-Inflammatory Role of Fruits, Vegetables, Herbs, Spices, and Cocoa in Humans. *Current Pharmaceutical Design*, *22*(44), 6701–6715. https://doi.org/10.2174/1381612823666161123094235.

Shahidi, F., Ambigaipalan, P. 2015. Phenolics and polyphenolics in foods, beverages, and spices: Antioxidant activity and health effects: A review. *J. Funct. Foods*, *18*, 820–897. https://doi.org/10.1016/j.jff.2015.06.018.

Shakoor, H., Feehan, J., Apostolopoulos, V., Platat, C., Al Dhaheri, A.S. and Ali, H.I. 2021. Immunomodulatory Effects of Dietary Polyphenols. *Nutrients*, *13*, 728. https://doi.org/10.3390/nu13030728.

Sharma, A.R., Lee, Y.-H., Bat-Ulzii, A., Chatterjee, S., Bhattacharya, M. and Chakraborty, C. 2023. Bioactivity, Molecular Mechanism, and Targeted Delivery of Flavonoids for Bone Loss. *Nutrients*, *15*, 919. https://doi.org/10.3390/nu15040919.

Simopoulos, A.P. 2002. Omega-3 Fatty Acids in Inflammation and Autoimmune Diseases. *J. Am. Coll. Nutr.*, *21*, 495–505. https://doi.org/10.1080/07315724.2002.10719248.

Singh, N. and Gaur, S. 2024. New insights into multifunctional aspects of milk derived bioactive peptides: A review. *Food Chemistry Advances*, *4*, 100628. https://doi.org/10.1016/j.focha.2024.100628.

Singh, R.H., Rohr, F., Frazier, D., Cunningham, A., Mofidi, S., Ogata, B., Splett, P.L., Moseley, K., Huntington, K., Acosta, P.B., Vockley, J. and Van Calcar, S.C. 2014. Recommendations for the nutrition management of phenylalanine hydroxylase deficiency. *Genetics in Medicine*, *16*, 121–131. https://doi.org/10.1038/gim.2013.179.

Slavin, J., Savarino, V., Paredes-Diaz, A. and Fotopoulos, G. 2009. A Review of the Role of Soluble Fiber in Health with Specific Reference to Wheat Dextrin. *Journal of International Medical Research*, *37*, 1–17. https://doi.org/10.1177/147323000903700101.

Sparg, S.G., Light, M.E. and van Staden, J. 2004. Biological activities and distribution of plant saponins. *J. Ethnopharmacol.*, *94*, 219–243. https://doi.org/10.1016/j.jep.2004.05.016.

Su, Y., Cappock, M., Dobres, S., Kucine, A.J., Waltzer, W.C. and Zhu, D. 2023. Supplemental mineral ions for bone regeneration and osteoporosis treatment. *Engineered Regeneration*, 4, 170–182. https://doi.org/10.1016/j.engreg.2023.02.003.

Tabrizi, A., Dargahi, R., Tehrani Ghadim, S., Javadi, M., Rasouli Pirouzian, H., Azizi, A. and Homayouni Rad, A. 2020. Functional laxative foods: Concepts, trends, and health benefits. In: *Bioactive Natural Products* (pp. 305–330). ResearchGate. https://doi.org/10.1016/B978-0-12-817907-9.00011-8.

Tanaka, T., Shimizu, M. and Moriwaki, H. 2012. Cancer Chemoprevention by Carotenoids. *Molecules*, *17*, 3202–3242. https://doi.org/10.3390/molecules17033202.

Terao, J. 2023. Revisiting carotenoids as dietary antioxidants for human health and disease prevention. *Food Funct.*, *14*, 7799–7824. https://doi.org/10.1039/D3FO02330C.

Thakur, S., Singh, A., Insa, B., Sharma, S. 2023. Food fortification in India as malnutrition concern: A global approach. *Sustainable Food Technology*, *1*, 681–695. https://doi.org/10.1039/D3FB00079F.

Valdes, A.M., Walter, J., Segal, E. and Spector, T.D. 2018. Role of the gut microbiota in nutrition and health. *BMJ, 361*, k2179. https://doi.org/10.1136/bmj.k2179.

Vauzour, D., Vafeiadou, K., Rodriguez-Mateos, A., Rendeiro, C. and Spencer, J.P.E. 2008. The neuroprotective potential of flavonoids: A multiplicity of effects. *Genes Nutr.*, *3*, 115–126. https://doi.org/10.1007/s12263-008-0091-4.

Walter, P. 2008. 10 Years of Functional Foods in Europe. *International Journal for Vitamin and Nutrition Research*, *78*, 253–260. https://doi.org/10.1024/0300-9831.78.6.253.

Wang, L., Zhang, P. and Yan, H. 2023. Functional foods and dietary supplements in the management of non-alcoholic fatty liver disease: A systematic review and meta-analysis. *Front. Nutr.*, *10*. https://doi.org/10.3389/fnut.2023.1014010.

Wang, W., Yang, W., Dai, Y., Liu, J. and Chen, Z.-Y. 2023. Production of Food-Derived Bioactive Peptides with Potential Application in the Management of Diabetes and Obesity: A Review. *J. Agric. Food Chem.*, *71*, 5917–5943. https://doi.org/10.1021/acs.jafc.2c08835.

Wang, X., Qi, Y. and Zheng, H. 2022. Dietary Polyphenol, Gut Microbiota, and Health Benefits. *Antioxidants*, *11*, 1212. https://doi.org/10.3390/antiox11061212.

Weaver, C.M., Gordon, C.M., Janz, K.F., Kalkwarf, H.J., Lappe, J.M., Lewis, R., O'Karma, M., Wallace, T.C. and Zemel, B.S. 2016. The National Osteoporosis Foundation's position statement on peak bone mass development and lifestyle factors: A systematic review and implementation recommendations. *Osteoporosis International, 27*, 1281–1386. https://doi.org/10.1007/s00198-015-3440-3.

Wei, B.-Z., Li, L., Dong, C.-W., Tan, C.-C. and Xu, W. 2023. The Relationship of Omega-3 Fatty Acids with Dementia and Cognitive Decline: Evidence from Prospective Cohort Studies of Supplementation, Dietary Intake, and Blood Markers. *Am. J. Clin. Nutr.*, *117*, 1096–1109. https://doi.org/10.1016/j.ajcnut.2023.04.001.

Williams, M., Bank, W. and Ragasa, C. 2006. Functional Foods : Opportunities and Challenges for Developing Countries. *Agriculture and Rural Development*, *19*. https://www.researchgate.net/publication/279437168.

Williamson, G. and Sheedy, K. 2020. Effects of Polyphenols on Insulin Resistance. *Nutrients*, *12*, 3135. https://doi.org/10.3390/nu12103135.

Yamamoto, Y., Aizawa, K., Mieno, M., Karamatsu, M., Hirano, Y. and Furui, K. 2017. The effects of tomato juice on male infertility. *Asia Pac. J. Clin. Nutr.*, *26*, 65–71. https://doi.org/10.6133/apjcn.102015.17.

Zabot, G.L., Schaefer Rodrigues, F., Polano Ody, L., Vinícius Tres, M., Herrera, E., Palacin, H., et al., 2022. Encapsulation of Bioactive Compounds for Food and Agricultural Applications. *Polymers* (*Basel*) *14*, 4194. https://doi.org/10.3390/polym14194194.

Zaky, A.A., Simal-Gandara, J., Eun, J.-B., Shim, J.-H., Abd El-Aty, A.M. 2022. Bioactivities, Applications, Safety, and Health Benefits of Bioactive Peptides from Food and By-Products: A Review. *Front Nutr.*, *8*. https://doi.org/10.3389/fnut.2021.815640.

Zhang, S.Q., Tian, D., Hu, C.Y. and Meng, Y.H. 2022. Chlorogenic Acid Ameliorates High-Fat and High-Fructose Diet-Induced Cognitive Impairment via Mediating the Microbiota-Gut-Brain Axis. *J. Agric. Food Chem.*, *70*, 2600–2615. https://doi.org/10.1021/acs.jafc.1c07479.

Zhao, Y., Jiang, Q. 2021. Roles of the Polyphenol-Gut Microbiota Interaction in Alleviating Colitis and Preventing Colitis-Associated Colorectal Cancer. *Adv. Nutr.*, *12*, 546–565. https://doi.org/10.1093/advances/nmaa104.

Zheng, Y., Qin, C., Wen, M., Zhang, L. and Wang, W. 2024. The Effects of Food Nutrients and Bioactive Compounds on the Gut Microbiota: A Comprehensive Review. *Foods*, *13*. https://doi.org/10.3390/foods13091345.

Zhong, J., Tan, L., Chen, M. and He, C. 2022. Pharmacological activities and molecular mechanisms of Pulsatilla saponins. *Chin. Med.*, *17*, 59. https://doi.org/10.1186/s13020-022-00613-8.

Zhong, X., Wang, G., Li, F., Fang, S., Zhou, S. and Ishiwata, A. 2023. Immunomodulatory Effect and Biological Significance of β-Glucans. *Pharmaceutics*, *15*, 1615. https://doi.org/10.3390/pharmaceutics15061615.

Zimmermann, M.B. 2009. Iodine deficiency in pregnancy and the effects of maternal iodine supplementation on the offspring: A review. *Am. J. Clin. Nutr.*, *89*, 668S–672S. https://doi.org/10.3945/ajcn.2008.26811C.

CHAPTER 14

Phenolic Bioactives in Functional Foods

Antioxidant and Anti-Inflammatory Health-Promoting Properties

Eleni Koutsia[1], Theodora Adamantidi[2] and Alexandros Tsoupras[3*]

1. Introduction

In recent decades, integrating bioactive compounds into food matrices and systems has drawn significant attention from academia and industry, driven by growing consumer demand for foods that promote health and prevent disease (Rana et al., 2022; Zhang et al., 2022). Among the most extensively studied and functionally diverse bioactive compounds are polyphenols, a notably large class of plant-derived secondary metabolites characterized by one or more aromatic rings bearing hydroxyl groups (Ciepei et al., 2024). These active compounds are ubiquitously found in fruits, vegetables, cereals, legumes, herbs, spices, tea, coffee, wine, and oils, contributing not only to the sensory attributes of foods, such as color, flavor, and bitterness, but also to their health-promoting properties (Khan et al., 2024; Güneş et al., 2019).

The successful utilization of polyphenols in functional foods largely depends on the efficiency of their extraction from plant matrices. The yield, purity, and activity of polyphenolic extracts are significantly influenced by several interconnected factors, including the choice of solvent, extraction temperature, duration, solid-to-solvent ratio, particle size of the raw material, and the specific method of extraction employed (e.g., ultrasound-assisted extraction (UAE), microwave-assisted extraction (MAE), pressurized liquid extraction (PLE), supercritical fluid extraction (SFE), or traditional solvent maceration) (Hu, 2022; Brglez et al., 2016). Moreover, the chemical structure and polarity of individual polyphenols determine their solubility and extractability, with green extraction technologies increasingly favored for their environmental friendliness and preservation of thermolabile compounds (Panja, 2018; Palos-H. et al., 2025)

Polyphenols are broadly classified into subgroups, including phenolic acids, flavonoids, lignans, stilbenes, and tannins, whereas multiple simple phenols have also displayed important beneficial properties (Ciepei et al., 2024). Each polyphenolic compound has distinct chemical structures, biological roles, and food-distribution profiles. Representative compounds include gallic acid, ferulic acid, caffeic acid, protocatechuic acid, and *p*-coumaric acid (phenolic acids: hydroxybenzoic

[1-3] Hephaestus Laboratory, School of Chemistry, Faculty of Sciences, Democritus University of Thrace, Kavala University Campus, St. Lucas 65404 Kavala, Greece.

* Corresponding author: atsoupras@chem.duth.gr

DOI: 10.1201/9781003659808-14

and hydrocinnamic acids) (Ficco et al., 2024), quercetin, catechins, apigenin, luteolin, naringenin, luteolin 7-*O*-glucoside, and cyanidin (flavonoids, flavonoid derivatives, and flavonoid subgroups) (Singla et al., 2019), secoisolariciresinol (lignans), resveratrol (stilbenes), gallotannins and ellagitannins (tannins), as well as curcumin, vanillin, and hydroxytyrosol (simple phenols) (Ciepei et al., 2024; Ficco et al., 2024; Singla et al., 2019). These compounds are abundant in dietary staples such as berries, grapes, tea, coffee, cocoa, olives, flaxseeds, and citrus fruits. At the same time, herbs and spices like rosemary, thyme, and turmeric are also rich in polyphenols (Manach et al., 2004; Li et al., 2023; Khan 2019).

Polyphenolic compounds are widely recognized for their antioxidant, anti-inflammatory, anti-diabetic, anti-obesity, cardioprotective, neuroprotective, and anti-carcinogenic effects (Ebrahimi, 2021; Rasoului et al., 2017), which make them ideal candidates for incorporation into functional foods –foods that go beyond basic nutrition to provide additional physiological benefits (Khan et al., 2024; Güneş et al., 2019; Ebrahimi, 2021; de Araújo et al., 2021). Increasing evidence suggests that polyphenol-rich diets contribute to the prevention and management of a plethora of chronic diseases, including cardiovascular disease, Type 2 diabetes (T2DM), neurodegenerative disorders (NDs) such as Alzheimer's and Parkinson's diseases (AD and PD, respectively), certain types of cancer, obesity, atherosclerosis, high blood sugar, metabolic syndrome, and skin diseases. These effects are attributed not only to their ability to neutralize oxidative stress and modulate inflammatory pathways, but also to their influence on gut microbiota composition, cell signaling, and gene expression (Rasouli et al., 2017; Zekrumal et al., 2023; El Oirdi 2024; Wu et al., 2025; Farhan, 2024).

However, the bioavailability and stability of polyphenols can be limited due to their sensitivity to heat, light, pH, and digestive conditions. To address these challenges, novel delivery systems such as nanoencapsulation, microencapsulation, liposomes, and biopolymeric carriers have been created, with encapsulation efficiency depending on the type of polyphenol, carrier system, and processing method (Enaru et al., 2021; Shi et al., 2021; Ali Redha et al., 2024). Recent studies have reported encapsulation efficiencies ranging from 60 to over 90% of compounds like resveratrol, ferulic acid, quercetin, catechins, and hydroxytyrosol, significantly enhancing their functional stability and targeted delivery (Ali Redha et al., 2024; Wang et al., 2024).

The bioactivity of polyphenols is strongly influenced by their chemical structure, source, degree of polymerization, and the extraction method employed. Hence, a comprehensive understanding of their classification, occurrence, extraction technologies, and mechanisms of action is essential for their successful application in functional food design (Ciepei et al., 2024; Khan et al., 2024; de Araújo et al., 2021; Ali Redha et al., 2024). This review aims to provide a detailed overview of the most prominent dietary polyphenols, mainly focusing on their chemical classification and representative compounds, their natural source, green and conventional extraction methods used to isolate them from raw materials, their unique role in plant and human health, their antioxidant and anti-inflammatory mechanisms supported by *in vivo* and *in vitro* studies, their role in the prevention and potential management of chronic non-communicable diseases, and their current and emerging applications in functional food systems (Ciepei et al., 2024; Khan et al., 2024; de Araújo et al., 2021; Ali Redha et al., 2024). By systemically presenting these aspects, this review will serve as a reference for food scientists, nutritionists, and product developers seeking to formulate polyphenol-rich functional foods with substantiated health claims.

2. Extraction Methods of Polyphenols and Factors Affecting the Extraction Yield

The extraction and separation of polyphenols, a structurally diverse group of multiple secondary metabolites including phenolic acids, flavonoids, lignans, stilbenes, tannins, and simple phenols from natural sources, represent key steps for their valorization, characterization, and application in many functional food, nutraceutical, and pharmaceutical applications (Panja 2018; Palos-H. et al.,

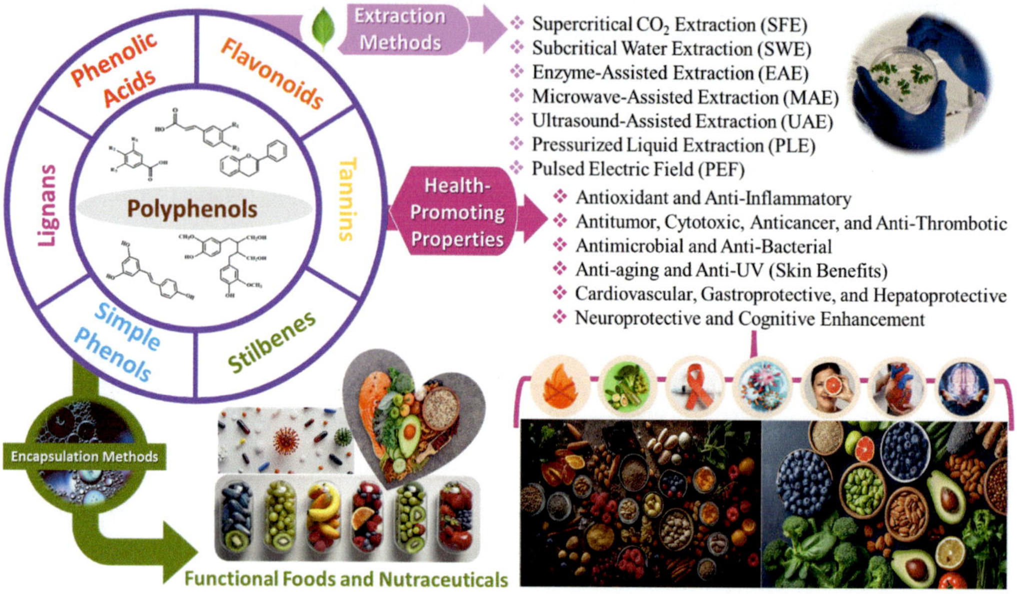

Fig. 1 Health-Promoting Properties of Polyphenols in Functional Food Applications. (Parts of this figure were obtained from https://www.freepik.com/ (accessed on May 13, 2025)).

2025). Given their wide range of polarity, molecular weight, and functional properties, optimized extraction protocols are required to achieve high yield while maintaining chemical integrity. Over the past two decades, intensive research efforts have been directed toward refining these protocols, transitioning from conventional solvent-based techniques to modern, green extraction strategies (Hu, 2022; Brglez et al., 2016; Panja, 2018; Palos-H et al., 2025; Khan et al., 2019).

2.1 *Conventional Extraction Methods*

Traditional extraction approaches, including maceration, Soxhlet extraction, and solid-liquid extraction, remain widely employed due to their simplicity, accessibility, and adaptability to various plant matrices. These methods generally use organic solvents such as methanol, ethanol, acetone, diethyl ether, and ethyl acetate, often in aqueous combinations, depending on the target class of polyphenols. The polarity of the solvent system critically affects extraction efficiency: for instance, methanol is suitable for extracting low-molecular-weight polyphenols such as simple phenols and phenolic acids, while aqueous acetone is more effective for flavonoids and tannins (Hu, 2022; Brglez et al., 2016; Lama-Muñoz and Contreras, 2022; Alara et al., 2021; Ćujić et al., 2016). Less polar solvents (e.g., dichloromethane, chloroform, hexane) are occasionally used to eliminate lipophilic non-polyphenolic compounds such as chlorophylls, waxes, oils, and sterols (Brglez et al., 2016). However, these solvents may pose environmental and health concerns and often necessitate subsequent steps of purification. Additionally, conventional methods may expose polyphenols to prolonged heat and oxidative conditions, resulting in degradation or isomerization, especially regarding thermolabile compounds like anthocyanins or stilbenes (Hu, 2022; Brglez et al., 2016; Lama-Muñoz and Contreras, 2022; Alara et al., 2021; Ćujić et al., 2016; Athanasiadis et al., 2023).

To mitigate degradation, extraction temperatures are typically maintained between 20–50°C, as temperatures above 70°C have been shown to compromise the stability of several polyphenols (Brglez et al., 2016). Parameters such as particle size, moisture content, solvent-to-solid ratio, and extraction time also play pivotal roles in extraction efficiency. Pre-treatment methods, including drying, grinding, and homogenization, enhance both surface area and mass transfer, while freeze-drying is preferred for preserving the structure and content of thermosensitive polyphenols. Additionally, pH adjustment of the extraction medium can improve selectivity and stability. For example, catechins are more stable at pH 4–5, whereas strong acidic or alkaline conditions accelerate their degradation (Hu, 2022; Brglez et al., 2016; Rasouli et al., 2017; Ajila et al., 2011).

2.2 Emerging Green Extraction Techniques

To overcome the limitations of conventional techniques, including high solvent usage, long processing times, and low selectivity, green extraction methods have gained significant attention. Among these, UAE and MAE, respectively, and especially SFE (mainly CO_2), represent more sustainable and efficient alternatives. Furthermore, PLE, pulsed electric field (PEF), subcritical water (SWE), and enzyme-assisted extraction (EAE) have also been pointed out as green, efficacious extraction techniques. These methods reduce the need for toxic solvents and energy input while maximizing recovery, enhancing bioavailability, and preserving bioactivity (Panja, 2018; Palos-H. et al., 2025; Ameer et al., 2017; Dalmoune et al., 2025; Dzah et al., 2020; Da Parto and Natolina, 2017; Gallo et al., 2019; Frontuto et al., 2019; Gligor et al., 2019).

UAE uses ultrasonic waves to enhance cell wall disruption and mass transfer, increasing extraction efficiency mostly at lower temperatures. It has been effectively applied, for instance, for oleuropein and hydroxytyrosol recovery from olive leaves and pomace (Chanioti and Tzia, 2017; Iraki et al., 2018). MAE involves rapid internal heating of plant tissues using microwave energy, strikingly reducing both time and solvent consumption; it is widely used for extracting flavonoids and phenolic acids from plant matrices like myrtle, chokeberries, and *Rosmarinus officinalis* herbs (Dalmoune et al., 2025; Simić et al., 2016; Švarc-Gajić et al., 2013). Plus, EAE employs cell wall-degrading enzymes like cellulases or pectinases to break down structural barriers, facilitating the release of bound polyphenols, including tannins and lignans, especially in by products like sweet cherry and grape pomace, as well as unripe apples and grape seed extract (Gligor et aL., 2019; Zheng et al ., 2009; Domínguez-R. et al., 2021; Chamorro et al., 2012).

PEF treatment improves the permeability of cell membranes through electroporation, therefore enhancing solvent penetration and extraction yield, particularly useful for stilbenes and simple phenols in matrices like pomegranate, potato, and orange peels (Athanasiadis et al., 2023; Frontuto et al., 2019; Luengo et al., 2013; Rajha et al., 2019). Additionally, PLE uses solvents at high pressure and moderate temperature to enhance solubility and diffusion, and is especially efficient for recovering flavonoids, caffeic acid, and ferulic acid from pomegranate, apple, and grape peels (Gallo et al., 2019; Alonso-S. et al., 2001; Sumere et al., 2018; Çam M, Hışıl, 2010). SWE is another green technique that utilizes water at high temperature and pressure, reducing water's polarity and allowing for efficient extraction of both polar and semi-polar polyphenols. SWE is particularly suited for complex matrices like grape skins and is attractive due to its low cost, scalability, and absence of hazardous solvents (Zhang et al., 2020; Duba et al., 2015).

Supercritical CO_2 extraction has emerged as a powerful approach due to the environmentally benign nature of CO_2, its ability to act as a tunable solvent, and the mild conditions under which it operates. In its supercritical state, CO_2 shows gas-like diffusivity and liquid-like solvation properties, which are ideal for extracting polyphenols, especially those with moderate-to-low polarity, such as flavonoids, stilbenes, and lignans, in contrast to conventional extraction methods. The extraction efficiency can be further enhanced by adding polar co-solvents like ethanol or water to facilitate the recovery of more polar polyphenols, including phenolic acids, flavonoids, and tannins (Da Parto and Natolina, 2017; Farías-C. et al., 2013; Espinosa-Petal., 2017; Santos et al., 2012). SFE has been successfully applied to plant-based residues and agro-industrial by products like, for example, orange pomace or grape seed extract, offering clean, solvent-free polyphenol extracts with minimal thermal degradation. Moreover, it supports the development of high-value ingredients for use in cosmetics, functional foods, and pharmaceuticals. Applications include decaffeination, aroma extraction, and the selective isolation of polyphenolic fractions with anti-inflammatory, antioxidant, anti-thrombotic, anti-aging, neuroprotective, and cardioprotective action (Da Parto and Natolina, 2017; Farías-C. et al., 2013; Espinosa-Petal, 2017; Santos et al., 2012).

2.3 Parameters Influencing Extraction Efficiency

The extraction efficiency of polyphenols is influenced by compound-specific properties (e.g., molecular weight, polarity, solubility) and process-related factors such as extraction time, solvent type and concentration, temperature, pH, and matrix structure. Polyphenols vary widely in solubility and reactivity. For example, hydrocinnamic acids and flavonoids are typically more soluble in polar aqueous-organic solvents, while lignans and stilbenes may require semi-polar media (Panja, 2018; Ajila et al., 2011; Sulaiman et al., 2011).

Pre-treatment techniques, including particle size reduction and enzyme-assisted hydrolysis, enhance the liberation of bound polyphenols from plant cell walls, where they are often linked to polysaccharides or proteins. Extraction kinetics are also highly affected by solvent viscosity, sample porosity, and agitation. While higher temperatures accelerate solute diffusion and solvent penetration, they must be carefully controlled to avoid thermal decomposition of sensitive polyphenols such as proanthocyanins or resveratrol (Panja, 2018; Ajila et al., 2011; Sulaiman et al., 2011). Finally, optimizing the solvent-to-solid ratio, extraction duration, and number of extraction cycles is critical for improving polyphenol recovery and lowering solvent consumption. The ideal conditions must strike a balance between maximizing yield and preserving the biological functionality of the polyphenolic compounds (Panja, 2018; Ajila et al., 2011; Sulaiman et al., 2011).

2.4 Antioxidant Activity Assessment and Polyphenol Separation

After completing the extraction procedure, the antioxidant potential of polyphenolic-abundant extracts is commonly evaluated using *in vitro* assays, with the 2,2-diphenyl-1-picrylhydrazyl (DPPH) radical scavenging method being among the most popular. This assay provides rapid and reliable results by measuring the decrease in absorbance at 515 nm upon reaction with antioxidants in the extract. The simplicity, reproducibility, and minimal thermal stress of the DPPH technique make it suitable for evaluating a wide range of polyphenols (Rusouli et al., 2017; Ajila et al., 2011; Frontuto et al., 2019; Chanioti and Tzia, 2017; Li et al., 2017; Jesinek et al., 2015; Pedan et al., 2016; Li et al., 2022; Chedea and Pop, 2019).

Complementary methods like 2,2'-azino-bis(3-ethylbenzothiazoline-6-sulfonic) acid (ABTS), ferric reducing antioxidant power (FRAP), and oxygen radical absorbance capacity (ORAC) assays offer additional insights into antioxidant mechanisms, including electron transfer and hydrogen atom donation capacity (Li et al., 2017; Pico-H. et al., 2020; Pascual-M. et al., 2018). To separate and quantify individual polyphenols from natural sources, advanced chromatographic techniques like high-performance liquid chromatography (HPLC), gas chromatography (GC), and capillary electrophoresis (CE) are employed. These techniques are often coupled with detectors, including ultraviolet (UV)-Vis, mass spectrometry (MS), or nuclear magnetic resonance (NMR) for accurate identification and quantification of specific compounds, including hydroxytyrosol, gallic acid, catechins, quercetin, cyanidin, and tannins (Pedan et al., 2016; Li et al., 2022; Pascal-M. et al., 2018; Lucas-T., et al., 2022; Tao et al., 2014; Gabaston et al., 2020; Barrulas et al., 2019).

3. Classification, Source, and Structural Profile of Polyphenols

Polyphenols constitute a diverse class of naturally occurring phytochemicals, widely distributed in the plant kingdom and commonly found in fruits, vegetables, cereals, legumes, tea, coffee, wine, and various medicinal herbs. They are secondary metabolites characterized by the presence of one or more hydroxylated aromatic rings and are broadly linked to defense mechanisms in plants, like protection against UV radiation, pathogens, and oxidative stress; (Zhang et al., 2022; Ciepei et al.,

2024; Khan et al., 2024; Khan et al., 2019; Rasouli et al., 2017; Abbas et al., 2017; Upadhyay and Dixit , 2015).

Structurally, polyphenols are classified based on their carbon skeleton and polymerization degree. The predominant categories include phenolic acids, flavonoids, stilbenes, lignans, tannins, and simple phenols. Among these, flavonoids represent the most abundant and structurally diverse subgroup, encompassing subclasses such as flavonols, flavones, flavanols, flavanones, isoflavones, anthocyanins, and their glycosylated forms. Phenolic acids are subdivided into hydroxybenzoic and hydrocinnamic acids, commonly present in berries, coffee, and whole grains (Ciepei et al., 2024; Li et al., 2023; Khan et al., 2019; Ebrahimi, 2021; Ćujić et al., 2016; Simić et al., 2016).

Each polyphenol subclass is associated with specific botanical sources and possesses distinct physicochemical characteristics that primarily influence their bioavailability and biological action. For instance, stilbenes, though less abundant, have gained attention due to compounds including resveratrol in grapes and red wine (Upadhyay and Dixit , 2015), while both hydrolysable (gallotannins and ellagitannins) and condensed tannins are prevalent in nuts, pomegranates, and oak-aged beverages (Okuda and Ito, 2011). Lignans, found in seeds and whole grains, are of particular interest for their phytoestrogenic properties (da Silva et al., 2025). Simple phenols, although structurally less complex, contribute significantly to the antioxidant potential of natural sources (Ciepei et al., 2024; Li et al., 2023; Khan et al., 2019; Ebrahimi, 2021; Ćujić et al., 2016; Simić et al., 2016). Table 1 presents a detailed classification of polyphenols, highlighting key subgroups, main compounds, dietary sources, and other characteristics such as molecular weight and formula, pKa, solubility, and $\log K_{OW}$. Figures 2, 3, and 4 represent the structures of the primary polyphenol subgroups and the most representative compounds of each.

4. Antioxidant, Anti-Inflammatory, and Other Health-Promoting Properties of Polyphenols

Numerous risk factors contribute to the onset of inflammation, oxidative stress, and age-affiliated diseases, such as genetic predisposition, advancing age, poor dietary habits, lack of physical activity, exposure to UV radiation and photo-oxidation, metabolic syndromes, and cardiovascular disorders. Under these circumstances, the overproduction of reactive oxygen species (ROS) and inflammatory mediators like thrombin, thromboxane A_2 (T_XA_2), eicosanoids, adenosine diphosphate (ADP), and collagen is promoted. Thus, both the platelet-activating factor (PAF) and oxidized phospholipids (ox-PLs) are activated, which in turn trigger multiple thrombo-inflammatory cascades. Ox-PLs and PAF contribute to the activation of nuclear factor kappa B (NF-κB) signaling, initiating the degradation of its inhibitor, IκB, and promoting pro-inflammatory gene transcription (Tsoupraset et al., 2024; Lordau et al., 2019; Wang et al., 2023; Soendjaja and Girard, 2024).

Polyphenols, including all major subgroups such as flavonoids (e.g., flavonols, isoflavones, flavones, flavanones, flavonoid glucosides, anthocyanins, and anthocyanidins), stilbenes, lignans, tannins, phenolic acids, and simple phenols are known to counteract these processes by effectively reducing pro-inflammatory cytokines, like tumor necrosis factor-α (TNF-α), interleukins-1β/6 (IL-1β/6), and interferon-γ (IFN-γ). Mechanistically, PAF binding to its receptor (PAF-R), stimulates thrombo-inflammatory signaling through both the mitogen-activated protein kinase (MAPK) and phospholipase Cβ (PLCβ) pathways. This cascade inhibits key enzymes like PAF-cholinetransferase (PAF-CPR) and lysophosphatidylcholine (LPC) acyltransferase, thereby suppressing NF-κB activity and the subsequent transcription of inflammation-associated genes (de Araújo et al., 2021; Wu et al., 2025; Tsoupraset et al., 2024; Caban and Lewandowska, 2022; Yahfoufiet al., 2018; Singh et al., 2020).

Additionally, the expression of cyclooxygenase-2 (COX-2) modulates the levels of transforming growth factor β (TGF-β), a cytokine involved in tumor progression and thrombo-inflammation driven by mediators such as thrombin, eicosanoids, T_XA_2, ADP, and collagen (Tsoupraset et al.,

Table 1 Bioactive Compounds Identified in Matricaria Chamomilla Extracts.

Bioactives			*Bioactive Source*	*Molecular Formula*	*Molecular Weight (g/mol)*	*pKa**	*Water Solubility (mg/L, 25°C)*	*logKow*	*References*
Phenolic Acids	Hydroxybenzoic Acids	p-Hydroxybenzoic acid	Red fruits, herbs, radishes, onions, raspberries, tea, blueberries, whole grains, cranberries, red wine, soybean, barley, purple rice	$C_7H_6O_3$	138.12	4.54	6000	1.58	(Ciepei et al., 2024; Manach et al., 2004; Rasouli et al., 2017; da Silva et al., 2025)
		Protocatechuic acid	Onions, Hibiscus sabdariffa, carrots, mushrooms, red wine, berries, rice bran, honey, salvia, soybean, Loquat, green tea, chia seed, strawberry, plum, soybean, barley, millet, olives, and oil	$C_7H_6O_4$	154.12	4.48; 8.83; 12.6	18200	0.86	(Ciepei et al., 2024; Manach et al., 2004; Silva et al., 2025; Kakkar and Bais, 2014)
		Gallic acid	Green tea, black currant, gallnut, mangoes, guava, onions, maize, coriander, garlic, barley, almond, chestnut, hazelnut, peanut, pecans, red wine, honey, walnut, grapes, pomegranate	$C_7H_6O_5$	170.12	4.40; 8.80; 10.00; 11.40	14,700	0.70	(Ciepei et al., 2024; Manach et al., 2004; Li et 2023; da Silva et al., 2025)
		Syringic acid	Strawberries, red wine, walnut, guava, purple rice, peanut, chestnut, almond, chia seed, honey, olive oil, berries, pumpkin	$C_9H_{10}O_5$	198.17	4.33	5780	1.04	(Ciepei et al., 2024; Manach et al., 2004; Li et 2023; da Silva et al., 2025);

Contd.

Table 1 *Contd.*

Bioactives			*Bioactive Source*	*Molecular Formula*	*Molecular Weight (g/mol)*	*pKa**	*Water Solubility (mg/L, 25°C)*	*logKow*	*References*
		Vanillic acid	Vanilla beans, mangoes, wheat, tomatoes, oats, coriander, purple rice, chestnut, red wine, beer, whole grains, green tea	$C_8H_8O_4$	168.15	4.16	1500	1.40	(Ciepei et al., 2024; Manach et al., 2004; Li et al., 2023; da Silva et al., 2025)
		Ellagic acid	Pomegranate, green tea, grapes, olives, walnuts, red wine, strawberries, chestnuts, pecans, chia seeds, dark chocolate, blackberries, raspberries	$C_{14}H_6O_8$	302.20	5.50	820	1.59	(Ciepei et al., 2024; Manach et al., 2004; Li et al., 2023; da Silva et al., 2025)
	Hydroxycinnamic Acids	Chlorogenic acid	Coffee, potatoes, eggplants, cherries, sunflower seeds, hawthorn, grapes, mangoes, guava, tomato, broccoli, coriander, garlic, red wine, oats, pears, spinach, maize, barley, purple rice, green tea, chestnut, black chokeberry, apples, blueberries	$C_{16}H_{18}O_9$	354.31	3.91	40,000	-0.40	(Ciepei et al., 2024; Manach et al., 2004; Li et al., 2023; da Silva et al., 2025)

Contd.

Table 1 *Contd.*

Bioactives			*Bioactive Source*	*Molecular Formula*	*Molecular Weight (g/mol)*	*pKa**	*Water Solubility (mg/L, 25°C)*	*logKow*	*References*
		Cinnamic acid	Cinnamon, honey, coriander, garlic, soybean, oats, whole grains, citrus, red wine	$C_9H_8O_2$	148.16	4.45	546	2.13	(Ciepei et al., 2024; Manach et al., 2004; Li et al., 2023; da Silva et al., 2025)
		Caffeic acid	Coffee, green tea, kiwi, oat, grapes, hawthorn, red wine, mangoes, tomato, hawthorn, celery, cocoa, red cabbage, garlic, honey, spinach, maize, soybean, sorghum, barley, millet, walnut, chestnut, hazelnut, chia seed, apples, pears, berries	$C_9H_8O_4$	180.16	4.62	1610	1.67	(Ciepei et al., 2024; Manach et al., 2004; Li et al., 2023; da Silva et al., 2025)
		p-Coumaric acid	Plum, guava, broccoli, celery, coriander, red cabbage, garlic, wheat, soybean, sorghum, millet, walnut, chestnut, hazelnut, almond, peanut, chia seed, red wine, honey, tomatoes, apples, beer, onions, garlic, carrots	$C_9H_8O_3$	164.16	4.65	1020	1.79	(Ciepei et al., 2024; Manach et al., 2004; Li et al., 2023; da Silva et al., 2025)

Contd.

Table 1 *Contd.*

Bioactives			*Bioactive Source*	*Molecular Formula*	*Molecular Weight (g/mol)*	*pKa**	*Water Solubility (mg/L, 25°C)*	*logKow*	*References*
		Ferulic acid	Cereal, aubergine, wheat grain, rice, oat, coffee, guava, mangoes, grapes, hawthorn, tomato, broccoli, celery, coriander, beet, red cabbage, wheat, maize, soybean, sorghum, barley, millet, chestnut, chia seed, green tea, red wine, honey, eggplant, orange juice, brans	$C_{10}H_{10}O_4$	194.18	4.61	780	1.58	(Ciepei et al., 2024; Manach et al., 2004; Li et al., 2023; da Silva et al., 2025)
		Sinapic acid	Apple, pear, oat, chicory, artichoke, potatoes, coffee, corn flour, wheat, rice, flour, cider, red cabbage, oats, wheat, maize, mustard seeds, canola oil, radish, broccoli, sprouts	$C_{11}H_{12}O_5$	224.21	3.41	630	1.63	(Ciepei et al., 2024; Manach et al., 2004; Li et al., 2023; da Silva et al., 2025)
Flavonoids	Flavonols	Quercetin	Curly kale, yellow onion, broccoli, strawberries, leeks, hawthorn, guava, coriander, beet, red cabbage, garlic, kale, spinach, barley, hazelnut, black chokeberry, green tea, honey, apples, capers	$C_{15}H_{10}O_7$	302.24	7.17; 8.26; 10.13; 12.30; 13.11	261	1.81	(Ciepei et al., 2024; Singla et al., 2019; Manach et al., 2004; Li et et al., 2023; da Silva et al., 2025; Abbas et al., 2017; Corcoran et al., 2012; Chaves et al., 2020; Haytowitz et al., 2013; Cione et al., 2013)

Contd.

Table 1 *Contd.*

Bioactives			*Bioactive Source*	*Molecular Formula*	*Molecular Weight (g/mol)*	*pKa**	*Water Solubility (mg/L, 25°C)*	*logKow*	*References*
		Myricetin	Yellow onion, cherry tomato, broccoli, black currant, apricot, blueberries, apple, beans, black grape, tomato, black tea, green tea, red, leek, wine, strawberries, chia seed, honey, walnut, sweet potatoes, cloves	$C_{15}H_{10}O_8$	318.23	6.37	Sparingly soluble	2.94	(Ciepei et al., 2024; Singla et al., 2019; Manach et al., 2004; Li et al., 2023; da Silva et al., 2025; Abbas et al., 2017; Corcoran et al., 2012; Chaves et al., 2020; Haytowitz et al., 2013)
		Kaempferol	Leek, yellow onion, broccoli, strawberries, red cabbage, guava, celery, coriander, barley, garlic, kale, honey, peanut, hazelnut, almond, chia seed, spinach, broccoli, tea	$C_{15}H_{10}O_6$	286.24	6.38	Insoluble	1.9	(Ciepei et al., 2024; Singla et al., 2019; Manach et al., 2004; Li et al., 2023; da Silva et al., 2025; Abbas et al., 2017; Corcoran et al., 2012; Chaves et al., 2020; Haytowitz et al., 2013)
	Flavones	Luteolin	Parsley, capsicum pepper, millet, spinach, wheat, peppers, celery, beet, garlic, oats, spinach, sorghum, green tea, honey, green pepper, thyme, chamomile tea, carrots	$C_{15}H_{10}O_6$	286.24	6.57	850	2.73	(Ciepei et al., 2024; Singla et al., 2019; Manach et al., 2004; Li et al., 2023; da Silva et al., 2025; Abbas et al., 2017; Corcoran et al., 2012; Chaves et al., 2020; Haytowitz et al., 2013)

Contd.

Table 1 *Contd.*

Bioactives			*Bioactive Source*	*Molecular Formula*	*Molecular Weight (g/mol)*	*pKa**	*Water Solubility (mg/L, 25°C)*	*logKow*	*References*
		Apigenin	Parsley, celery, millet, wheat, peppers, garlic, sorghum, oats, green tea, honey, chamomile tea, oranges, onions. grapefruit	$C_{15}H_{10}O_5$	270.24	6.38	183	2.46	(Ciepei et al., 2024; Singla et al., 2019; Manach et al., 2004; Li et al., 2023; da Silva et al., 2025; Abbas et al., 2017; Corcoran et al., 2012; Chaves et al., 2020; Haytowitz et al., 2013)
	Flavanols (Flavan-3-ols – Catechins)	(–)-Epicatechin	Beans, apricots, cherries, grapes, peaches, apples, blackberries, green tea, black tea, red wine, cider, beet, strawberries, fava beans, millet, chia seed, maize, barley, walnut, pistachio, pecans, dark chocolate, cocoa	$C_{15}H_{14}O_6$	290.27	9.00	450	0.4	(Ciepei et al., 2024; Singla et al., 2019; Manach et al., 2004; Li et al., 2023; da Silva et al., 2025; Abbas et al., 2017; Corcoran et al., 2012; Chaves et al., 2020; Haytowitz et al., 2013)
		(–)-Epicatechin gallate	Leguminous plants, grapes, green and black tea, strawberries, maize, barley, cocoa, pistachio, pecans, dark chocolate, apples	$C_{22}H_{18}O_{10}$	442.40	8.03	Slightly soluble	Nd	(Ciepei et al., 2024; Singla et al., 2019; Manach et al., 2004; Li et al., 2023; da Silva et al., 2025; Abbas et al., 2017; Corcoran et al., 2012; Chaves et al., 2020; Haytowitz et al., 2013)

Contd.

Table 1 *Contd.*

Bioactives			*Bioactive Source*	*Molecular Formula*	*Molecular Weight (g/mol)*	*pKa**	*Water Solubility (mg/L, 25°C)*	*logKow*	*References*
		(–)-Epigallocatechin	Leguminous plants, grapes, green and black tea, strawberries, maize, barley, pistachio, pecans, dark chocolate, cocoa, apples	$C_{15}H_{14}O_7$	306.27	8.73	Slightly soluble	0.71–1.49	(Ciepei et al., 2024; Singla et al., 2019; Manach et al., 2004; Li et al., 2023; da Silva et al., 2025; Abbas et al., 2017; Corcoran et al., 2012; Chaves et al., 2020; Haytowitz et al., 2013)
		(-)-Epigallocatechin gallate	Leguminous plants, grapes, green and black tea, strawberries, maize, barley, cocoa, pistachio, pecans, dark chocolate, oolong tea, cranberries	$C_{22}H_{18}O_{11}$	458.40	7.99	Slightly soluble	2.38	(Ciepei et al., 2024; Singla et al., 2019; Manach et al., 2004; Li et al., 2023; da Silva et al., 2025; Abbas et al., 2017; Corcoran et al., 2012; Chaves et al., 2020; Haytowitz et al., 2013; Cione et al., 2013)
	Flavanones	Hesperidin	Orange juice, oranges, grapefruit juice, citrus, walnut, tomato, lemon, tangerine, barley, pistachio, mandarins, lime	$C_{28}H_{34}O_{15}$	610.56	8.61	4.95	2.01	(Ciepei et al., 2024; Singla et al., 2019; Manach et al., 2004; Li et al., 2023; da Silva et al., 2025; Abbas et al., 2017; Corcoran et al., 2012; Chaves et al., 2020; Haytowitz et al., 2013)

Contd.

Table 1 *Contd.*

Bioactives			*Bioactive Source*	*Molecular Formula*	*Molecular Weight (g/mol)*	*pKa**	*Water Solubility (mg/L, 25°C)*	*logKow*	*References*
		Naringenin	Orange juice, lemon juice, garlic, grapefruit, citrus, lemons, tangerine, tomato, walnut, almond, peanut, tomatoes, cherries, cocoa, figs	$C_{15}H_{12}O_5$	272.25	7.05; 8.84	475	2.42	(Ciepei et al., 2024; Singla et al., 2019; Manach et al., 2004; Li et al., 2023; da Silva et al., 2025; Abbas et al., 2017; Corcoran et al., 2012; Chaves et al., 2020; Haytowitz et al., 2013)
	Isoflavones	Genistein	Soy flour, miso, soya, soya milk, fava beans. peas, barley, tofu, tempeh, chickpeas	$C_{15}H_{10}O_5$	270.24	7.63	Insoluble	2.70	(Ciepei et al., 2024; Singla et al., 2019; Manach et al., 2004; Li et al., 2023; da Silva et al., 2025; Abbas et al., 2017; Corcoran et al., 2012; Chaves et al., 2020; Haytowitz et al., 2013)
		Daidzein	Soy flour, boiled soybeans, soya, soya milk, pea, fava beans, tofu, tempeh, chickpeas, legumes, peanuts	$C_{15}H_{10}O_4$	254.24	6.48	Insoluble	2.50	(Ciepei et al., 2024; Singla et al., 2019; Manach et al., 2004; Li et al., 2023; da Silva et al., 2025; Abbas et al., 2017; Corcoran et al., 2012; Chaves et al., 2020; Haytowitz et al., 2013)

Contd.

Table 1 *Contd.*

Bioactives			*Bioactive Source*	*Molecular Formula*	*Molecular Weight (g/mol)*	*pKa**	*Water Solubility (mg/L, 25°C)*	*logKow*	*References*
	Flavonoid Glycosides	Quercetin-7-O-glucoside	Tangerine, lemon, strawberries, green tea, onions, black chokeberry, apples, kale, broccoli	$C_{21}H_{20}O_{12}$	464.39	5.73	3230	– 0.81	(Ciepei et al., 2024; Manach et al., 2004; Li et al., 2023; Corcoran et al., 2012; Xiao et al., 2016)
		Kaempferol-3-O-glucoside	Tangerine, lemon, strawberries, kale, onions, spinach, broccoli, green and black tea, apples	$C_{21}H_{20}O_{11}$	448.38	6.37	Insoluble	0.52	(Ciepei et al., 2024; Manach et al., 2004; Li et al., 2023; Corcoran et al., 2012; Xiao et al., 2016)
		Luteolin-7-O-glucoside	Tangerine, lemon, strawberries, green pepper, celery, onions, parsley, thyme, artichokes, chamomile tea	$C_{21}H_{20}O_{11}$	448.38	6.37	Insoluble	0.52	(Ciepei et al., 2024; Manach et al., 2004; Li et al., 2023; Corcoran et al., 2012; Xiao et al., 2016)
		Apigenin-7-O-glucoside	Tangerine, lemon, strawberries, onions, chamomile flowers, parsley, celery, oranges, oregano, cilantro	$C_{21}H_{20}O_{10}$	432.38	7.5	Insoluble	0.6	(Ciepei et al., 2024; Manach et al., 2004; Li et al., 2023; Corcoran et al., 2012; Xiao et al., 2016)
	Anthocyanidins	Cyanidin	Blackberries, aubergine, black grapes, beet, plum, cherries, red cabbage, black currant, raspberries	$C_{15}H_{11}O_6^+$	287.24	6.47	Insoluble	2.41	(Ciepei et al., 2024; Manach et al., 2004; Li et al., 2023; Zhang et al., 2019; Brooks MS-L and Celli, 2019; Krga and Milenkovic, 2019)

Contd.

Table 1 *Contd.*

Bioactives			*Bioactive Source*	*Molecular Formula*	*Molecular Weight (g/mol)*	*pKa**	*Water Solubility (mg/L, 25°C)*	*logKow*	*References*
		Delphinidin	Black grapes, aubergine, red grapes, blueberries, eggplants, black currants	$C_{15}H_{11}O_7^+$	303.25	5.98	Insoluble	Nd	(Ciepei et al., 2024; Manach et al., 2004; Li et al., 2023; Zhang et al., 2019; Brooks MS-L and Celli, 2019; Krga and Milenkovic, 2019)
		Malvidin	Cherries, rhubarbs, strawberries, red wine, red cabbage, plum, aubergine, black grapes, black rice, blueberries, plum, elderberries	$C_{17}H_{15}O_7^+$	331.30	5.99	Sparingly soluble	Nd	(Ciepei et al., 2024; Manach et al., 2004; Li et al., 2023; Zhang et al., 2019; Brooks MS-L and Celli, 2019; Krga and Milenkovic, 2019)
		Pelargonidin	Black currant, aubergine, red plums, cranberries, strawberries, red radishes	$C_{15}H_{11}O_5^+$	271.25	Nd	Sparingly soluble	Nd	(Ciepei et al., 2024; Manach et al., 2004; Li et al., 2023; Zhang et al., 2019; Brooks MS-L and Celli, 2019; Krga and Milenkovic, 2019)
		Peonidin	Blueberries, aubergine, plums, red onions, red grapes, cherries, black rice	$C_{16}H_{13}O_6^+$	301.27	Nd	Sparingly soluble	Nd	(Ciepei et al., 2024; Manach et al., 2004; Li et 2023; Zhang et al., 2019; Brooks MS-L and Celli, 2019; Krga and Milenkovic, 2019)

Contd.

Table 1 *Contd.*

Bioactives			*Bioactive Source*	*Molecular Formula*	*Molecular Weight (g/mol)*	*pKa**	*Water Solubility (mg/L, 25°C)*	*logKow*	*References*
		Petunidin	Aubergine, grapes, blackberries, sweet potatoes	$C_{16}H_{13}O_7^+$	317.27	Nd	Sparingly soluble	Nd	(Ciepei et al., 2024; Manach et al., 2004; Li et al., 2023; Zhang et al., 2019; Brooks MS-L and Celli, 2019; Krga and Milenkovic, 2019)
	Anthocyanins	Cyanidin-3-glucoside	Red wine, cereals, aubergine, beans, cabbage, onions, radishes, cherries, strawberries, pomegranate, blackberries, raspberry, maize, purple rice, pistachio, black chokeberries	$C_{21}H_{21}O_{11}^+$	449.40	Nd	600	0.39	(Ciepei et al., 2024; Manach et al., 2004; Li et al., 2023; Zhang et al., 2019; Brooks MS-L and Celli, 2019; Krga and Milenkovic, 2019)
		Delphinidin-3-galactoside	Red wine, cereals, aubergine, beans, cabbage, onions, radishes, cherries, strawberries, pomegranate, blueberries, raspberry, maize, purple rice, red grapes, pistachio, black currants	$C_{21}H_{21}O_{12}^+$	465.39	Nd	850	0.93	(Ciepei et al., 2024; Manach et al., 2004; Li et al., 2023; Zhang et al., 2019; Brooks MS-L and Celli, 2019; Krga and Milenkovic, 2019)

Contd.

Table 1 *Contd.*

Bioactives			*Bioactive Source*	*Molecular Formula*	*Molecular Weight (g/mol)*	*pKa**	*Water Solubility (mg/L, 25°C)*	*logKow*	*References*
		Malvidin-3,5-diglucoside	Red wine, cereals, aubergine, beans, cabbage, onions, radishes, cherries, strawberries, pomegranate, blackberries, raspberry, maize, purple rice, purple corn, pistachio, blueberries, sweet potatoes	$C_{15}H_{12}O_5$	272.25	Nd	Soluble	Nd	(Ciepei et al., 2024; Manach et al., 2004; Li et al., 2023; Zhang et al., 2019; Brooks MS-L and Celli, 2019; Krga and Milenkovic, 2019)
Lignans		Secoisolariciresinol	Linseed, algae, lentils, triticale, wheat, garlic, asparagus, carrots, pears, prunes, oats, barley, rye, sesame seed, flaxseeds, whole grains, berries, brassica, legumes	$C_{20}H_{26}O_6$	362.40	9.96	Sparingly soluble	2.05	(Ciepei et al., 2024; Manach et al., 2004; Li et al., 2023; Li et al., 2023; Rodríguez-G., et al., 2019; Jenkis et al., 2019)
		Matairesinol	Linseed, algae, lentils, triticale, wheat, garlic, asparagus, carrots, pears, prunes, oats, barley, rye, sesame seeds, whole grain, bran, legumes. pumpkin seeds	$C_{20}H_{22}O_6$	358.40	9.64	Sparingly soluble	2.79	(Ciepei et al., 2024; Manach et al., 2004; Li et al., 2023; Li et al., 2023; Rodríguez-G., et al., 2019; Jenkis et al., 2019)
Tannins	Gallotannins	Tannic Acid	Mangoes, tea leaves, oak bark, chestnuts, red wine, persimmons, pomegranate,	$C_{76}H_{52}O_{46}$	1701.20	7.61	407	4.73	Li et al., 2023; Li et al., 2023; Jourdes et al., 2021; Jafari et al., 2022)

Contd.

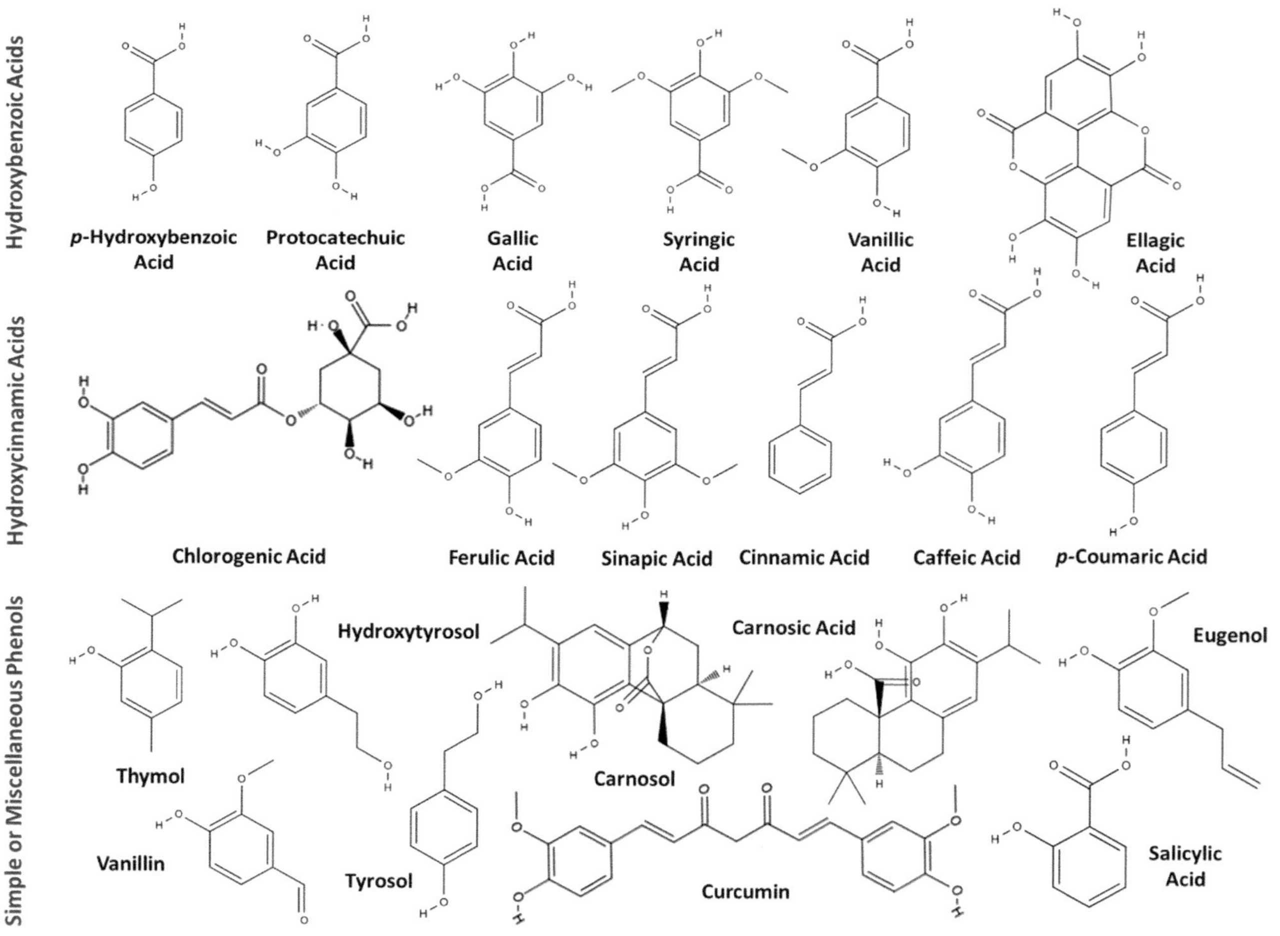

Fig. 3 Structural Presentation of the Main Phenolic (Hydroxybenzoic/Hydroxycinnamic) Acids and Simple Phenols (Parts of this figure were obtained from https://www.molview.org/ (accessed on May 14, 2025)).

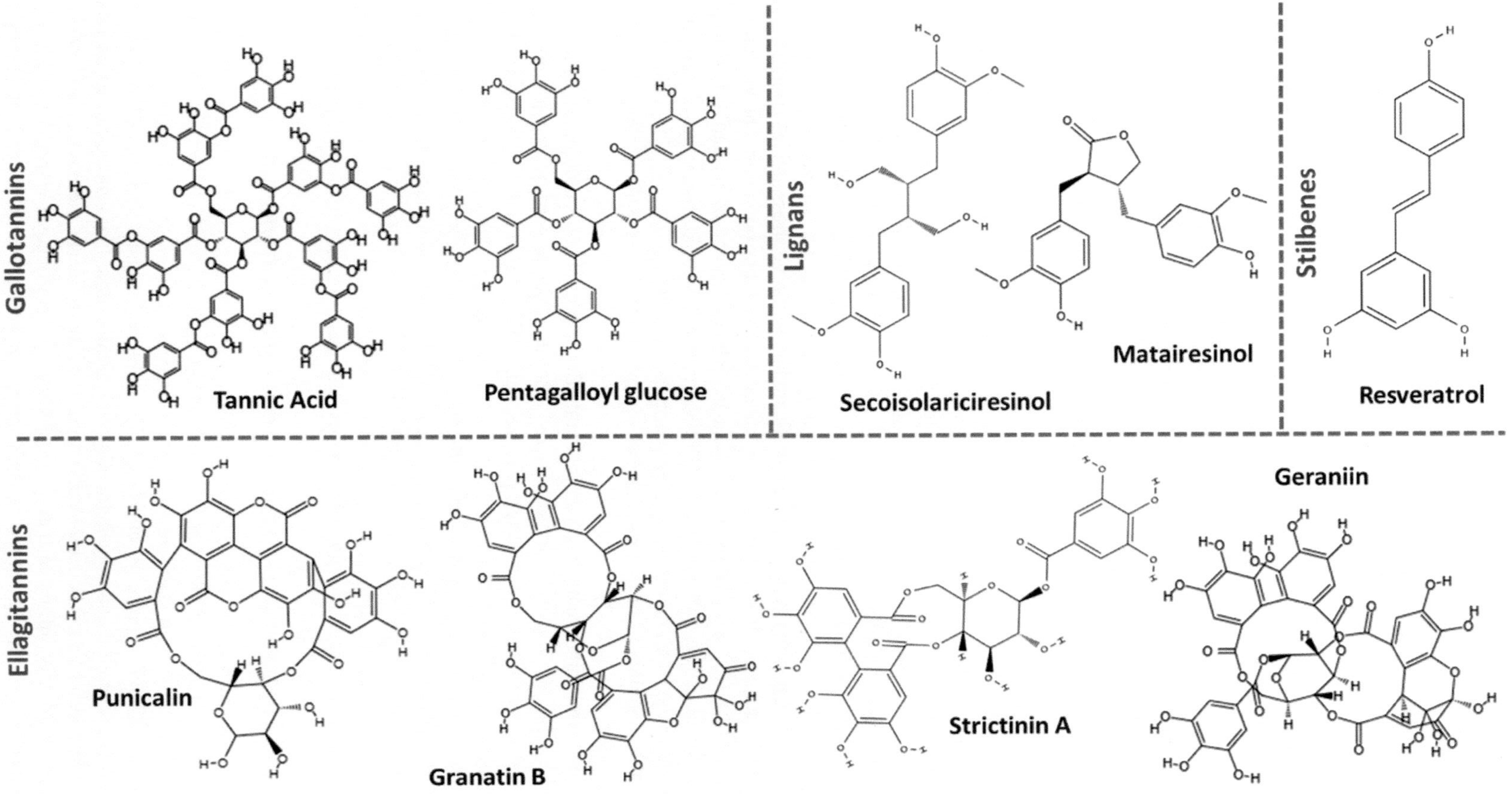

Fig. 4 Structural Presentation of the Main Tannins (Ellagitannins and Gallotannins), Lignans, and Stilbene. (Parts of this figure were obtained from https://www.molview.org/ (accessed on May 14, 2025)).

2024; Lordau et al., 2019; Wang et al., 2023; Soendjaja and Girard, 2024). Across polyphenol classes, various bioactives modulate these pathways to exert anti-inflammatory, antioxidant, and cytoprotective functions, offering protective effects against chronic conditions and age-associated disorders. ROS also causes oxidative damage, which activates the Kelch-like ECH-associated protein 1 (Keap1)/nuclear factor erythroid 2-related factor 2 (Nrf2) axis and the phosphoinositide 3-kinase (PI3K)/protein kinase B (Akt) pathway. Upon activation, Nrf2 translocates to the nucleus, where it binds to antioxidant response elements (AREs), inducing the expression of Phase II detoxifying enzymes (e.g., glutathione-S-transferase, nicotinamide adenine dinucleotide phosphate (NADPH)), thereby enhancing the cell's resistance to oxidative stress and inflammation (El Oirdi, 2024; Tsoupraset et al., 2024; Zhou et al., 2019). The aryl hydrocarbon receptor (AhR), another transcription factor responsive to environmental and dietary cues, further regulates detoxification and inflammatory processes. Once activated, AhR translocates to the nucleus, complexes with the AhR nuclear translocator (ARNT), and binds to dioxin response elements (DRE) on DNA, hence initiating the transcription of Phase I enzymes (e.g., esterases, amidases) that bio-transform xenobiotics and oxidative by products (Tsoupraset et al., 2024; Larigot et al., 2018; Amakura et al., 2008; Tang et al., 2023).

In general, polyphenols exhibit a broad spectrum of bioactivity owing to their structural diversity. Compounds including several flavonoids (e.g., apigenin, luteolin), phenolic acids (e.g., gallic acid, caffeic acid), stilbenes (e.g., resveratrol), tannins (e.g., tannic acid, punicalin), lignans (e.g., secoisolariciresinol, matairesinol), and simple phenols (e.g., curcumin, eugenol), are part of a wider group of polyphenols with a multitude of antioxidant, anti-inflammatory, anticancer, and neuroprotective effects (Singla et al., 2019; Khan et al., 2019; Rasouli et al., 2017; de Araújo et al., 2021; Abbas et al., 2017). Gallic and caffeic acid, for example, act by inhibiting NF-κB activation via the suppression of IκB degradation, downregulating PAF-mediated inflammation, COX-2 activity, and enhancing Nrf2 signaling, key processes implicated in oxidative stress and inflammation-related diseases. Phenolic acids contribute by chelating metal ions, and thus preventing the Fenton reaction and formation of highly reactive hydroxyl radicals, a major cause of oxidative stress-induced damage (Abbas et al., 2017; da Silva et al., 2925). Moreover, flavonoids like apigenin and quercetin, also directly scavenge ROS, neutralizing them before cellular damage occurs, while simultaneously modulating AhR signaling to further regulate inflammatory responses (Singla et al., 2019; Abbas et al., 2017; Corcoran et al., 2012).

Beyond their anti-inflammatory and antioxidant effects, polyphenols modulate multiple cellular processes relevant to chronic and aging-associated disorders such as cardiovascular complications, neurodegenerative and metabolic disorders (Zhang et al., 2022; Ciepei et al., 2024; Brglez-Mojzer et al., 2016; Manach et al., 2004; Khan et al., 2019; Ebrahimi, 2022). Due to these versatile biological activities, polyphenols serve as promising candidates for use in functional foods, nutraceuticals, as well as cosmetic, cosmeceutical and pharmaceutical applications, and dietary interventions targeting chronic inflammation and age-related pathologies. The health-promoting effects of polyphenol-rich extracts and the detailed mechanisms through which polyphenols contribute to health-promoting defenses, are explored in the following subsections and visually summarized in Figure 5.

4.1 Antioxidant and Pro-Oxidant Activity of Polyphenols

Polyphenols are renowned for their potent antioxidant properties, which stem from their ability to donate hydrogen atoms or electrons, neutralize free radicals, and chelate pro-oxidant metal ions. These compounds help stabilize ROS, thus preventing oxidative damage to vital cellular components such as lipids, proteins, and DNA. By reducing oxidative stress, polyphenols play a major role in the prevention and management of chronic diseases, like cancer, cardiovascular, and neurodegenerative disorders. Their molecular structures, particularly the number and position of hydroxyl groups, directly influence their radical-scavenging efficacy (Marino et al., 2014). For instance, gallic acid exhibited great radical scavenging activity towards OH and OOH radicals. In

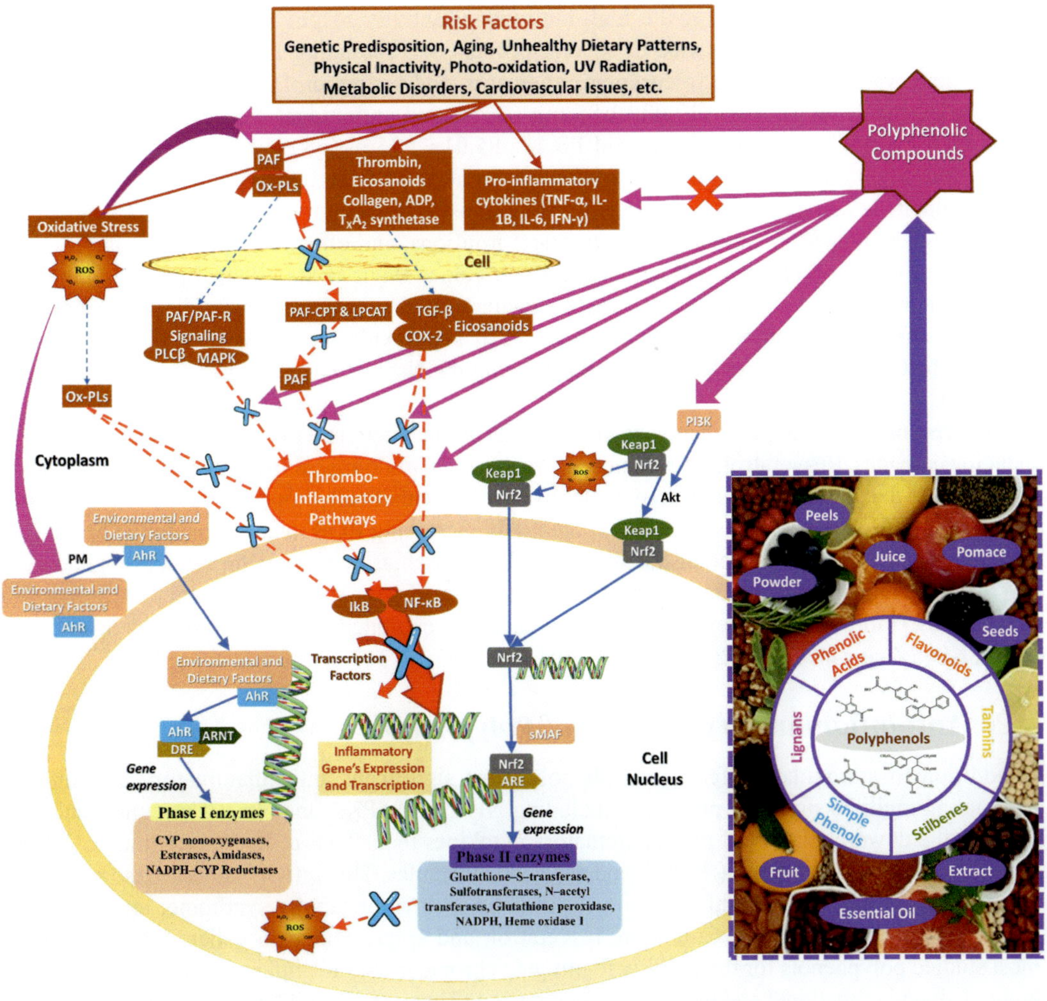

Fig. 5 Mechanistic Overview of the Health-Promoting Properties of Polyphenols. (Parts of this figure were obtained from https://www.freepik.com/ (accessed on May 20, 2025)).

functional food applications, polyphenols such as quercetin, catechins, tannins, and anthocyanins are utilized to boost the oxidative stability of products, while enhancing the consumer's antioxidant defense system (Zhang et al., 2022; Ciepei et al., 2024; Güneş Bayir et al., 2024; Hong et al., 2020).

The antioxidant activity of several polyphenols has been widely studied in a multitude of clinical and pre-clinical trials. In an in vitro study, simple phenols such as thymol and especially *Thymus serpyllum* essential oil displayed high growth inhibitory 50% values (GI_{50}= 7.02 – 52.69 µg/mL) (Nikolić et al., 2014). In contrast, tetraploid *Thymus vulgaris* essential oil was more potent with half-minimal inhibitory (IC_{50}) doses of 180.03 µg/mL (DPPH: 40.05 µg TE/mg) (Gupta et al., 2024). Moreover, the hydroethanolic extract of the plant *Anabasis aretioïdes* Coss. demonstrated mainly cinnamic acid, sinapic acid, and vanillin content and was rich in flavonoids (total flavonoid content (TFC): 8.52 ± 0.32 mg rutin equivalents (RE)/100 g dry weight (DW)), polyphenols (total phenolic content (TPC): 159.32 ± 0.63 mg gallic acid equivalents (GAE)/100 g DW), and condensed tannins (8.73 ± 0.23 mg catechin equivalents (CE)/100 g DW). It revealed a high total antioxidant capacity (117.29 ± 2.66 µg ascorbic acid equivalent (AA)/mg extract), in DPPH (IC_{50}: 51.28 ± 0.91 µg/mL), β-carotene (4.44 ± 0.21 mg/mL), ABTS (28.63 ± 1.02 TE µmol/mL) assays (Laaraj et al., 2024).

A representative example of the high antioxidant activity of polyphenols, is an olive leaf extract which showed the presence of several polyphenols including hydroxytyrosol, catechin, caffeic acid, vanillin, naringin, oleuropein, quercetin, and kaempferol and displayed high TPC (74.45 ± 1.2 mg GAE/g), TFC (17.08 ± 1.85 mg CE/g), and antioxidant capacity values (DPPH: 83.45 ± 0.89%, ABTS: 82.85 ± 1.52%, and FRAP: 85.01 ± 2.35 assays) (El Adnany et al., 2023). Curcumin, silymarin, and apigenin also showed great antioxidant capacity, tending to reduce malondialdehyde content, while catechins were the most effective for scavenging ABTS radicals, had the highest reduction equivalents for trivalent iron ions, and effectively protected against the oxidation of dihydrorhodamine (Zhang et al., 2022; Güneş Bayir et al., 2024). Lignans including secoisolariciresinol and matairesinol, and anthocyanins including cyanidin, petunidin, pelargonidin, malvidin, and peonidin upregulate the expression of superoxide dismutase (SOD) and catalase (CAT), effectively scavenging ROS through activating the Nrf2 cascade (Jang et al., 2022; Tena et al., 2020).

Interestingly, several polyphenols were pro-oxidant mediators in a plethora of oxidative stress-inducing cascades. Resveratrol, for instance, decreases the chronological lifespan of several fungi or yeast species due to its pro-oxidant activity. Resveratrol hypothetically impedes electron transport to complex II of the electron transport chain, potentially increasing ROS (leads to oxidative breakage of cellular DNA in the presence of transition metal ions such as copper and/or reduces γ-glutamic and α-aminoadipic protein semialdehydes in human serum albumin) (Canedo-S. et al., 2022; Aranjo et al., 2018). Gallic acid, daidzein, and curcumin have also exhibited significant pro-oxidant activity (ROS, NO) (Zhang et al., 2022; Lucas et al., 2021).

4.2 Anti-Inflammatory Activity of Polyphenols

In addition to antioxidant effects, polyphenols exert significant anti-inflammatory actions by modulating multiple signaling pathways, including NF-κB, MAPKs, and Nrf2. These mechanisms result in the downregulation of pro-inflammatory cytokines (e.g., TNF-α, IL-6), enzymes (COX-2, inducible nitric oxide synthase (iNOS)), and adhesion molecules. This activity underlies the potential of polyphenols in managing inflammatory conditions such as metabolic syndrome, arthritis, and inflammatory bowel diseases. Curcumin, resveratrol, and epigallocatechin gallate are among the most studied polyphenols for their ability to mitigate chronic low-grade inflammation, making them highly valued in functional and therapeutic food design (Zhang et al., 2022; Ciepei et al., 2024; Li et al., 2023; Güneş Bayir et al., 2024). For example, tetraploid thyme essential oils were more effective in inhibiting COX-2 than diploids at 50 μg/mL (Nikolić, et al., 2014). Vanillin also offers many anti-inflammatory and anti-neuroinflammatory benefits, as it lowers the production of pro-inflammatory mediators (TNF-α (128.50 ± 14.59 to 67.51 ± 10.88 pg/mL), IL-6 (531.5 ± 196.4 to 109.3 ± 24.14 pg/mL), and IL-1β (2569 ± 1648 to 731.8 ± 171.7 pg/mL)), iNOS, and COX-2, while inhibiting MAPKs and NF-κB signaling in lipopolysaccharide (LPS)-induced conditions (Guo et al., 2019; Kim et al., 2019).

Flavonoids, such as quercetin and galangin, have also demonstrated notable anti-inflammatory effects in LPS-stimulated RAW264.7 macrophages, downregulating NO, iNOS, IL-6, NF-κB, c-Jun N-terminal kinase (JNK), and extracellular signal-regulated kinase ½ ($ERK_{1/2}$) (Lee et al., 2018; Overman et al., 2011). Quercetin and catechins combination showed similar inflammation-suppressive results, while also restoring the impaired expression of the toll-like receptor 4, myeloid differentiation primary response gene 88 (TLR-4-MyD88), and halting the phosphorylation of NF-κB *p65* and *p50* subunits (Vazquez et al., 2015; Li et al., 2019). Curcuminoids and their analogs exhibited anti-inflammatory effects in hyperosmotic human corneal limbus epithelial cells, decreasing mRNA, TNF-α, IL-1β/6/17A, matrix metallopeptidase-9 (MMP-9), and intercellular adhesion molecule-1 (ICAM-1), while maintaining cell viability (Kasetsuwan et al., 2024). Similarly, resveratrol and hesperidin provide all afore-mentioned anti-inflammatory effects plus inhibit the Janus kinase/signal transduction and activator of transcription (JAK/STAT) cascade (resveratrol) (De Sá Coutinho et al.,

2018) and modulate Forkhead box proteins O1 and O3 (Foxo1/3) and Nrf2 signaling in H_2O_2-exposed chondrocytes (hesperidin) (Tsai, et al., 2019). Furthermore, dietary lignans efficiently suppress NF-κB, MAPK, Nrf2, and JAK/STAT3 signaling (Hassanein et al., 2024), whereas anthocyanins in several *in vivo* and *in vitro* studies have shown notable anti-inflammatory value against NF-κB, TLRs, MAPKs, NO, and ROS (Ma et al., 2021).

4.3 Other Health-Promoting Benefits of Polyphenols

Beyond antioxidant and anti-inflammatory activities, polyphenols contribute to a broader range of health benefits. Their cardioprotective effects include improvement of endothelial function, blood pressure regulation, and lipid profile modulation. Notably, polyphenols exhibit anti-platelet activity, which leads to cardiovascular protection by inhibiting platelet aggregation and thrombus formation. Polyphenolic compounds can interfere with platelet activation pathways (e.g., T_XA_2, collagen, and ADP-mediated signaling), thereby reducing the risk of atherothrombosis. Additionally, polyphenols show great antimicrobial, antidiabetic, anticancer, and neuroprotective effects, supporting their inclusion in functional foods targeting multiple chronic conditions (Zhang et al., 2022; Ciepei et al., 2024; Li et al., 2023; Güneş Bayir et al., 2024).

Simple phenols like thymol have displayed notable antimicrobial activity. *Thymus serpyllum* essential oil had the most potent antimicrobial profile (minimum inhibitory concentration (MIC)) 2.5-5 μg/mL, minimum bactericidal concentration (MBC) 5–10 μg/mL; MIC 1–2 μg/mL, minimum fungicidal concentration (MFC) 2–4 μg/mL)), while showing no toxicity (Nikolić et al., 14). Flavonoids, especially, have significantly higher antimicrobial (antibacterial) activity than other polyphenolic compounds. Quercetin, which has fewer free-OH groups than other phenolics, increases the chemical affinity for the microbial lipid membrane due to its lower hydrophilicity, thus, stronger antimicrobial action (Da Rossi et 2025). Curcumin, silymarin, catechins, tannins, anthocyanins (e.g., delphinidin, cyanidin glucosides), kaempferol, ellagic acid, gallic acid, chlorogenic and caffeic acid, and resveratrol have demonstrated vast antibacterial activity against *S. aureus, A. baumannii, P. aeruginosa, B. subtilis, K. pneumoniae, E. coli, Salmonella* sp., *Bacillus* sp., *Klebsiella* spp., and many other microorganisms (Manso et al., 2021).

Certain polyphenols exhibit selective cytotoxicity against cancer cells while sparing normal cells, making them promising agents in cancer prevention and therapy. Polyphenols' anticancer mechanisms include apoptosis induction, angiogenesis inhibition, suppression of cell proliferation, and modulation of oncogenic signaling pathways such as the phosphatidylinositol 3-kinase (PI3K)/ protein kinase B (Akt) and MAPKs. Hence, polyphenols have been extensively documented for their anticancer, antitumor, anti-thrombotic, chemopreventive, and cytotoxic action. Thus, incorporating these compounds into functional foods or using them as adjuncts to conventional therapies represents a growing area of research in integrative oncology (Zhang et al., 2022; Ciepei et al., 2024; Li et al., 2023; Güneş Bayir et al., 2024). Ouercetin has revealed a great ability to prevent angiogenesis by targeting the vascular endothelial growth factor receptor 2 (VEGFR-2)-mediated angiogenesis pathways and thus, impeding tumor growth, while it also inhibits epithelial-mesenchymal transition (EMT) and STAT3 signaling, thereby preventing metastasis (Zhang et al., 2022; Singaravalen and Tollefsbol, 2025). Curcumin supplementation prevented cell proliferation, altering mammalian target of rapamycin (mTOR), PI3K/Akt, caspase-3, MAPK, and NF-κB signaling cascades, with limited sub-chronic toxicity (Zhang et al., 2022; Singaravalen and Tollefsbol, 2025). Also, resveratrol exerts its anticancer impact by downregulating Akt and NF-κB and upregulating B-cell lymphoma 2 (Bcl-2) and its associated X protein (Bax), sirtuin 1 (SIRT-1), *p53* gene, and transcription factor EB (TFEB). Epigallocatechin gallate, on the other hand, increases miRNA, protein kinase C (PKC), RAS, adenosine monophosphate-activated protein kinase (AMPK), and phosphatase and tensin homolog (PTEN). Interestingly, several combinations of the afore-mentioned polyphenols and well-documented cancer medications have been utilized for colon, colorectal, prostate, breast, or cervical cancer and leukemia or lymphoma management (Zhang et al., 2022; Singaravalen and Tollefsbol, 2025).

Polyphenols contribute to cardiovascular health through several antioxidant, anti-inflammatory, anti-hypertensive, anti-atherogenic, and lipid-lowering actions. Consuming a variety of foods and supplements, including olive oil and cherries (rich in polyphenols like hydroxytyrosol, quercetin, resveratrol, and catechins), patients in a systematic review displayed improvements in inflammatory markers and lipid profiles, as well as decreased cholesterol levels (LDL), with no reported adverse effects or iron status interference (Speer et al., 2019). Moreover, flavonoids, flavanols, anthocyanins, phenolic acids, tannins, resveratrol, and other polyphenols improved vasodilation and blood pressure control, benefiting hypertensive patients (especially curcumin + amlodipine) (Zhang et al., 2022; Singaravalen and Tollefsbol, 2025; Lee et al., 2021). Epigallocatechin gallate, one of the main constituents in green tea, also lowered lipid levels, enhanced plaque stability and gut microbiota, promoted endothelial function recovery, inhibited apoptosis, and regulated Nrf2, autophagy, and inflammatory signaling cascades, offering anti-atherosclerotic benefits (Zhang et al., 2022; Wang et al., 2025). As expected, polyphenols (e.g., catechins, curcumin, quercetin) were ideal anti-diabetic agents, able to inhibit carbohydrate-hydrolyzing enzymes like α-amylase and α-glucosidase, hence slowing glucose absorption and reducing postprandial glycemic spikes in T2DM patients (Zhang et al., 2022; Wang et al., 2025; Wen et al., 2022; Ansari et al., 2022; Marton et al., 2021). Additionally, polyphenols positively affect weight regulation and obesity prevention via inhibiting adipogenesis and stimulating lipolysis by activating the AMPK pathway (Zhang et al., 2022; Ciepei et al., 2024; Li et al., 2023; Güneş Bayir et al., 2024).

One of the most significant properties of polyphenols is their well-known neuroprotective activity in a plethora of NDs, such as Alzheimer's disease (AD), Parkinson's disease (PD), cognitive decline, and age-related neurodegeneration. Polyphenols like epigallocatechin gallate play a major role in the complex interplay between gut microbiota, nervous, and immune system, heavily influencing the gut-brain-immune axis. They exert their neuroprotective abilities by inhibiting ROS, suppressing pro-inflammatory cytokines, regulating Nrf2/ARE or NF-κB cascades, impeding $ERK_{1/2}$ pathway, exhibiting free radical scavenging effect by activating PI3K/Akt/Nrf2 signaling, enhancing mitochondrial function, promoting synaptic plasticity, and protecting against the induction of tau hyperphosphorylation and amyloid-beta (Aβ) aggregation, key ND features (Uddin et al., 2020; Annunziata et al., 2021). Finally, such bioactive polyphenolic compounds, because of all their antioxidant, anti-inflammatory, and cytotoxic properties, activate key longevity SIRT-1, AMPK, and Foxo1/3 signaling cascades, implicated in mitochondrial function, DNA repair, and autophagy, protecting against UV-induced photo-aging, improving collagen synthesis, reducing MMP activity, and promoting overall skin health (Zhang et al., 2022; Luo et al., 2021).

5. Polyphenol-based Applications in Functional Foods and Nutraceuticals

The broad spectrum of bioactivities attributed to polyphenols, particularly their antioxidant, anti-inflammatory, cardioprotective, neuroprotective, antidiabetic, anticancer, and/or antimicrobial impact, has led to their widespread application in the development of functional foods and nutraceutical products. By leveraging their health-promoting potential, polyphenols are increasingly being utilized either as natural additives, health-enhancing ingredients, or active components in personalized nutrition strategies and preventive healthcare solutions (Zhang et al., 2022; Ciepei et al., 2024; Güneş Bayir et al., 2024; de Araújo et al., 2021; Taouzin et al., 2023).

5.1 Functional Dairy Products

Dairy matrices such as yogurt, cheese, and milk-based beverages provide an excellent medium for the incorporation of polyphenols, mainly due to their emulsion structure, buffering capacity, and consumer familiarity. The addition of several polyphenols and polyphenolic extracts has been shown to enhance the antioxidant capacity of dairy consumables, while maintaining or even improving both microbial stability and shelf-life. Challenges related to bitterness and stringency are often addressed through encapsulation, ensuring improved sensory acceptance (de Araújo et al., 2021; Taouzin et al., 2023; Pandey et al., 2021; Tzima et al., 2020). Broadly consumed dairy products like ice cream, yogurt, and buttermilk, for instance, lack anthocyanins and therefore, their antioxidant activity. Black carrots are, among others, the cheapest source of anthocyanins, which are mainly consumed as a natural fermented drink. The incorporation of black carrot concentrate at the level of 7.5% resulted in dairy products with high acceptability, nutraceutical efficacy, mineral and flavonoid content increase, stability, and antioxidant capacity (26.39–35.87 times higher) (Pandey et al., 2021).

Curcumin has been widely used in functional dairy products. A turmeric extract nanoemulsion powder incorporated into milk increased its stability (21 days), resulted in its yellowing but without affecting its aroma, and increased its digestion capacity (Park et al., 2019). Moreover, curcumin nanoemulsions improved the organoleptic properties of milk-based soft cheese by 150%, improving its shelf life, antioxidant, and antimicrobial activity against *S. aureus* and *E. coli* (Bagale et al., 2023), whereas Manchengo-type cheeses enriched with curcumin nanoemulsions, revealed high TPC (245.65 ± 8.81-307.91 ± 4.41 mg GAE/100 g), and antioxidant action (11.90 ± 0.39-13.73 ± 0.64 mg TE/g) (Sardiña-Valdés et al., 2021). Similarly, quercetin and catechins in cyclodextrin inclusions used in fortified cheeses and yogurt demonstrated excellent outcomes when encapsulated in such food matrices (Pereira et al., 2020; Ho et al., 2019). Finally, green coffee extracts and especially extracted chlorogenic acids, increased the antioxidant activity and organoleptic properties in caramel cottage cheese (Budryn et al., 2016), whereas yogurt enriched by olive leaf phenolics within liposomes, revealed elevated antioxidant and nutritional value, as well as public acceptance (Tavakoli et al., 2018).

5.2 Functional Beverages and Bio-preservatives

Polyphenol-enriched beverages represent one of the most successful categories in the functional food market. This includes fortified teas, fruit and vegetable juices, fortified water or smoothies, and plant-based milk alternatives. Extracts from several natural sources are commonly used to increase phenolic content and associated health benefits, while encapsulated forms of curcumin, quercetin, and hydroxytyrosol, for instance, are being utilized to overcome solubility issues and ensure stability in aqueous systems (Li et al., 2023; Da et al., 2019; Bińkowska et al., 2024; Lai and Wong, 2022). Anthocyanins are natural colorants that, when incorporated in model beverages, enhance stability and shelf-life, especially in conjunction with heat-denatured whey protein isolates (hydrogen bonding), serving as natural pigments (Chung et al., 2015).

Curcumin nanoparticles exhibited antibacterial activity against *Staphylococcus aureus* (minimum inhibitory concentration (MIC 125 μg/mL) and *Alicyclobacillus acidoterrestris* (MIC 62.5 μg/mL), antioxidant capacity, and stability when used in orange juice (Dutra et al., 2021). Nanoencapsulated Egyptian prickly pear peel fruit extracts in chitosan and chitosan-sodium alginate displayed high reducing power activity (462.49–481.76 μg GAE/100 g), thermal stability, and bioactive compound quality when integrated in guava juice (F. Mahmoud et al., 2018). Moreover, golden apple and red grape pomace extracts incorporated in chitosan and soy protein nanocapsules enhanced both the antioxidant activity and longevity of apple and pineapple juice as food edible materials (Gaber Ahmed et al., 2020). Interestingly, pomegranate-derived vanillin and geraniol reduced the native microflora, extending the microbiological shelf-life of strawberry juice and decreasing its *E. coli* content (Tomadoni et al., 2016). In a similar study, these added fibers also increased the antioxidant capacity and improved the sensory quality of strawberry juice, acting as natural bio-preservatives and increasing the TPC (Moon et al., 2006).

Considering white and red wine, catechin-rich extracts from green tea and grape seeds increased the TPC, antioxidant action, and consumer preference (Yeo et al., 2012; Li et al., 2020). Three types of oenological tannins derived from grape skin, seed, and French oak improved wine aroma, its sensorial properties, higher alcohols, and anthocyanin content (Chen et al., 2016). Plus, the commercial enological Tara tannin "Vitanil B" was added at different concentrations in white wine from grapes of *Vitis vinifera* L. *var.* Malagousia, and at a concentration of 300 mg/L, it enhanced its TPC, antioxidant stability, antiradical activity, and sensory character, while preventing color deterioration for 100 days and offering high aroma quality and overall acceptance of Malagousia white wine (Strati et al., 2021).

5.3 Functional Fruit- and Vegetable-Based Foods

Fruit- and vegetable-based foods represent an important and consumer-friendly platform for delivering polyphenols in their native or fortified forms. Formulations such as dried fruits, fruit bars, vegetable chips, jams, purées, and fermented vegetables retain significant amounts of polyphenols or can be further enriched through natural extracts. Apples, berries, plums, tomatoes, beets, carrots, and spinach are frequently utilized for their inherent phenolic content and antioxidant properties. Additionally, fortification with external polyphenol sources (e.g., grape seed extract, green tea, or olive polyphenols) is being explored to increase functionality (Khan et al., 2024; Abbas et al., 2017; Wang et al., 2023; Da et al., 2019; Bińkowska et al., 2024). For example, unripe grape extract was utilized to fortify three food models: carbohydrates/acidic pH/sweet-beetroot purée, proteins/neutral pH sweet-pea purée, and starch/neutral pH potato purée, where significant positive variations in sourness, saltiness, bitterness, and astringency were observed (Bucalossi et al., 2020).

Snack bars with added apple dietary fiber (DF) revealed high total DF (~5.3% w/w). At the same time, apple DF or inulin-enhanced snack bars had higher TPC levels (2.87 and 2.22 mg CT/g bar), with apple polyphenol extract (APE) in the bar filling, without showing extra browning and presenting great longevity, practically being convenient functional foods (Sun-W. et al., 2010). Plus, cabbage chips developed from a cabbage powder (12%), maize flour (10%), and rice flour (27.97%) blend, revealed the positive influence of plant-derived polyphenols in crispiness, taste, and acceptance. Rice flour had the most notable impact on the acceptance, aroma, and overall appearance, and all outcomes pointed out cabbage chips' balanced nutrition and potential as healthier snacks (Deshmukh et al., 2024).

5.4 Functional Bakery Products

Bakery products such as bread, muffins, cookies, and energy bars are ideal vehicles for polyphenol delivery. Incorporating flavonoid-rich plant powders (e.g., matcha, grape pomace, or olive pomace) or polyphenol extracts (e.g., from cocoa or turmeric) into the dough or butter can significantly enhance the antioxidant activity of the final product. However, thermal degradation during baking necessitates the utilization of heat-stable polyphenols or protective carriers, including microcapsules, to preserve bioactivity (Khan et al., 2024; Abbas et al., 2017; Wang et al., 2023; Da et al., 2019; Bińkowska et al., 2024) Adding matcha green tea powder full of the four major catechins to cake dough significantly suppressed the accumulation of reactive carbonyl species (RCSs), providing heat stability during cake baking, as a food additive (Sugimoto et al., 2021). In addition, (+)-catechin-based leavening agents (mucic or tartaric acids), increased both TPC (tartaric acid: 0.00175 meq catechin (CT)/g) and total antioxidant capacity values (tartaric acid: 0.0298 ± 0.021 and 0.0081 ± 0.0009 meq CT/g, respectively, and mucic acid: 0.518 ± 0.015 (DPPH) and 0.112 ± 0.010 (ABTS) mg/mL) in muffins obtained by both powders (Carullo et al., 2020).

Fortified cookies with liposomal curcumin or α-tocopherol revealed unaltered sensory appeal and retained antioxidant action ($92.70 \pm 3.34\%$ and $95.33 \pm 0.67\%$, respectively) during baking (Pamunuwa et al., 2023). Furthermore, tannic acid as a dough additive demonstrated great results, as it increased the volumes and specific volumes of different bread types, enhanced their antioxidant capacity by four- and three-fold in FRAP and ABTS assays, reduced bread crumb hardness, gumminess, and chewiness without affecting crust color, and improved their sensory attributes (Guan et al., 2023). Moreover, the dough's rheological properties were notably dose-dependently impacted by quercetin (0.2%) fortification, which also enhanced the antioxidant activity and inhibited advanced glycation end-products (AGEs) formation by 46–52%, positively altering the quality attributes of baked bread (Lin and Zhou, 2018). Sorghum flour addition increased resistant starch content, free phenolic acid, and antioxidant capacity of durum wheat pasta (Khan et al., 2013). Finally, black glutinous rice crisp rice bars revealed a high content of anthocyanins, leading to an increased nutritional value and storage stability of rice bars (Kasote et al., 2021).

5.5 Functional Meat and Fish Products and Storage Stability

The application of polyphenols in animal-based products focuses primarily on their antioxidant and antimicrobial roles. Natural extracts rich in rosmarinic acid, tannins, or phenolic acids can reduce lipid oxidation and microbial spoilage in processed meats, thereby extending shelf-life and reducing the need for synthetic additives. These ingredients may also be integrated into edible coatings or marinades. Nonetheless, organoleptic balance remains critical to prevent off-flavors or discoloration (Khan et al., 2024; Abbas et al., 2017; Wang et al., 2023; Da et al., 2019; Bińkowska et al., 2024). Curcumin, garlic, thyme, cinnamon, peppermint, and silymarin extracts nanoencapsulated in several formulations enhanced the growth performance, antimicrobial action, and antioxidant capacity, as feed additives in broiler chickens. Plus, flaxseed oil, pomegranate, red onion, garlic, fish oil, thyme, eugenol, and clove essential oil extracts enhanced the antibacterial and antioxidant action, and the nutritional value of chicken, beef, turkey, and pork meat (Taouzin et al., 2023).

For instance, tea (TP), apple (AP), and cinnamon (CP) polyphenols could perform as natural antioxidant additives for dry-fried bacon, as they significantly reduced pH, lipid oxidation, and residual nitrite content. AP exhibited a protective effect against protein oxidation-induced damage in bacon, while it also led to reduced thiobarbituric acid reactive substance (0.59 malondialdehyde (MDA)/kg), carbonyl contents (2.30 nmol/mg protein), as well as *N*-nitroso-methyl phenylamine (NMPhA) formation (1.211 ng/kg) (Deng et al., 2022). Moreover, a pectin-polyphenol combination inhibited oxidation, protein aggregation/crosslinking, and heterocyclic amine formation, while it improved particle size, zeta potential, and bound water with proteins in meat-based snacks (Nawaz et al., 2024). Non-conventional poultry patties finally benefited from the combined antimicrobial effects of thymol and carvacrol, which enhanced their quality and storage stability (Mastromatteo et al., 2009).

Pacific white shrimp (*Litopenaeus vannamei*) treated with catechin were monitored during iced storage and hindered the growth of psychrophilic bacteria and spoilage microorganisms, including H_2S-producing bacteria and Enterobacteriaceae. Catechin addition lowered melanosis incidence, lipid oxidation, loss of freshness, and polyphenol-oxidase (PPO), dose-dependently (Nirmal and Benjakul, 2009). Similarly, hydroxycinnamic acid (caffeic, chlorogenic, *o*-coumaric, and ferulic acid) and catechin (epicatechin, epicatechin gallate, epigallocatechin, and epigallocatechin gallate) powders were examined for their antioxidant capacity in minced fish muscle. As a result, 10 ppm caffeic acid most effectively inhibited lipid oxidation in fish muscle, epicatechin revealed the highest antioxidant activity, and an efficacy increment was shown using 10-100 ppm of both caffeic acid and epicatechin (Medina et al., 2007). Tea polyphenols, mainly catechins, retarded microbial growth and prevented protein denaturation, therefore leading to meat fish, and fish by product preservation, as well as reduced fish spoilage (Kamal et al., 2025).

5.6 Functional Confectionery and Snacks

The incorporation of polyphenols into chocolate, candy, granola bars, and extruded snacks is gaining traction, particularly for flavonoid-rich compounds. Dark chocolate, mainly, is an excellent carrier because of its intrinsic polyphenol content and fat matrix, which enhances solubility and stability.

Functional gummies and lozenges are also being developed with encapsulated polyphenols for several health applications (Khan et al., 2024; Abbas et al., 2017; Wang et al., 2023; Da et al., 2019; Bińkowska et al., 2024). Procyanidins and xanthones in mangosteen pericarp (3%) powder added to dark and compound chocolate significantly expanded their bioactive profile and total phenolic content (13% in dark and 50% in compound chocolates), without affecting their sensory quality (Sim et al., 2016). Additionally, whey protein isolates, along with polyphenol-rich cranberry juice spray-dried particles, used as a functional ingredient in high protein bars, inhibited bar hardening, increased their elasticity, and enhanced their nutritional content (Diaz et al., 2021).

Aiming to enhance the nutritional value of gummy candies, pistachio green hull extract (PGHE, 2%), stevia (0.026%), and gelatin-to-starch ratio (8:2) were incorporated, resulting in desirable physicochemical properties, enhanced bioactivity, increased both TPC (680.31 ± 0.6 mg GAE/100 g gummy candy) and antioxidant activity ($IC_{50} = 277$ μg/mL), and reduced sugar content (50%, equal to ~12% sucrose), eliminating the need for artificial flavors and synthetic colorants (Roudhari et al., 2024). Similarly, antioxidant gummies based on gelatin enriched with citrus fruit peel (red and blond oranges, lemons) extracts, normally treated as waste, showed excellent scavenger abilities (IC_{50} = 0.201–0.454 mg/mL) and rheological properties for ~14 days (lemon-based gummies returned the best results) (Aiello et al., 2024). In addition, honey candies in the form of lozenges or drops demonstrated higher antioxidant activity (49.5 ± 1.9 vs. 55.8 ± 0.5% (DPPH)), TPC (2090 ± 10 vs. 1730 ± 10 mg gallic acid/kg honey), and hydroxymethylfurfural components, compared to raw honey (Larsen and Ahmed, 2022).

5.7 Nutraceutical Supplements, Probiotic and Prebiotic Formulations

Polyphenols are widely formulated as dietary supplements in the form of capsules, powders, tablets, granules, and liquid drops. These supplements often contain purified or standardized forms of individual polyphenols (e.g., resveratrol, epigallocatechin gallate) or synergistic blends derived from botanical extracts (e.g., grape seed, green tea, turmeric, olive leaf). Moreover, a co-formulation of polyphenols with probiotics (e.g., *Lactobacillus, Bifidobacterium*) is of emerging interest due to their synergistic effects on gut health. Hence, fortified foods enriched with polyphenols are being developed to promote gut-brain axis balance, immune support, and metabolic health (Zhang et al., 2022, Khan et al., 2024; Muntaha et al., 2025). In a comprehensive review, collectively, resveratrol, curcumin, ferulic, gallic, and *p*-coumaric acid, luteolin, quercetin, hesperidin, genistein, and anthocyanidins integrated in nutraceuticals, reduced fed blood glucose levels, increased insulin, elevated high-density lipoprotein (HDL) cholesterol, alleviated diabetic neuropathy and cognitive decline, decreased weight and body mass index (BMI), improved hepatic and pancreatic antioxidant capacity, and lowered pro-inflammatory cytokines (e.g., TNF-α, IL-6), hyperglycemia, and blood pressure in diabetic individuals and animals (Fernandas et al., 2022).

A recent study highlighted that combining different probiotics and berry-derived polyphenols (from blueberry and cranberry) modulated immune response in dendritic cells (DCs), enhancing anti-inflammatory (IL-10) and at the same time suppressing pro-inflammatory activities (IL-12) (Taverniti et al., 2022). A second-generation rice-based snack prepared with 60% high-protein quinoa flour as well as *Bacillus coagulans* probiotics, achieved a 17% protein content, 75% viability towards artificial gastroenteric juice, 35.29, 5.52, and 13.74 μmol ascorbic acid (AA)/g antioxidant capacity measured via ABTS, DPPH, and FRAP assays, respectively (Pabon et al., 2024). Similarly, the prebiotic effect of dietary polyphenols, especially catechins, anthocyanins, and proanthocyanins, is strongly supported (Alves-S. et al., 2020).

5.8 Food Packaging and Preservation Applications

Polyphenols are increasingly being embedded into active food packaging systems, including biodegradable films, edible coatings, or nanocomposite materials to inhibit lipid peroxidation, delay microbial growth, and extend the shelf life of perishable products (Khan et al., 2024; Taouzin et al., 2023; Bian et al., 2024; Khanzada et al., 2024; Robledo et al., 2018; Benloch-T. et al., 2025). Gallic acid might be a great biocontrol tool for improving shelf-life and preventing food spoilage, as it exhibited vast antimicrobial activity (pore formation/membrane rupture) against *Pseudomonas* spp. Strains (4°C), decontaminated truffles, diminished off-flavors, and preserved their positive features (using 2.5 mg/ml gallic acid, storage for 28 days, at 4°C) (Sorrentino et al., 2018). Active essential oil thymol and carvacrol added to a vinegar-based marinade for controlling spoilage of marinated beef (Shawarma) stored in air or vacuum packaging, at 0.8% extended the microbiological shelf-life by 6 days and achieved higher antimicrobial activity, while at 0.4% increased the sensorial shelf-life by 9 and > 12 days (Karan et al., 2020). A turmeric extract, finally, acted as a natural antioxidant on modified atmosphere-packaged fresh lamb sausages with fat replacement by tiger nut during storage (2°C), as it hindered lipid oxidation and peroxidation, without notably affecting their physicochemical parameters (de Carvalho et al., 2020).

Thymol and carvacrol added to starch edible films and coatings revealed enhanced antimicrobial action against *Colletotrichum gloeosporioides*, with a MIC of 1500 mg/L and a delay in firmness, maturity index, and color change, therefore, they were able to reduce anthracnose incidence on mango and papaya (Ochoa-U. et al., 2021). Similarly, thymol encapsulation in ethyl cellulose-based microspheres revealed high antimicrobial activity, thermal stability, and a gradual release profile, while retaining pH and titratable acidity for 21 days of storage during cherry tomato packaging Giotopoulou et al., 2024). Another example underlying polyphenol's high food packaging value, are gelatin, chitosan, and methylcellulose films, which when integrated with tannic acid (15% w/w), displayed increased transparency, (methylcellulose: 0.572 to 4.73 A/mm), reduced weight loss for cherry tomatoes wrapped with chitosan (21.3–19.6%), gelatin (22.1–15.5%), and methylcellulose (26.2–20.5%), decreased browning index of fruits, and improved antibacterial impact of biopolymers (Halim et al., 2018). Chitosan-based films blended with tannic acid and Moringa Oleifera seeds (10% w/w), exhibited high antifungal and antimicrobial action against *S. aureus, E. coli, Aspergillus niger (A. niger),* and *Candida albicans* (*C. albicans*) and preserved strawberry freshness (12 days) (Venkatesan et al., 2024). Similarly, multifunctional tannic acid/chitosan films showed enhanced anti-UV, antioxidant, antimicrobial, and mechanical properties, reduced enzymatic browning and water loss, and increased spoilage-free storage period of bananas, as active food packaging materials (Lee et al., 2023).

6. Incorporation and Encapsulation Delivery Systems of Polyphenols for Functional Food and Nutraceutical Applications

Polyphenols are vital bioactive compounds in plants, known for their strong antioxidant, anti-inflammatory, cardioprotective, and neuroprotective effects. However, despite their promising therapeutic potential, their application in functional foods and nutraceuticals is often hindered by their low bioavailability, instability during processing and storage, poor water solubility, and limited absorption in the gastrointestinal tract. To address these limitations, innovative incorporation and encapsulation strategies have been developed, aiming to protect, stabilize, and enhance the delivery of polyphenols in complex food matrices or supplement formulations (Zhang et al., 2022; Khan et al., 2024; Enaru et al., 2021; Shi et al., 2021; Ali Redha et al., 2024; Wang et al., 2024).

Incorporation typically refers to the direct addition of polyphenolic compounds, either as crude extracts, purified molecules, or standardized formulations, into food products. This approach has been successfully applied in beverages, dairy products, cereals, baked goods, and functional snacks (Ficco et al., 2024; de Araújo et al., 2021; Da et al., 2019; Pandey et al., 2021; Ou et al., 2019). Nevertheless, direct incorporation might induce the degradation of sensitive polyphenols due to exposure to oxygen, light, temperature, and varying pH during processing and storage. Moreover, bitterness or astringency can negatively affect sensory properties (Ciepei et al., 2024; de Araújo et al., 2021; Ou et al., 2009).

To overcome these challenges, encapsulation technologies have emerged as powerful tools, offering protection, targeted delivery, controlled release, and improved bioaccessibility as well as bioavailability of polyphenols (Zhang et al., 2022; Ali Redha et al., 2024). Lipid-based carriers (delivery systems), like liposomes, solid lipid nanoparticles, and nano-emulsions, are widely utilized to encapsulate flavonoids (e.g., quercetin, kaempferol), stilbenes (e.g., resveratrol), and other hydrophobic polyphenols. These systems not only enhance the stability and solubility but also accelerate cellular uptake and provide a larger surface-to-mass ratio, making them ideal for nutraceuticals (de Araújo et al., 2021; Halevas et al., 2022; Veiko et al., 2020; Caddeo et al., 2008). Particularly, nano-emulsions are increasingly applied to incorporate polyphenols into lipid-containing functional foods. These systems are mostly effective for lipophilic polyphenols such as thymol, carnosic acid, and curcumin, providing enhanced dispersion and bioaccessibility (Ciepei et al., 2024; de Araújo et al., 2021; Robledo et al., 2018; Zheng et al., 2021; Jiang et al., 2020). Moreover, inclusion complexes, mainly with cyclodextrins (β-CD, hydroxypropyl (HP)-β-CD), form molecular cages that effectively entrap polyphenols like catechins, hydroxytyrosol, or eugenol, significantly improving water solubility, protection from oxidation, and sustained release in the gut (Cid-S. et al., 2022; Li et al., 2019; Zyze et al., 2018; Ntuli et al., 2022).

Biopolymer-based nanoformulations (e.g., nano-particles, -gels, -films, -fibers, -emulsions, -micelles), composed of natural polysaccharides (e.g., starch, chitosan, cellulose, lignin, glycogen, alginate) and proteins (e.g., whey, casein, zein, gelatin, albumin, soy or rice protein, ferritin) are widely popular due to their biocompatibility and ability to form stable structures via ionic gelation, emulsification, or coacervation. These have shown excellent results in delivering anthocyanins, phenolic acids, and tannins, while they have also demonstrated high biocompatibility, sustainability, biodegradability, eco-consciousness, and non-toxicity, with both hydrophobic and hydrophilic properties (Zhang et al., 2022; de Araújo et al., 2021; Shi et al., 2021; Alara et al., 2021; Muntaha et al., 2025; Bian et al., 2024; Khanzada et al., 2024; Benloch-T et al., 2025; Liu et al., 2023; Liu et al., 2016; Rodríguez et al., 2015). Interestingly, commercialized functionalized formulations incorporating nano-sized polyphenols (e.g., nano-curcumin or nano-resveratrol) have been introduced to the global market, already offering beneficial outcomes for consumers (Zhang et al., 2022).

At this point, it must be noted that spray-drying and freeze-drying techniques are extensively utilized in the food industry to encapsulate polyphenolic-rich extracts or isolated compounds into microparticles (microencapsulation) employing delivery carriers including maltodextrin, starch, or modified celluloses. This approach offers protection during storage and can be easily incorporated into powdered or dry functional food systems and nutraceuticals, bettering food manufacturing, safety, quality, and packaging technologies (Bińkowska et al., 2024; Zheng et al., 2011; Wang et al., 2017). Additionally, the most common methods used to produce nanoencapsulation systems include emulsification, coacervation, extraction, inclusion complexation, and emulsification solvent evaporation. Emerging technologies like electrospinning, electro-spraying, layer-by-layer assembly, nanoprecipitation, and supercritical fluid processing have also gained attention for designing nano- and micro-encapsulated systems with high efficiency, sustained release, and environmental trigger responsiveness (Figure 6) (Taouzinet et al., 2023; Rostamabadi et al., 2020; Elbaz et al., 2023; Martínez Rivas et al., 2017; Soh and Lee, 2019).

Encapsulation not only stabilizes polyphenols against several environmental stimuli and sensors but also allows for targeted delivery to specific sites within the gastrointestinal tract, modulating absorption and enhancing bioactivity. In food systems, encapsulated polyphenols are increasingly used in the formulation of fortified yogurts, cheese, probiotic beverages, cereal bars, and plant-based alternatives, without significantly altering the sensory profile or texture (Ficco et 2024; Li et al., 2023; Krga and Milenkwic, 2019; Da et al., 2019; Bińkowska et al., 2024). Moreover, encapsulation can mask undesirable tastes or odors often associated with certain polyphenols, such as the bitterness of catechins or the pungency of eugenol (Cione et al., 2020; Aburel et al., 2021; Zyze et al., 2018; Liu et al., 2016). Ultimately, the selection of encapsulation materials and methods is dictated by a plethora of factors, like the chemical profile of the polyphenol, its solubility and stability, as well as the desired release site (e.g., mouth, stomach, intestine) and the final product's processing conditions (Zhang et al., 2022; Khan et al., 2024; Enaru et al., 2021; Shi et al., 2021; Ali Redha et al., 2024; Wang et al., 2024). Advances in micro/nano-encapsulation, responsive delivery systems, and sustainable biopolymer matrices are believed to effectively aid in overcoming common problems during functional food and nutraceutical formulation, such as the bioavailability of a polyphenolic blend (Zhang et al., 2022; Khan et al., 2024; Enaru et al., 2021; Shi et al., 2021; Ali Redha et al., 2024; Wang et al., 2024).

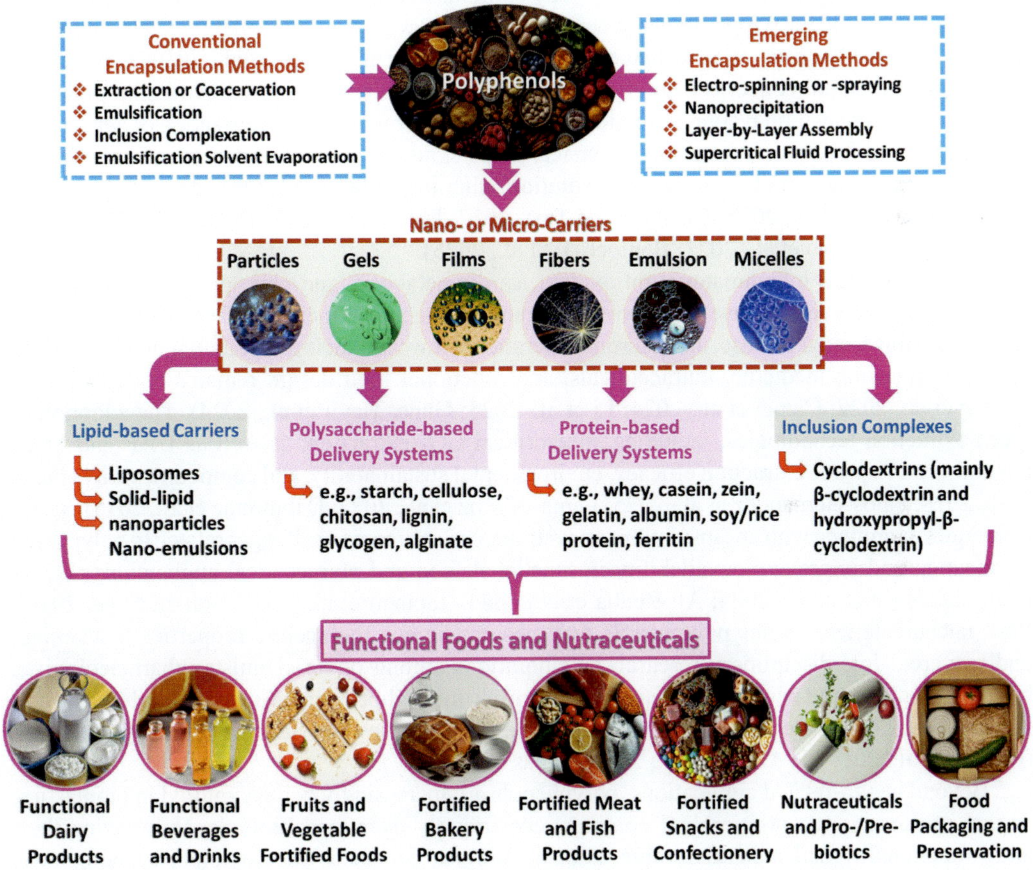

Fig. 6 Incorporation and Encapsulation Technologies of Polyphenols for Functional Food Applications. (Parts of this figure were obtained from https://www.freepik.com/ (accessed on May 15, 2025)).

7. Limitations, Challenges, and Future Perspectives

Despite their potent bioactivity, many polyphenolic compounds exhibit low bioavailability, largely due to poor absorption, rapid metabolism, and instability in physiological environments. To address these limitations, advanced delivery systems, such as liposomes, nanoformulations, and microencapsulation, are being explored to enhance stability, targeted delivery, and controlled release, ultimately improving their therapeutic efficacy. Large-scale production of high-quality polyphenol-rich extracts is challenged by variability in raw material cultivation, driven by factors such as plant genotype, cultivation conditions, geographical origin, harvesting time, and post-harvest processing. Hence, the standardization of agricultural and extraction practices is critical to ensure reproducibility in bioactive content and functional properties (Ali Redha et al., 2024; Alara et al., 2021; Ajila et al., 2011; Chaves et al., 2020; Haytowitz et al., 2013).

In parallel, comprehensive toxicological evaluations and long-term clinical trials are needed to validate the safety and efficacy of polyphenols, particularly in high-dose nutraceutical applications and functional food products. Additionally, potential interactions with pharmaceutical drugs must be investigated, especially as polyphenol-containing products are frequently consumed alongside many conventional medications (Zhang et al., 2022; Khan et al., 2024; Abbas et al., 2017; Güneş Bayir et al., 2024; Muntaha et al., 2025). Some polyphenol-rich sources, especially those from allergenic plant families (e.g., Asteraceae, Fabaceae), may stimulate hypersensitivity reactions in susceptible individuals. Thus, identifying allergenic constituents and developing hypoallergenic alternatives is essential for broad consumer acceptance (Wu et al., 2023; Demisow-W. et al., 2019; Christensen, 2018). Environmental concerns also arise from unsustainable harvesting and extraction practices, which may deplete natural resources or alter phytochemical profiles. The valorization of agricultural by-products and food waste (e.g., olive pomace, grape skins, tea leaves) as sources of polyphenols presents a sustainable and cost-effective solution, aligning with circular bioeconomy principles (Palos-Hernandez et al., 2025; Chanioti and Tzia, 2017; Iraki et al., 2018; Duba et al., 2015; Chen et al., 2016; Sugimoto et al., 2021; Benloch-T. et al., 2025).

Given their anti-inflammatory, antioxidant, cardiometabolic, and neuroprotective properties, polyphenols hold great promise in the prevention and management of chronic diseases like diabetes, obesity, cardiovascular disease, and neurodegeneration. However, their full potential in personalized nutrition, precision medicine, nutraceuticals, and functional food design remains underexploited (Zhang et al., 2022; Ciepei et al., 2024; Li et al., 2023: Güneş Bayir et al., 2024). The adoption of green extraction technologies, including supercritical CO_2 extraction, as well as UAE and MAE methods, can improve extraction efficacy, environmental sustainability, and compound purity (Panja et al., 2018; Palos-Hernandez et al., 2025; Lama-Muñoz et al., 2022; Dahmoune et al., 2015). These techniques, combined with encapsulation innovations, can overcome challenges related to polyphenol degradation and improve bioavailability in functional food and nutraceutical applications (Zhang et al., 2022; Khan et al., 2024; Ali Redha et al., 2024; Taouzin et al., 2023; Soh and Lee, 2019). Plus, uncharacterized minor polyphenols may possess unique therapeutic properties that remain undiscovered. Interdisciplinary research collaborations among phyto-chemists, pharmacologists, biotechnologists, and clinicians, supported by high-throughput screening, metabolomics, and systems biology tools, could potentially uncover new bioactives and improve formulation design (Rasouli et al., 2017; de Araújo et al., 2021; Abbas et al., 2017).

Beyond health-related applications, polyphenols are increasingly being explored in bioplastics, natural preservatives, antimicrobial coatings, eco-friendly dyes, and biofuels, showcasing their multi-functionality and industrial value. Biotechnological advances such as plant tissue culture, metabolic engineering, and synthetic biology can facilitate sustainable and scalable production of targeted polyphenols with consistent quality (de Araújo et al., 2021; Tomadoni et al., 2016; Robledo et al., 2018; Sorrentino et al., 2018; Yong et al., 2024; Yan et al., 2021; X. Shi, 2025). Lastly, the implementation of globally harmonized regulatory frameworks is essential to provide clear guidance on sourcing, standardization, labeling, safety, and efficacy evaluation. Enhanced regulatory oversight will support the safe, ethical, and effective integration of polyphenols into functional foods and nutraceuticals across international markets (Palos-Hernandez et al., 2025; Chanioti and Tzia, 2017; Iraki et al., 2018; Duba et al., 2015; Chen et al., 2016; Sugimoto et al., 2021; Benlloch-T. et al., 2025).

8. Conclusions

Polyphenols have transitioned from being traditionally recognized phytochemicals to becoming scientifically validated bioactives with a wide range of health-promoting benefits, including their antioxidant, anti-inflammatory, anticancer, antimicrobial, anti-aging, antiplatelet, cardioprotective, and neuroprotective properties. Today, polyphenols are incorporated into nutraceuticals, functional foods and beverages, as well as pharmaceuticals and cosmeceuticals, highlighting their growing relevance in chronic disease prevention and health maintenance.

Despite their promising potential, challenges like polyphenols' low bioavailability, variability in composition, and standardization difficulties remain. Advances in encapsulation technologies, green extraction methods, and sustainable delivery systems offer practical solutions to improve efficacy, stability, and scalability. Ongoing interdisciplinary research and innovation will be crucial in refining polyphenol-based products and ensuring their successful integration into modern health and wellness strategies, aimed at developing safe and efficacious functional foods and nutraceuticals.

Author Contributions: Conceptualization, A.T.; methodology, A.T.; software, all authors; valida tion, A.T.; investigation, A.T., E.K., and T.A.; writing—original draft preparation, all authors; writing—review and editing, A.T., E.K., and T.A.; visualization, A.T., E.K., and T.A.; supervision, A.T.; project administration, A.T. All authors have read and agreed to the published version of the manuscript.

Funding: This research received no external funding.

Acknowledgments: The authors would like to thank the School of Chemistry of the Faculty of Sciences of Democritus University of Thrace for their continuous support.

Conflicts of Interest: The authors declare no conflicts of interest.

References

Abbas, M., Saeed, Farhan, Anjum, Faqir Muhammad, et al., 2017. Natural polyphenols: An overview. *Int. J. Food Prop.*, *20*, 1689–1699. https://doi.org/10.1080/10942912.2016.1220393.

Aburel, O.M., Pavel, I.Z., Dănilă, M.D., et al., 2021. Pleiotropic Effects of Eugenol: The Good, the Bad, and the Unknown. *Oxid. Med. Cell Longev.*, *2021*, 3165159. https://doi.org/10.1155/2021/3165159.

Aiello, F., Caputo, P., Oliviero Rossi, C., et al., 2024. Formulation of Antioxidant Gummies Based on Gelatin Enriched with Citrus Fruit Peels Extract. *Foods*, *13*, 320. https://doi.org/10.3390/foods13020320.

Ajila, C.M., Brar, S.K., Verma, M., et al., 2011. Extraction and Analysis of Polyphenols: Recent trends. *Crit. Rev. Biotechnol.*, *31*, 227–249. https://doi.org/10.3109/07388551.2010.513677.

Alara, O.R., Abdurahman, N.H. and Ukaegbu, C.I. 2021. Extraction of phenolic compounds: A review. *Curr. Res. Food Sci.*, *4*, 200–214. https://doi.org/10.1016/j.crfs.2021.03.011.

Ali Redha, A., Kodikara, C. and Cozzolino, D. 2024. Does Encapsulation Improve the Bioavailability of Polyphenols in Humans? A Concise Review Based on *In Vivo* Human Studies. *Nutrients*, *16*, 3625. https://doi.org/10.3390/nu16213625.

Alonso-Salces, R.M., Korta, E., Barranco, A., et al., 2001. Pressurized liquid extraction for the determination of polyphenols in apple. *J. Chromatogr. A.*, *933*, 37–43. https://doi.org/10.1016/S0021-9673(01)01212-2.

Alves-Santos, A.M., Sugizaki, C.S.A., Lima, G.C. and Naves, M.M.V. 2020. Prebiotic effect of dietary polyphenols: A systematic review. *J. Funct. Foods*, *74*, 104169. https://doi.org/10.1016/j.jff.2020.104169.

Amakura, Y., Tsutsumi, T., Sasaki, K.,, et al., 2008. Influence of food polyphenols on aryl hydrocarbon receptor-signaling pathway estimated by *in vitro* bioassay. *Phytochemistry*, *69*, 3117–3130. https://doi.org/10.1016/j.phytochem.2007.07.022.

Ameer, K., Shahbaz, H.M. and Kwon, J-H. 2017. Green Extraction Methods for Polyphenols from Plant Matrices and Their Byproducts: A Review. *Compr. Rev. Food Sci. Food Saf.*, *16*, 295–315. https://doi.org/10.1111/1541-4337.12253.

Annunziata, G., Sureda, A., Orhan, I.E., et al., 2021. The neuroprotective effects of polyphenols, their role in innate immunity, and the interplay with the microbiota. *Neurosci. Biobehav. Rev.*, *128*, 437–453. https://doi.org/10.1016/j.neubiorev.2021.07.004.

Ansari, P., Choudhury, S.T., Seidel, V., et al., 2022. Therapeutic Potential of Quercetin in the Management of Type-2 Diabetes Mellitus. *Life*, *12*, 1146. https://doi.org/10.3390/life12081146.

Arcanjo, N.M.O., Luna, C., Madruga, M.S. and Estévez, M. 2018. Antioxidant and pro-oxidant actions of resveratrol on human serum albumin in the presence of toxic diabetes metabolites: Glyoxal and methyl-glyoxal. *Biochim. Biophys. Acta BBA - Gen Subj.*, *1862*, 1938–1947. https://doi.org/10.1016/j.bbagen.2018.06.007.

Athanasiadis, V., Chatzimitakos, T., Kotsou, K., et al., 2023. Polyphenol Extraction from Food (by) Products by Pulsed Electric Field: A Review. *Int. J. Mol. Sci.*, *24*, 15914. https://doi.org/10.3390/ijms242115914.

Bagale, U., Kadi, A., Abotaleb, M., et al., 2023. Prospect of Bioactive Curcumin Nanoemulsion as Effective Agency to Improve Milk Based Soft Cheese by Using Ultrasound Encapsulation Approach. *Int. J. Mol. Sci.*, *24*, 2663. https://doi.org/10.3390/ijms24032663.

Barrulas, R.V., Paiva, T.G. and Corvo, M.C. 2019. NMR methodology for a rational selection of ionic liquids: Extracting polyphenols. *Sep. Purif. Technol.*, *221*, 29–37. https://doi.org/10.1016/j.seppur.2019.03.077.

Bellumori, M., Cecchi, L., Innocenti, M., et al., 2019. The EFSA Health Claim on Olive Oil Polyphenols: Acid Hydrolysis Validation and Total Hydroxytyrosol and Tyrosol Determination in Italian Virgin Olive Oils. *Molecules*, *24*, 2179. https://doi.org/10.3390/molecules24112179.

Benlloch-Tinoco, M., Gentile, P., Taylor, L. and Girón-Hernández, J. 2025. Alginate edible films as delivery systems for green tea polyphenols. *Food Hydrocoll.*, *158*, 110518. https://doi.org/10.1016/j.foodhyd.2024.110518.

Bian, H., Luo, D., Shi, J., et al., 2024. Cellulose nanofibrils-based packaging films synergistically reinforced by lignin and tea polyphenols for strawberry preservation. *Chem. Eng. J.*, *502*, 157843. https://doi.org/10.1016/j.cej.2024.157843.

Bińkowska, W., Szpicer, A., Stelmasiak, A., et al., 2024. Microencapsulation of Polyphenols and Their Application in Food Technology. *Appl Sci.*, *14*, 11954. https://doi.org/10.3390/app142411954.

Bostanghadiri, N., Pormohammad, A., Chirani, A.S., et al., 2017. Comprehensive review on the antimicrobial potency of the plant polyphenol Resveratrol. *Biomed. Pharmacother.*, *95*, 1588–1595. https://doi.org/10.1016/j.biopha.2017.09.084.

Brglez Mojzer, E., Knez Hrnčič, M., Škerget, M., et al., 2016. Polyphenols: Extraction Methods, Antioxidative Action, Bioavailability, and Anticarcinogenic Effects. *Molecules*, *21*, 901. https://doi.org/10.3390/molecules21070901 1.

Brooks, MS-L. and Celli, G.B. 2019. *Anthocyanins from Natural Sources: Exploiting Targeted Delivery for Improved Health.* Royal Society of Chemistry, Cambridge.

Bucalossi, G., Fia, G., Dinnella, C., et al., 2020. Functional and sensory properties of phenolic compounds from unripe grapes in vegetable food prototypes. *Food Chem.*, *315*, 126291. https://doi.org/10.1016/j.foodchem.2020.126291.

Budryn, G., Zaczyńska, D. and Oracz, J. 2016. Effect of addition of green coffee extract and nanoencapsulated chlorogenic acids on aroma of different food products. *LWT*, *73*, 197–204. https://doi.org/10.1016/j.lwt.2016.06.019.

Caban, M. and Lewandowska, U. 2022. Polyphenols and the potential mechanisms of their therapeutic benefits against inflammatory bowel diseases. *J. Funct. Foods*, *95*, 105181. https://doi.org/10.1016/j.jff.2022.105181.

Caddeo, C., Teskač, K., Sinico, C. and Kristl, J. 2008. Effect of resveratrol incorporated in liposomes on proliferation and UV-B protection of cells. *Int. J. Pharm.*, *363*, 183–191. https://doi.org/10.1016/j.ijpharm.2008.07.024.

Çam, M. and Hışıl, Y. 2010. Pressurised water extraction of polyphenols from pomegranate peels. *Food Chem.*, *123*, 878–885. https://doi.org/10.1016/j.foodchem.2010.05.011.

Canedo-Santos, J.C., Carrillo-Garmendia, A., Mora-Martinez, I., et al., 2022. Resveratrol shortens the chronological lifespan of Saccharomyces cerevisiae by a pro-oxidant mechanism. *Yeast 39*, 193–207. https://doi.org/10.1002/yea.3677.

Carullo, G., Scarpelli, F., Belsito, E.L., et al., 2020. Formulation of New Baking (+)-Catechin Based Leavening Agents: Effects on Rheology, Sensory and Antioxidant Features during Muffin Preparation. *Foods*, *9*, 1569. https://doi.org/10.3390/foods9111569.

Chamorro, S., Viveros, A., Alvarez, I., et al., 2012. Changes in polyphenol and polysaccharide content of grape seed extract and grape pomace after enzymatic treatment. *Food Chem.*, *133*, 308–314. https://doi.org/10.1016/j.foodchem.2012.01.031.

Chanioti, S. and Tzia, C. 2017. Optimization of ultrasound-assisted extraction of oil from olive pomace using response surface technology: Oil recovery, unsaponifiable matter, total phenol content, and antioxidant activity. *LWT - Food Sci. Technol.*, *79*, 178–189. https://doi.org/10.1016/j.lwt.2017.01.029.

Chaves, J.O., de Souza, M.C., da Silva, L.C., et al., 2020. Extraction of Flavonoids from Natural Sources Using Modern Techniques. *Front. Chem*, 8. https://doi.org/10.3389/fchem.2020.507887.

Chedea, V.S. and Pop, R.M. 2019. Chapter 11 - Total Polyphenols Content and Antioxidant DPPH Assays on Biological Samples. In: Watson, R.R. (Ed.), *Polyphenols in Plants* (2nd Edn., pp 169–183). Academic Press.

Chen, K., Escott, C., Loira, I., et al., 2016. The Effects of Pre-Fermentative Addition of Oenological Tannins on Wine Components and Sensorial Qualities of Red Wine. *Molecules*, *21*, 1445. https://doi.org/10.3390/molecules21111445.

Christensen, L.P. 2018. Chapter 29 - Polyphenols and Polyphenol-Derived Compounds from Plants and Contact Dermatitis. In: Watson, R.R., Preedy, V.R. and Zibadi, S. (Eds.), *Polyphenols: Prevention and Treatment of Human Disease* (2nd Edn., pp. 349–384). Academic Press.

Chung, C., Rojanasasithara, T., Mutilangi, W. and McClements, D.J. 2015. Enhanced stability of anthocyanin-based color in model beverage systems through whey protein isolate complexation. *Food Res. Int.*, *76*, 761–768. https://doi.org/10.1016/j.foodres.2015.07.003.

Cid-Samamed, A., Rakmai, J., Mejuto, J.C., et al., 2022. Cyclodextrins inclusion complex: Preparation methods, analytical techniques and food industry applications. *Food Chem.*, *384*, 132467. https://doi.org/10.1016/j.foodchem.2022.132467.

Cione, E., La Torre, C., Cannataro, R., et al., 2020. Quercetin, Epigallocatechin Gallate, Curcumin, and Resveratrol: From Dietary Sources to Human MicroRNA Modulation. *Molecules*, *25*, 63. https://doi.org/10.3390/molecules25010063.

Ciupei, D., Colişar, A., Leopold, L., et al., (202.) Polyphenols: From Classification to Therapeutic Potential and Bioavailability. *Foods*, *13*, 4131. https://doi.org/10.3390/foods13244131.

Corcoran, M.P., McKay, Diane, L. and Blumberg, J.B. 2012. Flavonoid Basics: Chemistry, Sources, Mechanisms of Action, and Safety. *J. Nutr. Gerontol. Geriatr.*, *31*, 176–189. https://doi.org/10.1080/21551197.2012.698219.

Ćujić, N., Šavikin, K., Janković, T., et al., 2016. Optimization of polyphenols extraction from dried chokeberry using maceration as traditional technique. *Food Chem.*, *194*, 135–142. https://doi.org/10.1016/j.foodchem.2015.08.008.

Da Porto, C. and Natolino, A. 2017. Supercritical fluid extraction of polyphenols from grape seed (*Vitis vinifera*): Study on process variables and kinetics. *J. Supercrit. Fluids.*, *130*, 239–245. https://doi.org/10.1016/j.supflu.2017.02.013.

da Silva, A.P.G., Sganzerla, W.G., John, O.D. and Marchiosi, R. 2025. A comprehensive review of the classification, sources, biosynthesis, and biological properties of hydroxybenzoic and hydroxycinnamic acids. *Phytochem. Rev.*, *24*, 1061–1090. https://doi.org/10.1007/s11101-023-09891-y.

Dahmoune, F., Nayak, B., Mouss, K., et al., 2015. Optimization of microwave-assisted extraction of polyphenols from *Myrtus communis* L. leaves. *Food Chem.*, *166*, 585–595. https://doi.org/10.1016/j.foodchem.2014.06.066.

Das, A.B., Goud, V.V. and Das, C. 2019. 9 - Phenolic Compounds as Functional Ingredients in Beverages. In: Grumezescu, A.M. and Holban, A.M. (Eds.), Value-Added Ingredients and Enrichments of Beverages (pp. 285–323). Academic Press.

de Araújo, F.F., de Paulo Farias, D., Neri-Numa, I.A.,and Pastore, G.M. 2021. Polyphenols and their applications: An approach in food chemistry and innovation potential. *Food Chem.*, *338*, 127535. https://doi.org/10.1016/j.foodchem.2020.127535.

De Rossi, L., Rocchetti, G., Lucini, L. and Rebecchi, A. 2025. Antimicrobial Potential of Polyphenols: Mechanisms of Action and Microbial Responses—A Narrative Review. *Antioxidants*, *14*, 200. https://doi.org/10.3390/antiox14020200.

de Carvalho, F.A.L., Munekata, P.E.S., Lopes de Oliveira, A., et al., 2020. Turmeric (Curcuma longa L.) extract on oxidative stability, physicochemical, and sensory properties of fresh lamb sausage with fat replacement by tiger nut (*Cyperus esculentus* L.) oil. *Food Res. Int. Ott. Ont.*, *136*, 109487. https://doi.org/10.1016/j.foodres.2020.109487.

De Sá Coutinho, D., Pacheco, M.T., Frozza, R.L. and Bernardi, A. 2018. Anti-Inflammatory Effects of Resveratrol: Mechanistic Insights. *Int. J. Mol. Sci.*, *19*, 1812. https://doi.org/10.3390/ijms19061812.

de Oliveira, M.R. 2016. The Dietary Components Carnosic Acid and Carnosol as Neuroprotective Agents: A Mechanistic View. *Mol. Neurobiol.*, *53*, 6155–6168. https://doi.org/10.1007/s12035-015-9519-1.

Deng, S., Shi, S. and Xia, X. 2022. Effect of plant polyphenols on the physicochemical properties, residual nitrites, and *N*-nitrosamine formation in dry-fried bacon. *Meat Sci.*, *191*, 108872. https://doi.org/10.1016/j.meatsci.2022.108872.

Denisow-Pietrzyk, M., Pietrzyk, Ł and Denisow, B. 2019. Asteraceae species as potential environmental factors of allergy. *Environ. Sci. Pollut. Res.*, *26*, 6290–6300. https://doi.org/10.1007/s11356-019-04146-w.

Deshmukh, P., Patil, D., Bhalerao, P.P., et al., 2024. Optimization & characterization of chips developed from cabbage powder, maize, and rice flour using simplex lattice mixture design. *Food Chem. Adv.*, 5, 100787. https://doi.org/10.1016/j.focha.2024.100787.

Diaz, J.T., Foegeding, E.A. and Lila, M.A. 2021. Whey protein-polyphenol aggregate particles mitigate bar hardening reactions in high protein bars. *LWT*, *138*, 110747. https://doi.org/10.1016/j.lwt.2020.110747.

Domínguez-Rodríguez, G., Marina, M.L. and Plaza, M. 2021. Enzyme-assisted extraction of bioactive non-extractable polyphenols from sweet cherry (*Prunus avium* L.) pomace. *Food Chem.*, *339*, 128086. https://doi.org/10.1016/j.foodchem.2020.128086.

Duba, K.S., Casazza, A.A., Mohamed, H.B., et al., 2015. Extraction of polyphenols from grape skins and defatted grape seeds using subcritical water: Experiments and modeling. *Food Bioprod. Process*, *94*, 29–38. https://doi.org/10.1016/j.fbp.2015.01.001.

Dutra, T.V., de Menezes, J.L., Mizuta, A.G., et al., 2021. Use of nanoencapsulated curcumin against vegetative cells and spores of *Alicyclobacillus* spp. in industrialized orange juice. *Int. J. Food Microbiol.*, *360*, 109442. https://doi.org/10.1016/j.ijfoodmicro.2021.109442.

Dzah, C.S., Duan, Y., Zhang, H., et al., 2020. The effects of ultrasound assisted extraction on yield, antioxidant, anticancer, and antimicrobial activity of polyphenol extracts: A review. *Food Biosci.*, *35*, 100547. https://doi.org/10.1016/j.fbio.2020.100547.

Ebrahimi, E. 2021. Polyphenols: A Comprehensive Review of their Nutritional Properties. *Open Biotechnol. J.*, *15*, 164–172. https://doi.org/10.2174/1874070702115010164.

El Adnany, E.M., Elhadiri, N., Mourjane, A., et al., 2023. Impact and Optimization of the Conditions of Extraction of Phenolic Compounds and Antioxidant Activity of Olive Leaves (Moroccan picholine) Using Response Surface Methodology. *Separations*, *10*, 326. https://doi.org/10.3390/separations10060326.

Elbaz NM, Tatham LM, Owen A, et al., 2023. Layer by layer self-assembly for coating a nanosuspension to modify drug release and stability for oral delivery. *Food Hydrocoll.*, *144*, 108908. https://doi.org/10.1016/j.foodhyd.2023.108908.

El Oirdi., M. 2024. Harnessing the Power of Polyphenols: A New Frontier in Disease Prevention and Therapy. *Pharmaceuticals*, *17*, 692. https://doi.org/10.3390/ph1706069.

Enaru, B., Socaci, S., Farcas, A., et al., 2021. Novel Delivery Systems of Polyphenols and Their Potential Health Benefits. *Pharm Basel Switz.*, *14*, 946. https://doi.org/10.3390/ph14100946.

Espinosa-Pardo, F.A., Nakajima, V.M., Macedo, G.A., et al., 2017. Extraction of phenolic compounds from dry and fermented orange pomace using supercritical CO2 and cosolvents. *Food Bioprod. Process.*, *101*, 1–10. https://doi.org/10.1016/j.fbp.2016.10.002.

Fache, M., Boutevin, B. and Caillol, S. 2016. Vanillin Production from Lignin and Its Use as a Renewable Chemical. *ACS Sustain. Chem. Eng.*, *4*, 35–46. https://doi.org/10.1021/acssuschemeng.5b01344.

Fache, M., Boutevin, B. and Caillol, S. 2015. Vanillin, a key-intermediate of biobased polymers. *Eur. Polym,* J., *68*, 488–502. https://doi.org/10.1016/j.eurpolymj.2015.03.050.

Farhan, M. 2024. The Promising Role of Polyphenols in Skin Disorders. *Molecules*, *29*, 865. https://doi.org/10.3390/molecules29040865.

Farías-Campomanes, A.M., Rostagno, M.A. and Meireles, M.A.A. 2013. Production of polyphenol extracts from grape bagasse using supercritical fluids: Yield, extract composition, and economic evaluation. *J. Supercrit. Fluids*, *77*, 70–78. https://doi.org/10.1016/j.supflu.2013.02.006.

Fernandes, I., Oliveira, J., Pinho, A. and Carvalho, E. 2022. The Role of Nutraceutical Containing Polyphenols in Diabetes Prevention. *Metabolites*, *12*, 184. https://doi.org/10.3390/metabo12020184.

Ficco, D.B.M., Petroni, K., Mistura, L. and D'Addezio, L. 2024. Polyphenols in Cereals: State of the Art of Available Information and Its Potential Use in Epidemiological Studies. *Nutrients*, *16*, 2155. https://doi.org/10.3390/nu16132155.

Frontuto, D., Carullo, D., Harrison, S.M., et al., 2019. Optimization of Pulsed Electric Fields-Assisted Extraction of Polyphenols from Potato Peels Using Response Surface Methodology. *Food Bioprocess. Technol.*, *12*, 1708–1720. https://doi.org/10.1007/s11947-019-02320-z.

Gabaston, J., Leborgne, C., Waffo-Téguo, P., et al., 2020. Separation and isolation of major polyphenols from maritime pine (Pinus pinaster) knots by two-step centrifugal partition chromatography monitored by LC-MS and NMR spectroscopy. *J. Sep. Sci.*, *43*, 1080–1088. https://doi.org/10.1002/jssc.201901066.

Gaber Ahmed GH, Fernández-González A, Díaz García ME (2020) Nano-encapsulation of grape and apple pomace phenolic extract in chitosan and soy protein via nanoemulsification. *Food Hydrocoll.*, *108*, 105806. https://doi.org/10.1016/j.foodhyd.2020.105806.

Gallardo-Fernández, M., Gonzalez-Ramirez, M., Cerezo, A.B., et al., 2022. Hydroxytyrosol in Foods: Analysis, Food Sources, EU Dietary Intake, and Potential Uses. *Foods*, *11*, 2355. https://doi.org/10.3390/foods11152355.

Gallo, M., Formato, A., Giacco, R., et al., 2019. Mathematical optimization of the green extraction of polyphenols from grape peels through a cyclic pressurization process. *Heliyon*, *5*. https://doi.org/10.1016/j.heliyon.2019.e01526.

Giotopoulou, I., Stamatis, H. and Barkoula, N-M. 2024. Encapsulation of Thymol in Ethyl Cellulose-Based Microspheres and Evaluation of Its Sustained Release for Food Applications. *Polymers*, *16*, 3396. https://doi.org/10.3390/polym16233396.

Gligor, O., Mocan, A., Moldovan, C., et al., 2019. Enzyme-assisted extractions of polyphenols: A comprehensive review. *Trends Food Sci. Technol.*, *88*, 302–315. https://doi.org/10.1016/j.tifs.2019.03.029.

Guan, Y., Yang, X., Pan, C., et al., 2023. Comprehensive Analyses of Breads Supplemented with Tannic Acids. *Foods*, *12*, 3756. https://doi.org/10.3390/foods12203756.

Güneş Bayir, A., Aksoy, A.N. and KoçyiĞiT, A. 2019. The Importance of Polyphenols as Functional Food in Health. *Bezmialem Sci.*, *7*, 157–163. https://doi.org/10.14235/bas.galenos.2018.2486.

Güneş Bayir, A., Nur Aksoy, A.N., Kocyigit, A., Bezmialem Vakıf University. 2024. The Importance of Polyphenols as Functional Food in Health. *ResearchGate*, *7*, 157–163. https://doi.org/10.14235/bas.galenos.2018.2486.

Guo, W., Liu, B., Hu, G., et al., 2019. Vanillin protects the blood–milk barrier and inhibits the inflammatory response in LPS-induced mastitis in mice. *Toxicol. Appl. Pharmacol.*, *365*, 9–18. https://doi.org/10.1016/j.taap.2018.12.022.

Gupta, N., Bhattacharya, S., Dutta, A., et al., 2024. Synthetic polyploidization induces enhanced phytochemical profile and biological activities in *Thymus vulgaris* L. essential oil. *Sci. Rep.*, *14,* 5608. https://doi.org/10.1038/s41598-024-56378-7.

Halevas, E.G., Avgoulas, D.I., Katsipis, G. and Pantazaki, A.A. 2022. Flavonoid-liposomes formulations: Physico-chemical characteristics, biological activities, and therapeutic applications. *Eur. J. Med. Chem. Rep.*, *5*, 100059. https://doi.org/10.1016/j.ejmcr.2022.100059.

Halim, A.L.A., Kamari, A. and Phillip, E. 2018. Chitosan, gelatin, and methylcellulose films incorporated with tannic acid for food packaging. *Int. J. Biol. Macromol.*, *120*, 1119–1126. https://doi.org/10.1016/j.ijbiomac.2018.08.169.

Hassanein, E.H.M., Althagafy, H.S., Baraka, M.A., et al., 2024 The promising antioxidant effects of lignans: Nrf2 activation comes into view. *Naunyn Schmiedebergs Arch. Pharmacol.*, *397*, 6439–6458. https://doi.org/10.1007/s00210-024-03102-x.

Haytowitz, D.B., Bhagwat, S. and Holden, J.M. 2013. Sources of Variability in the Flavonoid Content of Foods. *Procedia Food Sci.*, *2*, 46–51. https://doi.org/10.1016/j.profoo.2013.04.008.

Hazafa, A., Iqbal, M.O., Javaid, U., et al., 2022. Inhibitory effect of polyphenols (phenolic acids, lignans, and stilbenes) on cancer by regulating signal transduction pathways: A review. *Clin. Transl. Oncol.*, *24*, 432–445. https://doi.org/10.1007/s12094-021-02709-3.

Ho, S., Thoo, Y.Y., Young, D.J. and Siow, L.F. 2019. Stability and recovery of cyclodextrin encapsulated catechin in various food matrices. *Food Chem.*, *275*, 594–599. https://doi.org/10.1016/j.foodchem.2018.09.117.

Hong, S., Pangloli, P., Perumal, R., et al., 2020. A Comparative Study on Phenolic Content, Antioxidant Activity and Anti-Inflammatory Capacity of Aqueous and Ethanolic Extracts of Sorghum in Lipopolysaccharide-Induced RAW 264.7 Macrophages. *Antioxidants*, *9*, 1297. https://doi.org/10.3390/antiox9121297.

Hu, Y. 2022. Recent Technologies for the Extraction and Separation of Polyphenols in Different Plants: A Review. *J. Renew. Mater.*, *10*, 1471–1490. https://doi.org/10.32604/jrm.2022.018811.

Irakli, M., Chatzopoulou, P. and Ekateriniadou, L. 2018. Optimization of ultrasound-assisted extraction of phenolic compounds: Oleuropein, phenolic acids, phenolic alcohols, and flavonoids from olive leaves, and evaluation of its antioxidant activities. *Ind. Crops Prod.*, *124*, 382–388. https://doi.org/10.1016/j.indcrop.2018.07.070.

Jafari, H., Ghaffari-Bohlouli, P., Niknezhad, S.V., et al., 2022. Tannic acid: A versatile polyphenol for design of biomedical hydrogels. *J. Mater. Chem. B.*, *10*, 5873–5912. https://doi.org/10.1039/D2TB01056A.

Jang, W.Y., Kim. M-Y. and Cho, J.Y. 2022. Antioxidant, Anti-Inflammatory, Anti-Menopausal, and Anti-Cancer Effects of Lignans and Their Metabolites. *Int. J. Mol. Sci.*, *23*, 15482. https://doi.org/10.3390/ijms232415482.

Jenkins, D.J.A., Kendall, C.W.C, and Sievenpiper, J.L. 2021. Plant Polyphenols Lignans and Cardiovascular Disease. *J. Am. Coll. Cardiol.*, *78*, 679–682. https://doi.org/10.1016/j.jacc.2021.06.014.

Jesionek, W., Majer-Dziedzic, Barbara and Choma, I.M. 2015. Separation, Identification, and Investigation of Antioxidant Ability of Plant Extract Components Using TLC, LC–MS, and TLC–DPPH•. *J. Liq. Chromatogr. Relat. Technol.*, *38*, 1147–1153. https://doi.org/10.1080/10826076.2015.1028295.

Jiang, T., Liao, W. and Charcosset, C. 2020. Recent advances in encapsulation of curcumin in nanoemulsions: A review of encapsulation technologies, bioaccessibility, and applications. *Food Res. Int.*, *132*, 109035. https://doi.org/10.1016/j.foodres.2020.109035.

Jourdes, M., Pouységu, L., Quideau, S., et al., 2012. Hydrolyzable Tannins: Gallotannins, Ellagitannins, and Ellagic Acid. In: *Handbook of Analysis of Active Compounds in Functional Foods*. CRC Press.

Jovanović, A.A., Djordjević, V.B., Petrović, P.M., et al., 2021. The influence of different extraction conditions on polyphenol content, antioxidant and antimicrobial activities of wild thyme. *J. Appl. Res. Med. Aromat. Plants*, *25*, 100328. https://doi.org/10.1016/j.jarmap.2021.100328.

Kakkar, S. and Bais, S. 2014. A Review on Protocatechuic Acid and Its Pharmacological Potential. *Int. Sch. Res. Not.*, *2014*, 952943. https://doi.org/10.1155/2014/952943.

Kamal, G.M., Uddin, J., Asmari, M., et al., 2025. Natural Polyphenols as a promising aquatic food preservative: A concurrent review. *J. Agric. Food Res.*, 102046. https://doi.org/10.1016/j.jafr.2025.102046.

Karam, L., Chehab, R., Osaili, T.M. and Savvaidis, I.N. 2020. Antimicrobial effect of thymol and carvacrol added to a vinegar-based marinade for controlling spoilage of marinated beef (Shawarma) stored in air or vacuum packaging. *Int. J. Food Microbiol.*, *332*, 108769. https://doi.org/10.1016/j.ijfoodmicro.2020.108769.

Karković Marković, A., Torić, J., Barbarić, M. and Jakobušić Brala, C. 2019. Hydroxytyrosol, Tyrosol and Derivatives, and Their Potential Effects on Human Health. *Molecules*, *24*, 2001. https://doi.org/10.3390/molecules24102001.

Kasetsuwan, N., Reinprayoon, U., Uthaithammarat, L., et al., 2024. Anti-inflammatory effect of curcuminoids and their analogs in hyperosmotic human corneal limbus epithelial cells. *BMC Complement Med. Ther.*, *24*, 172. https://doi.org/10.1186/s12906-024-04448-8.

Kasote, D., Tiozon, R.N., Sartagoda, K.J.D., et al., (2021.) Food Processing Technologies to Develop Functional Foods with Enriched Bioactive Phenolic Compounds in Cereals. *Front. Plant Sci.*, *12*. https://doi.org/10.3389/fpls.2021.771276.

Khan, I., Yousif, A., Johnson, S.K. and Gamlath, S. 2013. Effect of sorghum flour addition on resistant starch content, phenolic profile, and antioxidant capacity of durum wheat pasta. *Food Res. Int.*, *54*, 578–586. https://doi.org/10.1016/j.foodres.2013.07.059.

Khan, M.K., Paniwnyk, L. and Hassan, S. 2019. Polyphenols as Natural Antioxidants: Sources, Extraction and Applications in Food, Cosmetics and Drugs. In: Li. Y. and Chemat, F. (Eds.), *Plant Based "Green Chemistry 2.0": Moving from Evolutionary to Revolutionary* (pp. 197–235). Springer, Singapore.

Khan, M.U., Rohilla, S., Joshi, S. and Bora, J. 2024. Polyphenols as Functional Foods and Nutraceuticals. In: Bashir, K., Jan, K. and Ahmad, F.J. (Eds.), *Functional Foods and Nutraceuticals: Chemistry, Health Benefits, and the Way Forward* (pp. 145–171). Springer International Publishing, Cham.

Khanzada, B., Akhtar, N., ul Haq, I., et al., 2024. Polyphenol assisted nano-reinforced chitosan films with antioxidant and antimicrobial properties. *Food Hydrocoll.*, *153*, 110010. https://doi.org/10.1016/j.foodhyd.2024.110010.

Kim, M.E., Na, J.Y., Park, Y-D. and Lee, J.S. 2019. Anti-Neuroinflammatory Effects of Vanillin through the Regulation of Inflammatory Factors and NF-κB Signaling in LPS-Stimulated Microglia. *Appl. Biochem. Biotechnol.*, *187*, 884–893. https://doi.org/10.1007/s12010-018-2857-5.

Krga, I. and Milenkovic, D. 2019. Anthocyanins: From Sources and Bioavailability to Cardiovascular-Health Benefits and Molecular Mechanisms of Action. *J. Agric. Food Chem.*, *67*, 1771–1783. https://doi.org/10.1021/acs.jafc.8b06737.

Laaraj, S., Tikent, A., Chebaibi, M., et al., 2024. A Study of the Bioactive Compounds, Antioxidant Capabilities, Antibacterial Effectiveness and Cytotoxic Effects on Breast Cancer Cell Lines Using an Ethanolic Extract from the Aerial Parts of the Indigenous Plant Anabasis aretioïdes Coss. & Moq. *Curr. Issues Mol. Biol.*, *46*, 12375–12396. https://doi.org/10.3390/cimb46110735.

Lai, W-F. and Wong, W-T. 2022. Design and optimization of quercetin-based functional foods. *Crit. Rev. Food Sci. Nutr.*

Lama-Muñoz, A. and Contreras M. del M. 2022. Extraction Systems and Analytical Techniques for Food Phenolic Compounds: A Review. *Foods*, *11*, 3671. https://doi.org/10.3390/foods11223671.

Landete, J.M. 2011. Ellagitannins, ellagic acid and their derived metabolites: A review about source, metabolism, functions, and health. *Food Res. Int.*, *44*, 1150–1160. https://doi.org/10.1016/j.foodres.2011.04.027.

Larigot, L., Juricek, L., Dairou, J. and Coumoul, X. 2018. AhR signaling pathways and regulatory functions. *Biochim Open*, *7*, 1–9. https://doi.org/10.1016/j.biopen.2018.05.001.

Larsen, P. and Ahmed, M. 2022. Evaluation of antioxidant potential of honey drops and honey lozenges. *Food Chem. Adv.*, *1*, 100013. https://doi.org/10.1016/j.focha.2022.100013.

Lee, H.N., Shin, S.A., Choo, G.S., et al., 2018. Anti-inflammatory effect of quercetin and galangin in LPS-stimulated RAW264.7 macrophages and DNCB-induced atopic dermatitis animal models. *Int. J. Mol. Med.*, *41*, 888–898. https://doi.org/10.3892/ijmm.2017.3296.

Lee, S.J., Gwak, M.A., Chathuranga, K., et al., 2023. Multifunctional chitosan/tannic acid composite films with improved anti-UV, antioxidant, and antimicrobial properties for active food packaging. *Food Hydrocoll.*, *136*, 108249. https://doi.org/10.1016/j.foodhyd.2022.108249.

Lee, S., Jo, C., Choi, H-Y. and Lee, K. 2021. Effect of Co-Administration of Curcumin with Amlodipine in Hypertension. *Nutrients*, *13*, 2797. https://doi.org/10.3390/nu13082797.

Li, L., Li, Z., Wei, Z., et al., 2020. Effect of tannin addition on chromatic characteristics, sensory qualities, and antioxidant activities of red wines. *RSC Adv.*, *10*, 7108–7117. https://doi.org/10.1039/C9RA09846A.

Li, L., Zhao, J., Yang, T. and Sun, B. 2022. High-speed countercurrent chromatography as an efficient technique for large separation of plant polyphenols: A review. *Food Res. Int.*, *153*, 110956. https://doi.org/10.1016/j.foodres.2022.110956.

Li, T., Li, F., Liu, X., et al., 2019. Synergistic anti-inflammatory effects of quercetin and catechin via inhibiting activation of TLR4–MyD88-mediated NF-κB and MAPK signaling pathways. *Phytother. Res.*, *33*, 756–767. https://doi.org/10.1002/ptr.6268.

Li, W., Chen, H., Xu, B., et al., 2023. Research progress on classification, sources, and functions of dietary polyphenols for prevention and treatment of chronic diseases. *J. Future Foods*, *3*, 289–305. https://doi.org/10.1016/j.jfutfo.2023.03.001.

Li, W., Ran, L., Liu, F., et al., 2019. Preparation and Characterisation of Polyphenol-HP-β-Cyclodextrin Inclusion Complex that Protects Lamb Tripe Protein against Oxidation. *Molecules*, *24*, 4487. https://doi.org/10.3390/molecules24244487.

Li, Y., Li, L., Cui, Y., et al., 2017. Separation and purification of polyphenols from red wine extracts using high speed counter current chromatography. *J. Chromatogr. B.*, *1054*,105–113. https://doi.org/10.1016/j.jchromb.2017.03.006.

Lin, J. and Zhou, W. 2018. Role of quercetin in the physicochemical properties, antioxidant, and antiglycation activities of bread. *J. Funct. Foods*, *40*, 299–306. https://doi.org/10.1016/j.jff.2017.11.018.

Liu, C., Dong, S., Wang, X., et al., 2023. Research progress of polyphenols in nanoformulations for antibacterial application. *Mater. Today Bio.*, *21*, 100729. https://doi.org/10.1016/j.mtbio.2023.100729.

Liu, C., Ge, S., Yang, J., et al., 2016. Adsorption mechanism of polyphenols onto starch nanoparticles and enhanced antioxidant activity under adverse conditions. *J. Funct. Foods*, *26*, 632–644. https://doi.org/10.1016/j.jff.2016.08.036.

Lordan, R., Tsoupras, A. and Zabetakis, I. 2019. The Potential Role of Dietary Platelet-Activating Factor Inhibitors in Cancer Prevention and Treatment. *Adv. Nutr.*, *10*, 148–164. https://doi.org/10.1093/advances/nmy090.

Lorenzo, J.M., Munekata, P.E., Putnik, P., et al., 2019. Chapter 6 - Sources, Chemistry, and Biological Potential of Ellagitannins and Ellagic Acid Derivatives. In: Atta-ur-Rahman (Ed.), *Studies in Natural Products Chemistry* (pp. 189–221). Elsevier.

Lucas, M., Freitas, M., Xavier, J.A., et al., 2021. The scavenging effect of curcumin, piperine, and their combination against physiological relevant reactive pro-oxidant species using *in vitro* non-cellular and cellular models. *Chem. Pap.*, *75*, 5269–5277. https://doi.org/10.1007/s11696-021-01710-y.

Lucas Tenório, C.J., Assunção Ferreira, M.R. and Lira Soares, L.A. 2022. Recent advances on preparative LC approaches for polyphenol separation and purification: Their sources and main activities. *Trends Food Sci. Technol.*, *128*, 129–146. https://doi.org/10.1016/j.tifs.2022.08.004.

Luengo, E., Álvarez, I. and Raso, J. 2013. Improving the pressing extraction of polyphenols of orange peel by pulsed electric fields. *Innov. Food Sci. Emerg. Technol.*, *17*, 79–84. https://doi.org/10.1016/j.ifset.2012.10.005.

Luo, J., Si, H., Jia, Z. and Liu, D. 2021. Dietary Anti-Aging Polyphenols and Potential Mechanisms. *Antioxidants*, *10*, 283. https://doi.org/10.3390/antiox10020283.

Ma Z, Du B, Li J, et al (2021) An Insight into Anti-Inflammatory Activities and Inflammation Related Diseases of Anthocyanins: A Review of Both *In Vivo* and *In Vitro* Investigations. *Int. J. Mol. Sci.*, *22*, 11076. https://doi.org/10.3390/ijms222011076.

Mahmoud, F, K., S. Ali, H., A. Amin, A. (2018) Nanoencapsulation of Bioactive Compounds Extracted from Egyptian Prickly Pears Peel Fruit: Antioxidant and Their Application in Guava Juice. *Asian J. Sci. Res.*, *11*, 574–586. https://doi.org/10.3923/ajsr.2018.574.586.

Manach, C., Scalbert, A., Morand, C., et al., 2004. Polyphenols: Food sources and bioavailability. *Am. J. Clin. Nutr.*, *79*, 727–747. https://doi.org/10.1093/ajcn/79.5.727.

Manso, T., Lores, M. and de Miguel, T. 2021. Antimicrobial Activity of Polyphenols and Natural Polyphenolic Extracts on Clinical Isolates. *Antibiotics*, *11*, 46. https://doi.org/10.3390/antibiotics11010046.

Mantzourani, C., Tarantilis, P.A. and Kokotou, M.G. 2023. Carnosic Acid and Carnosol: Analytical Methods for Their Determination in Plants, Foods and Biological Samples. *Separations*, *10*, 481. https://doi.org/10.3390/separations10090481.

Marino, T., Galano, A. and Russo, N. 2014. Radical Scavenging Ability of Gallic Acid toward OH and OOH Radicals: Reaction Mechanism and Rate Constants from the Density Functional Theory. *J. Phys. Chem. B.*, *118*, 10380–10389. https://doi.org/10.1021/jp505589b.

Martínez Rivas, C.J., Tarhini, M., Badri, W., et al., 2017. Nanoprecipitation process: From encapsulation to drug delivery. *Int. J. Pharm.*, *532*, 66–81. https://doi.org/10.1016/j.ijpharm.2017.08.064.

Marton, L.T., Pescinini-e-Salzedas, L.M., Camargo, M.E.C., et al., 2021. The Effects of Curcumin on Diabetes Mellitus: A Systematic Review. *Front. Endocrinol.*, 12. https://doi.org/10.3389/fendo.2021.669448.

Mastromatteo M., Lucera, A., Sinigaglia, M. and Corbo, M.R. 2009. Combined effects of thymol, carvacrol and temperature on the quality of non conventional poultry patties. *Meat Sci.*, *83*, 246–254. https://doi.org/10.1016/j.meatsci.2009.05.007.

Mateus, A.R.S., Serrano, C., Almeida, C., et al., 2024. Beyond Thymol and Carvacrol: Characterizing the Phenolic Profiles and Antioxidant Capacity of Portuguese Oregano and Thyme for Food Applications. *Appl. Sci.*, *14*, 8924. https://doi.org/10.3390/app14198924.

Medina, I., Gallardo, J.M., Gonzalez, M.J., et al., 2007. Effect of molecular structure of phenolic families as hydroxycinnamic acids and catechins on their antioxidant effectiveness in minced fish muscle. *J. Agric. Food Chem.*, *55*, 3889–3895. https://doi.org/10.1021/jf063498i.

Moon, K-D., Delaquis, P., Toivonen, P. and Stanich, K. 2006. Effect of vanillin on the fate of *Listeria monocytogenes* and *Escherichia coli* O157:H7 in a model apple juice medium and in apple juice. *Food Microbiol.*, *23*, 169–174. https://doi.org/10.1016/j.fm.2005.02.005.

Muntaha, S.T., Rakha, A., Rasheed, H., et al., 2025. Polyphenol-protein particles: A nutraceutical breakthrough in nutrition and food science. *J. Agric. Food Res.*, *19*, 101641. https://doi.org/10.1016/j.jafr.2025.101641.

Nawaz, A., Irshad, S., Walayat, N., et al., 2024. Effect of structurally different pectin combined with polyphenol on water distribution, protein oxidation and heterocyclic amines formation in meat-based snacks. *LWT*, *193*, 115754. https://doi.org/10.1016/j.lwt.2024.115754.

Nikolić, M., Glamočlija, J., Ferreira, I.C.F.R., et al., 2014. Chemical composition, antimicrobial, antioxidant, and antitumor activity of *Thymus serpyllum* L., *Thymus algeriensis* Boiss. and Reut and *Thymus vulgaris* L. essential oils. *Ind. Crops Prod.*, *52*, 183–190. https://doi.org/10.1016/j.indcrop.2013.10.006.

Nirmal, N.P. and Benjakul, S. 2009. Melanosis and Quality Changes of Pacific White Shrimp (*Litopenaeus vannamei*) Treated with Catechin during Iced Storage. *J. Agric. Food Chem.*, *57*, 3578–3586. https://doi.org/10.1021/jf900051e.

Ntuli, S., Leuschner, M., Bester, M.J and Serem, J.C. 2022. Stability, Morphology, and Effects of *In Vitro* Digestion on the Antioxidant Properties of Polyphenol Inclusion Complexes with β-Cyclodextrin. *Mol. Basel Switz.*, *27*, 3808. https://doi.org/10.3390/molecules27123808.

Ochoa-Velasco, C.E., Pérez-Pérez, J.C., Varillas-Torres, J.M., et al., 2021. Starch Edible Films/Coatings Added with Carvacrol and Thymol: *In Vitro* and *In Vivo* Evaluation against Colletotrichum gloeosporioides. *Foods*, *10*, 175. https://doi.org/10.3390/foods10010175.

Tavakoli, H., Hosseini, O., Jafari, S.M. and Katouzian, I. 2018. Evaluation of Physicochemical and Antioxidant Properties of Yogurt Enriched by Olive Leaf Phenolics within Nanoliposomes. *J. Agric. Food Chem.*, *66*, 9231–9240. https://doi.org/10.1021/acs.jafc.8b02759.

Taverniti, V., Del Bo', C., Fiore, W., et al., 2022. Combination of different probiotics and berry-derived (poly)phenols can modulate immune response in dendritic cells. *J. Funct. Foods*, *94*, 105121. https://doi.org/10.1016/j.jff.2022.105121.

Tena, N., Martín, J. and Asuero, A.G. 2020. State of the Art of Anthocyanins: Antioxidant Activity, Sources, Bioavailability, and Therapeutic Effect in Human Health. *Antioxidants*, *9*, 451. https://doi.org/10.3390/antiox9050451.

Tian, B. and Liu, J. 2020. Resveratrol: A review of plant sources, synthesis, stability, modification, and food application. *J. Sci. Food Agric.*, *100*, 1392–1404. https://doi.org/10.1002/jsfa.10152.

Tsai, Y-F., Chen, Y-R., Chen, J-P., et al., 2019. Effect of hesperidin on anti-inflammation and cellular antioxidant capacity in hydrogen peroxide-stimulated human articular chondrocytes. *Process Biochem.*, *85*, 175–184. https://doi.org/10.1016/j.procbio.2019.07.014.

Tomadoni, B., Viacava, G., Cassani, L., et al., 2016. Novel biopreservatives to enhance the safety and quality of strawberry juice. *J. Food Sci Technol.*, *53*, 281–292. https://doi.org/10.1007/s13197-015-2068-9.

Tsoupras, A., Adamantidi, T., Finos, M.A., et al., 2024 Re-Assessing the Role of Platelet Activating Factor and Its Inflammatory Signaling and Inhibitors in Cancer and Anti-Cancer Strategies. *Front. Biosci. Landmark Ed.*, *29*, 345. https://doi.org/10.31083/j.fbl2910345.

Uddin, M.S., Al Mamun, A., Kabir, Md. T., et al., 2020. Neuroprotective role of polyphenols against oxidative stress-mediated neurodegeneration. *Eur. J. Pharmacol.*, *886*, 173412. https://doi.org/10.1016/j.ejphar.2020.173412.

Tzima, K., Brunton, N.P., Choudhary, A. and Rai, D.K. 2020. Potential Applications of Polyphenols from Herbs and Spices in Dairy Products as Natural Antioxidants. In: *Herbs, Spices, and Medicinal Plants* (pp. 283–299). John Wiley & Sons, Ltd.

Upadhyay, S. and Dixit, M. 2015. Role of Polyphenols and Other Phytochemicals on Molecular Signaling. *Oxid. Med. Cell Longev.*, *2015*, 504253. https://doi.org/10.1155/2015/504253.

Vazquez Prieto, M.A., Bettaieb, A., Rodriguez Lanzi, C., et al., 2015. Catechin and quercetin attenuate adipose inflammation in fructose-fed rats and 3T3-L1 adipocytes. *Mol. Nutr. Food Res.*, *59*, 622–633. https://doi.org/10.1002/mnfr.201400631.

Veiko, A.G., Sekowski, S., Lapshina, E.A., et al., 2020. Flavonoids modulate liposomal membrane structure, regulate mitochondrial membrane permeability, and prevent erythrocyte oxidative damage. *Biochim. Biophys. Acta BBA - Biomembr.*, *1862*, 183442. https://doi.org/10.1016/j.bbamem.2020.183442.

Venkatesan, R., Vetcher, A.A., Al-Asbahi, B.A. and Kim, S-C. 2024. Chitosan-Based Films Blended with Tannic Acid and Moringa Oleifera for Application in Food Packaging: The Preservation of Strawberries (*Fragaria ananassa*). *Polymers*, *16*, 937. https://doi.org/10.3390/polym16070937.

Wang, D,, Xiao, H,, Lyu, X., et al., 2023. Lipid oxidation in food science and nutritional health: A comprehensive review. *Oil Crop Sci.*, *8*, 35–44. https://doi.org/10.1016/j.ocsci.2023.02.002.

Wang, W., Jung, J. and Zhao, Y. 2017. Chitosan-cellulose nanocrystal microencapsulation to improve encapsulation efficiency and stability of entrapped fruit anthocyanins. *Carbohydr. Polym.*, *157*, 1246–1253. https://doi.org/10.1016/j.carbpol.2016.11.005.

Wang, X., Chen, C., Bao, Y., et al., 2024. Encapsulation of three different types of polyphenols in casein using a customized pH-driven method: Preparation and characterization. *Food Res. Int.*, *189*, 114547. https://doi.org/10.1016/j.foodres.2024.114547.

Wen, C., Dechsupa, N., Yu, Z., et al., 2023. Pentagalloyl Glucose: A Review of Anticancer Properties, Molecular Targets, Mechanisms of Action, Pharmacokinetics, and Safety Profile. *Molecules*, *28*, 4856. https://doi.org/10.3390/molecules28124856.

Wen, L., Wu, D., Tan, X., et al., 2022. The Role of Catechins in Regulating Diabetes: An Update Review. *Nutrients*, *14*, 4681. https://doi.org/10.3390/nu14214681.

Wang, L., Tang, C. and Pan, Q. 2025. Mechanisms Underlying the Anti-Atherosclerotic Effects of EGCG. *Curr. Mol. Med.* https://doi.org/10.2174/0115665240354839241113044604.

Wu, T., Li, Z., Wu, Y., et al., 2023. Exploring plant polyphenols as anti-allergic functional products to manage the growing incidence of food allergy. *Front. Nutr.*, *10*, 1102225. https://doi.org/10.3389/fnut.2023.1102225.

Wu, W., Mi, Y, Meng, Q., et al., 2025. Natural polyphenols as novel interventions for aging and age-related diseases: Exploring efficacy, mechanisms of action, and implications for future research. *Chin. Herb. Med.*, *17*, 279–291. https://doi.org/10.1016/j.chmed.2024.09.001.

Xiao, J., Capanoglu, Esra, Jassbi ,Amir Reza and Miron, A. 2016. Advance on the Flavonoid C-glycosides and Health Benefits. *Crit. Rev. Food Sci. Nutr.*, *56*, S29–S45. https://doi.org/10.1080/10408398.2015.1067595.

Yahfoufi, N., Alsadi, N., Jambi, M. and Matar, C. 2018. The Immunomodulatory and Anti-Inflammatory Role of Polyphenols. *Nutrients*, *10*, 1618. https://doi.org/10.3390/nu10111618.

Yan, B., Chen, Z.S., Hu, Y. and Yong, Q. 2021. Insight in the Recent Application of Polyphenols from Biomass. *Front. Bioeng. Biotechnol.*, *9*. https://doi.org/10.3389/fbioe.2021.753898.

Yong, Y., Ahmad, H.N., Gu, Y., et al., 2024. The synergistic effect of polyphenols and polypeptides for plant-based bioplastic film: Enhanced UV resistance, antioxidant, and antibacterial performance. *Food Chem.*, *460*, 140746. https://doi.org/10.1016/j.foodchem.2024.140746.

Yoo, Y.J., Saliba, A.J., Prenzler, P.D. and Ryan, D. 2012. Total Phenolic Content, Antioxidant Activity, and Cross-Cultural Consumer Rejection Threshold in White and Red Wines Functionally Enhanced with Catechin-Rich Extracts. *J. Agric. Food Chem.*, *60*, 388–393. https://doi.org/10.1021/jf203216z.

Zekrumah, M., Begua, P., Razak, A., et al., 2023. Role of dietary polyphenols in non-communicable chronic disease prevention, and interactions in food systems: An overview. *Nutrition*, *112*, 112034. https://doi.org/10.1016/j.nut.2023.112034.

Zhang, J., Celli, G.B. and Brooks, M.S. 2019. Chapter 1. Natural Sources of Anthocyanins. In: Brooks, MS-L., and Celli, G.B. (Eds.), *Food Chemistry, Function, and Analysis* (pp 1–33). Royal Society of Chemistry, Cambridge.

Zhang, J., Wen, C., Zhang, H., et al., 2020. Recent advances in the extraction of bioactive compounds with subcritical water: A review. *Trends Food Sci. Technol.*, *95*, 183–195. https://doi.org/10.1016/j.tifs.2019.11.018.

Zhang, Z., Li, X., Sang, S., et al., 2022. Polyphenols as Plant-Based Nutraceuticals: Health Effects, Encapsulation, Nano-Delivery, and Application. *Foods*, *11*, 2189. https://doi.org/10.3390/foods11152189.

Zheng, H., Hwang, I-W. and Chung, S-K. 2009. Enhancing polyphenol extraction from unripe apples by carbohydrate-hydrolyzing enzymes. *J. Zhejiang Univ. Sci. B.*, *10*, 912–919. https://doi.org/10.1631/jzus.B0920186.

Zheng, H., Li, J., Ning, F., et al., 2021. Improving *in vitro* bioaccessibility and bioactivity of carnosic acid using a lecithin-based nanoemulsion system. *Food Funct.*, *12*, 1558–1568. https://doi.org/10.1039/d0fo02636k.

Zheng, L., Ding, Z., Zhang, M. and Sun, J. 2011. Microencapsulation of bayberry polyphenols by ethyl cellulose: Preparation and characterization. *J. Food Eng.*, *104*, 89–95. https://doi.org/10.1016/j.jfoodeng.2010.11.031.

Zhou, Y., Jiang, Z., Lu, H., et al., 2019. Recent Advances of Natural Polyphenols Activators for Keap1-Nrf2 Signaling Pathway. *Chem. Biodivers.*, *16*, e1900400. https://doi.org/10.1002/cbdv.201900400.

Żyżelewicz, D., Oracz, J., Kaczmarska, M., et al., 2018. Preparation and characterization of inclusion complex of (+)-catechin with β-cyclodextrin. *Food Res. Int.*, *113*, 263–268. https://doi.org/10.1016/j.foodres.2018.07.018.

Index

About the Editors

Dr. Surajbhan Sevda is currently serving as an Associate Professor in the Department of Biotechnology at the National Institute of Technology (NIT) Warangal, India. His research focuses on microbial fuel cells, bioelectrochemical systems, and sustainable wastewater treatment, with a strong emphasis on resource recovery and circular bioeconomy approaches. He obtained his Ph.D. in Biochemical Engineering and Biotechnology from the Indian Institute of Technology (IIT) Delhi in 2013. Prior to his current position, he worked as a Scientist at IIT Guwahati and served as an Assistant Professor at NIT Jalandhar. He also gained international research experience as a Postdoctoral Fellow at Qatar University and Virginia Tech, USA, and as a Visiting Doctoral Fellow at VITO, Belgium. Dr. Sevda holds an M.Tech in Fermentation Technology from the Institute of Chemical Technology (ICT), Mumbai, and a B.E. in Biotechnology from the University of Rajasthan. He was also a visiting scientist at the University of Calgary, Canada, in 2018. He has authored over 50 research articles in reputed SCI journals, along with several book chapters and edited books. His research contributions span industrial wastewater treatment, bioenergy and biofuels, life cycle assessment, metal recovery, biosensor development, green chemistry, functional foods, and microbial electrosynthesis.

Dr. Meena Krishania is a Scientist at the National Agri-Food and Biomanufacturing Institute (BRIC-NABI), formerly the Center of Innovative and Applied Bioprocessing (CIAB), Mohali, India. She is an expert with over a decade of experience in Food Bioprocessing. She has a PhD from IIT Delhi in 2013, MTech in Chemical Engineering from MNIT Jaipur in 2009, and B Tech in Biotechnology Engineering from Rajasthan University in 2007. She joined CIAB in 2014. Her research group focuses on food processing, nutraceuticals, biomanufacturing, and scale-ups, resulting in more than 8 patents filed, many publications in reputed SCI journals, and 3 technologies licenses. She completed the national and international projects in her area. She has received several prestigious awards. She was a visiting research fellow at the University of Southampton, UK, in 2011. Dr. Meena is a life member of several professional societies, including BRSI, SEEC, and IFIbiop, and web-coordinator of prestigious professional societies Indian National Young Academy of Science (INSA-INYAS).

For Product Safety Concerns and Information please contact our
EU representative GPSR@taylorandfrancis.com Taylor & Francis
Verlag GmbH, Kaufingerstraße 24, 80331 München, Germany